PEACH

In Memoriam: Dra. Maria Luisa Badenes Catalá (1963–2022)

Maria Luisa Badenes was a research professor and coordinator of the Center for Citriculture and Plant Production of the Valencian Institute of Agricultural Research (IVIA), Spain, and one of the most important scientists in the long history of the institute. Marisa devoted her career to horticultural crops, including temperate and minor species fruits, after finishing her PhD in Agricultural Engineering from the Polytechnic University of Valencia in 1991. She was responsible for the plant breeding programmes of apricot, peach, loquat, persimmon and kiwifruit for many years and a leading authority in international fruit growing. After 2 years of postdoctoral training at the University of California, Davis, she rejoined the Fruit Growing Unit of the IVIA with Dr Gerardo Llácer. They started one of the first two Spanish apricot breeding programmes, in which cultivars resistant to sharka (plum pox) were obtained, facilitating identification of the gene responsible for resistance to this virus in apricot: a milestone of great international impact. Along with her colleagues, she participated in obtaining new varieties of apricot, peach, nectarine, loquat and persimmon, thus diversifying and improving the crop portfolio with cultivars with low-chill, high-fruit quality traits and expansion of the ripening calendar. Her scientific career was extensive; she coordinated many international, national and public–private projects. She was also a professor for students seeking Master's Degrees in Plant Breeding, and Plant Molecular and Cellular Biotechnology. She served as the research supervisor for PhD and MS students at the IVIA and the Polytechnic University of Valencia. She was a proficient writer. Her scientific productivity is one of the most voluminous and impactful, including numerous articles for the industry. She served on the editorial review boards for several scientific journals and was highly respected for her scientific expertise on adaptation of temperate fruit crops. Dra. Badenes was very active in several scientific organizations, particularly the International Society for Horticultural Science (ISHS). She was Chair of the ISHS Working Group Persimmon, convened two symposia in Valencia, the VI International Symposium on Persimmon (2016) and the IV International Symposium on Pomegranate and Minor Mediterranean Fruits (2017), and convened the X International Symposium on Temperate Fruits in the Tropics and Subtropics at IHC2018 in Turkey. She also served as General Secretary of the European Association for Research on Plant Breeding (EUCARPIA). In Spain, she served as the Coordinator of Agriculture of the Spanish Agency of Evaluation and Prospective (ANEP) and the Spanish Research Agency (AEI). Marisa was beloved for being a positive colleague and workmate. 'Don't worry' was her favourite expression. She was a kind and loving friend, who appreciated everything. She enjoyed and balanced her ambitious career and her personal life as a wife, mother and grandmother.

PEACH

Edited by

George A. Manganaris

Cyprus University of Technology, Cyprus

Guglielmo Costa

University of Bologna, Italy

Carlos H. Crisosto

University of California, Davis, USA

CROP PRODUCTION SCIENCE IN HORTICULTURE SERIES

This series examines economically important horticultural crops selected from the major production systems in temperate, subtropical and tropical climatic areas. Systems represented range from open field and plantation sites to protected plastic and glass houses, growing rooms and laboratories. Emphasis is placed on the scientific principles underlying crop production practices rather than on providing empirical recipes for uncritical acceptance. Scientific understanding provides the key to both reasoned choice of practice and the solution of future problems.

Students and staff at universities and colleges throughout the world involved in courses in horticulture, as well as in agriculture, plant science, food science and applied biology at degree, diploma or certificate level, will welcome this series as a succinct and readable source of information. The books will also be invaluable to progressive growers, advisers and end-product users requiring an authoritative, but brief, scientific introduction to particular crops or systems. Keen gardeners wishing to understand the scientific basis of recommended practices will also find the series very useful.

The authors are all internationally renowned experts with extensive experience of their subjects. Each volume follows a common format covering all aspects of production, from background physiology and breeding, to propagation and planting, through husbandry and crop protection, to harvesting, handling and storage. Selective references are included to direct the reader to further information on specific topics.

Titles Available:
1. **Ornamental Bulbs, Corms and Tubers** A.R. Rees
2. **Citrus** F.S. Davies and L.G. Albrigo
3. **Onions and Other Vegetable Alliums** J.L. Brewster
4. **Ornamental Bedding Plants** A.M. Armitage
5. **Bananas and Plantains** J.C. Robinson
6. **Cucurbits** R.W. Robinson and D.S. Decker-Walters
7. **Tropical Fruits** H.Y. Nakasone and R.E. Paull
8. **Coffee, Cocoa and Tea** K.C. Willson
9. **Lettuce, Endive and Chicory** E.J. Ryder
10. **Carrots and Related Vegetable *Umbelliferae*** V.E. Rubatzky, C.F. Quiros and P.W. Simon
11. **Strawberries** J.F. Hancock
12. **Peppers: Vegetable and Spice Capsicums** P.W. Bosland and E.J. Votava
13. **Tomatoes** E. Heuvelink
14. **Vegetable Brassicas and Related Crucifers** G. Dixon
15. **Onions and Other Vegetable Alliums, 2nd Edition** J.L. Brewster
16. **Grapes** G.L. Creasy and L.L. Creasy
17. **Tropical Root and Tuber Crops: Cassava, Sweet Potato, Yams and Aroids** V. Lebot

CABI is a trading name of CAB International

CABI
Nosworthy Way
Wallingford
Oxfordshire OX10 8DE
UK

CABI
200 Portland Street
Boston
MA 02114
USA

Tel: +44 (0)1491 832111
E-mail: info@cabi.org
Website: www.cabi.org

Tel: +1 (617)682-9015
E-mail: cabi-nao@cabi.org

A catalogue record for this book is available from the British Library, London, UK.

Library of Congress Cataloging-in-Publication Data

Names: Manganaris, G. A. (George A.), editor. | Costa, G. (Guglielmo), editor. | Crisosto, Carlos H., editor.
Title: Peach / edited by George Manganaris, Guglielmo Costa, and Carlos Crisosto.
Other titles: Crop production science in horticulture.
Description: Boston, MA, USA : CAB International, [2023] | Series: Crop production science in horticulture | Includes bibliographical references and index. | Summary: "Provides comprehensive, fundamental and up-to-date information on the production processes for fresh market and canning peach and nectarine, including orchard establishment, production, postharvest handling and uses. Highly illustrated in full colour"-- Provided by publisher.
Identifiers: LCCN 2022055741 (print) | LCCN 2022055742 (ebook) | ISBN 9781789248432 (paperback) | ISBN 9781789248449 (ebook) | ISBN 9781789248456 (epub)
Subjects: LCSH: Peach. | Peach--Diseases and pests. | Peach--Breeding. | Peach--Handling. | Peach industry.
Classification: LCC SB371 .P418 2023 (print) | LCC SB371 (ebook) | DDC 634/.25--dc23/eng/20230215
LC record available at https://lccn.loc.gov/2022055741
LC ebook record available at https://lccn.loc.gov/2022055742

ISBN-13: 9781789248432 (paperback)
        9781789248449 (ePDF)
        9781789248456 (ePub)

DOI: 10.1079/9781789248456.0000

Commissioning Editor: Rebecca Stubbs
Editorial Assistant: Emma McCann
Production Editor: Shankari Wilford

Typeset by SPi, Pondicherry, India
Printed and bound in the USA by Integrated Books International, Dulles, Virginia

CONTENTS

CONTRIBUTORS

J.E. Adaskaveg, Department of Microbiology and Plant Pathology, University of California, Riverside, California, USA. E-mail: jim.adaskaveg@ucr.edu

B.M. Anthony, Department of Horticulture and Landscape Architecture, Colorado State University, Fort Collins, CO 80523, USA. E-mail: Brendon. Anthony@colostate.edu

M.L. Badenes†, Instituto Valenciano de Investigaciones Agraria, 46113, Moncada, Valencia, Spain.

D. Bassi, Faculty of Agriculture and Food Science, University of Milan, 20133 Milan, Italy. E-mail: daniele.bassi@unimi.it

B.R. Blaauw, Department of Entomology, University of Georgia, Athens, Georgia, USA. E-mail: bblaauw@uga.edu

C. Bonghi, Department of Agronomy, Food, Natural Resources, Animals and Environment, University of Padua, Agripolis – Legnaro (PD), Italy. E-mail: claudio.bonghi@unipd.it

A. Botton, Department of Agronomy, Food, Natural Resources, Animals and Environment, University of Padua, Padua, Italy. E-mail: alessandro.botton@ unipd.it

M. Cirilli, Faculty of Agriculture and Food Science, University of Milan, Milan, Italy. E-mail: marco.cirilli@unimi.it

L. Cisneros-Zevallos, Department of Horticultural Sciences, Texas A&M University, College Station, Texas, USA. E-mail: L.Cisneros-Zevallos@ag.tamu. edu

G. Costa, Department of Agri-Food Sciences and Technologies, University of Bologna, Bologna, Italy. E-mail: guglielmo.costa@unibo.it

C.H. Crisosto, Department of Plant Sciences, University of California, Davis, California, USA. E-mail: chcrisosto@ucdavis.edu

M. Christofi, Department of Agricultural Sciences, Biotechnology & Food Science, Cyprus University of Technology, 3603 Lemesos, Cyprus. E-mail: Man.christofi@edu.cut.ac.cy

B. Dichio, Dipartimento delle Culture Europee e del Mediterraneo: Architettura, Ambiente, Patrimoni Culturali – University of Basilicata, Potenza, Italy. E-mail: bartolomeo.dichio@unibas.it

M.F. Drincovich, Centro de Estudios Fotosintéticos y Bioquímicos, Consejo Nacional de Investigaciones Científicas y Técnicas, Facultad de Ciencias Bioquímicas y Farmacéuticas, Universidad Nacional de Rosario, Rosario, Argentina. E-mail: drincovich@cefobi-conicet.gov.ar

P. Drogoudi, Department of Deciduous Fruit Trees, Institute of Plant Breeding and Genetic Resources, Hellenic Agricultural Organization 'Demeter', 38 R.R. Station, Naoussa 59035, Greece. E-mail: pdrogoudi@elgo.gr

G. Echeverria, Postharvest Program, Institute of Agrifood Research and Technology (IRTA), 25003 Lleida, Spain. E-mail: gemma.echeverria@irta.cat

H. Förster, Department of Microbiology and Plant Pathology, University of California, Riverside, California, USA. E-mail: helgaf@ucr.edu

S. Foschi, Rinova, Cesena, Italy. E-mail: sfoschi@rinova.eu

K. Gasic, Department of Plant and Environmental Sciences, Clemson University, SC 29631, USA. E-mail: kgasic@clemson.edu

D. Giovannini, CREA, Research Centre for Olive, Fruit and Citrus Crops, 47121 Forlì, Italy. E-mail: daniela.giovannini@crea.gov.it

T.M. Gradziel, Department of Plant Sciences, University of California, Davis, California, USA. E-mail: tmgradziel@ucdavis.edu

I. Iglesias, Agromillora Group, Plaça M. Raventós, 3. 08770 Sant Sadurní d'Anoia, Spain. E-mail: iiglesias@agromillora.com

G. Lang, Michigan State University, East Lansing, Michigan, USA. E-mail: langg@msu.edu

G.A. Manganaris, Department of Agricultural Sciences, Biotechnology & Food Science, Cyprus University of Technology, 3603 Lemesos, Cyprus. E-mail: George.manganaris@cut.ac.cy

J.C. Melgar, Department of Plant and Environmental Sciences, Clemson University, USA. E-mail: jmelgar@clemson.edu

P. Milonas, Benaki Phytopathological Institute, Kifissia, Greece. E-mail: p.milonas@bpi.gr

I.S. Minas, Department of Horticulture and Landscape Architecture, Colorado State University, Fort Collins, CO 80523, USA. E-mail: Ioannis.Minas@colostate.edu

B. Morandi, Department of Agri-Food Sciences and Technologies, University of Bologna, Bologna, Italy. E-mail: brunella.morandi@unibo.it

M.Á. Moreno, Pomology Department, Estación Experimental de Aula Dei - CSIC, 50080 Zaragoza, Spain. E-mail: mmoreno@eead.csic.es

A.L. Nielsen, Department of Entomology, Rutgers University, New Brunswick, New Jersey, USA. E-mail: nielsen@njaes.rutgers.edu

N.T. Papadopoulos, Department of Agriculture, Crop Production and Rural Environment, University of Thessaly, Volos, Greece. E-mail: nikopap@uth.gr

J.R. Pieper, Department of Horticulture and Landscape Architecture, Colorado State University, Fort Collins, CO 80523, USA. E-mail: jeff.pieper@colostate.edu

G. Reig, Fruit Production Programme, Institute of Agrifood Research and Technology (IRTA), Edifici Fruitcentre, Parc Agrobiotech Lleida, Parc de Gardeny, 25003 Lleida, Spain. E-mail: reiggemma@gmail.com

G.L. Reighard, Department of Plant and Environmental Sciences, Clemson University, SC 29634, USA. E-mail: grghrd@clemson.edu

D.F. Ritchie, Department of Entomology and Plant Pathology, North Carolina State University, Raleigh, North Carolina, USA. E-mail: david_ritchie@ncsu.edu

S. Sansavini, Department of Agri-Food Sciences and Technologies, University of Bologna, Bologna, Italy. E-mail: silviero.sansavini@unibo.it

G. Schnabel, Department of Plant and Environmental Sciences, Clemson University, Clemson, South Carolina, USA. E-mail: schnabe@clemson.edu

P. Tonutti, Crop Science Research Center, Scuola Superiore Sant'Anna, Pisa, Italy. E-mail: pietro.tonutti@santannapisa.it

L. Trainotti, Department of Biology, University of Padua, Padua, Italy. E-mail: livio.trainotti@unipd.it

A.R. Vicente, University of La Plata, Calle 60 y 119, La Plata, CP 1900, Argentina. E-mail: arielvicente@gmail.com

C. Xiloyannis, Dipartimento delle Culture Europee e del Mediterraneo: Architettura, Ambiente, Patrimoni Culturali – University of Basilicata, Potenza, Italy. E-mail: cristosxiloyannis15@gmail.com

1

PEACH: AN INTRODUCTION

George A. Manganaris[1]*, Silviero Sansavini[2], Tom M. Gradziel[3], Daniele Bassi[4] and Carlos H. Crisosto[3]

[1]*Department of Agricultural Sciences, Biotechnology & Food Science, Cyprus University of Technology, Lemesos, Cyprus; [2]Department of Agri-Food Sciences and Technologies, University of Bologna, Bologna, Italy; [3]Department of Plant Sciences, University of California, Davis, California, USA; [4]Faculty of Agriculture and Food Science, University of Milan, Milan, Italy*

1.1 HISTORY

Peach (*Prunus persica* (L.) Batsch) as well as nectarine (which is a natural mutation of peach that lacks fuzz) belong to the botanical family Rosaceae (Faust and Timon, 1995). Peach originated in China and came to Persia (modern-day Iran) via silk trading routes in the second or third century BC where they were widely cultivated and became known as 'Persian apple'. Alexander the Great encouraged the spread of peach into the Mediterranean region with the Roman army bringing it to Italy by the first century BC where images of peaches can still be found on the walls of Herculaneum and Paestum (Italy), preserved despite the destruction of Vesuvius (Bassi and Monet, 2008).

Although introduced from Iran, the origin of peach from China is well supported in the *Origin of Cultivated Plants* (de Candolle, 1883), with further support in *The Peaches of New York* (Hedrick, 1917). Chinese literature references peach more than 1000 years before it first appeared in any European writings. Furthermore, there is documented evidence of peach cultivation in China more than 3000 years ago. In Taoist mythology, the peach is a symbol of immortality (Layne and Bassi, 2008). The Queen Mother (goddess) of the West (Xi Wang Mu) had a jade palace that was surrounded by a beautiful garden containing the peach trees of immortality. In the classic Chinese novel *The Journey to the West* (Wu Ch'eng-en, 1590 AD, translated by Anthony C. Yu), Sun Wukong, or the Monkey King, attained immortality because of a memorable visit to this garden (Layne and Bassi, 2008):

* Email: george.manganaris@cut.ac.cy

© CAB International 2023. *Peach* (G. Manganaris, G. Costa and C. Crisosto eds)
DOI: 10.1079/9781789248456.0001

'I have been authorized by the Jade Emperor,' said the Monkey King, 'to look after the Garden of Immortal Peaches. The local spirit hurriedly saluted him and led him inside.' The Monkey King then asked the local spirit, 'How many trees are there?' 'There are three thousand six hundred,' said the local spirit. 'In the front are one thousand two hundred trees with little flowers and small fruits. These ripen once every three thousand years, and after one taste of them a man will become immortal with healthy limbs and a lightweight body. In the middle are one thousand two hundred trees of layered flowers and sweet fruits. They ripen once every six thousand years. If a man eats them, he will ascend to Heaven with the mist and never grow old. At the back are one thousand two hundred trees with fruits of purple veins and pale yellow pits. These ripen once every nine thousand years and, if eaten, will make a man's age equal to that of Heaven and Earth, the sun, and the moon.' Highly pleased by these words, the Monkey King made a thorough inspection of the trees and a list of the arbors and pavilions before returning to his residence. One day he saw that more than half of the peaches on the branches of the older trees had ripened, and he wanted very much to eat one and sample its novel taste. Closely followed, however, by the local spirit of the garden, the stewards, and the divine attendants, he found it inconvenient to do so. He therefore devised a plan on the spur of the moment and said to them, 'Why don't you all wait for me outside and let me rest a while in this arbor?' The various immortals withdrew accordingly. The Monkey King then took off his cap and robe and climbed up onto a big tree. He selected the large peaches that were thoroughly ripened and plucking many of them, ate to his heart's content right on the branches.

After the establishment of European colonies on the East Coast of North America (~1650s), the English, French and Spanish also brought and planted peaches in the New World. In the 16th century, peach was introduced to Mexico, most likely by the Spanish, and from there to St Augustine, Florida, and subsequently to Virginia. In the early 18th century, Spanish missionaries introduced the peach to California, which eventually became one of the most important peach (fresh and canning) production regions. About the same time, the Russians apparently brought peach seeds or trees by ship to western North America, planting them at Fort Ross near what is now Jenner, California. During and following the Gold Rush, early settlers introduced additional peach selections from their homelands. Regional selection and further genetic improvements resulted in the development of new cultivars adapted to the climate, soils and availability of water for irrigation (Crisosto *et al.*, 2020).

1.2   WORLD PEACH PRODUCTION AND CULTIVATED AREA

Currently, there are approximately 24.6 million t of peaches produced yearly, with China accounting for almost 60%, followed by the European Union (EU; Spain, Italy and Greece) at 15% and the USA at approximately 3%. A significant growth of production over the last 20 years has been seen due to advancements

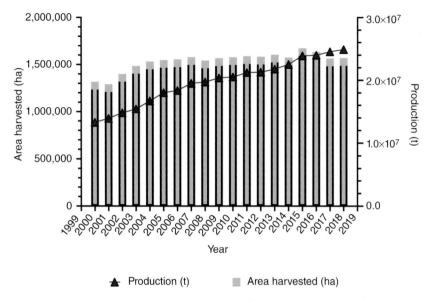

Fig. 1.1. World peach production and harvested area during the period 2000–2018. Data from FAOSTAT (www.fao.org/faostat; accessed 1 December 2020).

in fruit production systems (Fig. 1.1). However, a descending trend has been registered over recent years in several key peach-producing countries, mainly due to the increased labour costs and reduced revenue for the farmer (Manganaris *et al.*, 2022).

China produces around 15 million t on over 780,000 ha. Traditionally, the main peach-producing province is Shandong, followed by the provinces of Hebei, Henan, Hubei and Shanxi; such areas are characterized by their favourably dry and thus low-disease growing conditions. Lately, peach acreage seems to be increasing in Sichuan and Hunan provinces where citrus was overproduced, and in Yunnan, Guizhou, Fujian, Jiangxi and Guangxi provinces where peach production is developing at higher elevations and so with higher chill requirements. In addition, peach greenhouse production and high-density cultivation have recently emerged in limited areas of northern China. Due to the addition of these new production areas combined with enhanced production efficiency, China is steadily increasing production, having increased from 40% to 60% of the total world peach production (Fig. 1.2).

The EU is the second largest producer of peach after China, with an average annual production of 3,612,000 t in the period 2018–2020 and a total harvested area of 206,660 ha in 2019 (Iglesias and Echeverria, 2022). Spain is the first country in the ranking, with 77,464 ha and 1,480,000 t year[-1], followed by Italy and Greece (Europêch, 2021). Annual exports for the 2018–2020 period amounted to 55% of total production, corresponding to 826,100 t. Nectarine represents 41% of total annual production, followed by peach (21% flat and 18%

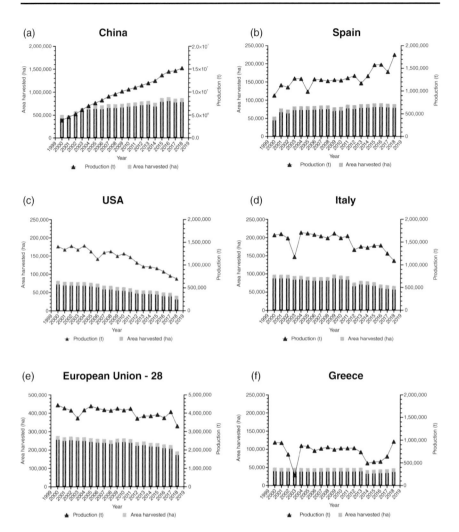

Fig. 1.2. Peach production and harvested area in the main peach-producing countries during the period 2000–2018. Data from FAOSTAT (www.fao.org/faostat; accessed 1 December 2020).

round) and clingstone cultivars (20%). The provinces of Catalonia, Aragón and Murcia, all regions located in the Mediterranean Basin, are the most important areas of peach production in Spain (Iglesias and Echeverria, 2022).

Italy, with 1.2 million t produced on approximately 67,000 ha, has most production in the northern regions, especially Emilia-Romagna (the country's top producing region) and Piedmont, with other production in more southern areas, primarily Campania.

Greece is in fourth position with 900,000 t on about 41,000 ha. The bulk of Greek production is in the northern part of the country, with Imathia and

Pella being the key producing counties. Peach cultivation has additionally expanded to central Greece, in addition to the southern part for the early-ripening cultivars.

The USA ranks as the fifth largest producer with 824,000 t on approximately 45,000 ha. Over the past 12 years, production fell from 1,100,000 t to 651,500 t, representing almost a 40% reduction (fresh and processed), although total market value remains relatively unchanged. Production is confined mainly to California, South Carolina and Georgia. California produces nearly 56% of the USA fresh peach crop and about 96% of canning peaches (USDA/NASS, 2020). South Carolina produces approximately 75,000 t, Georgia around 39,000 t, Pennsylvania and New Jersey around 19,500 t, Colorado 14,300 t and Washington 11,150 t. California fresh peach and nectarine production is estimated at 60 million packages of 11.4 kg from more than 450 cultivars (CDFA, 2022). In the San Joaquin Valley, harvest of early cultivars starts in late April, and is completed for late cultivars by early October. In recent years, there has been a large increase in the production of white-fleshed and yellow-fleshed, low-acid peach and nectarine cultivars. However, California is characterized by the lowest farm gate prices among all areas of the USA where peach is being produced.

In general, world peach production is gradually increasing, mainly due to China. Overall, peach production in the EU has been relatively steady with a descending trend (FAO, 2020), compensating for Italy's decline. With the aim of advancing peach fruit production and consumption, there is an urgent need to dissect solutions to valorize on the market the exceptional diversity and flavour potential of peach, already present in the varietal landscape (Manganaris *et al.*, 2022).

Innovative companies that are reducing production costs, selecting more flavourful cultivars and producing ripe fruit for improved eating quality are still profiting from producing fresh peaches. The current approach for increased consumption includes greater releases of white- and yellow-fleshed peach cultivars with enhanced flavour, including high total soluble solids, a high aromatic profile, a desirable texture and attractive appearance. Low acidity combined with high total soluble solids and flat peach (easy-to-eat) cultivars have been launched from active breeding programmes around the world (Iezzoni *et al.*, 2020). These new cultivars that are currently being introduced represent the continuing efforts to provide improved flavour with enhanced nutritional and eating qualities, with recent consumer feedback being very promising.

1.3   TAXONOMY AND BOTANY

Peach is member of the family Rosaceae and is grouped within the genus *Prunus* and is included in the section *Euamygdalus* Schneid of the subgenus *Amygdalus* (Hedrick, 1917). Peach can be distinguished from almond (*Prunus dulcis* (Mill.) D.A. Webb) because the mesocarp of the latter becomes dry and splits at maturity and the leaves are serrulate (Rehder, 1990).

Peach is a diploid species ($2n = 16$) with a medium tree height (up to 8 m); the leaves are lanceolate, glabrous and serrate, broadest near the middle, often with a glandular petiole. The flowers are generally pink but can also be white or red. The fruit skin is pubescent or glabrous, while the mesocarp is fleshy and does not split. The stony endocarp is very deeply pitted, wrinkled and very hard (Bassi and Monet, 2008).

Peach has a perigynous perfect flower, which develops into fruit and contains both male and female reproductive structures. Fruit develops from a single carpel (the ovary, containing two ovules) and bears one or two seeds (Fig. 1.3). Flower and fruit anatomy are tightly linked; basal portions of the petals, calyx and stamen are fused into hypanthium tissue and form a cup-like structure around the ovary (perigynous).

The ovary is in an intermediate position relative to the petals, calyx and stamen, characterized by lack of fuzz in the case of nectarine (Fig. 1.4). The

Fig. 1.3. Peach flower morphology.

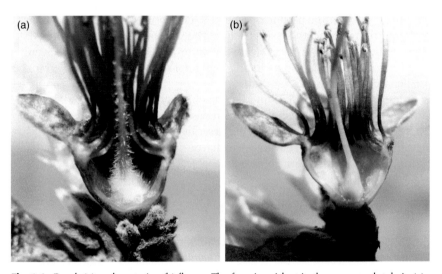

Fig. 1.4. Peach (a) and nectarine (b) flower. The fuzz is evident in the ovary and style in (a).

fruit has no sepals, petals or stamen residues, and the hypanthium tissue is shed after fruit set.

The overall tree structure (growth habit) is determined by the number and location of lateral shoots produced per year, the strength of apical control and branch angles and internode length. Most commercial peach cultivars have a relatively strong apical dominance, thus producing fewer lateral shoots in the current year growth. There are distinct bearing and growth habits, depending on branch angle and internode length. In most cultivars, flower buds are formed laterally at the leaf base, while the vegetative buds are in the middle (Fig. 1.5). Details on tree structure and bearing habits are discussed in Chapters 2 and 3.

Botanically, stone fruits are drupes with a fleshy texture showing: (i) a thin, edible outer skin (exocarp or epicarp) derived from the ovary; (ii) an edible flesh of varying thickness beneath the skin (mesocarp); and (iii) a lignified inner wall (endocarp), commonly referred to as the stone or pit, enclosing one seed (or very rarely two) (Fig. 1.6). The seed includes the following parts, starting from the outside: testa, cotyledons and embryo.

Fig. 1.5. One-year old shoots (black arrows) form the main fruiting body of a peach tree.

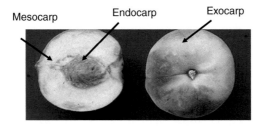

Fig. 1.6. Structure of a typical peach fruit, composed of the exocarp, mesocarp and endocarp.

The skin (epicarp) is a protective layer composed of cuticle, epidermis cells and some hypodermal cell layers. The cuticle has a thin coating of wax to reduce water loss and to protect the fruit against mechanical injury and pathogens. The epidermis is responsible for most of the skin's mechanical strength and consists of thick-walled, flexible cells, and surface trichomes or hairs (known as 'fuzz', and absent in nectarines), being an extension of some epidermal cells.

Anthocyanins, the glycoside derivatives of the anthocyanidins, are responsible for the colours ranging from purple to red and are localized mainly within the epidermis cell vacuoles or in the flesh parenchyma cells. Anthocyanin pigments are synthesized from flavonoids via phenylalanine. The presence of anthocyanins in skin and/or flesh can be quantitative or qualitative and is independent of the skin/flesh ground colour (yellow or white). The quantitative trait is influenced positively by light exposure with maximum concentration occurring at the full ripe stage, while the qualitative trait, expressed only in the epidermis, is not related to light or to ripening (Bassi and Monet, 2008). When anthocyanins are present in the flesh, their localization is mainly just under the skin and/or close to the pit. The amount and distribution of anthocyanins throughout the mesocarp depends on the cultivar (quantitative trait), and may be variably expressed among the fruit on a given tree (Lesley, 1957). However, one more red fruit pigmentation (i.e. the 'red flesh', or 'blood' flesh) trait should be mentioned. In this phenotype, almost all the flesh is heavily stained by anthocyanins (Gallesio, 2003), independently of the basic skin/flesh colour, either white or yellow, while showing a dull purple skin. At least two Mendelian 'blood' flesh traits are known: (i) *Bf*, which induces the early development of anthocyanins in the fruit flesh beginning at pit hardening and is associated with red midrib leaf veins (often on small trees); and (ii) *DBF*, where the mesocarp is fully red around 2 weeks before ripening, with no red on the midrib leaf vein.

The flesh (mesocarp), which is the main edible portion of the fruit, consists mainly of storage parenchyma tissue of large and relatively thin-walled cells with a high content of water, sugars, acids and nutrients, including numerous compounds showing health-promoting benefits (Crisosto and Valero, 2008). Carotenoids localized in chloroplasts (chromoplasts) are liposoluble pigments that belong to the subgroup of isoprenoids with typically 40 carbons in their polyene backbone with conjugated double bonds and rings at the ends. The extensively conjugated double rings allow carotenoids to absorb visible light, yielding yellow, orange and occasionally red colours (Vicente *et al.*, 2009). Yellow-fleshed fruit shows higher carotenoid levels including carotenes (orange pigments: α-carotene, β-carotene) and xanthophylls (yellow pigments: antheraxanthin, luteoxanthin, zeaxanthin, lutein) than white-fleshed cultivars. Xanthophylls are synthesized via hydroxylation from carotenoids: lutein from carotene, and zeaxanthin, antheraxanthin and violaxanthin from β-carotene (Demmig-Adams and Adams, 2002). In a survey of 25 Californian cultivars, total carotenoid concentrations varied considerably (Gil *et al.*, 2002). Among the carotenoids, β-carotene and β-cryptoxanthin are the primary provitamin A factors, with concentrations reaching 2000 µg kg^{-1} of fresh weight for the former and up to 3400 µg kg^{-1} of fresh weight for the latter (Tourjie *et al.*, 1998). High carotene levels mask oxidation from bruising or other blemishes reducing harvesting losses on yellow-fleshed cultivars. White peaches are attractive to consumers for their distinct flavour and/or aroma (Robertson *et al.*, 1990; Crisosto *et al.*, 2001), although some of them are either too soft or too susceptible to skin bruising and flesh browning to be commercially successful. Carotenoids are rather heat stable relative to anthocyanins, which are very labile and subject to browning in canning operations; this has led to the selection of yellow-flesh canning peaches that are anthocyanin free. As localization of anthocyanins in skin is independent of that in the flesh, commercial canned peaches may or may not develop a red overcolour, as the anthocyanins are removed when the fruit is peeled before canning (Vicente *et al.*, 2009).

1.3.1   Flesh texture

At least four distinct peach phenotypes exist based on ripening/softening patterns, although their biological basis is not fully understood. The two best-characterized phenotypes are the melting (M) and non-melting (NM) types. A wide range of diverse metabolic pathways during fruit ripening of fresh fruits with melting-flesh and non-melting-flesh characteristics have been reported (Manganaris *et al.*, 2006). The M texture shows a prominent softening during the late-ripening phase, resulting in a delicate texture and increased juiciness. Within cultivars in the M group, members are classified according to their rate of softening ranging from soft to medium to firm (Yoshida, 1976; Ramming, 1991). The freestone 'firm' subgroup (FM) softens slowly, and is firm near commercial harvest, being less prone to bruising during handling, allowing easier management of harvest timing, grading and shipping operations, which further results in a potentially longer shelf life.

The NM phenotype (the so-called 'canning peach') has a firm texture when fully mature and softens very slowly but when overripe may become rubbery as the flesh never fully melts. Some of these cultivars also display a distinctive off-flavour (Sherman *et al.*, 1990; Crisosto *et al.*, 1999, 2008). However, it has been demonstrated that it is possible to select for the absence of off-flavours within breeding progeny (Beckman and Sherman, 1996). The lack of full softening (melting) in the NM phenotype is related to the loss of endopolygalacturonase (endoPG) activity, as this enzyme is responsible for cleaving pectins (polygalacturonic acid chains) from the cell wall in M fruits (Lester *et al.*, 1996; Brummell *et al.*, 2004; Peace *et al.*, 2005; Iezzoni *et al.*, 2020). The M and NM phenotypes develop high levels of ethylene between fruit growing stages III and IV (Tonutti *et al.*, 1991; Mignani *et al.*, 2006). The NM flesh is also much less susceptible to mealiness, a common chilling injury storage disorder, as it does not go through the melting phase (Brovelli *et al.*, 1998; Crisosto *et al.*, 2008), but some of these cultivars are highly susceptible to flesh browning and other symptoms of chilling injury (Manganaris *et al.*, 2019). Notably, a large quantitative trait locus (a DNA zone containing several genes responsible for a given quantitative trait) for mealiness was detected near the endoPG locus, confirming the previous observation that this disorder occurs particularly in M freestone phenotypes (Peace *et al.*, 2005; Martínez-García *et al.*, 2012).

1.3.2   Freestone versus clingstone trait

Based on the separation or adherence of endocarp and mesocarp, peaches can be segregated into two groups: freestone (where the endocarp does not adhere to the flesh) and clingstone (where flesh adheres to the endocarp). Intermediate types (semi-freestone or semi-clingstone) with varying degrees of adhesion have also been observed, particularly in early-maturing cultivars (Weinberger, 1950). Because the rapid flesh maturation, due to early ripening, delays the appearance of the freestone phenotype, the very early semi-freestone phenotypes should be regarded as 'physiologically clingstone' but 'genetically freestone' (Beckman and Sherman, 1996). As the intensity of these traits changes during fruit ripening, flesh adherence to the endocarp in early-ripening cultivars should be assessed at the fully ripe stage, or even at the early stage of senescence, to assure reliable phenotyping. Based on endocarp adherence and flesh texture (M and NM), three alleles at a single locus (F) for the endoPGase enzymes were found to control the three flesh-adherence phenotypes: freestone/melting, clingstone/melting and clingstone/non-melting. A diagnostic polymerase chain reaction (PCR) test has been made available to detect all four alleles (Lester *et al.*, 1996; Peace *et al.*, 2005; Iezzoni *et al.*, 2020).

1.3.3   Stony-hard trait

A third flesh texture, the stony-hard (SH) phenotype is described as very firm and crispy (Yoshida, 1976). However, this type never melts, as in 'Yumyeong', a white-fleshed peach from Korea. SH texture resembles the NM phenotype but does not become rubbery when overripe, always keeping the crispy structure. A remarkable difference of SH compared with NM is the almost complete lack of ethylene production in the SH phenotype during ripening (Goffreda, 1992; Haji *et al.*, 2001, 2003; Tatsuki *et al.*, 2006), although ethylene can be a trigger for a stress response (Tatsuki *et al.*, 2006; Begheldo *et al.*, 2008). The lack of ethylene evolution is due to the transcription suppression (and not to mutation) of the 1-aminocyclopropane-1-carboxylic acid synthase isogene (*Pp-ACS1*), a key gene of the ethylene enzymatic pathway (Tatsuki *et al.*, 2006). Independent inheritance of SH flesh from the *M/NM* locus has been demonstrated, suggesting an epistatic effect of SH (*hd* gene), as, when exogenous ethylene is applied, the SH/M phenotype (*hdhd/F-*) is induced to melt, while the SH/NM phenotype (*hdhd/f1f1*) remains firm and crisp (Haji *et al.*, 2005). From a practical point of view, SH fruits are often very difficult to distinguish from NM or very firm/unripe M phenotypes. Therefore, evaluation of progenies from controlled crosses segregating for SH is puzzling. SH flesh identification on the tree is very time consuming (several measurements over time are required to characterize firmness evolution), and sensory evaluation is not always reliable. So far, the only sound method for SH texture phenotyping is to measure ethylene production (Goffreda, 1992). However, a molecular marker is now available to score the trait.

1.3.4   The slow-softening trait

A fourth Mendelian flesh texture trait resembles the SH flesh in texture, firmness and crispiness but when fully ripe becomes melting and evolves ethylene (Bassi and Monet, 2008; Ciacciulli *et al.*, 2018a,b). This flesh texture, firmer than the 'firm' M type, is found in many cultivars (including nectarines such as 'Big Top' and standard peaches such as 'Rich Lady' and 'Diamond Princess') and was developed from the end of the 1990s. It was first commercially introduced by private breeders from California, and is sometimes (e.g. 'Big Top' and others) associated with the low-acid trait. This phenotype shows a slow softening on the tree and a remarkable keeping quality that make it very desirable to growers. Detailed biochemical (physiological) and genetic studies are in progress to identify and understand this trait, as it seems to be dominant over the standard M type (Ciacciulli *et al.*, 2018b).

1.3.5   Endocarp (stone, pit) and seed

The endocarp has lignin deposited in the cell walls, with the outer surface being deeply wrinkled and pitted. In very early-ripening cultivars, lignification is limited and the endocarp may be rather soft. A pronounced funiculus ridge may be present near the ventral suture, with an undesirably sharp and lignified tip. Endocarp splitting (at the carpel suture) or shattering (radial fractures) may affect either early- or late-ripening cultivars, depending on orchard conditions. It is reported that cultivation practices to improve fruit size (e.g. supplemental irrigation, girdling, mineral nutrition) may also increase the incidence of endocarp splitting or shattering (Crisosto and Costa, 2008). These two undesired endocarp structural failures present commercial problems because of the potential consumer hazard of biting into stony fragments. In the canning industry, susceptible cultivars are not processed because the resulting fragments are difficult to eliminate from processed fruits.

The stone shape changes according to the fruit shape, from globose (in round fruit) to elliptic (in ovate or elliptic fruit) to roundoblate (in flat fruits). Fruit contains one seed (occasionally two), whose content makes them taste very bitter, but they are not unsafe unless eaten in quantity. The bitter taste is Mendelian, being dominant over the non-bitter. The seeds show a very high germination rate after stratification, their chilling requirement being related to the chilling requirement of the mother tree (Perez, 1990). Embryo viability can be limited in very early-ripening cultivars, and aseptic embryo culture is required for seed recovery. This type of embryo rescue of otherwise abortive, immature embryos can occur as early as 50 days after full bloom (Ramming, 1985). *In vitro* embryo rescue is also a valuable technique that allows seedlings to grow from very early ripening parents used as seed parents in breeding for early-ripening cultivars (Cantabella *et al.*, 2020).

1.4   SPECIES RELATED TO PEACH

Peach relatives show very poor fruit-eating quality but are of horticultural interest for disease resistance traits and/or as rootstocks (Bassi and Monet, 2008). The most interesting species that are considered close relatives of peach are briefly described below.

1.4.1   *Prunus davidiana* (Carr.) Franch.

Prunus davidiana (Carr.) Franch. is a wild species native to central China, where it is often used as a seedling rootstock. It shows tolerance to drought but is very sensitive to nematodes. The tree is tall (up to 10 m), with a reddish-brown bark. The leaves are long and glabrous, and are ovate-lanceolate and broadest near the base. The flower is white or light pink. The pit is small and pitted, and the flesh is freestone. Accessions of this species have been hybridized with peach to improve disease resistance on scion cultivars to plum pox, powdery mildew, leaf curl and other diseases (Moing *et al.*, 2003; Fresnedo-Ramírez *et al.*, 2015), or to breed interspecific rootstocks adaptable to marginal soils or to avoid replant problems (Pisani and Roselli, 1983).

1.4.2   *Prunus ferganensis* (Kost. and Rjab) Kov. and Kost.

This is a wild form found in western China classified as a subspecies of *P. persica*. A wide variability of fruit types can be found (e.g. yellow- and white-fleshed, fuzzless). It shows resistance to powdery mildew. The leaves have parallel veins and there are parallel grooves in the stone, both of which are single Mendelian traits (Okie and Rieger, 2003). The seed can be cyanogenic glycoside-free (not bitter).

1.4.3   *Prunus kansuensis* Rehd.

Prunus kansuensis Rehd. is a wild species found in north-east China, where it is used as a seedling rootstock. It produces a bushy tree with glabrous winter buds and is early blooming, and the flowers are somewhat resistant to frost (Meader and Blake, 1939). The leaves are villous along the midrib near the base and broadest below the middle, and the style is longer than the stamens. The fruit quality is poor (astringent), and the pit is furrowed (with parallel grooves) but not pitted.

1.4.4   *Prunus mira* Koehne

This is a wild species found in far-west China and eastern Tibet. The tree is tall (up to 20 m) and lives for up to 1000 years. The leaves are lanceolate, villous along the midrib beneath and rounded at base, and the flowers are white. The fruit is highly variable in shape, colour and size. The pit surface is smooth, although in some types it resembles *P. persica*. Some forms are cultivated in Tibet, and it is also used as a seedling rootstock in some regions of India. It is presumed to be an ancestor of *P. persica*, having spread south and east from the Himalayan mountains (Yoshida, 1987).

REFERENCES

Bassi, D. and Monet, R. (2008) Botany and taxonomy. In: Layne, D. and Bassi, D. (eds) *The Peach: Botany, Production and Uses*. CAB International, Wallingford, UK, pp. 1–36.

Beckman, T.G. and Sherman, W.B. (1996) The non-melting semi-freestone peach. *Fruit Varieties Journal* 50, 189–193.

Bevheldo, M., Manganaris, G.A., Bonghi, C. and Tonutti, P. (2008) Different postharvest conditions modulate ripening and ethylene biosynthetic and signal transduction path ways in stony hard peaches. *Postharvest Biology and Technology* 48, 84–91.

Brovelli, E.A., Brecht, J.K., Sherman, W.B. and Sims, C.A. (1998) Anatomical and physiological responses of melting-flesh and nonmelting-flesh peaches to postharvest chilling. *Journal of the American Society for Horticultural Science* 123, 668–674.

Brummell, D.A., Cin, V.D., Crisosto, C.H. and Labavitch, J.M. (2004). Cell wall metabolism during maturation, ripening and senescence of peach fruit. *Journal of Experimental Botany* 56, 2029–2039.

Cantabella, D., Dolcet-Sanjuan, R., Casanovas, M., Solsona, C., Torres, R. and Teixidó, N. (2020) Inoculation of *in vitro* cultures with rhizosphere microorganisms improve plant development and acclimatization during immature embryo rescue in nectarine and pear breeding programs. *Scientia Horticulturae* 273: 109643.

CDFA (2022) California Agricultural Production Statistics. California Department of Food and Agriculture, Sacramento, California. Available at: www.cdfa.ca.gov/Statistics/ (accessed 30 November 2022).

Ciacciulli, A., Chiozzotto, R., Attanasio, G., Cirilli, M. and Bassi, D. (2018a) Identification of a melting type variant among peach (*P. persica* L. Batsch) fruit textures by a digital penetrometer. *Journal of Texture Studies* 49, 370–377.

Ciacciulli A., Cirilli M., Chiozzotto, R., Attanasio G., da Silva C. *et al.* (2018b) Linkage and association mapping for the slow softening (SwS) trait in peach (*P. persica* L. Batsch) fruit. *Tree Genetics and Genomes* 14: 93.

Crisosto, C.H. and Costa, G. (2008) Preharvest factors affecting peach quality. In: Layne, D. and Bassi, D. (eds) *The Peach: Botany, Production and Uses*. CAB International, Wallingford, UK, pp. 536–549.

Crisosto, C.H. and Valero, D. (2008) Harvesting and postharvest handling of peaches for the fresh market. In: Layne, D. and Bassi, D. (eds) *The Peach: Botany, Production and Uses*. CAB International, Wallingford, UK, pp. 575–596.

Crisosto, C.H., Mitchell, F.G. and Ju, Z. (1999) Susceptibility to chilling injury of peach, nectarine, and plum cultivars grown in California. *Horticultural Science* 34, 1116–1118.

Crisosto, C.H., Day, K.R., Crisosto, G.M. and Garner, D. (2001) Quality attributes of white flesh peaches and nectarines grown under California conditions. *Journal of the American Pomological Society* 55, 45–51.

Crisosto, C.H., Crisosto, G.M. and Day, K.R. (2008) Market life update for peach, nectarine, and plum cultivars grown in California. *Advances in Horticultural Science* 22, 201–204.

Crisosto, C.H., Echeverria, G., and Manganaris, G.A. (2020) Peach and nectarine postharvest handling. In: Crisosto, C.H. and Crisosto, G.M. (eds) *Mediterranean Tree Fruits and Nuts Postharvest Handling*. CAB International, Wallingford, UK, pp. 53–87.

de Candolle, A. (1883) *L'Origine Delle Piante Coltivate*. Fratelli Dumolard, Milan, Italy.

Demmig-Adams, B. and Adams, W.W. (2002) Antioxidants in photosynthesis and human nutrition. *Science* 5601, 2149–2153.

Europêch (2021) Synthese de la recolte Europeenne 2020 et previsions de recolte: Pêches-nectarines et pavies. Montpellier, France.

FAO (2020) FAOSTAT Database. Food and Agriculture Organization of the United Nations, Rome. Available at: www.fao.org/faostat/ (accessed 23 January 2020).

Faust, M. and Timon, B. (1995) Origin and dissemination of Peach. *Horticultural Reviews* 17, 331–379.

Fresnedo-Ramírez J., Bink M.C.A.M., van de Weg E., Famula T.R., Crisosto C.H. *et al.* (2015) QTL mapping of pomological traits in peach and related species breeding germplasm. *Molecular Breeding* 35: 166.

Gallesio, G. (2003) Il trattato del pesco di Giorgio Gallesio. In: Baldini, E. (ed.) *Gli inediti trattati def pesco e def ciliegio. Complementi scientifici della "Pomona Italiana" di Giorgio Gallesio*. Accademia dei Georgofili. Florence, Italy, pp. 9–146.

Gil, M.I., Tomás-Barberán, F.A., Hess-Pierce, B. and Kader, A.A. (2002) Antioxidant capacities, phenolics compounds, carotenoids, and vitamin C content of nectarine, peach, and plum cultivars from California. *Journal of Agricultural and Food Chemistry* 50, 4976–4982.

Goffreda, J.C. (1992) Stony hard gene of peach alters ethylene biosynthesis, respiration and other ripening-related characteristics. *Horticultural Science* 27: 610.

Haji, T., Yaegaki, H. and Yamaguchi, M. (2001) Changes in ethylene production and flesh firmness of melting, non-melting and stony hard peaches after harvest. *Journal of the Japanese Society for Horticultural Science* 70, 458–459.

Haji, T., Yaegaki, H. and Yamaguchi, M. (2003) Softening of stony hard peach by ethylene and the induction of endogenous ethylene by 1-aminocyclopropane 1-carboxylic acid (ACC). *Journal of the Japanese Society for Horticultural Science* 72, 212–217.

Haji, T., Yaegaki, H. and Yamaguchi, M. (2005) Inheritance and expression of fruit texture melting, non-melting and stony hard in peach. *Scientia Horticulturae* 105, 241–248.

Hedrick, U.P. (1917) *The Peaches of New York*. J.B. Lyon, Albany, New York. Available at: https://archive.org/details/peachesofnewyork0002uphe/page/n7/mode/2up (accessed 18 January 2023).

Iezzoni, A.F., McFerson, J., Luby, J., Gasic, K., Whitaker, V. *et al.* (2020) RosBREED: bridging the chasm between discovery and application to enable DNA-informed breeding in rosaceous crops. *Horticulture Research* 7: 177.

Iglesias, I. and Echeverria, G. (2022) Current situation, trends and challenges for efficient and sustainable peach production. *Scientia Horticulturae* 296: 110899.

Layne, D. and Bassi, D. (2008) Preface. In: Layne, D. and Bassi, D. (eds) *The Peach: Botany, Production and Uses*. CAB International, Wallingford, UK, pp. xiii–xiv.

Lesley, J.W. (1957) A genetic study of inbreeding and of crossing inbred lines in peaches. *Proceedings of the American Society for Horticultural Science* 70, 93–103.

Lester, D.R., Sherman, W.B. and Atwell, B.J. (1996) Endopolygalacturonase and the melting flesh (*M*) locus in peach. *Journal of the American Society for Horticultural Science* 121, 231–235.

Manganaris, G.A., Vasilakakis, M., Diamantidis, G. and Mignani, I. (2006) Diverse metabolism of cell wall components of melting and non-melting peach genotypes during ripening after harvest or cold storage. *Journal of the Science of Food and Agriculture* 86, 243–250.

Manganaris, G.A., Vincente, A.R., Martinez, P. and Crisosto, C.H. (2019) Postharvest physiological disorders in peach and nectarine. In: Tonetto de Freitas, S. and Pareek, S. (eds) *Physiological Disorders in Fruits and Vegetables*. CRC Press, Boca Raton, Florida, pp. 253–264.

Manganaris, G.A., Minas, I., Cirilli, M., Torres, R., Bassi, D. and Costa, G. (2022) Peach for the future: a specialty crop revisited. *Scientia Horticulturae* 305: 111390.

Martínez-García, P., Peace, C.P., Parfitt, D., Ogundiwin, E., Fresnedo-Ramirez, J. *et al.* (2012) Influence of year and genetic factors on chilling injury susceptibility in peach (*Prunus persica* (L.) Batsch). *Euphytica* 185, 267–280.

Meader, E.M. and Blake, M.A. (1939) Some plant characteristics of the second-generation progeny of *Prunus persica* and *Prunus kansuensis* crosses. *Proceedings of the American Society for Horticultural Science* 37, 223–231.

Mignani, I., Ortugno, C. and Bassi, D. (2006) Biochemical parameters for the evaluation of different peach flesh types. *Acta Horticulturae* 713, 441–448.

Moing, A., Poessel, J.L., Svanella-Dumas, L., Loonis, M. and Kervella, J. (2003) Biochemical basis of low fruit quality of *Prunus davidiana*, a pest and disease donor for peach breeding. *Journal of the American Society for Horticultural Science* 128, 55–62.

Okie, W.R. and Rieger, M. (2003) Inheritance of venation pattern in *Prunus ferganensis* × *persica* hybrids. *Acta Horticulturae* 622, 261–264.

Peace, C.P., Crisosto, C.H and Gradziel, T.M. (2005) Endopolygalacturonase: a candidate gene for *Freestone* and *Melting Flesh* in peach. *Molecular Breeding* 16, 21–31.

Perez, G.S. (1990) Relationship between parental blossom season and speed of seed germination. *Horticultural Science* 25, 958–960.

Pisani, P.L. and Roselli, G. (1983) Interspecific hybridization of *Prunus persica* x *P. davidiana* to obtain new peach rootstocks. *Genetica Agraria* 37, 197–198.

Ramming, D.W. (1985) *In ovulo* embryo culture of early maturing *Prunus*. *Horticultural Science* 20, 419–420.

Ramming, D.W. (1991) Genetic control of a slow-ripening fruit trait in nectarine. *Canadian Journal of Plant Science* 71, 601–603.

Rehder, A. (1990) *Manual of Cultivated Trees and Shrubs Hardy in North America*. Discords Press, Portland, Oregon.

Robertson, J.A., Horvat, R.J., Lyon, B.G., Meredith, F.I., Senter, S.D. and Okie, W.R. (1990) Comparison of quality characteristics of selected yellow- and white-fleshed peach cultivars. *Journal of Food Science* 55, 1308–1311.

Sherman, W.B., Topp, B.L. and Lyrene, P.M. (1990) Non-melting flesh for fresh market peaches. *Proceedings of the Florida State Horticultural Society* 103, 293–294.

Tatsuki, M., Haji, T. and Yamaguchi, M. (2006) The involvement of 1-aminocyclopropane 1-carboxylic acid synthase isogene, *Pp-ACS 7*, in peach fruit softening. *Journal of Experimental Botany* 57, 1281–1289.

Tonutti, P., Casson, P. and Ramina, A. (1991) Ethylene biosynthesis during peach fruit development. *Journal of the American Society for Horticultural Science* 116, 274–279.

Tourjie, K.R., Barret, D.M., Romero, M.V. and Gradziel, T.M. (1998) Measuring flesh colour variability among processing clingstone peach genotypes differing in carotenoid composition. *Journal of the American Society for Horticultural Science* 123, 433–437.

USDA/NASS (2020) 2020 State Agriculture Overview. US Department of Agriculture, National Agricultural Statistics Service, Washington, DC. Available at: www.nass. usda.gov/Quick_Stats/Ag_Overview/stateOverview.php?state=CALIFORNIA (accessed 1 April 2020).

Vicente, A.R., Manganaris, G.A., Sozzi, G.O. and Crisosto, C.H. (2009) Nutritional quality of fruits and vegetables. In: Florkowski, W., Shewfelt, R.L., Brueckner, B. and Prussia, S.E. (eds) *Postharvest Handling: A Systems Approach*. Elsevier, Academic Press, Oxford, UK, pp. 57–106.

Weinberger, J.H. (1950) Chilling requirement of peach varieties. *Proceedings of the American Society for Horticultural Science* 56, 122–128.

Yoshida, M. (1976) Genetical studies on the fruit quality of peach varieties. 3: texture and keeping quality. *Bulletin of the Tree Research Station* 3, 1–16.

Yoshida, M. (1987) Peach germplasm. *Kajitsu Nippon* 42, 70–74.

2

PEACH TREE ARCHITECTURE: TRAINING SYSTEMS AND PRUNING

Ignasi Iglesias[1], Gregory L. Reighard[2] and Gregory Lang[3]*

[1]Agromillora Group, Sant Sadurní d'Anoia, Barcelona, Spain; [2]Department of Plant and Environmental Sciences, Clemson University, South Carolina, USA; [3]Department of Horticulture, Michigan State University, East Lansing, Michigan, USA

2.1 TRAINING SYSTEMS

2.1.1 Orchard systems and production economics

The main cost in peach/nectarine production is labour, mainly for harvest, thinning and pruning (Fig. 2.1). Labour represents around half the total production cost for a mid-season harvest cultivar trained to a Spanish gobelet shape. This varies significantly among regions, mainly due to the huge differences in labour costs among countries, which currently ranges from €0.8 to €16 h^{-1} for Tunisia and California (USA), respectively. Although crop protection, fertilization and soil management together comprise the highest cost category, harvest is nearly as much and comprises the greatest single labour cost, followed by thinning. Regardless of country, the trends over the last two decades have included increased labour costs, a lack of specialized workers and a scarcity of labour (Neri *et al.*, 2015; Iglesias, 2022).

Flower and fruit thinning, winter and summer pruning, and harvest and crop-protection costs can be reduced significantly by changing the tree architecture by moving to smaller and two-dimensional (planar) canopies that are more accessible to labour, machines and light, resulting in a more efficient use of inputs and increased fruit quality (Chalmers *et al.*, 1978; Barthélémy *et al.*, 1991). Changing the canopy from the traditional three-dimensional volume to a planar structure facilitates a significant increase in efficiency. For 'Luciana', the harvest rate ranged from 120 kg (person h)$^{-1}$ for mature Spanish gobelet trees to 225 kg (person h)$^{-1}$ for planar canopies assisted by platforms. Both canopy architecture and dimensions affect the total cost of production.

---

* Email: langg@msu.edu

© CAB International 2023. *Peach* (G. Manganaris, G. Costa and C. Crisosto eds)   17
DOI: 10.1079/9781789248456.0002

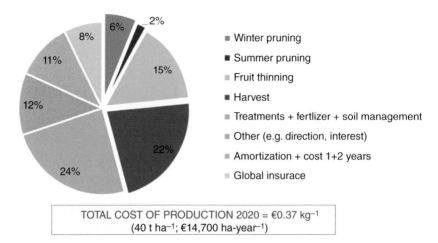

Fig. 2.1. The 2020 cost of production for 'Luciana', a mid-season nectarine with a 40 t ha⁻¹ yield, trained in a Spanish gobelet, spacing 3 × 5 m and expected lifespan of 12 years in the Ebro Valley, Spain. Adapted from Iglesias and Echeverria (2022).

Developing planar canopies and using mechanization for pruning, thinning and harvest can reduce the total cost of production by 25–30%, saving €2957 ha⁻¹ year⁻¹ for a mid-season cultivar in north-east Spain (Iglesias and Echeverría, 2022). For intensive orchards, the cost of planting is about twice that compared with standard open-vase trees; however, the reduced annual cost of production results in better economic sustainability (grower profits with reasonable prices), as well as better environmental sustainability (improved efficiency and use of inputs such as pesticides) (Iglesias, 2021, 2022; Iglesias and Echeverría, 2022).

2.1.2   Rootstocks and training systems

Modern peach orchards should be designed to be labour and input efficient. An efficient orchard design begins with the cultivar and rootstock combination. Cultivars vary in fruiting habit, which directly affects pruning (Iglesias and Echeverría, 2009; Sutton *et al.*, 2020). Most cultivars produce the best fruit quality on 1-year-old proleptic shoots, particularly early- and mid-season cultivars. Mid- or late-season cultivars, such as 'Elegant Lady' or 'O'Henry', tend to produce more on old wood, requiring specific pruning techniques. In addition, inherent canopy vigour can vary by cultivar; for example, the 'Rich' cultivar series usually has higher vigour than standard cultivars such as 'O'Henry'.

Orchard design is highly dependent on the selection of the rootstock, which, when combined with the cultivar's vigour and fruiting habit, is crucial for optimizing efficiencies throughout the productive life of the orchard. The range of rootstocks available for peach production includes an extensive list of different *Prunus* spp. and/or interspecific hybrids, but typically only a few are used in each region of a country. The choice of rootstock can confer important and different traits to the tree, including vigour, fruit quality, adaptation to soil conditions (sensitivity to iron chlorosis, tolerance to water logging, replanting situations) and cold hardiness. Interspecific hybrids (e.g. INRA® 'GF-677', 'Garnem®' and 'Cadaman®'), various selected seedlings and cultivars of plum (mainly *Prunus insititia*) are the main rootstocks used in European countries (Iglesias and Echeverría, 2022). In the USA, the most common rootstocks are peach seedlings of 'Nemaguard', 'Nemared', 'Lovell', 'Halford', 'Bailey' and 'Guardian®', depending on the region.

Other peach seedlings, plums and interspecific *Prunus* hybrids, such as 'Controller™ 5', 'Controller™ 6', 'Montclar®', 'Adesoto® 101', 'Ishtara®', 'Penta', or some of the Rootpac® series, have been used as peach rootstocks, especially for vigour control (DeJong *et al.*, 2005; Iglesias, 2018; Iglesias *et al.*, 2020; Iglesias and Echeverría, 2022). Typically, adaptation to specific soil conditions and rootstock-modulated tree vigour are the most important traits to be considered in the design of efficient orchards and selection of appropriate training systems. The range of vigour provided by currently available rootstocks is illustrated in Fig. 2.2. Currently, the most common rootstocks used are vigorous, as is the case of 'GF-677' and 'Garnem®' in Europe or 'Nemaguard' and 'Guardian®' in the USA. In addition to vigour control, some new rootstocks confer better yield efficiency and fruit quality, in particular fruit size and colour.

For modern peach production to be profitable, growers must achieve an optimum balance between tree vigour and yield, with the former being adequate to support the crop load without being excessive, and the latter being more than adequate to recover the costs of production. When selecting a training system and determining tree spacing for future peach orchards, it is important to recognize the global trends over the past two decades for other deciduous tree fruits, such as apples and sweet cherries, for which productivity and efficiency have increased tremendously through progressive orchard intensification on size-controlling rootstocks. This provides important benefits such as reducing the juvenility period, obtaining earlier yields, improving fruit quality and consistency, and increasing input efficiencies (labour, pesticides and fertilizers) (Iglesias, 2022). The use of planar and/or smaller canopies that facilitate better accessibility for labour and machines for thinning, pruning and harvest can reduce the total cost of production by 20–30% (around €2000 ha^{-1} year^{-1}) in early- and mid-season peach cultivars (Iglesias and Echeverría, 2022).

Training system decisions will depend on the markets and priorities of the growers in each country or region. In many fruit-producing regions of the world, the increasing cost and scarcity of skilled labour has become a major concern. Traditional open-vase trees usually have large, semi-organized, three-dimensional canopies that require extensive labour to manage and harvest, and adoption of mechanization solutions for orchard tasks is difficult. Transitioning

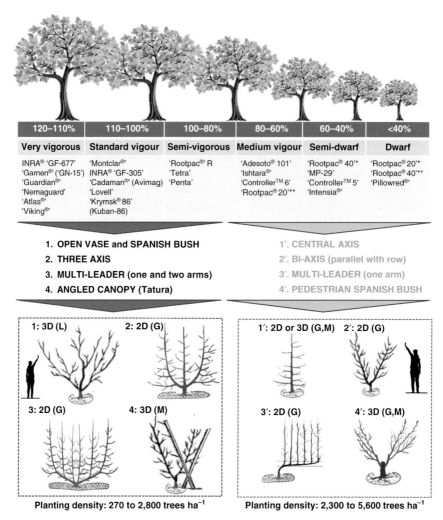

120–110%	110–100%	100–80%	80–60%	60–40%	<40%
Very vigorous	**Standard vigour**	**Semi-vigorous**	**Medium vigour**	**Semi-dwarf**	**Dwarf**
INRA® 'GF-677'	'Montclar®'	'Rootpac®' R	'Adesoto® 101'	'Rootpac® 40'*	'Rootpac® 20'*
'Garnem®' ('GN-15')	INRA® 'GF-305'	'Tetra'	'Ishtara®'	'MP-29'	'Rootpac® 40'**
'Guardian®'	'Cadaman®' (Avimag)	'Penta'	'Controller™ 6'	'Controller™ 5'	'Pillowred®'
'Nemaguard'	'Lovell'		'Rootpac® 20'**	'Intensia®'	
'Atlas®'	'Krymsk® 86'				
'Viking®'	(Kuban-86)				

1. OPEN VASE and SPANISH BUSH	**1′. CENTRAL AXIS**
2. THREE AXIS	**2′. BI-AXIS (parallel with row)**
3. MULTI-LEADER (one and two arms)	**3′. MULTI-LEADER (one arm)**
4. ANGLED CANOPY (Tatura)	**4′. PEDESTRIAN SPANISH BUSH**

1: 3D (L) **2: 2D (G)**	**1′: 2D or 3D (G,M)** **2′: 2D (G)**
3: 2D (G) **4: 3D (M)**	**3′: 2D (G)** **4′: 3D (G,M)**
Planting density: 270 to 2,800 trees ha⁻¹	**Planting density: 2,300 to 5,600 trees ha⁻¹**

Fig. 2.2. Tree vigour induced by a range of peach rootstocks (top) and the most appropriate training systems (bottom) for respective rootstock vigour levels, including three-dimensional (3D) and two-dimensional (2D) canopies. The potential for mechanization of some tasks is indicated (low: L, medium: M, good: G) for each canopy architecture. Adapted from Iglesias and Echeverria (2022).

to orchard systems with canopy architectures that are more planar can partially ameliorate this problem. Additionally, the design of competitive modern orchards should facilitate easy access for pruning, thinning and harvesting, as well as promoting earlier yields for a more rapid return on investment. Intensive training systems for peach can be achieved by using size-controlling rootstocks and/or training techniques that diffuse vigour (Lang, 2022).

Once a canopy architecture and training system for a specific cultivar/rootstock combination is selected, the optimum planting distances can be determined relative to the inherent vigour of the orchard site, as conferred by soil fertility and climate (Table 2.1). Training systems are defined by the architecture of the mature canopy and its technological development and management in terms of pruning. As training and pruning techniques affect tree vigour, the training system, rootstock/cultivar combination and inherent site vigour must all be considered when deciding within-row planting distance. Row-spacing decisions are a function of both adequate access for orchard equipment down tractor alleys and anticipated mature canopy architecture. Tree height influences the length and duration of the daily shadow cast on lower portions of canopies in adjacent rows, and canopy depth and structure influence not only light penetration and distribution within the tree's own canopy but also the potential diffusion of light through the canopy to adjacent row canopies. Tree and row spacing directly define tree density per orchard area, thereby affecting orchard establishment costs via tree numbers. As noted above and in Table 2.1, the training system directly affects the potential for mechanization and canopy accessibility by labour and machines for pruning, thinning and harvesting, which together represent nearly half of the costs of production (Fig. 2.1).

As tree size and vigour is decreased through the use of size-controlling rootstocks and/or vigour-diffusing training systems, it becomes easier to establish and maintain planar canopies, creating 'fruiting walls' that facilitate the imposition of more intensive, precise and uniform tree-management techniques. Planar canopy architectures minimize the potential for severely shaded interior portions of the canopy, thereby increasing the overall proportion of sun-exposed leaves within the canopy and consequently photosynthetic efficiency. Similarly, exposure of developing fruit to light improves fruit photosynthesis and transpiration, drawing more nutrients such as calcium to the fruit via transpiration, increasing carbohydrate availability from nearby leaves and improving fruit coloration (anthocyanin biosynthesis) during ripening. Greater and more uniform light distribution to fruit within the canopy improves ripening uniformity and, similarly, canopy penetration and distribution of protective sprays is greater and more uniform for disease and insect control. As contemporary orchard technologies such as digital imaging and mapping of canopy and crop loads develop rapidly, planar canopy architectures will facilitate their utilization and enhance their precision (Lang and Whiting, 2021).

The training system has a direct effect on the cost of orchard establishment and production, beginning with the cost of trees, any associated trellis support structures and labour for canopy training, plus the costs at maturity of annual maintenance and harvest. The recurring annual costs are dependent mainly on the potential for the mechanization and the accessibility of the canopy for labour and machines to conduct pruning, thinning and harvesting. The range of available rootstocks based on tree vigour conferred, and the training systems associated with each for optimal canopy and orchard

Table 2.1. Factors to be considered in the design of modern peach orchards: canopy and row architecture, canopy volume, training system, rootstocks and recommended tree and row spacing (ranges accommodate lower to higher vigour orchard sites, respectively, as well as shorter to taller ultimate tree heights).

Training system	Canopy/row architecture, volume	Rootstock	Tree × row spacing (m)	Potential mechanization options
Open vase (e.g. gobelet)	Three-dimensional tree row, high	Vigorous	2.5 × 5.0 to 4.5 × 6.0	Limited (top hedging)
Structured Y multi-leader (e.g. perpendicular V, quad-V, hex-V)	Three-dimensional y-tree row, high	Vigorous to semi-vigorous	1.8 × 5.0 to 2.2 × 5.0 to 3.0 × 5.4	Partial (top and side hedging, some blossom/fruit thinning)
Central leader (e.g. spindle, fusetto)	Three-dimensional tree row, medium	Semi-vigorous to semi-dwarfing	2.5 × 4.0 to 3.5 × 5.0	Partial (top and some side hedging, some blossom/fruit thinning)
Planar multi-leader (e.g. UFO, palmette)	Two-dimensional fruiting wall, medium–low	Vigorous to semi-dwarfing	2.0 × 3.0 to 3.0 × 3.5	Extensive (top and side hedging, blossom/fruit thinning)
Planar tri-leader (e.g. trident or tri-axis)	Two-dimensional fruiting wall, medium–low	Semi-vigorous to semi-dwarfing	1.5 × 3.0 to 2.2 × 4.0	Extensive (top and side hedging, blossom/fruit thinning)
Planar dual leader (e.g. bi-axis)	Two-dimensional fruiting wall, medium–low	Semi-dwarfing to dwarfing	1.0 × 3.0 to 1.5 × 4.0	Extensive (top and side hedging, blossom/fruit thinning)
Planar central leader (e.g. espalier, super spindle)	Two-dimensional fruiting wall, medium–low	Dwarfing	0.5 × 3.0 to 1.0 × 4.0	Extensive (top and side hedging, blossom/fruit thinning)
Dual planar espalier (e.g. Tatura)	Two-dimensional Y or V dual fruiting wall, medium	Semi-vigorous to semi-dwarfing	0.5 × 3.5 to 1.5 × 4.5	Partial (top and some side hedging, blossom/fruit thinning)

management, are illustrated in Fig. 2.2. Most of the rootstocks traditionally used are in the vigorous to semi-vigorous group because the open vase has been the most common training system used around the world. Although there are fewer rootstocks in the semi-dwarfing to dwarfing group, the commercial availability of rootstocks in this group is increasing with recent new selections from the USA and Europe.

Currently, the most popular training system in all peach-producing countries is the open vase (Fig. 2.2), with variations in planting distances, tree architecture and canopy volume, such as the quad-V and delayed vasette (Anthony and Minas, 2021) or Spanish gobelet (Montserrat and Iglesias, 2011). Open-vase canopies, including the Spanish gobelet, are usually associated with vigorous rootstocks such as 'GF-677', 'Garnem®', 'Nemaguard' or 'Guardian®' planted at low to medium densities (270–700 trees ha^{-1}) and achieving a large canopy volume.

The second most common training system used, albeit to a much lesser extent, is the central-leader or central-axis canopy architecture, usually with semi-vigorous or semi-dwarfing rootstocks (Figs 2.2 and 2.3, Table 2.1). The goal is a somewhat conical or spindle tree shape, narrower at the top and broader at the base. This is the case in Italy, Greece and Spain. As vigour is contained within the single leader, the system is best accomplished using rootstocks with some level of vigour control and/or for orchard sites with less

Fig. 2.3. The gobelet (a), central axis (b), bi-axis (c) and tri-axis (d) peach training systems used in Spain.

inherent vigour, otherwise it becomes difficult to control vigour at the top of the canopy (and increases the risk of losing productive fruiting wood at the bottom). Central-leader trees also can be planted at high densities on dwarfing to semi-dwarfing rootstocks, or with two leaders on semi-dwarfing to semi-vigorous rootstocks at half the orchard density and trained as a super spindle canopy with severe annual dormant pruning and summer hedging to control vigour. The closer tree spacing also contributes to vigour control through root competition. The formation of many fruiting lateral shoots and early cropping of young orchards also contributes to reducing annual vigour. Without these multiple factors for controlling the innate vigour of most peach cultivars, resulting shoot growth tends to be excessively strong with low fruitfulness.

Increasing the number of the leaders per tree diffuses and moderates the vigour of each leader somewhat proportionally, even though it increases the overall vigour of the composite canopy due to greater leaf area and a more extensive root system occupying a greater volume of soil per tree for water and nutrient uptake when the trees are spaced more widely (Lang, 2022). For example, compared with a single leader tree, 'Fantasia'/'Lovell' nectarine trees with eight leaders were 105% more vigorous as measured by trunk cross-sectional area, but each leader was 22% shorter and 63% less vigorous as measured by leader cross-sectional area, with 38% fewer lateral shoots and 15% less dense lateral shoot formation (Table 2.2). As vigour is moderated by its diffusion among more leaders, the growth of lateral shoots arising from each leader also is moderated. This allows spacing of the leaders closer together for more efficient light interception and fruitwood formation with a reduced presence of excessively vigorous (and shade-inducing) shoots. These interactive factors can be utilized to create a more uniform, structured canopy

Table 2.2. The effect of vigour diffusion among multiple leaders for 5-year-old 'Fantasia' nectarine trees on Lovell rootstock with one, two, four, six or eight leaders per tree (planted at 1, 2, 2, 3, or 4 m spacing, respectively), with each leader trained similarly with a vertical orientation and short-pruning renewal of every lateral shoot (Benton Harbor, Michigan).

Tree or individual leader vigour parameters[a]	Number of leaders per tree				
	1	2	4	6	8
Leader height (m)	100	98	98	80	78
Trunk cross-sectional area (cm²), 15 cm above graft union	100	124	141	162	205
Leader cross-sectional area (cm²), 1.5 m above the ground	100	74	57	39	37
Lateral shoots (no. per leader)	100	98	83	69	62
Lateral shoot density (no. m⁻¹)	100	104	89	90	84

[a]Results are given as parameter values relative to a one-leader tree (%).

that facilitates improved precision for optimizing crop load management, particularly in two-dimensional canopy architectures.

Consequently, peach training system experimentation over the past 5–10 years has focused on increasing leader numbers on semi-vigorous to vigorous rootstocks, or increasing the density of single- or dual-leader trees on dwarfing to semi-dwarfing rootstocks, to achieve a high number of moderate-vigour leaders per hectare, each with moderate lateral shoot vigour, creating narrow, two-dimensional planar canopies that can be uniformly structured into a fruiting wall that facilitates crop load quantification. This follows the development of such canopy systems (e.g. upright fruiting offshoots (UFO)) for sweet cherry and other tree fruits (Lang, 2019). By utilizing canopy vigour diffusion in proportion to rootstock vigour, the same canopy training techniques can be applied across a wide range of rootstocks that are most suitable for each orchard site and region.

These planar canopy architectures with an optimized distribution of moderate-vigour lateral shoots can simplify canopy management such that only 20–30 fruits per leader are needed to achieve sustainable economic yields. This can simplify previously variable decisions regarding fruit thinning to a goal of retaining a single fruit per shoot, optimizing its leaf area and sunlight exposure. This also simplifies the thinning decision to be made by hand labour or, possibly in the near future, by orchard machines that can image the initial blossom or fruit crop load and selectively apply a thinning compound spray or mechanical force (e.g. an individual-activated string thinner or jet of air or water) to retain the target number of fruit and remove all others. Such technologies are currently under study to achieve the next logical step of mapping the simplified planar canopy for operation of a mechanical fruit-removal apparatus (i.e. a robotic harvester).

2.1.3 Low- and medium-density orchard systems: three-dimensional canopies

Three-dimensional canopies, in particular the open vase, have been the most common training system for peaches for more than a century due to the lack of efficient size-controlling, productive rootstocks comparable to M9 in apple (Iglesias, 2022). Open-vase training varies by country, depending on the endemic climatic and edaphic conditions and the rootstocks available (e.g. Fig. 2.2), and continues to be widely used in the world's main producing countries such as China, Spain, Italy and the USA. In these countries, the use of vigorous or semi-vigorous rootstocks is associated with low-density plantings (~300–700 trees ha^{-1}) because three-dimensional canopy architectures diffuse tree vigour across a larger canopy volume. This results in more difficult canopy access for labour and machines, with a higher annual cost of production compared with planar systems of intensive or semi-intensive orchards (Iglesias and Echeverría, 2022).

Low-density plantings generally have wider tractor alleys (4–6 m) and can facilitate taller trees ranging from 2.2 to 5.0 m (Fig. 2.2). Open-vase or Spanish gobelet variations do not require trellising, as the permanent scaffolds and structure are strong enough to support the fruit load. This, paired with the reduced number of trees required at planting, can minimize orchard establishment costs compared with medium- or high-density plantings (Iglesias and Echeverría, 2022). However, canopy light interception on a per-hectare basis often is low (<50%) for open-vase trees in the early production years, and the lack of uniform light distribution can lead to suboptimal yields and inconsistent quality.

The open-vase peach canopy typically consists of three to five primary scaffolds that emanate from a short trunk and split into secondary/tertiary scaffolds derived from each primary scaffold. Open-vase planting spacing is typically 3.5–5.0 m between trees and 4.0–6.0 m between rows (Table 2.1, Fig. 2.2). During recent decades, several modifications of the traditional open vase have been developed in different countries based on tree vigour and available labour and mechanization (e.g. Fig. 2.4). The main differences are the number of scaffolds and the presence or absence of secondary structure. Reducing the number of scaffolds requires an increase in secondary structure for optimum light interception and productivity within the allotted orchard space.

Montserrat and Iglesias (2011) and Neri and Massetani (2011) described another three-dimensional option to reduce canopy volume and production costs and promote early yields: the Spanish gobelet or Catalonian vase, developed over the last decades mainly in Spain and other Mediterranean countries such as France, Italy and Greece. Today, ~92% of Spanish peach production utilizes the Spanish gobelet. The most commonly used rootstocks are all highly vigorous ('GF-677', 'Garnem®' and 'Cadaman®'), with spacing of 5 × 3 m (667 trees ha^{-1}) and tree height from 2.3 to 3.0 m (Fig. 2.2).

Partial mechanization of summer pruning, flower/fruit thinning and harvest is possible, but is not as efficient as in two-dimensional training systems. During the establishment year, the trees are headed at planting at 0.5 m high

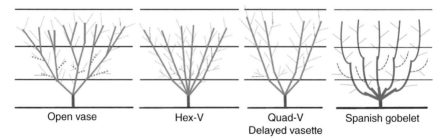

Fig. 2.4. Different options for three-dimensional gobelet-type training and variations based on tree canopy architecture. Adapted from Anthony and Minas (2021).

and then mechanically top pruned through years 2 and 3 at additional 0.5 m intervals of growth until a mature height of 2–3 m is achieved (Fig. 2.5). During the second year and thereafter, four to six open scaffolds are maintained by removal of the central branches to maintain an open centre of the vase. Paclobutrazol, a powerful plant growth inhibitor, is applied to prevent extensive growth once the mature canopy is achieved. The combination of vigorous rootstocks with the use of paclobutrazol in adult trees allows growers to fill the orchard space allotted to each tree and then control tree vigour and canopy volume, compared with the standard open vase (Iglesias and Echeverría, 2022). The widespread use of this canopy system has been tightly linked to the availability of paclobutrazol, which is not labelled for peach in many peach-producing countries. Increasingly strict European Union regulations for the use of such growth regulators could limit future utilization of vigorous rootstocks for Spanish gobelet training.

Production across years with cultivars having different harvest times is illustrated in Fig. 2.6. Full yields, ranging from 33 to 61 t ha^{-1}, are achieved in year 4 for most cultivars.

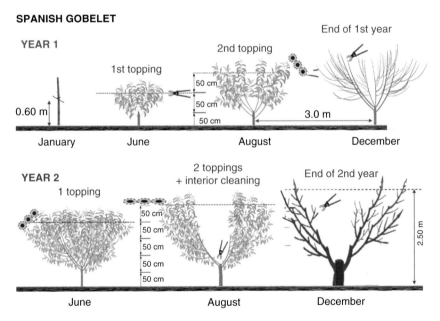

Fig. 2.5. Training and pruning of peach trees with the Spanish gobelet architecture during the first 2 years. The periphery of the three-dimensional canopy is continuous around an open centre, but for the final two drawings, portions of the bowl-shaped canopy are not depicted for clarity to illustrate pruning cuts to open the centre of the canopy. Adapted from Neri and Massetani (2011) and Iglesias and Echeverría (2022).

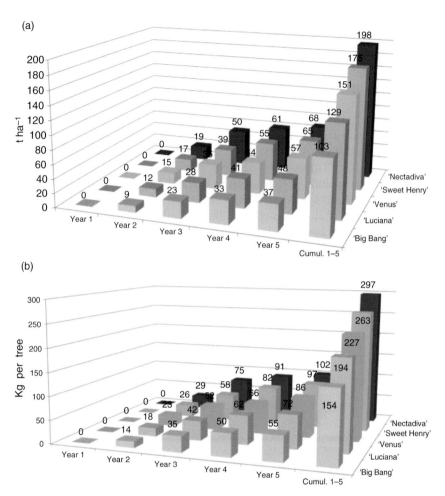

Fig. 2.6. Annual and cumulative harvest yields as t ha⁻¹ (a) and kg per tree (b) up to year 5 after planting for different peach/nectarine cultivars 'Big Bang®' (June) to 'Nectadiva®' (September) trained in the Spanish gobelet or Catalonian vase shape on 'GF-677' rootstock at the Institute of AgriFood Research and Technology (IRTA, Gimenells-Lleida). Adapted from Iglesias and Echeverría (2021).

Over the last decade, in both Spain and Italy, both the traditional open-vase and Spanish gobelet training systems have been modified to decrease canopy size and increase intensification with medium-density planting systems using size-controlling rootstocks such as 'Rootpac® 40', 'Rootpac® 20', 'Adesoto® 101' or 'Ishtara®' (Table 2.1, Fig. 2.2). The objective is the development of semi-intensive, fully pedestrian orchards with reduced volume canopies compared with traditional systems, which can achieve early yields without a requirement for trellising.

2.1.4 Medium- and high-density orchard systems: two-dimensional canopies

Two-dimensional (planar) canopies, independent of the number of leaders per tree, have been used for decades in all deciduous trees, particularly apple, pear and peach. In the last decade, central axis and bi-axis tree training have been used most often, as training labour is simplified for the first 2 or 3 years. These orchard systems are medium- or high-density plantings, typically with tree spacings of 0.8–2.5 m (depending on rootstock vigour) and row spacings of 3.0–4.0 m (600–3000 trees ha⁻¹), with heights of 2.5–3.5 m (Fig. 2.2). The objectives of these orchard systems include the reduction of canopy complexity, better canopy accessibility by labour and machines, decreased tree height, increased uniformity and density of simplified fruiting structures (trees or leaders), and, in the case of more than one leader, diffusion of vigour when vigorous or semi-vigorous rootstocks are used.

Planar canopies can be achieved with rootstocks ranging from vigorous to dwarfing. Different training system options can be selected depending on tree vigour, primarily central axis (leader), bi-axis, three-leader or multi-leader (Figs 2.2 and 2.7). These differ mainly in the cost of canopy training during the unproductive period when the canopy is filling its allotted orchard space. Trees on vigorous and semi-vigorous rootstocks require the development of more leaders to diffuse vigour, and thus more time and labour are required to train each tree. Consequently, most two-dimensional orchards in Europe utilize single- or bi-axis nursery trees so that the ultimate number of future leaders at maturity is achieved at the time of planting. In most cases, this requires size-controlling rootstocks (Fig. 2.2). Canopy training systems with more than two leaders are useful when vigour-limiting rootstocks are not available or when initial tree costs must be minimized, but the labour required for ini-

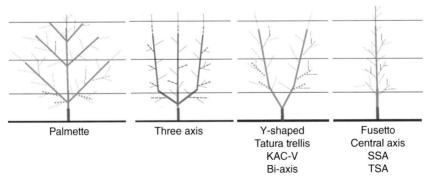

| Palmette | Three axis | Y-shaped
Tatura trellis
KAC-V
Bi-axis | Fusetto
Central axis
SSA
TSA |

Fig. 2.7. Several options for two-dimensional planar peach orchards based on tree canopy architecture. Adapted from Anthony and Minas (2021). KAC-V, Kearney Agricultural Center V system; SSA, super slender axe; TSA, tall spindle axe.

tial tree training is increased three- to five-fold. However, all these intensive planar canopies result in earlier yields compared with standard open-vase trees (Anthony and Minas, 2021; Iglesias and Echeverría, 2022).

High-density plantings that combine vigour-limiting rootstocks and different central-leader (e.g. fusetto, central axis, super slender axe, tall spindle axe) or dual-plane (e.g. Y or V systems such as the Tatura trellis) canopy iterations allow maximization of land-area production and light interception, while maintaining good light distribution throughout the canopy. Properly managed high-density plantings reduce the unproductive period and the cost of production by ~20% (Iglesias and Echeverría, 2022), achieve higher crop loads, and improve fruit size, fruit colour, soluble solids content and overall quality due to improved light distribution in the tree (Anthony and Minas, 2021; Iglesias and Echeverría, 2022). The establishment cost for higher tree numbers and required trellising may be a potential financial barrier to entry for some growers, but the return on these investments may be recouped quickly due to increased precocity and lower annual cost of production in these systems (Iglesias and Echeverría, 2022). Furthermore, high-density plantings require more intensive management and horticultural knowledge at the orchard management level, but confer simplicity at the orchard labour level, including the potential for increased mechanization (e.g. thinning, pruning, harvesting with platforms or small ladders) and robotics to reduce labour numbers and cost. Some high-density plantings with two-dimensional canopies can be developed to be fully or almost fully pedestrian.

Canopy development and training, from planting to maturity, is illustrated in Fig. 2.8 for central-leader, bi-axis and three-leader trees. Once the leaders are formed, annual maintenance hedging and pruning to maintain short fruiting lateral shoots is similar across these systems. Consequently, such pruning begins almost immediately for single-axis trees to promote many weak lateral shoots and prevent the formation of strong branches on the leader. In contrast, two- and three-leader trees are headed at planting and two strong leaders are developed at the desired spacing (closer for two leaders, wider for the eventual three leaders). While annual maintenance hedging and pruning begins in year 2 for these systems, the delayed centre leader is also developed at this point for the three-leader trees. This delayed formation of the third leader helps to create a more stable balance among the leaders, as the middle leader would tend to exhibit the greatest inherent vigour if allowed to form at the same time as the outlying leaders.

In addition to the above planar training systems, alternative multi-leader planar systems currently are being explored for peach in the USA, Spain, Italy, Greece and Australia (Fig. 2.9). These canopy architectures were developed in cherries (e.g. UFOs; Lang, 2019) during the last two decades and have more recently been applied to apples and other tree fruits (e.g. Dorigoni and Micheli, 2018; Tustin *et al.*, 2018), using semi-vigorous to semi-dwarfing rootstocks. The main advantages of these UFO-type systems in spur-bearing tree fruits are as follows: (i) the very narrow planar canopy (20–30 cm wide), with fruits borne close to the narrowly spaced vertical leaders, has the highest labour efficiency for pruning and harvest; (ii) the simplified canopy is easily adapted to mechanization and future potential robotic harvest; and (iii) the canopy architecture is adaptable across rootstocks of all vigour levels by varying the number of leaders (UFOs) in proportion to the rootstock vigour. A challenge of adopting UFO-type training to peach is that the fruiting habit of peach on

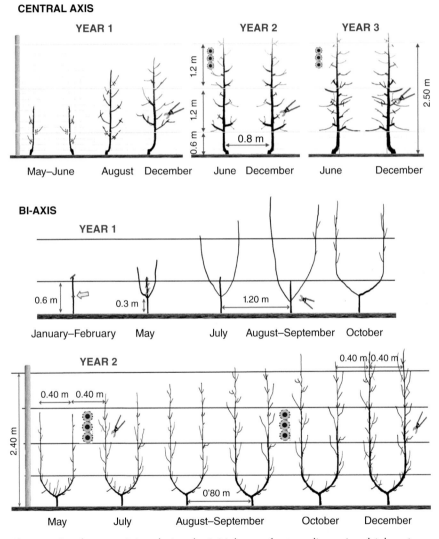

Fig. 2.8. Peach tree training during the initial years for two-dimensional (planar) single-, bi- and delayed three-axis canopies.

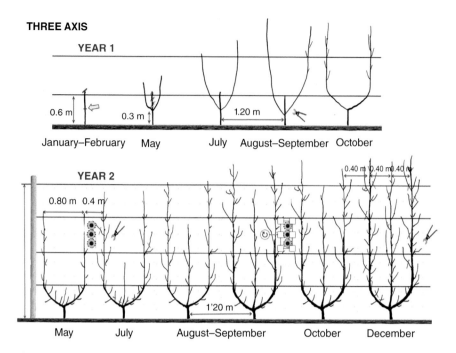

Fig. 2.8. Continued.

previous season shoots, rather than on spurs, requires greater spacing between the UFO leaders for the formation of fruit-bearing weak lateral shoots, creating a wider narrow canopy and necessitating modifications in annual pruning. Additionally, this system requires intensive trellising and specialized training (Fig. 2.10) during the first 2 years compared with single- or dual-leader trees, resulting in higher establishment costs for materials and labour.

The different steps for two UFO-type multi-leader planar peach training options are illustrated in Fig. 2.10. Spacing of the leaders can range from ~25 cm to ~50 cm each. The decision to utilize one or two horizontal 'cordons' is based on matching the diffusion of canopy vigour to rootstock vigour level and planting densities. Summer pruning is essential to train the trees during the first 2 years. Initial yields are obtained in the second year, by which time the tree will have achieved its almost final structural development.

2.2 PRUNING

Pruning is one of the most important and effective cultural techniques to create a canopy architecture that intercepts and distributes light efficiently for the leaf photosynthetic activity that will provide carbohydrates to developing fruit, and

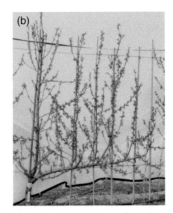

Fig. 2.9. Two-year-old trees of 'Boreal' nectarine grafted on 'Rootpac® R' in winter 2021 and spring 2022, trained to upright fruiting offshoot (UFO) like multi-leader planar canopies in a commercial orchard (Ebro Valley, Spain).

for the pigment biosynthetic activity that will achieve highly coloured fruit. However, other factors, such as nursery tree quality and availability, local soils and climate, rootstock vigour, available skilled labour and economics tend to preclude one 'recipe' for peach tree pruning to be the same across locations or regions, especially in the orchard establishment years. Therefore, this chapter provides a number of pruning concepts and techniques that will require some flexibility and deviation for developing and adapting pruning protocols suitable to optimizing peach tree growth on an orchard-by-orchard basis.

When planning an orchard, there may be a range of nursery tree types available to growers, including 1-year-old 'June-budded' or 2-year-old chip-budded, bare-root, field-grown trees, bench-grafted containerized trees,

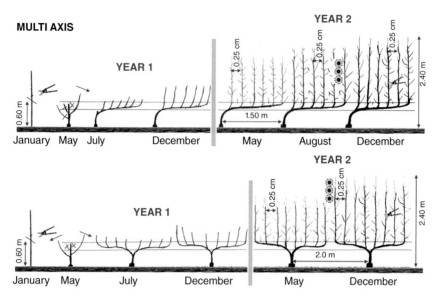

Fig. 2.10. The process of training peaches during the first 2 years into planar single-cordon, multi-leader (top) and double-cordon, multi-leader (bottom) trees.

or trees budded on tissue-cultured rootstocks or rooted cuttings grown in containers or in the nursery field. Nursery peach trees are usually single leader 'whips' or have a few lateral sylleptic branches, although dual-leader nursery trees may be feasible from custom nurseries. No matter what type of nursery tree is to be planted, balancing the top with the root system and stimulating new shoot production is often accomplished with pruning cuts to the leader and branches, and sometimes to the roots for bare-root trees. (e.g. to remove broken roots). Once the trees are established and growing, depending on the training system, pruning is practised in both summer ('green pruning') and winter ('dormant pruning'). The objective of pruning peach trees is to direct growth optimally to fit the desired canopy training system that maximizes light interception within the canopy, renews fruiting wood, increases labour efficiency for fruit thinning and harvest, accommodates mechanization, and improves fruit yield and quality.

2.2.1   Canopy structure establishment pruning

For training systems that create a three-dimensional canopy architecture, the newly planted nursery tree leader often is cut back to ~50–70 cm above the soil, depending on tree size and quality, to remove apical dominance for the formation of multiple leaders and to balance the shoot:root ratio (Figs 2.4 and 2.5). Any existing lateral branches also are cut back to ~3–5 cm long (i.e. stubbed). Retaining uncut lateral branches can lead to long, flimsy, flat-angled branches that could inhibit the strong new growth needed for ideal scaffold selection. The type of three-dimensional training system will dictate how the trees are pruned and/or managed for the remainder of the first growing season. For open-vase or other open-centre trees, tipping (i.e. 'pinching') or breaking vigorous shoots arising in the interior of the developing canopy will reduce apical dominance in the centre and promote the outside intact shoots (not tipped) from any stubbed branches to grow upward and outward at a more desirable angle (Fig. 2.5). This will produce more options for selecting evenly spaced and correctly angled scaffolds during the first winter.

For the quad-V training system (four evenly spaced scaffolds, ~90° apart with two on each side extending into the tractor alley), the scaffolds are usually chosen in the second winter to achieve better uniformity and spacing of these four permanent leaders (Fig. 2.4). Consequently, removing vigorous interior water sprouts (epicormic shoots) might be the only pruning needed during the first summer. Overall, very little summer pruning should occur during the first (and second) year except to trim back and/or remove vigorous shoots from inside the bowl of the intended vase-shaped tree.

For the Spanish gobelet training system, the objective is to reduce labour inputs by mechanizing pruning during the first 2 or 3 years (Fig. 2.5). In the year of planting, shoot-tip 'pinching' is done twice, in June and August, first by hand and then by hand or machine ('topping'). This is repeated mechanically in year 2 (Fig. 2.11). In August of year 2, the primary branches inside the canopy are cut off and by the end of that growing season, the main structure

Fig. 2.11. Training peaches in a modern Spanish gobelet. (a) Mechanical topping in June of year 1. (b) Fully established canopy after winter pruning. (c) Mature canopies in autumn of year 5.

of the tree has been established, to be finished in year 3. In winter of year 2, the final scaffolds will be selected from the lowest branch positions for a naturally open canopy (Fig. 2.4). This will result in an optimum light penetration (Figs 2.5 and 2.11). Significant yields begin in year 3 (Fig. 2.6) and the final canopy volume is achieved by the end of year 4.

For the perpendicular V or Y training systems, which are essentially dual-wall two-dimensional planar canopies, the newly planted tree is head pruned at about 50 cm or as low as feasible to promote two opposing scaffolds to arise within 20 cm of the cut and close to the ground to extend the linear fruit-bearing surface early during orchard establishment. Additional pruning of each leader to an outward-facing bud or lateral shoot promotes extension of each leader out and away from the central base trunk. The two scaffolds chosen are perpendicular to the row (i.e. one extending into each adjacent tractor alley). Upright interior regrowth is removed with manual hedging or pinching so that a 40–60° angle develops between the two eventual scaffolds (Fig. 2.7). This angle can be wider (up to 80°, DeJong *et al.*, 1994) if trees are supported with a trellis or the scaffolds are to be tied together at the top. Otherwise, the scaffolds tend to flatten as they get taller, which can lead to excessively strong upright shoots and sunburn on the interior of the V- or Y-trained canopy.

Trees trained to the two-dimensional hedgerow palmette or central-leader (slender-spindle or fusetto) systems (Fig. 2.7) are not headed back at planting unless necessary to balance the root:shoot ratio for bare-rooted trees. Containerized nursery trees are not pruned until the first winter after planting in the orchard. Thus, singular planar tree forms have minimal to no pruning during the establishment years.

2.2.2   Pruning precepts and shoot types

Following establishment during the first growing season, subsequent pruning should be focused on guiding the trees to quickly fill their allotted orchard space in the selected canopy architecture so that fruit production can begin to recover investment costs. Higher-density orchards require more trees to be planted per orchard area, and thus establishment costs are higher. Therefore, new peach plantings must promote early fruit production while filling their orchard space, preferably providing a harvestable crop by the second or third year depending on region (climate) and training system (tree density and canopy architecture). In general, the fewer the number of pruning cuts made in the first 2 or 3 years, the faster the tree will come into production. To maximize early yields, pruning should be applied judiciously when developing the desired tree architecture for the chosen training system.

After the tree canopy has reached its desired height and volume, and the desired scaffolds and subscaffolds (if any) have been established, pruning is key to maintaining annual fruitfulness and high fruit quality. Pruning approaches to maximize peach yields differ within each training system, but to maintain orchard sustainability and profitability, all methods should conceptualize how the pruning of specific branch types influences shoot growth and fruitwood structure. Pruning has long been considered a horticultural art form rather than a science, although many of the decisions regarding when and where to cut are (or should be) based on scientific principles. This is due to the complexity of tree growth, which involves continuous interactions between plant physiological processes and environmental cues. DeJong *et al.* (2012) and Allen *et al.* (2005) developed physiology-based computer model simulations to predict peach tree growth responses to pruning, which has aided the integration of physiological interactions and science-based precepts with the art of peach pruning.

The morphological foundations of these precepts centre on the three types of peach fruiting shoots – proleptic, sylleptic and epicormic – and how each type impacts the amount and quality of fruit buds (DeJong *et al.*, 2012; Prats-Llinas *et al.*, 2019). Proleptic shoots originate from visible preformed buds (i.e. vegetative shoot buds) that go through a dormancy phase and produce most of the quality fruiting wood on a peach tree for the crop in the following year. Sylleptic shoots originate from actively growing proleptic shoots (i.e. side branches on elongating proleptic shoots), as well as on epicormic shoots (i.e. water sprouts). Epicormic shoots are neoformed in that they do not arise from visible vegetative buds but originate from preventitious meristems that lie quiescent unless apical dominance control (i.e. correlative inhibition) has been lost due to heading cuts on older wood, branch breakage, or training branches to or below horizontal orientation. Severe pruning often stimulates the production of water sprouts. In contrast, no or little dormant pruning minimizes water-sprout production but favours production of smaller (sometimes less desirable) fruiting shoots due to the effects of apical dominance on lateral shoot growth.

In peach, short shoots usually terminate growth after 30 days, whereas epicormic shoots may grow until weather conditions stop growth (DeJong *et al.*, 2012). Typically, proleptic (and their associated sylleptic) shoots (i.e. future fruiting wood) stop growing or initiating new leaves between 60 and 100 days after budbreak for medium and long shoots. Short shoots, and especially epicormic shoots, generally produce poor fruiting wood; the sylleptic shoots arising from epicormic shoots have fewer flower buds, although they can set fruit almost as well as proleptic shoots (Fyhrie *et al.*, 2018). The overriding objective when pruning peach trees is to retain or maximize quality proleptic and sylleptic fruiting wood each year, while removing shoots that are problematic, inferior or non-productive.

2.2.3 Dormant pruning

After the first growing season, dormant pruning is required to 'train' the tree to the canopy architecture desired. The concept of growth reiteration following pruning (Barthélémy *et al.*, 1991), which was validated by Gordon and DeJong (2007), demonstrates that peach trees respond to pruning in a predictable way, and controlling vigour can be managed only partly with pruning. Reiteration growth is the tendency of the tree to replace what was lost due to pruning, breakage, etc. and the extent of regrowth depends on: (i) the timing of the loss of previous growth; and (ii) the available carbohydrate reserves in the tree when the loss occurs. Winter pruning increases the ratio of reserves to remaining vegetative buds, and with apical control removed, excessive growth can occur on larger branch cuts where preventitious meristems will be activated to grow. Similar vigorous regrowth occurs after large branch cuts or breaks during summer when the canopy photosynthetic activity is at a peak. Therefore, other means, including management of irrigation, fertility, root pruning and/or choice of rootstock at planting, are necessary to help keep the orchard within the system's designed architecture.

For peach, there are essentially two types of dormant pruning cuts: (i) heading cuts (e.g. bench) and (ii) thinning cuts. A heading cut removes the terminal section of the shoot's main axis, but retains the basal part of the shoot with its existing buds and nodes. A thinning cut removes the entire shoot, leaving no buds or meristems from which to regrow. Heading cuts are used primarily in the first and second year to encourage and divide vigorous growth to quickly develop a scaffold structure and reduce correlative inhibition (apical dominance) and its associated shoot epinasty (the formation of shorter lower shoots with wider branch angles). Heading cuts follow the reiteration rule, as tree regrowth tends to replace what was lost and the replacement shoots generally become longer than what the cut branch would have produced due to having fewer buds with access to available carbohydrates. This response is the basis for a pruning rule to 'cut what you want to grow' (T. DeJong, personal communication). Consequently, once the canopy has filled its allotted space for the desired training system, heading cuts produce too much unwanted vegetative growth, and other cultural practices must be implemented to control growth.

When the tree has filled its canopy space and reached full production, dormant thinning cuts are preferred over heading cuts to maintain canopy boundaries such as height, and to selectively remove unwanted or undesirable shoots without stimulating excessive epicormic shoot production. Thinning cuts also help remove excess cropping potential for better fruit size and quality, open the canopy for improved light distribution and increase partitioning of reserves to regenerate new fruit wood for next year. Knowing which type of cut is best to make for specific purposes will increase long-term orchard sustainability and productivity.

In three-dimensional training systems like the open vase, little has changed in how to train or prune this canopy architecture except for the trend to increase planting densities, which requires more precise scaffold placement and timely summer pruning. DeJong *et al.* (2012) described the system detailed by Micke *et al.* (1980) in terms of the reiteration concept. They noted that June-budded nursery trees have only neoform growth potential (i.e. epicormic shoots), so, under good growing conditions, they grow vigorously until autumn in the first summer. In the first winter, three or four of these epicormic shoots are selected to become the primary scaffolds, and other shoot growth is removed by thinning cuts. These unpruned new scaffolds are headed at ~0.5–0.8 m from the tree trunk to stimulate a reiteration response that will produce additional new water sprouts. With the traditional vase system, this type of pruning is repeated for another year to establish a set of two tertiary scaffolds on top of each secondary scaffold. At the end of the third year, the open-vase tree will have a strong structure to support a peach crop, but, depending on the rootstock and orchard management regime, the water sprouts produced can be excessive, with the risk of shading out fruiting wood. Thus, in the absence of summer pruning, the dormant removal of these shoots can repeat the cycle of vigorous, unproductive water sprouts. Without dwarfing rootstocks or control over other inputs, this can create an undesirable cycle of pruning and excessive growth.

While the tree structure is being formed to fill the canopy's allotted space, dormant pruning should be focused on using primarily thinning cuts to select and space each season's proleptic shoots. Avoiding heading cuts will reduce the invigoration of reiterative shoots that would negatively affect the quality of new fruiting wood during the summer. Heading should only be done if scaffolds need to be reoriented. Heading cuts at the top of the tree result in vigorous epicormic shoots, often referred to as 'crow's feet', which creates excessive shade and strong correlative inhibition of the development of fruiting wood along the scaffolds. Consequently, the top of each scaffold should be thinned to a single shoot oriented outwards (versus vertical) to lessen apical dominance. Removal of other shoots with thinning cuts should be focused on keeping the tree within its allotted space (both height and width), and removing water sprouts, misplaced (crowded or overlapping) and poorly oriented (vertical or pendent) shoots, and 1-year-old shoots or older branches that are damaged (broken), diseased (fungal/bacterial infections), shaded or weak (<20 cm extension growth). Mechanical topping (i.e. hedging) to reduce tree height should be avoided if possible, as it creates excessive epicormic shoot growth in the top of the canopy that may cause a number of other problems, including the requirement for further hand pruning to correct.

Once the undesirable wood is removed and the canopy is dormant-pruned to contain it within its allotted space, additional pruning may be necessary to renew poorly aligned or damaged secondary or tertiary scaffolds, and to increase sunlight penetration into the lower canopy. In hot climates, it can be strategic to leave some fruiting wood on lower scaffolds (even if not high quality) to protect against sunburn and reduce water-sprout growth. When excessively long (>50 cm), vertically oriented proleptic shoots exist on the interior scaffolds, a thinning cut that retains a side (sylleptic) shoot can be used to avoid a heading cut that would cause a vigorous reiterative growth response. The goal should be to leave a fixed number (based on cultivar, age, tree size and crop load) of proleptic and sylleptic shoots between 20 and 50 cm in length, well distributed throughout the canopy.

In two-dimensional training systems, light interception and quality depend on regenerative (winter) and timely (summer) pruning, as these planar canopies can quickly fill their space. These narrow two-dimensional canopies, whether vertical single planes or V or Y dual planes, facilitate improved cropping precision, as the number of proleptic (fruit-bearing) shoots can be quantified and adjusted to modify the crop load via pruning, thereby reducing fruit-thinning costs. If slender-spindle or central-leader systems are not grown on vigour-limiting rootstocks, the trees are prone to too much upper canopy growth, particularly if too many heading cuts are used. Maintaining canopy 'windows' for adequate light penetration and renewal of fruiting wood requires removal of water sprouts and previous-year fruitwood with thinning cuts and trimming of fruitwood to less than 50 cm in length. Without timely and selective pruning, these trees can become 'top heavy' in structure, and the lower canopy fruitwood can become unproductive and/or of poor quality. Once the permanent tree structure for planar canopies has been created, dormant pruning becomes similar to that for three-dimensional orchards.

The most common two-dimensional orchard systems are dual-plane V and Y canopies, as these increase the canopy volume per orchard area (and therefore yield potential) while maintaining the light and efficiency advantages of narrow planar structures. A true V canopy creates the dual-angled planes by alternating tree inclination, providing access down the centre of the tree row for labour. In contrast, a Y canopy creates the dual-angled planes by heading each tree to form two opposing scaffolds, one inclined into each plane; labour has no access to the middle of the planes if the trees are trellised. Otherwise, for establishment and maintenance training, both are similar. An example of the former is the open Tatura trellis, a trellised V canopy formed with alternating trees. An example of the latter would be the original Tatura trellis (Chalmers *et al.*, 1978) formed with dual-leader trees, or the perpendicular Kearney Agricultural Center V system (KAC-V system) (DeJong *et al.*, 1994), which is actually a Y canopy without a trellis. The KAC-V system requires a heading cut at planting and then selection of two opposite-facing scaffolds perpendicular to the row, which can be done the first winter after planting. Although selecting scaffolds at planting or soon after is ideal, wind, sunburn, herbivores and/or lack of well-oriented shoots often make that option unfeasible. Scaffolds can be selected in the first winter if tree growth was good. However, if trees grew poorly, heading cuts would be needed to repeat the process to develop potential scaffolds in the second year, thus delaying fruit production for a year. At maturity, tree ropes around the scaffolds about three-quarters of the tree height may be needed to support the crop load as there is no trellis.

Dormant pruning in the KAC-V system and similar two-dimensional systems should minimize heading cuts of strong shoots in the top (as would occur with top hedging) once the tree has attained the desired height, for the reasons discussed earlier about invigorating the upper canopy at the expense of shade and fruiting wood loss lower in the canopy. Pruning is focused on production of fruitwood (primarily proleptic shoots) by cutting back last year's fruiting shoots to basal buds or side shoots. In addition, fruit-thinning costs can be reduced by pruning off excess fruitwood to adjust the crop load, leaving only the number of fruiting shoots that the tree needs to size a full crop to commercial standards. The advantages of this system are two scaffolds of similar size, with the fruiting wood retained close to the scaffold trunks. Thus, the number of fruiting shoots (and fruit) can be accurately controlled by pruning.

The Tatura trellis system (Chalmers *et al.*, 1978) differs from the KAC-V primarily in that the tree is supported by trellis wires and originally had a wider crotch angle (~60° between two scaffolds) with lateral shoots that may or may not be tied down to the wires like espalier training. In recent years, the angles of most dual-plane trellised systems, including the Tatura, have become steeper, often with scaffolds closer to 70° from horizontal or even slightly more. Most of the same dormant pruning concepts can be applied, as for other V or Y systems. The Tatura system can carry higher crop loads due to its substantial trellis system. Like most two-dimensional canopy systems, both the Tatura and KAC-V systems are faster to winter prune as less decision making is required, and they are better adapted to potential mechanization (e.g. blossom thinning) than the three-dimensional open-vase systems.

Dormant pruning is fundamental for achieving a desirable balance between crop load and the supply of photosynthetic carbohydrates to achieve optimal yields and fruit quality. Selecting the best-positioned fruitwood in the canopy, and eliminating excess shoots, also promotes better light distribution and accomplishes the first important step in thinning of the potential crop load by eliminating unneeded fruiting wood. Dormant pruning will promote a progressive and continuous renewal of bearing branches year after year. This is achieved by stimulating new 1-year-old shoots that will bear flower buds the next year, while maintaining canopy volume control and new growth close to the structural scaffolds with stub cuts (Figs 2.12 and 2.13).

Dormant pruning is also influenced by the bearing habit of different cultivars. Most cultivars crop regularly on 1-year-old shoots, so dormant pruning should promote the progressive renewal of this type of wood, independent of the training system and canopy architecture (Fig. 2.14).

2.2.4 Summer pruning

One of the most important and often overlooked cultural practices in peach training systems is summer (or 'green') pruning. This is primarily due to time

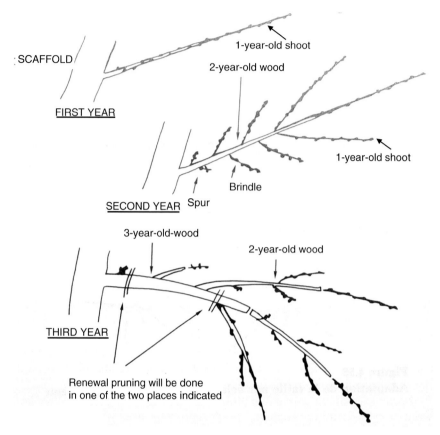

Fig. 2.12. Formation and evolution of a 1-year-old shoot during the second and third year, illustrating pruning options for maintenance and renewal.

and labour constraints during the summer harvest season. Summer pruning has the same objectives for most training systems, which include forming the initial scaffold framework, removing vigorous shoots and creating windows within the foliage to allow light penetration and distribution throughout the canopy to improve fruit quality and flower-bud formation for the next year's fruitwood. This is one of the most important training practices when the trees are young, as it can shorten the unproductive period by controlling tree development in a 'softer' way, avoiding dormant cuts and causing less reactive vegetative growth. As summer pruning removes foliage, it can have varied impacts including: (i) carbohydrate supply to current-season fruit that impacts size and quality; (ii) carbohydrate supply to current-season shoot extension growth that can impact shading and the loss of future fruit-bearing wood: (iii) the aforementioned light distribution windows that can impact fruit colour/quality and flower-bud formation; and (iv) carbohydrate reserves that can impact early

Fig. 2.13. (a) Successive year pruning for classic two-shoot peach fruiting shoot management to maintain growth close to scaffolds with stub cuts. Adapted from Brunner (1990). (b) A Spanish gobelet scaffold after dormant pruning with new 1-year-old shoots and brindles.

growth in the next season. Consequently, the timing of summer pruning is important to achieve the desired beneficial impacts. Summer pruning, either manual or mechanical, can improve fruit colour formation without a signifi-cant effect on available assimilates when timed appropriately, i.e. retaining ad-equate leaf area to support fruit growth while removing excessive leaf area that contributes to excessive shoot growth and shade. However, top hedging of peach trees in the summer can be of questionable value as, although it is labour efficient and reduces the overall tree vigour, it also promotes numerous water sprouts at the hedged (heading) cuts, which subsequently must be controlled with more selective pruning, usually by hand.

In the classic open-vase system, summer pruning in the first year is done primarily to keep the vertically growing, interior shoots pinched back so that they still produce carbohydrates but do not become dominant over the per-ipheral shoots that will form the scaffolds (Fig. 2.11). In successive years, depending on latitude (which can impact the time of vegetative growth and flower-bud initiation), summer pruning in warmer climates occurs in early summer (before flower-bud initiation in mid-summer) to remove water sprouts and other vigorous shoots inside the vase. Depending on regrowth, another pruning 3–4 weeks before harvest on late cultivars may be needed to increase fruit skin colour, soluble solids and flower-bud density. In cooler climates, summer pruning may occur 3–4 weeks before harvest to remove water sprouts and other excessively vigorous shoots while also improving fruit colour and flower-bud formation.

Fig. 2.14. (a, b) Dormant pruning of 1-year-old shoots to promote fruiting close to the scaffolds and short secondary structures in open-vase and bi-axis canopies. (c) One-year-old shoots arising close to the axis and from 1-year-old wood.

During the first two or three summers, water sprouts can be controlled by hand-breaking them in half, which removes their apical dominance but still provides photosynthates for the developing tree. Once the canopy has filled its allotted space, water sprouts can be removed completely with thinning cuts if there is adequate canopy to protect the scaffolds from sunburn. The subsequent goal is to get light infiltration throughout the middle and lower canopy to improve fruit quality and produce new fruitwood for the following year. Corelli-Grappadelli and Marini (2008) reported that a minimum of 23–40% full sun was necessary to increase fruit quality and promote flower-bud development on new shoots, so directing 30–35% of ambient sunlight to the lower canopy might be considered a minimum target when summer pruning fruiting trees.

Light quantity and penetration often are better for two-dimensional bi-axis and single-planar systems than for vase-trained trees (Grossman and DeJong, 1998; Corelli-Grappadelli *et al.*, 2003). These training systems require minimal summer pruning other than water-sprout control until they reach their intended mature height in about 2 years. Once their orchard space is filled, summer pruning follows the same principles of the three-dimensional systems in which vigorous growth is removed to improve fruit quality and flower-bud formation. However, more emphasis is focused on controlling the upper third of the canopy as it tends to quickly outgrow the lower canopy with vigorous shoots (due to apical dominance) while reducing lower canopy shoot elongation and accelerating fruitwood extinction due to shading. For early-ripening cultivars, which have a long growing season after harvest, initial summer pruning will be earlier (e.g. 2–3 weeks before harvest) and a second (or third) summer pruning is likely to be required. Mid- to late-season cultivars can be summer pruned 3–4 weeks preharvest, but they also might need summer pruning in early summer if growth is too vigorous and risks shading out next year's flower-bud development. Consequently, summer pruning is a critical cultural practice to maximize consistent peach yields and maintain the training system's optimal canopy structure. The greater the canopy complexity and/or tree density, the more summer pruning becomes essential.

In Mediterranean areas, with a long growing season and early-ripening cultivars, vase training systems (i.e. low open vase) are commonly managed with spring–summer pruning. In fact, small vase formation can be improved using summer shoot cuts to direct vegetative growth to well-placed lateral sylleptic and proleptic shoots. Modern systems such as the Spanish gobelet derived from the vase are characterized by a low scaffold (0.5 m above ground), low tree height (2.5 m) and a mostly free-growing canopy during the initial years, creating a bush type to enhance early bearing (Figs 2.5 and 2.11) (Neri and Massetani, 2011).

Spring and summer pruning is increasingly important as intensive plantings and two-dimensional training systems are adopted. If managed well, these cultural techniques can dwarf the tree canopy while maintaining the open habit of mature peach trees. During training of intensive orchards, spring pruning is therefore utilized more than summer (and especially winter) pruning to address the growth inclination of vigorously growing shoots and to anticipate formation of the foundational structure of the canopy (Neri and Massetani, 2011). The removal of poorly positioned water sprouts and stimulation of a greater selection of well-positioned shoots achieves a dwarf-statured tree (Fig. 2.15).

2.2.5   Manual versus mechanical pruning

Typically, pruning has been done manually with hand tools, but technological advancements, labour constraints and economics have increasingly incentivized growers to consider mechanical options and training systems that facilitate mechanization. Manual dormant pruning between leaf abscission and bloom represents 5–10% of the total cost of peach production. To increase labour efficiency, the use of hand-held pneumatically or electrically powered hand-pruning tools is increasingly common, particularly in countries with high production. Pneumatic pruning tools are powered by tractor or motorized platform engines, while electric pruning tools can also be powered by long-lasting lithium battery packs worn by each labourer for complete mobility. These can reduce the cost of pruning labour by 10–15%.

As most peach orchards are not pedestrian, the use of motorized platforms for labour instead of ladders is popular in two-dimensional mid- and high-density orchards for more efficient dormant and summer pruning, fruit thinning and harvest. The use of two-dimensional orchard systems and platforms provides better accessibility for labour to work in the upper canopy, improving efficiency. The combination of pneumatical pruners, motorized platforms and two-dimensional orchards can reduce the cost of production by 18% (Iglesias and Echeverría, 2022). Consequently, training systems and technological innovations have evolved in parallel, from traditional open-vase

Fig. 2.15. Different growth responses and timings of manual summer pruning. (a) Water sprouts to be pruned the first week of June. (b) The result (pictured during dormancy) of the early June-pruned scaffold top and laterals. (c) Pre-pruning of the top scaffold in September.

canopies to the reduced canopy volume of the Spanish bush for nearly pedestrian orchards, as well as two-dimensional canopies assisted by labourers on platforms almost comparable to working conditions in pedestrian orchards (Fig. 2.16).

Mechanical pruning has been used extensively in recent decades for important tree fruit crops such as olive, almond, apple and pear. In most cases, trees are pruned mechanically in summer with two objectives: (i) training the tree canopy in the establishment years; and (ii) controlling canopy volume

Fig. 2.16. Improvements in dormant pruning efficiency by changing tree canopy architecture and adopting new orchard technologies. (a) Traditional open-vase and manual ladder-based pruning. (b) Spanish gobelet with one-person hydraulic lifts. (c) Pedestrian Spanish gobelet with multi-person pneumatical pruners. (d) Near-pedestrian, two-dimensional, three-axis orchards with a motorized multi-person platform with pneumatical pruners. (e, f) Near-pedestrian two-dimensional orchards with two-axis (e) and central axis (f) trees with workers in both using one- or two-step ladder platforms and pneumatical or electric pruners.

in mature trees. It is utilized to complement and reduce manual dormant or summer pruning but not to replace those practices. It is increasingly common in countries where there is a scarcity and/or high cost of skilled orchard workers. Trees are pruned mechanically in summer to improve light penetration, reduce shading, increase fruit colour, avoid heavy cuts in winter and efficiently contribute to vigour control. In the Spanish gobelet system, this occurs in June and August of the first 2 years and then once in consecutive years until the final canopy volume is achieved. Mature trees may be pruned mechanically once a year or not, depending on tree vigour, usually after harvest. Mechanical pruning of mature trees in autumn, before full leaf abscission, often is followed by fungicide treatments with copper. Although mechanical pruning is currently used for some pruning tasks in certain open-vase training systems, it is most efficient for two-dimensional canopies of semi-intensive or intensive plantings due to the narrow, simplified, vertical canopies (Fig. 2.17).

Fig. 2.17. (a) Spring mechanical pruning of a late-ripening peach cultivar trained as a two-dimensional bi-axis canopy planted at 3.5 × 1.2 m. (b) Details of the cuts made to 1-year-old shoots. (c) The row view of the canopy after pruning.

The timing of mechanical pruning depends on the cultivar's harvest time and vigour. For cultivars that are semi-vigorous or vigorous, and/or that ripen early to mid-season, mechanical pruning should occur twice, first around the middle to the end of May and then with a follow-up before harvest, usually around the end of July (Fig. 2.18). For late-ripening cultivars, usually only the first pruning is required, as the crop load will help to control tree vigour.

Mechanization of pruning can provide positive benefits at a low cost. The time required in typical two-dimensional plantings (2000–2500 trees ha^{-1}) ranges from 1.0 to 1.5 h, depending on the orchard, cultivar and tree vigour. Usually, complementary manual pruning is done after mechanical pruning at about the time of fruit thinning or later. As with flower or fruit thinning, mechanization of winter/summer pruning can decrease the cost of production and increase fruit quality (Vittone *et al.*, 2010; Sutton *et al.*, 2020). In both cases, complementary follow-up manual thinning or pruning is required.

Mechanization can increase economic benefits for growers despite the higher capital investment for intensive orchard establishment and suitable equipment. This is because of earlier yields and significantly reduced production costs throughout the lifespan of the orchard (Iglesias and Echeverría, 2022). The various modern peach production techniques and seasons for application are illustrated in Fig. 2.19. Integrating tasks associated with winter pruning, flower and fruit thinning, summer pruning, harvest, fertigation and crop protection in an efficient manner is the key to achieve constant and high-quality yields for each combination of cultivar, rootstock and training system.

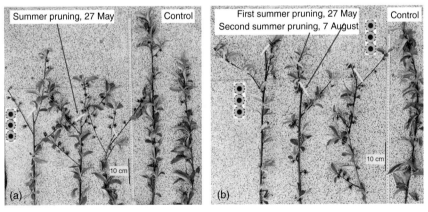

Fig. 2.18. The effect of mechanical summer pruning at bloom the following year: (a) after the summer pruning in May 2021; (b) after two summer prunings in May and August 2021.

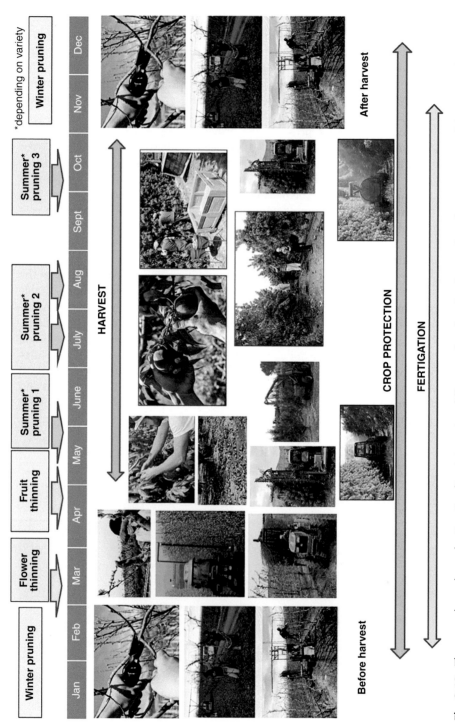

Fig. 2.19. The annual peach production timeline integrating different cultural orchard operations from pruning and thinning to harvest, fertigation and crop protection in the Ebro Valley (north-east Spain). Adapted from Iglesias and Echeverría (2022).

2.3 CONCLUSION

As in other deciduous tree fruit species, modern peach orchard training systems continue to evolve rapidly, from a regular geometrical approach to a more functional two-dimensional 'fruiting wall' approach that favours the natural growth habit of most common cultivars. Even though traditional geometric canopy shapes are very efficient for collecting light energy, the improvement of light distribution within the canopy requires intensive labour for tree training during the establishment years and canopy maintenance during the mature production years. High labour costs and labour shortages are increasingly common around the world. Skilled workers for pruning, thinning, etc. are more and more difficult to source, so future fruit production will require orchard operations that are easy to understand and easy to apply.

An increasingly common solution is to design modern orchards as a hedgerow or fruiting wall, using central axis, bi-axis, or multi-leader planar (2-D) canopy architecture and size-controlling rootstocks, as in apples, sweet cherries, and pears. Such orchards facilitate and improve the efficiency of manual and mechanical pruning (spring, summer, or autumn) to reduce the cost of production, better control tree vigour, and increase fruit quality. In these orchards, mechanical flower and/or fruit thinning also may be an option to achieve similar objectives. In addition, these 2-D canopies based on small trees result in more efficient use of inputs such as pesticides, fertilizers, irrigation and labour. Many of these factors open the door to future advances in robotics, with easier canopy digital imaging, mapping and quantification of critical organs like flower buds, floral clusters, and fruit and leaf populations. Both environmental and economic sustainability are required for successful future production, utilizing efficient peach orchards that combine the best genetics, optimized training systems and new technologies.

REFERENCES

Allen, M.T., Prusinkiewicz, P. and DeJong, T.M. (2005) Using L-systems for modeling source–sink interactions, architecture and physiology of growing trees: the L-PEACH model. *New Phytologist* 166, 869–880.

Anthony, B.M. and Minas, I.S. (2021) Optimizing peach tree canopy architecture for efficient light use, increased productivity and improved fruit quality. *Agronomy* 11: 1961.

Barthélémy, D., Edelin, C. and Hallé, F. (1991) Canopy architecture. In: Raghavendra, A.S. (ed.) *Physiology of Trees.* Wiley, New York, pp. 1–20.

Brunner, T. (1990) *Physiological Fruit Tree Training for Intensive Growing.* Akaeidai Kiado es Nyomda, Budapest, Hungary.

Chalmers, D., van den Ende, B. and van Heek, L. (1978) Productivity and mechanization of the Tatura Trellis orchard. *HortScience* 13, 517–521.

Corelli-Grappadelli, L. and Marini, R.P. (2008) Orchard planting systems. In: Layne, D.R. and D. Bassi (eds) *The Peach: Botany, Production and Uses.* CAB International, Wallingford, UK, pp. 193–220.

Corelli-Grappadelli, L., Ravaglia, G. and Morandi, B. (2003) Efficienza di conversione della radiazione in forme di allevamento di pesco. In: *Atti Convegno Macfrut, Cesena, Italy, 10 May 2003.*

DeJong, T.M., Day, K.R., Doyle, J.F. and Johnson, R.S. (1994) The Kearney Agricultural Center Perpendicular "V" (KAC-V) orchard system for peaches and nectarines. *HortTechnology* 4, 362–367.

DeJong, T.M., Johnson, R.S., Doyle, J.F. and Ramming, D. (2005) Research yields size-controlling rootstocks for peach production. *California Agriculture* 59, 80–83.

DeJong, T.M., Negron, C., Favreau, R., Day, K.R., Lopez, G. *et al.* (2012) Using concepts of shoot growth and architecture to understand and predict responses of peach trees to pruning. *Acta Horticulturae* 962, 225–232.

Dorigoni, A. and Micheli, F. (2018) Guyot training: a new system for producing apples and pears. *European Fruit Magazine* 2, 18–23.

Fyhrie, K., Prats-Llinàs, M.T., López, G. and DeJong, T.M. (2018) How does peach fruit set on sylleptic shoots borne on epicormics compare with fruit set on proleptic shoots? *European Journal of Horticultural Science* 83, 3–11.

Gordon, D. and DeJong, T.M. (2007) Current-year and subsequent-year effects of crop-load manipulation and epicormic-shoot removal on distribution of long, short and epicormic shoot growth in *Prunus persica. Annals of Botany* 99, 323–332.

Grossman, Y.L. and DeJong, T.M. (1998) Training and pruning system effects on vegetative growth potential, light interception, and cropping efficiency in peach trees. *Journal of the American Society for Horticultural Science* 123, 1058–1064.

Iglesias, I. (2018) Patrones de melocotonero: situación actual, innovación, comportamiento agronómico y perspectivas de futuro. *Revista de Fruticultura* 61, 6–43.

Iglesias, I. (2021) La intensificación sostenible como respuesta al Pacto Verde de la Unión Europea: retos y ejemplos en la producción agrícola y el consumo alimentario. *Revista de Fruticultura* 79, 45–87.

Iglesias, I. (2022) Situación actual e innovación tecnológica en fruticultura: una apuesta por la eficiencia y la sostenibilidad. *Revista de Fruticultura* 85, 6–45.

Iglesias, I. and Echeverría, G. (2009) Differential effect of cultivar and harvest date on nectarine colour, quality and consumer acceptance. *Scientia Horticulturae* 120, 41–50.

Iglesias, I. and Echeverría, G. (2021) Overview of peach industry in the European Union with special reference to Spain. *Acta Horticulturae* 1304, 163–176.

Iglesias, I. and Echeverría, G. (2022) Current situation, trends and challenges for efficient and sustainable peach production. *Scientia Horticulturae* 292: 110899.

Iglesias, I., Torrents, J., Moreno, M.A. and Ortíz, M. (2020) Actualización de los portainjertos utilizados en cerezo, duraznero y ciruelo. *Revista Frutícola* 42, 8–18.

Lang, G.A. (2019) Sustainable sweet cherry cultivation: a case study for designing optimized orchard production systems. In: Lang, G.A. (ed.) *Achieving Sustainable Cultivation of Temperate Zone Tree Fruits and Berries. Vol. 2: Case Studies.* Burleigh Dodds Science Publishing, Cambridge, UK, pp. 89–127.

Lang, G.A. (2022) Precision-Ready Peach Orchards Growing Into a Thing of Beauty. Available at: www.growingproduce.com/fruits/precision-ready-peach-orchards-growing-into-a-thing-of-beauty/ (accessed 1 December 2022).

Lang, G.A. and Whiting, M.D. (2021) Canopy architecture – optimizing the interface between fruit physiology, canopy management, and mechanical/robotic efficiencies. *Acta Horticulturae* 1314, 287–296.

Micke, W., Hewitt, A.A. and Clark, J.K. (1980) *Pruning Fruit and Nut Trees.* Leaflet 21171. Division of Agricultural Sciences, University of California, California.

Montserrat, R. and Iglesias, I. (2011) I sistemi di allevamento adottati in Spagna: l'esempio del vaso catalano. *Rivista di Frutticoltura* 7/8, 18–26.

Neri, D. and Massetani, F. (2011) Spring and summer pruning in apricot and peach orchards. *Advances in Horticultural Science* 25, 170–178.

Neri, D., Massetani, F. and Murri, G. (2015) Pruning and training systems: what is next? *Acta Horticulturae* 1084, 429–437.

Prats-Llinas, M.T., Lopez, G., Fyhrie, K., Pallas, B., Guedon, Y. *et al.* (2019) Long proleptic and sylleptic shoots in peach (*Prunus persica* L. Batsch) trees have similar, predetermined, maximum numbers of nodes and bud fate patterns. *Annals of Botany* 124, 993–1004.

Sutton, M., Doyle, J., Chavez, D. and Malladi, A. (2020) Optimizing fruit-thinning strategies in peach (*Prunus persica*) production. *Horticulturae* 6: 41.

Tustin, D.S., van Hooijdonk, B.M. and Breen, K.C. (2018) The Planar Cordon – new planting systems concepts to improve light utilisation and physiological function to increase apple orchard yield potential. *Acta Horticulturae* 1228, 1–12.

Vittone, G., Asteggiano, L. and Demaria, D. (2010) Buona pezzatura e costi minori diradando a macchina il pesco. *L'Informatore Agrario* 26, 50–53.

3

PEACH ROOTSTOCK DEVELOPMENT AND PERFORMANCE

Ioannis S. Minas[1]*, María Ángeles Moreno[2], Brendon M. Anthony[1], Jeff R. Pieper[1] and Gregory L. Reighard[3]

[1]*Department of Horticulture and Landscape Architecture, Colorado State University, Fort Collins, Colorado, USA; [2]Pomology Department, Estación Experimental de Aula Dei – CSIC, Zaragoza, Spain; [3]Department of Plant and Environmental Sciences, Clemson University, Clemson, South Carolina, USA*

3.1 INTRODUCTION

Rootstocks represent an invaluable genetic tool to optimize fruit tree adaptability in different growing regions, control tree vigour and facilitate cropping arrangements that can improve yield efficiency and productivity (Anthony and Minas, 2021). Previous experience from other fruit crops has taught us that, in orchard production systems, no revolution in grower profitability can result from cultivar breeding programmes alone but rather from rootstock breeding and cropping systems innovation (Minas *et al.*, 2022). The main soil-related challenges peach producers face worldwide are associated with texture, high pH, drought, waterlogging, and nematodes, fungal and bacterial pathogens that cause orchard replant disease syndrome (Reighard and Loreti, 2008; Anthony and Minas, 2021). These global soil challenges have shaped objectives for rootstock breeding programmes; thus, producers can overcome these limitations through targeted selection for specific pedoclimatic conditions and cropping systems.

Traditionally, peach (*Prunus persica* (L.) Batch) has been planted in low-density plantings utilizing vigorous peach seedling rootstocks. However, the use of such rootstocks has been increasingly discontinued due to their inability to withstand replant disease, waterlogging and high pH with increased levels of calcium carbonate, and to control tree vigour (Reighard and Loreti, 2008; Minas *et al.*, 2018). The main direction for breeders over the last decades in the development of new rootstocks – primarily from peach seedlings, domesticated and wild *Prunus* spp., and interspecific hybrids between peach,

* Email: Ioannis.Minas@colostate.edu

almond, plum, prune and apricot – is to overcome these abiotic and biotic challenges that limit peach cultivation (Byrne *et al.*, 2012; Anthony and Minas, 2021). Several new *Prunus* rootstocks have been selected from breeding programmes directed by universities, research institutions and private enterprises in countries such as the USA, Spain, Italy, Russia and France (Grasselly, 1987; Layne, 1987; Nicotra and Moser, 1997; Moreno, 2004; Reighard and Loreti, 2008; Felipe, 2009; Pinochet, 2010; Eremin and Eremin, 2014). These new *Prunus* rootstocks of varying vigour classification are available for use in the main peach production areas of the world (Table 3.1).

The increasing availability of new peach rootstocks depends largely on the various *Prunus* spp. or interspecific hybrids that can be used (Byrne *et al.*, 2012; Reighard *et al.*, 2013; Font i Forcada *et al.*, 2020; Reig *et al.*, 2020). These new rootstocks showcase a myriad of traits for a diverse array of abiotic/biotic conditions, which has allowed broad application and use. These rootstocks allow peach to be grown under limiting pathological and soil conditions, as peach shows low intraspecific variability to biotic and abiotic stresses. Furthermore, the use of *in vitro* propagation methods has resulted in increased genetic reproducibility, and allowed nurseries to propagate potted trees. Potted trees mitigate transplant shock and enhance soil rooting, which has improved survival and accelerated precocity over bare-rooted trees.

Nematode resistance was also a large research priority, in which popular peach rootstock cultivars such as 'Nemaguard' and 'Guardian®' were developed and released in the USA. Peach × almond hybrids, such as 'GF-677', which was bred in France, are by far the most successful interspecific rootstocks that are commonly used in Mediterranean countries because they tolerate calcareous (high-pH) soils and lime-induced iron chlorosis. They are also replant tolerant, graft compatible with peach cultivars and can be easily propagated in nurseries (Mestre *et al.*, 2015; Ben Yahmed *et al.*, 2016; Reig *et al.*, 2020). Peach × almond hybrids are characterized by their high vigour and adaptability in poor soils and dry conditions, such as the western USA and Mediterranean region (Moreno *et al.*, 1994a; Reig *et al.*, 2020; Reighard *et al.*, 2020). More recently, 'Garnem®', an almond ('Garfi') × peach ('Nemared') hybrid with similar vigour characteristics to 'GF-677' was selected in Spain for its root-knot nematode resistance and superior performance in the nursery (Pinochet *et al.*, 1999; Felipe, 2009). In the USA, the most recent peach × almond hybrid that was selected by the University of California at Davis (UC-Davis) rootstock breeding programme was 'Hansen® 536'. Similarly, private US entities such as Bright's Nursery, Inc. released their Bright's Hybrid® series and Zaiger Genetics, Inc. released other interspecific rootstocks, such as 'Viking®', and 'Atlas®' (Reighard and Loreti, 2008). In general, all of these rootstocks are characterized by their extreme vigour, which helps withstand issues with soil biotic and abiotic challenges. However, excessive peach tree vigour inhibits exploiting the advantages of higher-density plantings by creating the need for heavier pruning practices that

Table 3.1. The most important peach rootstock genotypes from various breeding programmes around the world and their genetic origin and vigour classification.

Rootstock	Breeder/country of origin	Species and interspecific hybrids	Vigour classification[a]
'Controller™ 5' (K146-43)	UC-Davis, USA	Plum × peach interspecific hybrid (*Prunus salicina* × *P. persica*)	Dwarfing
'Controller™ 6' (HBOK 27)	UC-Davis, USA	Peach × peach (*P. persica* × *P. persica*)	Semi-dwarfing
'Controller™ 7' (HBOK 32)	UC-Davis, USA	Peach × peach (*P. persica* × *P. persica*)	Semi-dwarfing
'Controller™ 8' (HBOK 10)	UC-Davis, USA	Peach × peach (*P. persica* × *P. persica*)	Semi-dwarfing
'MP-29'	USDA-Georgia, USA	Peach × interspecific plum hybrid (*P. umbellata* × plum species)	Dwarfing
'Lovell'	G.W. Thissell, USA	Peach seedling (*P. persica*)	Standard
'Hansen® 536'	UC-Davis, USA	Almond × peach interspecific hybrid (*P. dulcis* × *P. persica*)	Vigorous
'Nemaguard'	USDA, USA	Peach × wild peach interspecific hybrid (*P. persica* × *P. davidiana*)	Vigorous
'Guardian®'	Clemson-USDA, USA	Peach seedling (*P. persica*)	Vigorous
'Bright's Hybrid® 5'	Bright's Nursery, Inc., USA	Almond × peach interspecific hybrid (*P. dulcis* × *P. persica*)	Vigorous
'Viking®'	Zaiger Genetics, USA	Complex interspecific hybrid of peach, almond, plum or apricot (*P. persica, P. dulcis, P. cerasifera* or *P. mume*)	Vigorous
'Atlas®'	Zaiger Genetics, USA	Complex interspecific hybrid of peach, almond, plum or apricot (*P. persica, P. dulcis, P. cerasifera* or *P. mume*)	Vigorous
'Rootpac®R'	Agromillora Iberia, Spain	Plum and almond interspecific hybrid (*P. cerasifera* × *P. dulcis*)	Vigorous

Table 3.1. Continued.

			Vigour[a]
'Rootpac® 70' (Purplepac)	Agromillora Iberia, Spain	Almond × peach and peach × wild peach interspecific hybrid ((*P. dulcis* × *P. persica*) × (*P. persica* × *P. davidiana*))	Vigorous
'Rootpac® 40' (Nanopac)	Agromillora Iberia, Spain	Almond × peach interspecific hybrid ((*P. dulcis* × *P. persica*) × (*P. dulcis* × *P. persica*))	Dwarfing/semi-Dwarfing
'Rootpac® 20' (Densipac)	Agromillora Iberia, Spain	Plum interspecific hybrid (*P. besseyi* × *P. cerasifera*)	Dwarfing/semi-Dwarfing
'Krymsk® 1' (VVA-1)	KEBS, Russia	Cherry × plum interspecific hybrid (*P. tomentosa* × *P. cerasifera*)	Dwarfing
'Krymsk® 86' (Kuban 86)	KEBS, Russia	Plum × peach interspecific hybrid (*P. cerasifera* × *P. persica*)	Standard
'Empyrean® 2' (Penta)	CREA, Italy	Prune (*P. domestica*)	Semi-dwarfing
'Empyrean® 3' (Tetra)	CREA, Italy	Prune (*P. domestica*)	Semi-dwarfing
'Adesoto® 101'	CSIC, Spain	Plum (*P. insititia*)	Semi-dwarfing
'Cadaman®' (Avimag)	INRAE, France/Hungary	Wild peach × peach interspecific hybrid (*P. davidiana* × *P. persica*)	Vigorous
'Garnem®'	CITA, Spain	Almond × peach interspecific hybrid (*P. dulcis* × *P. persica*)	Vigorous
'Ishtara®' (Ferciana)	INRAE, France	Plum × peach interspecific hybrid ((*P. cerasifera* × *P. salicina*) × (*P. cerasifera* × *P. persica*))	Standard
'GF-677'	INRAE, France	Almond × peach interspecific hybrid (*P. dulcis* × *P. persica*)	Vigorous

UC-Davis, University of California, Davis; USDA, US Department of Agriculture; KEBS, Krymsk Experiment Breeding Station; CREA, Centro di Ricerca Alimenti e Nutrizione; INRAE, National Research Institute for Agriculture, Food and Environment; CITA, Agrifood Research and Technology Centre of Aragon; CSIC, Spanish National Research Council.

[a]Vigour classification is as follows: vigorous rootstocks are >110% the size of 'Lovell' with the size estimated by trunk cross-sectional area; standard size rootstocks are ~90–110% of 'Lovell'; semi-dwarfing rootstocks are ~60–89% of 'Lovell' and dwarfing rootstocks are <60% the size of 'Lovell' (Reighard et al., 2020).

can increase canker incidence and tree decline, and also lowers fruit quality (Font i Forcada *et al.*, 2012; Minas *et al.*, 2018; Pieper *et al.*, 2022).

Plum-based rootstocks have provided semi-dwarfing to vigorous classifications such as 'Adesoto® 101' and 'Ishtara®', which were selected in Spain and France, respectively, for their lower vigour compared with 'GF-677', higher fruit quality, good adaptation to heavy and calcareous soil conditions, tolerance to iron chlorosis and root asphyxia, and resistance to root-knot nematodes (Moreno, 2004; Iglesias *et al.*, 2019; Font i Forcada *et al.*, 2019, 2020; Reig *et al.*, 2020). Currently, 'Krymsk® 86', a Russian *P. cerasifera* × *P. persica* hybrid rootstock of standard size (80% of the size of vigorous almond hybrids such as 'Atlas®'), is dominating the new peach and almond planting decisions in the western USA due to the high graft compatibility with both species, good adaptation to heavy and calcareous soil conditions, increased anchorage, tolerance to iron chlorosis and root asphyxia due to waterlogging, cold hardiness, productivity, and good fruit size and quality (Reighard *et al.*, 2020; Minas *et al.*, 2022).

Less vegetative growth favours light distribution and interception in the canopy, consequently improving photosynthesis (Anthony *et al.*, 2021). Conversely, excessive shading in the canopy negatively affects fruit quality such as size, colour, sugar and phytochemical concentration, and antioxidant activity (Marini *et al.*, 1991; Font i Forcada *et al.*, 2012, 2019; Gullo *et al.*, 2014). Xylem anatomy and exchange of endogenous plant hormones among the plant organs are the primary mechanisms of rootstock–scion interactions that affect plant productivity and fruit quality, modifying the sink rate from the fruit to the shoot (Jackson, 1993; Tombesi *et al.*, 2010). The percentage of dry matter partitioned to fruit decreased with increasing rootstock vigour, even under increasing fruit sink (number of fruits) demand due to crop load (Caruso *et al.*, 1997; Inglese *et al.*, 2002). Dwarfing rootstocks can generally translocate more sugars (photosynthesis products) to the fruit because of the lower competition from the vegetative organs (Chalmers and Ende, 1975; Font i Forcada *et al.*, 2012, 2019; Gullo *et al.*, 2014).

Apple and sweet cherry production has been dramatically transformed due to the availability and use of precocious, dwarfing and efficient rootstocks that facilitate the development of high-density cropping systems (Robinson *et al.*, 2013; Musacchi *et al.*, 2015; Lang, 2019; Autio *et al.*, 2020). The primary focus of recent rootstock breeding efforts for peach is tree size control with a number of dwarfing and semi-dwarfing genotypes that have been selected. However, as the vigour control mechanism in most of these genotypes is governed by restrictions in xylem vessel diameter and sap flow, effective tree canopy size control is usually accompanied by reduced fruit size (DeJong *et al.*, 2014; Minas *et al.*, 2018). Two series of *Prunus* interspecific hybrid rootstocks that have demonstrated promising size-controlling genotypes are the Controller™ series from UC-Davis, USA, and the Rootpac® series from Agromillora Iberia S.L., Spain (Table 3.1). Extensive evaluation of these novel rootstock selections for their responses to different pedoclimatic conditions and intensive cropping systems that utilize simplified canopy architectures for improved productivity and labour efficiency is highly needed (DeJong *et al.*, 2005; Reighard and Loreti, 2008; Reig *et al.*, 2020). Several of these new rootstock genotypes are currently under evaluation across different peach-growing regions in the USA, under the guidance of the US Department of Agriculture (USDA) multi-state project NC-140 (Minas *et al.*, 2022), as well as in Canada (Cline and Bakker, 2022) and Mediterranean countries (Ben Yahmed *et al.*, 2016, 2020; Font i Forcada *et al.*, 2020; Reig *et al.*, 2020).

3.2 SELECTION CRITERIA FOR PEACH ROOTSTOCKS

3.2.1 Graft compatibility and mechanisms in peach

Graft compatibility between a scion and rootstock is fundamental to the orchard's economic, environmental and productive longevity (Reig *et al.*, 2019). Compatibility is defined as the successful graft union between a scion and rootstock, which allows the survivability and proper functioning of the grafted tree (Goldschmidt, 2014). Numerous scion–rootstock interactions must take place to solicit a compatible response, including callus formation, new vascular tissue differentiation and the development of an active vascular system across the graft interface (Martínez-Ballesta *et al.*, 2010). Furthermore, biochemical processes are occurring at this interface, such as the accumulation of proanthocyanidins or condensed tannins, which influence the lignification process that helps impart a successful graft union (Pina *et al.*, 2017). Unsuccessful grafts are due mainly to interspecific combinations that have distinct anatomical, morphological, physiological and/or biochemical differences between the grafted components (Darikova *et al.*, 2011; Amri *et al.*, 2021).

Two distinct types of graft incompatibility have previously been described as 'translocated' or 'localized'. Translocated incompatibility is usually characterized by visual symptoms, such as chlorosis (i.e. leaf yellowing, which becomes orange and then red), premature defoliation, leaf wilting, earlier cessation of growth, a reduction in vigour or an underdeveloped root system (Zarrouk *et al.*, 2006; Reig *et al.*, 2019). Translocated graft incompatibility can occur in the first year but can also occur later (Moreno *et al.*, 1993). In addition, some form of stress-exacerbated delayed graft incompatibility can appear when dwarfing rootstocks are exposed to some environmental stresses (e.g. higher temperatures in warmer areas) (Moreno *et al.*, 2001; Webster, 2004). These situations make it difficult to ascertain ideal genotypes, without long-term investigations (Reig *et al.*, 2019). Translocated graft incompatibility in peach is most evident when it is grafted on to myrobalan plum (*P. cerasifera* Ehrh.) (Moreno *et al.*, 1993). These symptoms arise due to impaired phloem transport from the shoots to the roots, as a result of degraded/reduced sieve-tube numbers at the graft union and cambium cell disorganization (Zarrouk *et al.*, 2010), and nitrogen starvation in the aerial parts (Moreno *et al.*, 1994b). This inhibits carbohydrate translocation to the roots, which starves the rootstock within a year or two (Reighard and Loreti, 2008). The differential expression of phenylalanine ammonia lyase (*PAL*) genes together with the antioxidant enzyme peroxidase, PAL and polyphenol oxidase involved in the phenylpropanoid pathway, as well as other biochemical factors (phenolic compounds) could be good markers for the 'translocated' peach–plum graft incompatibility (Amri *et al.*, 2021). Localized graft incompatibility is due to anatomical abnormalities in the callus bridge between the grafted components, contributing to degeneration of phloem and xylem cells and disruptions in the cambial and vascular connectivity (Zarrouk *et al.*, 2010). In contrast to translocated incompatibility, localized grafting issues can be ameliorated using mutually compatible interstocks (Hartmann *et al.*, 2002).

Peach scions have the capacity to be partially or completely compatible with various rootstock genetics within its taxonomic section: Euamygdalus. Peach is most compatible with itself (*P. persica*) and peach × almond hybrids, with less success with pure almond genotypes (*P. dulcis* (Mill.) D.A. Webb). Peach has also demonstrated good compatibility with other Euamygdalus section species, such as the 'wild' peach *P. davidiana* (Carr.) Franch and its hybrids 'Cadaman®' and 'Barrier 1', *P. mira* Koehne, *P. ferganensis* (Kost. and Rjab.) Kov. & Kost. and *P. kansuensis* Rehd. An additional taxonomic section, Euprunus, has also demonstrated grafting success with peach (Moreno *et al.*, 1995a,b; Salesses *et al.*, 1998; Duval, 2015). Notable species in this section include damson plums (*P. insititia* L.), sloe plums (*P. spinosa* L.), European plums (*P. domestica* L.), Japanese plums (*P. salicina* Lindl.) and myrobalan (or cherry) plums (*P. cerasifera*). Nevertheless, the development of new rootstocks with *P. besseyi* L.H. Bailey and *P. cerasifera* or *P. spinosa* in their genetic background will require graft-compatibility tests prior to their commercial release, as failure of peach and nectarine cultivars has been found when budded on these species (Moreno *et al.*, 1995b; Zarrouk *et al.*, 2006). Scion–rootstock compatibility is only the first benchmark evaluation criteria towards considering new rootstocks for peach production, as optimal rootstocks are expected to contribute to both abiotic and biotic adaptability/resistance.

3.2.2   Adaptability to abiotic conditions

Peach seedling rootstocks do not perform well in poorly drained and/or high pH (i.e. calcareous) soils that limit the bioavailability of iron, leading to iron-deficient chlorosis. When soil pH is above 7.5, these symptoms can appear, contributing to weak growth and low productivity of peach trees. Furthermore, waterlogged soils that have heavy clay attributes will inhibit respiration activities in the roots, which will contribute to root anoxia, tree decline and eventual death. For these abiotic stresses, new genotypes and *Prunus* hybrid rootstocks have been bred and used across the world to deal with suboptimal soil conditions with respect to pH and drainage (Byrne *et al.*, 2012).

An example is 'GF-677', a French-bred peach × almond hybrid rootstock that has been widely adopted in the Mediterranean region due to its ability to sustain vigour and productivity in calcareous soils, although it is quite susceptible to waterlogging and root-knot nematodes (Pinochet *et al.*, 1999; Mestre *et al.*, 2015; Reig *et al.*, 2020). Two Spanish peach × almond rootstocks, 'Adafuel' and 'Adarcias', are also highly tolerant to high-pH soils (Moreno *et al.*, 1994a). 'Adarcias' is also tolerant to waterlogged soils (Font i Forcada *et al.*, 2012, 2020). Other hybrid combinations including genotypes from *P. salicina*, *P. spinosa*, *P. insititia*, *P. davidiana*, *P. dulcis* and *P. cerasifera* have also demonstrated tolerance to high-pH and calcareous soils (Duval, 2015). However, compatibility with these genotypes must also be evaluated, as several *Prunus* hybrid rootstocks such as 'Jaspi®' ((*P. domestica* × *P. salicina*) × *P. spinosa*), 'Damas GF-1869' (*P. domestica* × *P. spinosa*), 'Myrobalan GF 3-1' (*P. cerasifera* × *P. salicina*) and 'Krymsk® 1' (*P. tomentosa* Thunb. × *P. cerasifera*) have demonstrated clear symptoms of graft incompatibility with peach and nectarine cultivars (Loreti and Massai, 2002; Zarrouk *et al.*, 2006, 2010).

Peach seedlings struggle to survive under root asphyxia conditions. In a study where various clonal plums and hybrid interspecific rootstocks were evaluated for waterlogging survival, plum and plum hybrid genotypes performed superiorly to peach and peach × almond genotypes (Pinochet *et al.*, 2012). In particular, young peach trees grafted on to 'GF-677' only withstood 6 days of flooding before succumbing (dying) (Pinochet *et al.*, 2012), whereas the plum and plum hybrids 'Evrica' (low vigour), 'Marianna 2624' (high vigour) and 'Pollizo AD.105' (medium vigour) could withstand 29, 32 and 34 days of flooding, respectively, before 100% tree mortality occurred (Pinochet *et al.*, 2012). 'Krymsk® 1', a *P. tomentosa* × *P. cerasifera* hybrid and dwarfing rootstock, is classified as waterlogging tolerant (Reighard and Loreti, 2008), but exhibited high levels of nematode infestation and iron-induced chlorosis in a previous trial (Pinochet *et al.*, 2012). This underscores the difficulty of identifying rootstocks that can address several abiotic and biotic factors simultaneously.

Additional abiotic factors to consider when selecting an appropriate rootstock are winter conditions and cold temperatures. Rootstocks can help improve the cold hardiness of peaches through various strategies inducing earlier acclimation, increased hardiness mid-winter and delaying de-acclimation (Minas and Sterle, 2020). These strategies are relevant to regions that experience early autumn frosts, excessive cold mid-winter and late spring freezes. The most relevant cold-hardy genotypes have historically been bred in Canada and Russia, regions that are typically colder and have marginal peach production due to these seasonal weather challenges (Reighard and Loreti, 2008; Cline and Bakker, 2022). However, where cold hardiness may be gained in breeding, limitations in their tolerance to other biotic and abiotic factors may be present. The most common cold-hardy peach seedling rootstocks in North America are 'Bailey', 'Siberian C' and 'Harrow Blood' (Reighard and Loreti, 2008; Cline and Bakker, 2022). More recent releases from Russia include the Krymsk® series, such as 'Krymsk® 1', 'Krymsk® 2' (*P. incana* (Pall.) Batsch × *P. tomentosa*) and 'Krymsk® 86' (*P. cerasifera* × *P. persica*), which have demonstrated improved cold hardiness of peach scions (Reighard and Loreti, 2008). However, it appears that the primary strategy in achieving 'cold hardiness' in rootstock breeding is through delaying bloom in spring (Reighard, 2000), which has not been an effective approach.

Recent concern has been expressed over the utilization of these and other *Prunus* hybrid rootstocks with respect to delaying acclimation in autumn (Layne *et al.*, 1977). The increased vigour of these rootstocks in fertile soils has been hypothesized to delay scion acclimation as vegetative growth is continued further into the season. This is especially an issue in relatively warm autumns where vegetative growth continues postharvest and an early frost occurs without any prior cold temperatures to prime acclimation in peach trees. Delayed acclimation can lead to severe woody tissue damage that may become susceptible to infection by canker causing pathogens (e.g. *Cytospora* canker or bacterial canker) in the seasons to come (Miller *et al.*, 2019, 2021). Furthermore, tissue damage may also inhibit the translocation of essential nutrients to the root system reserves prior to dormancy (e.g. nitrogen and carbon), as well as leading to vascular tissue collapse and subsequent fruit abscission in the following year.

3.2.3 Resistance to soil-borne pests and diseases

Parasitic nematodes and pathogenic soil fungi represent the primary pests and pathogens that negatively impact peach rootstocks and subsequent tree health, growth and productivity. There are four primary types of nematodes that impact peach tree roots: ring, root-knot, root-lesion and dagger nematodes (Reighard and Loreti, 2008). Rootstocks are often classified as: (i) resistant or tolerant, being a poor to fair host for nematode species; or (ii) susceptible, being a good to excellent host for nematode species. The distinction between resistant and tolerant are the number of genes involved. Tolerant genotypes might have many genes involved in this trait and will exhibit quantitative inheritance, while resistant genotypes may have few major genes or the involvement of a single-locus dominant/recessive model for inheritance of resistance (Esmenjaud *et al.*, 2015). Additionally, it is interesting to note that some rootstocks can be considered non-host species, highly resistant or immune to nematodes (Pinochet *et al.*, 1999, 2012), due to specific characteristics of the plant (physical or biochemical issues, hypersensitive-like reaction) preventing nematode infection (J. Pinochet, personal communication). A susceptible genotype provides a suitable host for nematode feeding and reproduction, which inhibits the scion's ability to receive minerals, hormones and water from the root system. This inhibits the tree's growth, fruiting and survival. Similarly, peach trees can be susceptible and/or tolerant to soil fungi species that are present primarily in fine-particle, poorly drained soils, where moisture is high and fungal pathogens can thrive. Notable fungal pathogens include crown rot (*Phytophthora* spp.) and oak root rot or *Armillaria* root rot (ARR) (*Armillaria mellea, Desarmillaria tabescens*). Cultural control/eradication for both nematode and fungal pathogens are difficult and time- and cost-intensive, and may be challenging in organic systems where particular chemical applications are not allowed. Therefore, genetic resistance to these biotic factors is desired for peach production systems. In addition, genetic resistance needs to be evaluated and 'field tested' in regions, countries and/or continents where environmental conditions will be different from where the breeding facilities and test plots were located (Reighard and Loreti, 2008).

A recent study evaluating nematode incidence in various *Prunus* rootstocks concluded that 'Krymsk® 1', 'Krymsk® 86' and 'Mayor' (a peach × almond hybrid) were all susceptible to root-knot nematodes (*Meloidogyne javanica*) (Pinochet *et al.*, 2012). 'Krymsk® 1' and 'Krymsk® 86' were also noted to be moderate hosts for root-lesion nematodes (*Pratylenchus vulnus*), while 'Krymsk® 2' and PAC 9801-02 ('Rootpac® 20') were poor hosts for this species (Pinochet *et al.*, 2012). 'Evrica' demonstrated increased tolerance to root-knot nematodes, while acting as a good host for root-lesion nematodes (Pinochet *et al.*, 2012). This underscores the difficulty in selecting a rootstock that demonstrates tolerance to all types of nematodes. Again, a strong understanding of the growing conditions should be gained before scion–rootstock selections are made, especially if planting in a replant site and warmer areas.

When peach trees are cultivated under intensive conditions, an unbalanced microbiome can develop in the rhizosphere, where microbes can become pathogenic and/or compete with trees for nutrients. This complex can be especially noticeable when fruit trees of the same species are replanted into the same soil and is usually referred to as orchard replant disease (Li *et al.*, 2019). Fungal and bacterial pathogens have been noted as the causal agent in apple replant disease, as well as in peach replant disease. Historically, chemical fumigation of the soil with broad-spectrum fumigant pesticides such as methyl bromide were the only cultural tool to mitigate the impacts of peach replant disease, but this pesticide has been banned by the US Environmental Protection Agency (Li *et al.*, 2019) and European Union. Therefore, the use of replant-tolerant rootstocks is necessary to overcome the problem, as peach seedlings are the most prone to replant disorders. Most of the currently available replant tolerant genotypes (i.e. peach × almond hybrids such as 'GF-677' or peach × *P. davidiana* hybrids as 'Cadaman®' and 'Barrier 1') (Massai and Loreti, 2004) demonstrate excessive vigour, which may interact with cold hardiness strategies (i.e. delayed acclimation) and/or decrease fruit quality due to intra-tree shading (Font i Forcada *et al.*, 2012; Minas *et al.*, 2018; Pieper *et al.*, 2022). Fortunately, several European studies carried out in replant sites reported size-controlling and tolerant rootstocks in plum genotypes, such as 'Adesoto® 101' (*P. insititia*) and 'Empyrean®3' (Tetra) (*P. domestica*), and other plum-based hybrid rootstocks, such as 'Replantpac' and 'Ishtara®' (Massai and Loreti, 2004; Jiménez *et al.*, 2011; Mestre *et al.*, 2015; Remorini *et al.*, 2015). However, these same plum-derived rootstocks were not tolerant to replant conditions in the south-eastern USA and died prematurely from replant-associated diseases (Reighard *et al.*, 2008).

Another disease complex affecting young peach trees (3–5 years old) stemming from cold damage and/or bacterial canker infections (*Pseudomonas syringae* pv. *syringae*) is known as peach tree short life (PTSL) (Ritchie and Clayton, 1981). PTSL is most notably detected after a sudden spring collapse of the trees, followed by subsequent death to the rootstock graft. Many factors contribute to PTSL including cultural practices, such as autumn pruning, orchard floor management, low soil pH and rootstock selection. Ring nematodes (*Mesocriconema xenoplax*) are associated with PTSL, being the primary pest that predisposes the tree to cold injury and bacterial infection (Beckman and Nyczepir, 2004). Research has demonstrated that rootstock selection is a key strategy to manage PTSL. The commercial rootstocks 'Nemaguard', 'Siberian C' and 'Elberta' were extremely susceptible to PTSL, especially where ring nematode was present (Zehr *et al.*, 1976; Yadava and Doud, 1989). 'Lovell' became the recommended rootstock for PTSL sites. However, when the soil fumigant DBCP (1,2-dibromo-3-chloropropane) was banned in 1979, 'Lovell' was less effective and was eventually replaced in the 1990s by 'Guardian®' (BY520-9), which was superior to 'Lovell' in PTSL tolerance (Okie *et al.*, 1994a,b; Beckman and Nyczepir, 2004). Additional genetic studies were conducted to identify PTSL-resistant gene analogues, which came from 'Guardian®' genotypes, to determine the best targets for breeding more tolerant rootstock genotypes (Blenda *et al.*, 2007), and are currently being employed in the breeding programme at Clemson University, South Carolina (Gasic and Saski, 2019). The current recommendation for PTSL control is preplant soil fumigation targeted at ring nematodes, paired with a ring nematode-tolerant rootstock (e.g. 'Guardian®', 'Viking®' or 'MP-29').

3.2.4 Size-controlling rootstocks and compatibility with training systems

Dwarfing mechanisms have been studied extensively in apple, with less information available in peach and other temperate fruit trees (Webster, 2004; Anthony and Musacchi, 2021). A primary vigour control mechanism in dwarfing apple rootstocks is the reduction of shoot length, neoformed nodes, internode length and the duration of growth in a season, therefore setting terminal buds earlier than more vigorous rootstocks (Seleznyova *et al.*, 2003, 2004, 2008). Changes in the production and movement of plant hormones (flows of auxins and cytokinins) between the rootstock and scion could also help to explain vigour mechanisms in dwarfing rootstocks (Webster, 2004). Furthermore, dwarfing apple rootstocks induce earlier flowering and fruiting (i.e. precocity), increase the number of flower buds on the primary axis and reduce the number of sylleptic shoots (Seleznyova *et al.*, 2008; van Hooijdonk *et al.*, 2010). As a result, the increased precocity and fertility indices of dwarfing rootstocks contribute largely to increased crop loads, which have beneficial impacts on production but deleterious effects on fruit quality (Anthony *et al.*, 2020).

In peach, the dwarfing mechanism has been hypothesized to be largely related to water relations within the tree, limiting stem water potentials and subsequent shoot growth (Basile *et al.*, 2003). This inhibition of xylem conductance has also contributed to reduced fruit size in dwarfing rootstocks. Understanding how the dwarfing rootstock influences the tree physiology and microclimate of the developing fruit is fundamental to evaluating their performance and comparing them among other genotypes of variable vigour. In other words, the influence of rootstock on fruit production and quality can easily be confounded by several variables. Therefore, to understand unbiased impacts of rootstock genotype on yield and fruit quality, the following confounding variables must be controlled for: crop load, canopy position and fruit maturity status (Anthony *et al.*, 2020; Anthony and Minas, 2021, 2022).

Although size-controlling rootstocks are more precocious than vigorous rootstocks in apple, that is not the situation in peach. Vigorous peach rootstocks still flower as early as dwarfing ones but do affect fruit quality due to interacting factors such as increased shading and water requirements. Recent peach rootstock studies sought to eliminate confounding variables when comparing the impact of rootstock on peach quality (Orazem *et al.*, 2011; Minas *et al.*, 2020, 2021, 2023; Pieper *et al.*, 2022). In certain cases, delayed maturity with vigorous rootstocks such as 'GF 677', 'Barrier 1' and 'Cadaman®' could provide at least a partial explanation for poor fruit quality (Orazem *et al.*, 2011). Nevertheless, it has been demonstrated that at uniform crop loads, canopy positions and maturity levels, fruit quality (dry matter concentration and soluble solids concentration) was enhanced in the most dwarfing rootstocks, when compared with the most vigorous ones, largely due to the increased levels of available light for the developing fruit (Orazem *et al.*, 2011; Font i Forcada *et al.*, 2019, 2020; Pieper *et al.*, 2022; Minas *et al.*, 2023). When rootstocks can control vigour, eliminating intra- and inter-tree canopy shading, fruit quality can be enhanced.

Long-term agronomical studies carried out in different peach-growing areas have shown that the size-controlling plum 'Adesoto® 101', which is well adapted to heavy-calcareous soil conditions, is conferring good fruit quality based on sugar and organic acid profiles and other biochemical traits (Orazem *et al.*, 2011; Mestre *et al.*, 2017; Font i Forcada *et al.*, 2019, 2020; Reig *et al.*, 2020). In fact, the higher content of organic acids, as well as sugars, induced by 'Adesoto® 101', will greatly increase the sweetness sensation of peach and nectarine fruits (Orazem *et al.*, 2011; Font i Forcada *et al.*, 2019). However, increased light intensity in a less shaded canopy of dwarfing rootstocks may lead to increased tree temperature and transpiration, which could lead to flower bud failure (blind wood), increased sunburn or reduced water available to size fruit in warmer areas (Ben Yahmed *et al.*, 2016). Thus, the amount of size control of the rootstock is important to consider when planning the orchard spacing.

In several rootstock trials, it has been well documented that larger, more vigorous canopies (e.g. increased trunk cross-sectional area) produce more fruit on a per-tree basis (Moreno *et al.*, 1995b; Mestre *et al.*, 2015; Font i Forcada *et al.*, 2020; Reighard *et al.*, 2020). Larger canopies can support higher crop loads per tree, with more abundant leaf area to support higher numbers of developing fruit. However, vigorous canopies limit planting densities, which are increasing in peach orchard systems as vigour-limiting rootstocks become more commercially available. Moreover, size-controlling rootstocks such as 'Rootpac® 40', 'Ishtara®' or 'Adesoto® 101', among others, resulted in better yield efficiency and improved fruit quality compared with 'GF 677' (Reig *et al.*, 2020). Unfortunately, these as well as many other European dwarfing rootstocks have not performed well in US trials (Johnson *et al.*, 2011). Thus, selecting a size-controlling rootstock adapted to different orchard environments in North America is still an important objective of US research programmes, with the Controller™ series and 'MP-29' showing the most promise.

Vigorous rootstocks could work in modern high-density planting systems if their innate vigour can be diffused and distributed across multiple leader training systems (Anthony and Minas, 2021; Iglesias and Echeverría, 2022; Manganaris *et al.*, 2022). These systems may include cordon or bi-cordon systems that are becoming increasingly popular in other tree fruit species such as apple and cherry (e.g. upright fruiting offshoot, bi-cordon upright fruiting offshoot) (Long *et al.*, 2015; Tustin *et al.*, 2018, 2022; Zhang *et al.*, 2015; Lang, 2019). New modern bi-cordon systems in peach are now under development across the globe, with varying numbers of uprights, in Spain, Greece and Colorado (Anthony and Minas, 2021; Iglesias and Echeverría, 2022; Manganaris *et al.*, 2022). However, these systems require a high level of upfront capital and infrastructure, given the number of trees and trellising at planting, as well as reliable vigour-limiting rootstocks. Fortunately, these are becoming more available, and several such as 'Rootpac® 20' are demonstrating promise in experimental and commercial high-density planting system orchards (Anthony and Minas, 2021; Iglesias and Echeverría, 2022).

3.3 MAIN PEACH ROOTSTOCK BREEDING AND EVALUATION PROGRAMMES

3.3.1 Western Europe

During recent decades, Western European countries in the Mediterranean Basin have been among the most important peach producers in the world, due to favourable climate conditions and technological developments. In fact, Spain ranks as the second highest producer in the world and is currently the highest producer and exporter in Europe, followed by Italy, Greece and France (FAOSTAT, 2022). In these Mediterranean countries, more advanced techno-logical tools (e.g. pruning, grafting, irrigation and fertilization practices) have facilitated the sharp increase in peach production and the higher productivity of new orchards. Moreover, the breeding and selection of new rootstock culti-vars have helped increase yields while decreasing management and chemical costs. Thus, rootstock breeding programmes in this region have been very active in the release of new *Prunus* genotypes (Byrne *et al.*, 2012).

Initial breeding efforts were focused on peach seedling rootstocks because of the availability of inexpensive wild types or commercial seeds, the ease of sexual propagation and the good compatibility with budded peach scion culti-vars. However, lack of uniformity in seedlings from open-pollinated trees and genetically heterogeneous seed sources guided the selection of specific peach lines such as 'GF-305', 'Montclar®', 'Rubira' and 'Higama' by the Institut National de la Recherche Agronomique (INRA) in France (Grasselly, 1985), and 'P.S.A5' and 'P.S.A6' in Italy (de Salvador *et al.*, 2002; Loreti and Massai, 2002). In certain cases, additional positive attributes were incorporated into these selections, such as moderate tolerance to waterlogging in 'GF-305' or resistance to nematodes in the 'Rubira' and 'Higama' selections (Layne, 1987). However, peach seedling rootstocks are poorly adapted to heavy clay or cal-careous soils where the pH is above 8.0. Therefore, rootstock breeding in these countries, including Spain, has been focused on peach × almond hybrids and plum-based rootstocks (Moreno, 2004; Felipe, 2009).

The peach × almond hybrids have been used primarily in calcareous soils as they tolerate iron chlorosis well and exhibit good graft compatibility with peaches and nectarines (Zarrouk *et al.*, 2005, 2006; Jiménez *et al.*, 2008). They are also vigorous and therefore appropriate for use in poor, dry soils and in fruit tree replanting situations (Massai and Loreti, 2004; Jiménez *et al.*, 2011; Mestre *et al.*, 2015; Remorini *et al.*, 2015). The early selection of the natural peach × almond hybrid 'GF-677' by INRA in France (Bernhard and Grasselly, 1981) has represented the best alternative for decades to prevent lime-induced iron deficiency in calcareous soils. Similarly, other natural peach × almond hybrids were selected locally such as 'Adafuel' and 'Adarcias' (Cambra, 1990; Moreno *et al.*, 1994a) by the Experimental Station of Aula Dei (EEAD-CSIC) and 'Mayor' by the Instituto Murciano de Investigación y Desarrollo Agrario y Medioambiental (IMIDA) (Cos *et al.*, 2004) in Spain. In addition, several selections ('PR204/84', 'AN 1/6', 'KID 1' and 'KID 2') from the Pomology Institute of Naoussa in Greece (Tsipouridis and Thomidis, 2005), as well as 'Sirio', 'Castore' and 'Polluce' from the University of Pisa in Italy (Loreti and Massai, 1994, 2006) were released, but most peach × almond hybrids are sus-ceptible to root-knot nematodes and crown gall (Pinochet *et al.*, 1999, 2002; Byrne *et al.*, 2012). Other peach × *P. davidiana* hybrids resistant to root-knot nematodes were selected by the Centro Nacionale della Recerca (CNR) in Italy such as 'Barrier 1' (de Salvador *et al.*, 1991, 2002) and by INRA of France and Hungary, such as 'Cadaman®' (Edin and Garcin, 1994), but both are less tolerant to iron chlorosis than 'GF-677' (Jiménez *et al.*, 2008).

Three new, very vigorous peach × almond hybrids, 'Garnem®', 'Felinem®' and 'Monegro®' (Felipe, 2009), were developed by the Centro de Investigación y Tecnología Agroalimentaria de Aragon (CITA) in Spain. They were selected from a controlled cross between the almond 'Garfi' and the peach 'Nemared', searching for red-coloured leaves and resistance to root-knot nematodes. Among these, 'Garnem®' has become the second most planted rootstock in Spain after 'GF-677' (Reig *et al.*, 2016, 2020), as has also happened in other Mediterranean countries. Similarly, a new hybrid rootstock, 'Rootpac® 70' (Purplepac) (Jiménez *et al.*, 2011), was also developed from a peach- and almond-based cross ((*P. persica* × *P. davidiana*) × (*P. amygdalus* × *P. persica*)) by the nursery company Agromillora Iberia S.L. in Spain. All of these rootstocks are highly vigorous, tolerant to iron chlorosis, and appropriate for infertile and droughty soils, as long as they are permeable and well drained. In addition, replant problems are ameliorated because of their vigour and resistance to nematodes. However, these rootstocks are not recommended for very fertile soils, non-replant soils or when high planting densities are required (Mestre *et al.*, 2015; Font i Forcada *et al.*, 2020). In fact, despite that 'GF-677' has been and still is the most commonly used hybrid rootstock across the Mediterranean area, it is becoming obsolete because of its susceptibility to root-knot nematodes and *Agrobacterium tumefaciens*, as well as its excess of vigour, which is not suitable in the current more intensive production systems (Reig *et al.*, 2020; Iglesias and Echeverría, 2022).

The group of 'St Julien d'Orleans' plums in France and 'Pollizo de Murcia' plums in south-east Spain are considered to be *P. insititia*, a subgroup within *P. domestica*. Clonal selections from 'St Julien' and 'Pollizo de Murcia' gave rise to 'St Julien GF-655/2' by INRA in France (Grasselly, 1988), 'St Julien A' by East Malling Research Station in the UK and 'Adesoto® 101' by EEAD-CSIC in Spain (Moreno *et al.*, 1995a). They have been the most widespread commercialized clonal rootstocks from this group, although other selections of 'St Julien' (St Julien hybrids No. 1 and No. 2) and 'Pollizo de Murcia' ('Montizo' and 'Monpol') (Felipe *et al.*, 1997) were also initially released. Other plum rootstocks, considered to be *P. domestica*, were clonally developed at the Centro di Ricerca per la Frutticoltura (CRA-FRU) in Italy (e.g. 'Empyrean® 2' (Penta) and 'Empyrean®3' (Tetra)) (Nicotra and Moser, 1997), at EEAD-CSIC in Spain (e.g. 'Constantí') (Mestre *et al.*, 2017) and at INRA in France (e.g. 'GF-43' (*P. domestica*) and 'Julior' (*P. insititia* × *P. domestica*)) (Grasselly, 1987, 1988). The plums induce an intermediate level of vigour, higher fruit quality and good productivity to peach scions (Jiménez *et al.*, 2011; Orazem *et al.*, 2011; Mestre *et al.*, 2017; Font i Forcada *et al.*, 2019) and their selections are highly resistant or immune to root-knot nematodes (Pinochet *et al.*,

1999). Several additional notable selections are the INRA French rootstocks including 'Damas GF-1869' (*P. domestica* × *P. spinosa*), 'Jaspi®' ((*P. domestica* × *P. salicina*) × *P. spinosa* L.) and 'Ishtara®' ((*P. cerasifera* × *P. salicina*) × (*P. cerasifera* × *P. persica*)) (Renaud *et al.*, 1988; Salesses *et al.*, 1998), and the Italian rootstocks 'Mr. S. 2/5' (*P. cerasifera*) and 'Mr. S. 2/8' (*P. cerasifera*), developed by the University of Pisa (Reighard and Loreti, 2008).

Collaborative projects supported by the European Union were focused on developing several hybrid progenies that originate from interspecific crosses among different Euamygdalus/Amygdalus and Euprunus/Prunophora species (Duval, 2015). These hybrids were developed through the collaboration of different groups from this region, including France, Italy and Spain (Salesses *et al.*, 1998). Some of these hybrids were selected for their tolerance/resistance to biotic and/or abiotic stresses (Dichio *et al.*, 2004; Duval, 2015) and were initially evaluated in orchard conditions with promising results for some of them in Italy (Giovannini *et al.*, 2015). Different interspecific *P. cerasifera* × (*P. dulcis* × *P. persica*) ('PADAC 99-05', 'PADAC 04-01', 'PADAC 04-03') and *P. besseyi* × *P. cerasifera* ('Rootpac® 20') hybrids have been also developed in Spain searching for multi-tolerance and/or resistance to abiotic and biotic stresses (Moreno, 2004; Pinochet *et al.*, 2012). Results from orchard trials have already demonstrated their commercial interest as rootstocks for peaches based on fruit quality and productive characteristics (Font i Forcada *et al.*, 2019, 2020; Reig *et al.*, 2020), although control of tree vigour in fertile soils compared with single plum genotypes remains to be achieved. Further studies of these interspecific hybrids should be performed with more peach cultivars to better understand their agronomical and fruit quality performance under different growing conditions in long-term orchard experiments.

3.3.2   Eastern Europe

Eastern Europe has continued to develop rootstocks, with the primary focus of improving the genetics of wild-type seedling rootstocks (Zec *et al.*, 2013). Research stations in Romania have been breeding rootstocks over the past 40 years, and rootstock selection has focused on several traits including graft compatibility, vigour, tolerance to calcareous soils, disease, severe cold and drought (Indreias, 2011, 2013). Since 2006, several seedling rootstock selections ('Ordea 2', 'Ordea 3', 'Ordea 5', 'Titan', 'Bucur', 'Cosmin' and 'Ordea 4') have been released, as well as hybrids with different plum species in their genetic background ('Adaptabil' and 'Miroper') (Stănică *et al.*, 2021). Rootstock selection and production in Bulgaria has primarily been through seeds. More recently, clonally propagated Western rootstocks have been used to compare traits of domestic selections. For example, rootstocks were evaluated in other stone fruit species for vigour and yield efficiency assessments of plum (Tabakov *et al.*, 2021) and apricot (Tabakov and Yordanov, 2011). In Poland, the Research Institute of Horticulture has been working to identify wild-type seedling rootstocks suitable for a cold climate that produce a high volume of viable seed (Szymajda *et al.*, 2019). In Ukraine, the main approach in breeding is selecting for pathogen resistance and adaptation to local environmental conditions to reduce production costs. One example of this approach appears to be 'Pamirskij 5', a rootstock resistant to powdery mildew (Pascal *et al.*, 2010).

In Russia, many new clonal rootstocks have been selected at the Krymsk Experimental Breeding Station (KEBS) (e.g. 'Krymsk® 1' (VVA-1), 'Krymsk® 2' (VSV-1), 'Krymsk® 86' (Kuban 86) or 'Fortuna') that could be suitable for peach in certain cases and colder areas, although they might be better adapted for plum and sweet cherry cultivation (e.g. 'Krymsk® 1', 'Krymsk® 2', 'Krymsk® 5' (VSL-2) or 'Krymsk® 6' (LC-52)). The origins of the Krymsk® germplasm from wild *Prunus* spp. dates to the 1960s, when Gennadiy Eremin began breeding for peach and cherry rootstocks for the cold weather and heavy soils of the region near the Black Sea and the foothills of the Caucasus Mountains. The main focus of the KEBS breeding programme was to select for resistance against a complex of harmful biotic and abiotic stresses, providing high adaptability to adverse environmental factors in southern Russia. Selected genotypes for peach span from dwarf ('Krymsk® 1') to standard or semi-vigorous ('Krymsk® 86'). 'Krymsk® 1' has been the most dwarfing rootstock in many orchard trials in warmer growing areas, probably due to poor adaptation to the growing conditions and/or to some form of stress-exacerbated delayed graft incompatibility (Ben Yahmed *et al.*, 2016; Reig *et al.*, 2020). It is known that incompatibility problems can occur more frequently in warmer areas, especially when more distant species are used as rootstocks (Moreno *et al.*, 2001). These genotypes exhibit an extensive root system that provides good tree anchorage and precocity (Eremin and Eremin, 2014). The genotype originally developed for peach, 'Krymsk® 86', is also increasingly being adopted by almond growers due to the easiness of propagation and high anchorage compared with peach roots in California.

3.3.3   North America

In the USA, rootstock development by the USDA and public university researchers occurred mainly between the 1930s and 1990s. Since then and until now, breeding has been dominated by private nurseries and breeding companies (Reighard and Loreti, 2008). Original breeding efforts focused on the incorporation of root-knot nematode resistance from various sources (e.g. 'Okinawa' and 'Nemaguard') (Reighard and Loreti, 2008), which resulted in the release of various selections such as 'Nemared' and 'Flordaguard' (Ramming and Tanner, 1983; Sherman *et al.*, 1991). Following this, a serious effort was put into the development of peach × almond hybrid rootstocks for improved peach performance in calcareous soils by D. Kester of UC-Davis (e.g. 'Hansen®

536'), by Burchell Nursery, Inc. (Oakdale, California) (e.g. 'Cornerstone') and Bright's Nursery, Inc. (Reedley, California) (e.g. Bright's Hybrid® series). Widely adapted, complex interspecific *Prunus* rootstocks (e.g. 'Viking®' and 'Atlas®') have been released by Zaiger Genetics, Inc. (Modesto, California) for multiple types of soil and replant problems.

Another important objective of the US breeding efforts was the development of size-controlling rootstocks, which was led by the USDA and UC-Davis (DeJong *et al.*, 2004). In 1987, David Ramming of the USDA and Ted DeJong and Scott Johnson of UC-Davis collaborated on a rootstock evaluation project for peaches and plums. In 2004, two plum × almond hybrids of the original 100 selections were patented as the slightly dwarfing rootstock 'Controller™ 9' (P30-135) and the very dwarfing stock 'Controller™ 5' (K146-43) (DeJong *et al.*, 2005). In the early 1990s, Fred Bliss from UC-Davis, who was succeeded by DeJong, collaborated with Craig Ledbetter of the USDA for the evaluation of USDA-bred 'Harrow Blood' (HB) peach × 'Okinawa' (OK) peach hybrids. Four more rootstocks were released: 'Controller™ 7' (HBOK 32), 'Controller™ 8' (HBOK 10), 'Controller™ 9.5' (HBOK 50) and 'Controller™ 6' (HBOK 27) (DeJong *et al.*, 2004, 2014).

Considerable efforts have been undertaken to develop tolerant rootstocks for PTSL syndrome, which is limiting orchard longevity in the south-eastern USA. These efforts led to the released 'Guardian®' (BY520-9), a vigorous rootstock with acceptable survival in field tests in South Carolina and Georgia (Okie *et al.*, 1994b; Reighard *et al.*, 1997). Following this release, breeders began to focus on determining anatomical, biochemical and genetic tolerance to ARR (Devkota *et al.*, 2020; Gasic, 2020) through marker-assisted genomic selections developed as part of the RosBREED project. The interdisciplinary, multi-university team at RosBREED targeted diseases that industry stakeholders across the country had identified as key challenges for advanced cultivar and rootstock breeding. For peach, this is being utilized to investigate ARR-tolerant peach rootstocks. Plum, plum hybrids and cherry (*P. avium* and *P. maackii*) genotypes exhibit the ability to implement ARR prevention mechanisms, unlike peach genotypes (Devkota *et al.*, 2020; Gasic, 2020). These efforts were further continued at the USDA in Byron, Georgia, for the development of rootstocks that are tolerant to ARR, PTSL and bacterial canker (Beckman *et al.*, 1997). A product of this work is 'MP-29', a dwarfing plum × peach interspecific hybrid rootstock resulting from a cross made in 1994 between 'Edible Sloe', an apparent natural plum hybrid thought to include *P. umbellata* Elliott, and a red-leafed peach rootstock selection from the Byron programme, 'SL0014' (*P. persica*) (Beckman *et al.*, 2012).

Most other peach rootstock development and evaluations in the USA are of imported rootstocks licensed by commercial nurseries (Reighard, 2000). Many of the above rootstocks are evaluated in North America through the NC-140 national multi-state, multi-disciplinary rootstock project (www.nc140.org; accessed 15 December 2022) prior to or shortly after being available to fruit growers. The NC-140 research project was established under the authority of the USDA, and its main endeavour is rootstock evaluations under diverse pedo-climatic conditions across multiple sites in North America. The NC-140 scientific coordination team is considered the international leader for temperate zone tree fruit rootstock science. Several NC-140 trials on peach rootstocks have been conducted, the most recent of which were the 2009 NC-140 peach rootstock trial (Reighard *et al.*, 2013, 2020) and the 2017 NC-140 semi-dwarf peach rootstock trial (Minas *et al.*, 2022).

The 2009 NC-140 peach rootstock trial encompassed 16 replicated orchard trials across 13 states in the USA and Chihuahua, Mexico. The rootstock cultivars included the peach seedlings 'Lovell', 'Guardian®', 'KV 10123' and 'KV 10127'; the self-rooted peaches 'Controller™ 8' (HBOK 10) and 'Controller™ 7' (HBOK 32); the interspecific *Prunus* hybrids 'Bright's Hybrid #5' (BH-5), 'Fortuna', 'Rootpac® R' (Replantpac), 'Krymsk® 86' (Kuban 86), 'Krymsk® 1' (VVA-1), 'Controller™ 5' (K146-43), 'Viking®' and 'Atlas®'; and the *P. americana* and *P. domestica* cultivars 'Empyrean®2' (Penta), 'Empyrean®3' (Tetra) and 'Imperial California'. This trial included many of the new rootstocks being sold in North America and the results (Reighard *et al.*, 2018, 2020) showed that peach seedlings on average survived better than complex hybrids after 9 years. New rootstocks such as 'Krymsk® 1', 'Krymsk® 86', 'Empyrean® 2', 'Empyrean® 3', 'Controller™ 5', 'Imperial California' and 'Rootpac® R' were the most susceptible to dying from bacterial canker in the four south-eastern states. The most vigorous rootstocks were the *Prunus* interspecific hybrids 'Viking®', 'Atlas®', 'Bright's Hybrid®5' and 'Krymsk® 86', and the peach seedlings 'Guardian®' and 'Lovell'. Cumulative yields were highest for the peach seedling rootstocks 'Guardian®', 'Lovell' and 'KV 10127', and the hybrids 'Atlas®' and 'Viking®'. The lowest yields were from trees on plum hybrids and plum species. These data suggested that there was no demonstrated advantage to increase yield per hectare by using clonal interspecific *Prunus* hybrids for peach production under current cultural practices, but the potential to increase productivity per hectare exists with higher planting densities, as all rootstocks were planted at the same density in this trial. Moreover, on high-pH soils in the western states (Utah and Colorado), peach seedlings were not the superior rootstocks for production, so continuing evaluation of non-peach rootstocks is warranted (Reighard *et al.*, 2020; Black *et al.*, 2021; Minas *et al.*, 2023).

A follow-up trial of semi-dwarf peach rootstocks was established as a high-density orchard system at ten sites across nine states in the USA and Ontario, Canada. The rootstock cultivars included the peach seedlings 'Lovell' and 'Guardian®' (i.e. standards); peach clonal hybrids 'Controller™ 6', 'Controller™ 7' and 'Controller™ 8' from UC-Davis; the interspecific *Prunus* hybrids 'Rootpac® 20' (Densipac) and 'Rootpac® 40' (Nanopac) from Agromillora Iberia, Spain; and 'MP-29' from the USDA, Georgia. During the first three years of establishment the largest trees were on 'Guardian®', and the smallest trees were on 'Rootpac® 40' and 'MP-29' (both 41% of 'Guardian®'). The highest yields were on the most vigorous rootstocks such as 'Guardian®', 'Lovell' and 'Rootpac® 20', while the lowest yield was on 'Rootpac® 40'. However, the rootstock with the highest yield efficiency was 'MP-29', which indicates that a productive dwarfing rootstock could be advantageous for higher-density plantings in peach (Minas *et al.*, 2022).

3.4 BASIC CHARACTERISTICS OF PEACH ROOTSTOCKS THAT ARE CURRENTLY COMMERCIALLY AVAILABLE

3.4.1 Peach cultivars

'Lovell'
'Lovell' is a longtime standard peach seedling rootstock propagated in the USA. It was selected in California in 1882. It gives standard-vigour peach trees with good root anchorage. It is not widely used in the major peach-producing regions any more as it is not suitable for replanting without fumigation, as well as because of its root-knot nematode susceptibility. 'Lovell' is susceptible to ARR. However, it is considered one of the best rootstocks in terms of resistance to bacterial canker. 'Lovell' prefers well-drained soils, as it shows low tolerance to waterlogging and poor tolerance to calcareous soil, expressed as iron chlorosis. 'Lovell' exhibits good acclimation performance in autumn and moderate mid-winter hardiness.

'Halford'
'Halford' is a peach seedling rootstock that is considered a sibling of 'Lovell', and is equal to it in vigour and many of its other characteristics. It has largely replaced 'Lovell' over the past 20 years as an inexpensive rootstock, simply because it was available as cannery pits. It is still used extensively in the eastern USA, but its availability may become limited as it is being replaced as a processing peach cultivar in California.

'Bailey'
'Bailey' is a peach seedling rootstock discovered in Iowa in 1890 and has been popular for sites with sandy soils and colder climates. It has some lesion-nematode and winter-cold tolerance. It produces a tree about 90–95% the size of 'Lovell' and is susceptible to root-knot nematodes and wet soils, as with 'Lovell' and 'Halford'. It is also susceptible to PTSL in the southern climates and is therefore mostly planted in northern regions.

'Nemaguard'
'Nemaguard' is a vigorous peach seedling rootstock (originally thought to also be partially *P. davidiana*, but molecular studies indicate it is primarily *P. persica*) that

was released by the USDA in 1959. It is the main rootstock used in California because it produces excellent nursery trees in 1 year, is not patented and, unlike 'Lovell', is resistant to root-knot nematodes. It supports a vigorous, productive tree that produces large fruit. However, it does not tolerate waterlogged or calcareous soils and is susceptible to many soil pests including ring nematode, bacterial canker, ARR and *Verticillium* spp.

'Nemared'

'Nemared' is a vigorous peach seedling rootstock that was released by the USDA in 1983 (Ramming and Tanner, 1983). It is very similar to 'Nemaguard' in most of its characteristics, except that it has red leaves, which is very convenient for tree propagation (grafting) and management in the nursery, but it is more sensitive to powdery mildew than 'Nemaguard'.

'Guardian®'

'Guardian®' (BY520-9) is a vigorous peach seedling rootstock that was released by Clemson University and the USDA in 1993 (Okie *et al.*, 1994b). It was selected for its tolerance to PTSL and has root-knot nematode resistance but slightly less than 'Nemaguard'. Like most peaches, it is susceptible to ARR (Beckman *et al.*, 1997), but has greater tolerance to ring nematode than 'Nemaguard'. Although it is similar to 'Nemaguard' in many ways, it is substantially less susceptible to bacterial canker (*Pseudomonas syringae*) and PTSL, and thus it is widely planted in the south-eastern USA, as well as being used for almonds on sandy sites in California.

'GF-305'

'GF-305' is a standard, vigorous peach seedling rootstock that was introduced by INRA in France in 1945. It shows a good level of resistance/tolerance to leaf curl, shot hole and powdery mildew in nursery conditions (Grasselly, 1985, 1988). However, it is susceptible to crown gall, as well as to root-knot (*M. javanica* and *M. incognita*) and root-lesion (*Pratylenchus vulnus*) nematodes. It is rarely used in Mediterranean regions because, like peach seedlings, it does not tolerate waterlogged or calcareous soils. Nevertheless, it is interesting to note that this rootstock is widely used as a good indicator of *Ilarvirus* and *Nepovirus* diseases for *Prunus* virus indexing.

'Montclar®'

'Montclar®' is a standard, vigorous peach seedling rootstock that was introduced by INRA in France (Grasselly, 1985). It is more vigorous than 'GF-305', but approximately 80–90% as vigorous as 'GF-677'. It is susceptible to crown gall, as well as to root-knot and root-lesion nematodes. It is used mainly in France, and to a lesser extent in other Mediterranean countries. It performs better in calcareous soils than most peach seedlings as it is moderately tolerant to iron-induced chlorosis.

'Controller™ 6'

'Controller™ 6' (HBOK 27) is a semi-dwarfing intraspecific hybrid originating from the F_2 of an open pollination of 'Harrow Blood' × 'Okinawa' peach (*P. persica*) cultivars. In the first 4 years after establishment in the 2017 NC-140 semi-dwarf peach rootstock trial, it produced trees that were 63% and 82% the size of 'Lovell' and 'Guardian®', respectively, across ten different orchard sites in the USA. It is moderately tolerant to calcareous soils and has performed very well in the calcareous soils of Colorado, without any symptoms of iron chlorosis. 'Controller™ 6' is moderately resistant to root-knot nematodes. The semi-dwarfing trees of 'Controller™ 6' in a perpendicular V training system produced extensive numbers of laterals per scaffold (i.e. high feathering) and the largest fruit in the first couple of cropping seasons in the 2017 NC-140 rootstock trial (Minas *et al.*, 2022).

'Controller™ 7'

'Controller™ 7' (HBOK 32) is a semi-dwarfing intraspecific hybrid originating from the F_2 of an open pollination of 'Harrow Blood' × 'Okinawa' peach cultivars. In the first 4 years after establishment in the 2017 NC-140 semi-dwarf peach rootstock trial, it produced trees that were 48% and 64% the size of 'Lovell' and 'Guardian®', respectively, across ten different orchard sites in the USA. It is resistant to root-knot nematodes, is very susceptible to iron deficiency in calcareous soils and poorly performed in the calcareous soils of Colorado with severe iron chlorosis symptoms. However, it performed well in the first 4 years of the same rootstock trial in the eastern USA (Minas *et al.*, 2022).

'Controller™ 8'

Controller™8 (HBOK 10) is a semi-dwarfing, intraspecific hybrid originating from the F_2 of an open pollination of 'Harrow Blood' × 'Okinawa' peach cultivars. In the first 4 years of the 2017 NC-140 semi-dwarf peach rootstock trial, it produced trees that were 53% and 70% of the size of 'Lovell' and 'Guardian®', respectively, across ten different orchard sites in the USA. It has good root anchorage and shows root-knot nematode resistance levels equal to that of 'Nemaguard'. It is moderately tolerant to calcareous soils and performed well in the calcareous soils of Colorado without symptoms of iron chlorosis. It also performed well in the first 4 years of the 2017 NC-140 rootstock trial in the eastern USA (Minas *et al.*, 2022).

'Flordaguard'

'Flordaguard' is a low-chill peach rootstock released in 1991 by the University of Florida (Sherman *et al.*, 1991). It has *P. davidiana* in its distant pedigree but is primarily cultured as a peach. It is very popular in low-chill regions, although it is not as low chill as 'Okinawa', which is also in its pedigree. The original clone is red leaved and has excellent resistance to three major root-knot nematode species. It is a very vigorous rootstock that does well on sandy soils. However, it is very susceptible to iron chlorosis in calcareous soils with high pH. Some believe there are different genotypes of the original cultivar, so it is also propagated by cuttings to ensure it is the correct genotype and to obtain more uniformity.

3.4.2   Plum cultivars

'Adesoto® 101'

'Adesoto® 101' is a semi-dwarf plum (*P. insititia*) rootstock developed at the Estación Experimental de Aula Dei, Zaragoza, Spain (Moreno *et al.*, 1995a). 'Adesoto® 101' adapts well to highly calcareous and compact soils, being highly tolerant to root asphyxia and iron chlorosis. It is immune or highly resistant to root-knot nematodes and susceptible to lesion nematodes (Pinochet *et al.*, 1999, 2012). In Mediterranean countries, it has demonstrated higher peach fruit quality based on higher content of sugars, organic acids and antioxidant compounds (Orazem *et al.*, 2011; Mestre *et al.*, 2017; Font i Forcada *et al.*, 2019, 2020; Reig *et al.*, 2020) and has proved to be one of the most interesting in terms of yield traits and tree survival, even under replanting conditions (Massai and Loreti, 2004; Orazem *et al.*, 2011; Mestre *et al.*, 2015). The production of suckers is a common drawback with most plum genotypes (Mestre *et al.*, 2017), mainly when they are propagated by *in vitro* techniques. Nevertheless, in long-term orchard trials, 'Adesoto® 101' showed less suckering compared with the 'GF-655/2' (Mestre *et al.*, 2017) and 'Krymsk® 1' rootstocks (Reig *et al.*, 2020). The performance of 'Adesoto® 101' in the NC-140 trials across the USA indicated that it is precocious, but it is highly susceptible to bacterial canker in the south-eastern states and exhibited excessive suckering.

'Empyrean® 3'

'Empyrean® 3' (Tetra) is a European plum (*P. domestica*) and a semi-dwarf rootstock, which is resistant to root-knot nematodes and tolerant to *Phytophthora* spp., waterlogging and heavy soils; it is graft compatible with peach, plum, almond and apricot. It was released by the Istituto Sperimentale per la Frutticoltura (ISF) of Rome, Italy, in 1997 (Nicotra and Moser, 1997). In a long-term rootstock trial, 'Empyrean® 3', together with 'Adesoto® 101' and 'Adarcias', proved to be the most dwarfing rootstocks, while 'Cadaman®' and 'Rootpac® R' were the most vigorous (Mestre *et al.*, 2015) in the heavy and calcareous soils of the Ebro Valley in Spain. In addition, 'Empyrean® 3' showed the most balanced leaf nutritional/mineral values, especially when compared with 'Barrier 1' and 'Cadaman®', although it did not differ significantly from 'GF-677' (Mestre *et al.*, 2015) and showed similar tolerance to iron chlorosis as 'GF-677' (Reig *et al.*, 2020). In all of the NC-140 trials where it was tested, it survived poorly. However, the trees that survived were significantly less vigorous than the standard rootstock, 'Lovell', being 48% of the size of 'Lovell' 9 years after planting (Reighard *et al.*, 2020).

'Empyrean® 2'

'Empyrean® 2' (Penta) is a European plum (*P. domestica*) and a semi-dwarf rootstock, which is resistant to root-knot nematodes and tolerant to *Phytophthora* spp., waterlogging and heavy soils. It appears to be very susceptible to ring nematode, so it could be highly susceptible to bacterial canker as well. It was released by the ISF in 1997 (Nicotra and Moser, 1997). In Europe, it is reported to be very vigorous, similar to peach × almond hybrids and peach × *P. davidiana*, such as 'Cadaman®' (Reig *et al.*, 2020). In all the NC-140 trials where it was tested, it survived well. The trees that survived were significantly less vigorous than the standard rootstock, 'Lovell', and were approximately 70% of the size of 'Lovell' 9 years after planting (Reighard *et al.*, 2020).

3.4.3   Interspecific hybrid cultivars

3.4.3.1   Peach × almond hybrids and similar complex interspecific hybrids within the subgenus Amygdalus

'GF-677'

'GF-677' (Paramount® in the USA), is a vigorous almond × peach (*P. dulcis* × *P. persica*) hybrid that was introduced by INRA in France (Grasselly, 1985, 1988). Due to its tolerance of calcareous soils and ease of micropropagation it has been the standard, most successful, rootstock for European peach production. However, it has many drawbacks including inducing excessive tree vigour and water sprout production, as well as susceptibility to most soil pests and diseases, and it does not tolerate waterlogged soils. The consolidated tendency to use 'GF-677' in peach replanting has been validated by data resulting from different European experimental trials established under replant situations (Massai and Loreti, 2004; Mestre *et al.*, 2015; Remorini *et al.*, 2015).

'Cadaman®'

'Cadaman®' is a vigorous wild peach (*P. davidiana*) × peach (*P. persica*) interspecific hybrid rootstock that was selected in Hungary and introduced by INRA in France (Edin and Garcin, 1994). In the NC-140 trials, its vigour was similar to that of 'Nemaguard'. However, in Europe, it has been reported to start off very vigorous and then slow down after several years, exhibiting a slight vigour control compared with 'GF-667' (Iglesias, 2013). In general, it produces similar or more vigorous trees than 'GF-677', as reported in different long-term European trials (Font i Forcada *et al.*, 2012, 2020; Mestre *et al.*, 2015; Reig *et al.*, 2020). 'Cadaman®' is widely used in Europe because of its tolerance to calcareous soils and moderate tolerance to waterlogging conditions (Font i Forcada *et al.*, 2012, 2020). It has good root anchorage and is resistant to root-knot nematodes.

'Hansen® 536'

'Hansen® 536' is a vigorous almond (*P. dulcis*) × peach (*P. persica*) interspecific hybrid that came from a cross between 'Almond B' and 'Peach 1-8-2' and was released by UC-Davis. It is a popular, high-vigour almond rootstock and is compatible with peach, with good anchorage and drought tolerance. It was found to be more tolerant of saline, alkaline and high boron soils. It has good resistance to root-knot nematodes but is susceptible to *Phytophthora* spp. and ARR. It is also very susceptible to waterlogging and requires very well-drained soils.

'Atlas®'

'Atlas®' is a vigorous complex almond (*P. dulcis* × *P. blireiana*) × 'Nemaguard' peach (*P. persica*) interspecific hybrid that was introduced by Zaiger Genetics in the USA in 1994. In the 2009 NC-140 'Redhaven' peach rootstock trial, it showed good survival and produced very vigorous trees and high cumulative yields after 9 years of evaluation across 16 different orchard sites in the USA (Reighard *et al.*, 2020). In the same trial and in two sites with calcareous soils in Utah and Colorado, this rootstock was among the most vigorous and most productive (Black *et al.*, 2021). It is not cold hardy, acclimates slowly in autumn and shows low tolerance to waterlogging. It is resistant to root-knot nematodes and susceptible to lesion and ring nematodes.

'Viking®'

'Viking®' is a vigorous complex almond (*P. dulcis* × *P. blireiana*) × 'Nemaguard' peach (*P. persica*) interspecific hybrid rootstock that was introduced by Zaiger Genetics in the USA in 1994. 'Viking®' tolerates saline and alkaline soil conditions and has low tolerance to waterlogging. It is intolerant of dehydration during transplanting, so container-grown trees are often planted. In the 2009 NC-140 'Redhaven' peach rootstock trial, it showed good survival and produced very vigorous trees and high cumulative yields after 9 years of evaluation across 16 different orchard sites in the USA (Reighard *et al.*, 2020). In the same trial and at a site with calcareous soils and a desert dry environment in Colorado, this rootstock did not survive well the first season after planting, but the trees that did survive were among the most vigorous and most productive (Reighard *et al.*, 2020). It is not cold hardy and acclimates slowly in autumn (I.S. Minas, personal communication). It is also resistant to root-knot nematodes and less susceptible to bacterial canker than 'Atlas®'. In other parts of the world, 'Viking' has performed well and has multiple advantages over peach rootstocks.

'Bright's hybrid® 5'

'Bright's hybrid® 5' is a vigorous almond (*P. dulcis*) × peach (*P. persica*) interspecific hybrid rootstock that was introduced by Bright's Nursery. It is very vigorous, nematode resistant, deep rooting, well anchored, and drought and replant tolerant. This rootstock needs deep, well-drained soil. The selection #5 was a survivor in an orchard with high pH due to excess calcium carbonate. It shows poor resistance to ARR. In the 2009 NC-140 'Redhaven' peach rootstock trial, it showed moderate survival and produced very vigorous trees after 9 years of evaluation across 16 different orchard sites in the USA (Reighard *et al.*, 2020).

'Rootpac®40'

'Rootpac®40' (Nanopac) is a standard-vigour almond × peach ((*P. dulcis* × *P. persica*) × (*P. dulcis* × *P. persica*)) interspecific hybrid rootstock that was developed by Agromillora Iberia, S.L. in Barcelona, Spain. It has been reported to give a tree that is around 25–30% smaller than 'GF-677' (Jiménez *et al.*, 2011), but with a well-developed root system. It adapts very well to all climates, but especially to warm conditions (low- and medium-chilling areas) (Reig *et al.*, 2020). It is moderately tolerant to calcareous and saline soils, and resistant to root-knot nematodes. However, it is susceptible to root-lesion nematodes and sensitive to crown gall and ARR. The results from the first 4 years after establishment in the 2017 NC-140 'Cresthaven' semi-dwarf peach rootstock trial across 10 different orchard sites in North America are inconsistent with what has been reported for this rootstock in Europe (Minas *et al.*, 2022). It has been dwarfing and short-lived in other rootstock trials in the south-east USA (G.L. Reighard, personal communication). However, in a more recent rootstock × training system trial that was established in 2019 in Colorado, it performed similarly to the European reports regarding its vigour.

'Cornerstone'

'Cornerstone' is a vigorous almond (*P. dulcis*) × peach (*P. persica*) interspecific hybrid rootstock that was introduced by Burchell Nursery in the USA. It was slightly more vigorous than 'Nemaguard' in the 2001 NC-140 trial. As a peach × almond hybrid, it is tolerant to calcareous soils with high pH and consequently to iron chlorosis. It is more drought tolerant than 'Nemaguard' and tolerates heavier soils, but it needs good drainage. It is similar to 'Nemaguard' in root-knot nematode resistance, is more tolerant of *Phytophthora* spp. than 'Hansen® 536' and is less susceptible to crown gall than most almond × peach interspecific hybrids.

'Garnem®'

'Garnem®' is a vigorous almond × peach interspecific hybrid rootstock that was selected for replant tolerance, nematode resistance and good nursery characteristics by CITA in Spain, as well as the closely related peach × almond hybrids 'Felinem®' and 'Monegro®' (Felipe, 2009). These three clones were selected from the progeny obtained in a cross between the Spanish almond 'Garfi' (*P. dulcis*) as the female parent and the North American peach 'Nemared' (*P. persica*) as the pollen donor. They were released to address the problems of *Prunus* spp. growing in Mediterranean conditions. 'Garnem®' is characterized by red leaves, good vigour, easy clonal propagation, resistance to root-knot nematodes, adaptation to calcareous soils and other Mediterranean agroecological conditions, and good graft compatibility with peach. It is tolerant to water stress but requires well-drained soils to avoid tree losses by waterlogging (Font i Forcada *et al.*, 2012, 2020). It is susceptible to crown gall and root-lesion nematodes (Felipe, 2009; Iglesias, 2013). It has not performed (survived) well in several trials in South Carolina where bacterial canker is a problem (G.L. Reighard, personal communication).

3.4.3.2 *Peach × plum hybrids and similar complex plum hybrids*

'Controller™ 5'

'Controller™ 5' (K146-43) is a dwarfing plum × 'Favorcrest' peach (*P. salicina × P. persica*) hybrid rootstock that was released by UC-Davis in 2004. It is compatible with peach and nectarine, with no root suckering. In the 2009 NC-140 peach rootstock trial, it produced trees that were 60% the size of 'Lovell' after 9 years of evaluation across 16 different orchard sites in the USA (Reighard *et al.*, 2020). In the same trial and in two sites with calcareous soils in Utah and Colorado, this rootstock did not perform well, but the surviving trees did not show iron chlorosis. It also shows low tolerance to waterlogging and has performed poorly in root-knot, lesion and ring nematode tests at a bacterial canker field site. 'Controller™ 5' has been reported to produce smaller fruit, which may be due to high crop loads (heavy flowering) and restricted water conductance in the xylem (even under well-irrigated conditions).

'Controller™ 9'

'Controller™ 9' (P30-135) is a vigorous Japanese plum (*P. salicina*) × peach (*P. persica*) interspecific hybrid. It is slightly less vigorous than 'Nemaguard' and produces a similar-sized tree but with fewer water sprouts and with good root anchorage. 'Controller™ 9' is susceptible to bacterial canker and root-knot and ring nematodes, but is more resistant to lesion nematode than 'Nemaguard'.

'Krymsk® 86'

'Krymsk® 86' (Kuban 86) is a standard-vigour myrobalan plum (*P. cerasifera*) × peach (*P. persica*) interspecific hybrid rootstock that was introduced by KEBS in Russia. 'Krymsk® 86' exhibits increased root strength and outstanding anchorage. The tree size is similar to 'Lovell' with a wide-spreading root system. 'Krymsk® 86' is tolerant to calcareous soil with a high pH, and to waterlogging and heavy-textured soils. 'Krymsk® 86' is tolerant to dry conditions and is cold hardy in mid-winter, with slower acclimation rates in autumn than 'Lovell' (I.S. Minas, personal communication). It has been reported to have poor performance with the flat peach cultivars 'UFO 3' and 'Subirana' in south-eastern Spain and northern Tunisia, respectively, probably due to its higher chilling requirements in these typical warm Mediterranean regions with low chilling (Legua *et al.*, 2012; Ben Yahmed *et al.*, 2016). However, it performed excellently in the Fresno area in California within the frame of the 2009 NC-140 'Redhaven' peach rootstock trial, which is also a moderate- to low-chilling region. Unlike other rootstocks with plum genes, 'Krymsk® 86' is compatible with 'Nonpareil', which made it the most popular almond rootstock for the Sacramento Valley, California. It is susceptible to root-knot and lesion nematodes, crown gall (Pinochet *et al.*, 2012) and *Phytophthora* spp. but has exhibited good levels of tolerance to ARR. 'Krymsk® 86' grows to 80% the size of vigorous almond hybrids such as 'Atlas®' and is dominating new peach and almond planting decisions in the USA due to its high graft compatibility with both species, good adaptation to heavy and calcareous soil conditions, increased anchorage, tolerance to iron chlorosis, minimal root asphyxia due to waterlogging, cold hardiness, productivity, and good fruit size and quality (Pieper *et al.*, 2022; Minas *et al.*, 2023).

'Krymsk® 1'

'Krymsk® 1' (VVA-1) is a dwarfing myrobalan plum (*P. cerasifera* × *P. tomentosa*) interspecific hybrid rootstock that was introduced by KEBS in Russia. It produces a tree that is generally about 50% the size of 'Nemaguard'. In an NC-140 trial, this rootstock demonstrated consistently smaller trees with high yield efficiency and no reduction in fruit size, while also showing scion incompatibility (with the 'Redhaven' scion), suckering and susceptibility to bacterial canker in some sites (Reighard *et al.*, 2011). Many other California cultivars have been grafted on to 'Krymsk® 1' and many of these show signs of poor compatibility. Zarrouk *et al.* (2006) reported 'translocated' or 'localized' graft incompatibilities when 'Krymsk® 1' was budded with 29 peach cultivars. Similarly, peach trees budded on 'Krymsk® 1' experienced the highest mortality rates in orchard conditions in northern Tunisia (Ben Yahmed *et al.*, 2016). Besides scion incompatibility, some *Prunus* rootstock studies (Reighard *et al.*, 2011, 2016) also reported high root suckering on 'Krymsk® 1' in California and other US states, which supports the similar findings of Reig *et al.* (2020) in loamy and calcareous soil under the hot climate conditions of the Ebro River Basin in Spain. 'Krymsk® 1' has been shown to be moderately tolerant of waterlogging and iron chlorosis in regions with calcareous soils. It did not survive well in a bacterial canker hot spot in South Carolina (Reighard *et al.*, 2020). 'Krymsk® 1' has shown good tolerance to ARR but is susceptible to root-knot nematodes and slightly resistant to lesion nematodes. However, Pinochet *et al.* (2012) classified it as a host to a mixture of five isolates of *Pratylenchus vulnus*. In the 2009 NC-140 'Redhaven' peach rootstock trial at the Colorado site, 'Krymsk® 1' was very impressive with healthy-looking dwarf trees with little suckering, good production and large fruit size with superior internal quality characteristics (Pieper *et al.*, 2022; Minas *et al.*, 2023). However, many other California cultivars have been grafted on this rootstock, and most produce weak trees with rolled leaves and poor graft unions due to compatibility issues.

'MP-29'

'MP-29' is a dwarfing 'Edible Sloe' plum × red-leafed peach (SL0014, *P. persica*) interspecific hybrid that was developed and jointly released by the USDA Agricultural Research Service at Byron, Georgia, and the Florida Agricultural Experiment Station (Beckman *et al.*, 2012). 'MP-29' has been reported to possess good resistance to root-knot nematodes (*M. incognita* and *M. floridensis*) and to ARR (*D. tabescens*) in field trials. 'MP-29' is tolerant to PTSL, comparable to 'Guardian®' peach seedling rootstocks. 'MP-29' also provides a marked reduction in tree vigour compared to peach seedling-type rootstocks and may prove useful as a dwarf rootstock in high-density cropping systems (Minas *et al.*, 2022).

'Rootpac® 20'

'Rootpac® 20' (Densipac) is a dwarfing, plum (*P. besseyi* × *P. cerasifera*) interspecific hybrid rootstock that was developed by Agromillora Iberia, S.L. in Barcelona, Spain. In Europe, it has been reported to produce a tree 40–50% the size of 'GF-677' (Duval, 2015), although the differences were not significant when budded with 'Big Top' in an orchard trial established under the growing conditions of the Ebro River Basin (Spain) in the 11th year after planting (Reig *et al.*, 2020). It is reported to be cold and waterlogging tolerant and moderately tolerant to salinity and calcareous-soil-induced iron chlorosis. It is moderately tolerant to root-knot nematodes and sensitive to crown gall (Pinochet *et al.*, 2012). It has good adaptation to heavy soils and colder areas. Based on its reported characteristics, it could be a good option for high-density plantings, but more data are required from different regions prior to final recommendations. Its vigour performance and classification from the first 4 years after establishment in the 2017 NC-140 semi-dwarf peach rootstock trial across 10 different orchard sites in North America are inconsistent (i.e., exhibits standard vigour) with what has been reported for this rootstock in Europe (Minas *et al.*, 2022). However, in a more recent rootstock × training system trial that was established in 2019 in Colorado, it performed similarly to the European reports regarding its reduced vigour and high precocity (I.S. Minas, personal communication).

'Rootpac® R'

'Rootpac® R' (Replantpac or Mirobac) is an open-pollinated hybrid of a myrobalan plum (*P. cerasifera* Ehr.), probably the female parent, and almond, both of unknown origin, and was developed by Agromillora Iberia, S.L. in Barcelona, Spain (Pinochet, 2010). It has high vigour, similar to 'Cadaman®' and 'GF-677' (Mestre *et al.*, 2015; Font i Forcada *et al.*, 2020) and 'Nemaguard' rootstocks.

It provides a good option for replant sites (Pinochet, 2010; Mestre *et al.*, 2015). It is tolerant to waterlogging and heavy-textured soils (Pinochet *et al.*, 2012). 'Rootpac® R' is well anchored and highly resistant to root-knot nematodes, and tolerant to ARR and *Phytophthora* spp. It has been shown to be highly susceptible to bacterial canker in the south-eastern USA (Reighard *et al.*, 2020). In the 2009 NC-140 'Redhaven' peach rootstock trial, and at the western sites of Colorado and Utah that are characterized by colder and dry conditions and calcareous soils, 'Rootpac® R' behaved as a standard vigour rootstock, little suckering, without any symptoms of iron chlorosis (Black *et al.*, 2021; Minas *et al.*, 2023).

'Ishtara®'

'Ishtara®' (Ferciana) is a standard to vigorous, complex myrobalan plum × peach ((*P. cerasifera* × *P. salicina*) × (*P. cerasifera* × *P. persica*)) interspecific hybrid rootstock (Grasselly, 1988; Renaud *et al.*, 1988). In Italy, it is reported to maintain good fruit size, even when grown in calcareous soils (Loreti and Massai, 2006). Results from long-term trials in Spain indicate that it may represent a good compromise between canopy size control, yield, yield efficiency, fruit size and tolerance to iron chlorosis (Reig *et al.*, 2020; Iglesias and Echeverría, 2022). It shows good adaptation to heavy and waterlogged soils, and is resistant or tolerant to root-knot nematodes and susceptible to ring and bacterial canker. In South Carolina rootstock trials, 'Ishtara®' was very susceptible to bacterial canker (Reighard *et al.*, 1997) but was resistant to the greater peach tree borer (*Synanthedon exitiosa*) in the 1994 NC-140 trial (Reighard *et al.*, 2004).

REFERENCES

Amri, R., Font i Forcada, C., Giménez, R., Pina, A. and Moreno, M.A. (2021) Biochemical characterization and differential expression of *PAL* genes associated with 'translocated' peach/plum graft-incompatibility. *Frontiers in Plant Science* 10: 622578.

Anthony, B. and Musacchi, S. (2021) Dwarfing mechanisms and rootstock–scion relationships in apple. *Italus Hortus* 28, 22–36.

Anthony, B.M. and Minas, I.S. (2021) Optimizing peach tree canopy architecture for efficient light use, increased productivity and improved fruit quality. *Agronomy* 11: 1961.

Anthony, B.M. and Minas, I.S. (2022) Redefining the impact of preharvest factors on peach fruit quality development and metabolism: a review. *Scientia Horticulturae* 297: 110919.

Anthony, B.M., Chaparro, J.M., Prenni, J.E. and Minas, I.S. (2020) Early metabolic priming under differing carbon sufficiency conditions influences peach fruit quality development. *Plant Physiology and Biochemistry* 157, 416–431.

Anthony, B.M., Chaparro, J.M., Sterle, D.G., Prenni, J.E. and Minas, I.S. (2021) Metabolic signatures of the true physiological impact of canopy light environment on peach fruit quality. *Environmental and Experimental Botany* 191: 104630.

Autio, W., Robinson, T., Blatt, S., Cochran, D., Francescato, P. *et al.* (2020) Budagovsky, Geneva, Pillnitz, and Malling apple rootstocks affect 'Honeycrisp' performance over eight years in the 2010 NC-140 'Honeycrisp' apple rootstock trial. *Journal of the American Pomological Society* 74, 182–195.

Basile, B., Marsal, J. and DeJong, T.M. (2003) Daily shoots extension growth of peach trees growing on rootstocks that reduce scion growth to daily dynamics of stem water potential. *Tree Physiology* 23, 695–704.

Beckman, T.G. and Nyczepir, A.P. (2004) Peach tree short life. In: *Southeastern Peach Growers Handbook*. Cooperative Extension Service, University of Georgia, Athens, Georgia, pp. 199–205.

Beckman, T.G., Okie, W.R., Nyczepir, A.P., Reighard, G.L., Zehr, E.I. and Newall, W.C. (1997) History, current status and future potential of Guardian® (BY520-9) peach rootstock. *Acta Horticulturae* 451, 251–258.

Beckman, T.G., Chaparro, J.X. and Sherman, W.B. (2012) MP-29, a clonal interspecific hybrid rootstock for peach. *HortScience* 47, 128–131.

Ben Yahmed, J., Ghrab, M., Moreno, M.A., Pinochet, J. and Ben Mimoun, M. (2016) Performance of 'Subirana' flat peach cultivar budded on different *Prunus* rootstocks in a warm production area in North Africa. *Scientia Horticulturae* 206, 24–32.

Ben Yahmed, J., Ghrab, M., Moreno, M.A., Pinochet, J. and Ben Mimoun, M. (2020) Leaf mineral nutrition and tree vigor of 'Subirana' flat peach cultivar grafted on different *Prunus* rootstocks in a warm Mediterranean area. *Journal of Plant Nutrition* 43, 811–822.

Bernhard, R. and Grasselly, C. (1981) Les pêchers × amandiers. *L'Arboriculture Fruitière* 328, 37–42.

Black, B.L., Minas, I.S., Reighard, G.L. and Beddes, T. (2021) Alkaline soil tolerance of rootstocks included in the NC-140 'Redhaven'. Peach Trial. *Journal of the American Pomological Society* 75, 9–16.

Blenda, A.V., Verde, I., Georgi, L.L., Reighard, G.L., Forrest, S.D. *et al.* (2007) Construction of a genetic linkage map and identification of molecular markers in peach rootstocks for response to peach tree short life syndrome. *Tree Genetics and Genomes* 3, 341–350.

Byrne, D.H., Raseira, M.C., Bassi, D., Piagnani, M.C., Gasic, K. *et al.* (2012) Peach. In: Badenes, M.L. and Byrne, D.H. (eds) *Fruit Breeding, Handbook of Plant Breeding 8*. Springer Science + Business Media, Düsseldorf, Germany, pp. 505–565.

Cambra, R. (1990) 'Adafuel', an almond × peach hybrid rootstock. *HortScience* 25: 584.

Caruso, T., Inglese, P., Sidari, M. and Sotille, F. (1997) Rootstock influences seasonal dry matter and carbohydrate content and partitioning in above-ground components of 'Flordaprince' peach trees. *Journal of the American Society for Horticultural Science* 122, 673–679.

Chalmers, D.J. and van den Ende, B. (1975) Productivity of peach trees: factors affecting dry-weight distribution during tree growth. *Annals of Botany* 39, 423–432.

Cline J.A. and Bakker C.J. (2022) Early performance of several *Prunus* interspecific hybrid rootstocks for Redhaven peach in Southern Ontario. *Canadian Journal of Plant Science* 102, 385–393.

Cos, J., Frutos, D., García, R., Rodríguez, J. and Carrillo, A. (2004) *In vitro* rooting study of the peach-almond hybrid 'Mayor'. *Acta Horticulturae* 658, 623–627.

Darikova, J.A., Savva, Y.V., Vaganov, E.A., Grachev, A.M. and Kuznetsova, G.V. (2011) Grafts of woody plants and the problem of incompatibility between scion and rootstock (a review). *Journal of Siberian Federal University Biology* 1, 54–63.

de Salvador, F.R., Liverani, A. and Fideghelli, C. (1991) La scelta dei portinnesti delle piante arboree da frutto: Pesco. *L'Informatore Agrario (Suppl.)* 36, 43–50.

de Salvador, F.R., Ondradu, B. and Scalas, B. (2002) Horticultural behaviour of different species and hybrids as rootstocks for peach. *Acta Horticulturae* 592, 317–322.

DeJong, T., Johnson, R.S., Doyle, J.F., Weibel, A., Solari, L. *et al.* (2004) Growth, yield and physiological behaviour of size-controlling peach rootstocks developed in California. *Acta Horticulturae* 658, 449–455.

DeJong, T., Almehdi, A., Johnson, S. and Day, K. (2005) Improved rootstocks for peach and nectarine. In: *California Tree Fruit Agreement 2004 Annual Research Report.* California Tree Fruit Agreement, Reedley, California, pp. 98–107.

DeJong, T., Grace, L.D., Almehdi, A. and Johnson, R.S. (2014) Performance and physiology of the Controller™ series of peach rootstocks. *Acta Horticulturae* 1058, 523–529.

Devkota, P., Iezzoni, A., Gasic, K., Reighard, G. and Hammerschmidt, R. (2020) Evaluation of the susceptibility of *Prunus* rootstock genotypes to *Armillaria* and *Desarmillaria* species. *European Journal of Plant Pathology* 158, 177–193.

Dichio, B., Xiloyannis, C., Celano, G., Vicinanza, L., Gómez-Aparisi, J. *et al.* (2004) Performance of new selections of *Prunus* rootstocks, resistant to root knot nematodes, in waterlogging conditions. *Acta Horticulturae* 658, 403–405.

Duval, H. (2015) Use of *Prunus* genetic diversity for peach rootstocks. *Acta Horticulturae* 1084, 277–282.

Edin, M. and Garcin, A. (1994) Un nouveau porte-greffe du pêcher Cadaman®-Avimag. *L'Arboriculture Fruitière* 475, 20–23.

Eremin, V.G. and Eremin, G.V. (2014) Clonal rootstocks of stone cultures for intensive orchards of the South of Russia. *Horticulture and Viticulture* 6, 24–29 (in Russian).

Esmenjaud, D., van Ghelder, C., Polidori, J., Khallouk, S. and Duval, H. (2015) Biological features, positional cloning and functional validation of the *Ma* gene for complete-spectrum resistance to root-knot-nematodes in *Prunus*. *Acta Horticulturae* 1084, 33–38.

FAOSTAT (2022) Value of Agricultural Production. Food and Agriculture Organization of the United Nations, Rome. Available at: https://www.fao.org/faostat/en/#data/QV (accessed 5 December 2022).

Felipe, A. (2009) 'Felinem', 'Garnem' and 'Monegro' almond × peach hybrid rootstock. *HortScience* 44, 196–197.

Felipe, A., Carrera, M. and Gómez-Aparisi, J. (1997) 'Montizo' and 'Monpol', two new plum rootstocks for peaches. *Acta Horticulturae* 451, 273–276.

Font i Forcada, C., Gogorcena Y. and Moreno M.Á. (2012) Agronomical and fruit quality traits of two peach cultivars on peach-almond hybrid rootstocks growing on Mediterranean conditions. *Scientia Horticulturae* 140, 157–163.

Font i Forcada, C., Reig, G., Giménez. R., Mignard. P., Mestre, L. and Moreno, M.Á. (2019) Sugars and organic acids profile and antioxidant compounds of nectarine fruits influenced by different rootstocks. *Scientia Horticulturae* 248, 145–153.

Font i Forcada, C., Reig, G., Mestre, L., Mignard, P., Betrán, J.Á. and Moreno, M.Á. (2020) Scion × rootstock response on production, mineral composition and fruit quality under heavy-calcareous soil and hot climate. *Agronomy* 10: 1159.

Gasic, K. (2020) Advances in cultivar and rootstock breeding: a case study in peach. *Acta Horticulturae* 1281, 1–8.

Gasic, K. and Saski, C. (2019) Advances in fruit genetics. In: Lang, G.A. (ed.) *Achieving Sustainable Cultivation of Temperate Zone Tree Fruits and Berries, Vol. 1: Physiology,*

Genetics and Cultivation. Burleigh Dodds Science Publishing, Cambridge, UK, pp. 135–162.

Giovannini D., Sirri, S., Dichio, B., Tuzio, A.C. and Xiloyannis, C. (2015) First results of the evaluation of six *Prunus* interspecific rootstock selections in comparative trials set up in two different environments. *Acta Horticulturae* 1084, 263–269.

Goldschmidt, E.E. (2014) Plant grafting: new mechanisms, evolutionary implications. *Frontiers in Plant Science* 5: 727.

Grasselly, C. (1985) Selection of peach seedling rootstocks. *Acta Horticulturae* 173, 245–250.

Grasselly, C. (1987) New French stone fruit rootstocks. *Fruit Varieties Journal* 41, 65–67.

Grasselly, C. (1988) Les porte-greffes du pêcher: des plus anciens aux plus récents. *L'Arboriculture Fruitière* 409, 29–34.

Gullo, G., Motisi, A., Zappia, R., Dattola, A., Diamanti, J. and Mezzetti, B. (2014) Rootstock and fruit canopy position affect peach [*Prunus persica* (L.) Batsch] (cv. Rich May) plant productivity and fruit sensorial and nutritional quality. *Food Chemistry* 153, 234–242.

Hartmann, H.T., Kester, D.E., Davies, F.T. and Geneve, R.L. (2002) *Plant Propagation. Principles and Practices,* 7th edn. Prentice Hall, Upper Saddle River, New Jersey.

Iglesias, I. (2013) Peach production in Spain: current situation and trends, from production to consumption. In: *Proceedings of the 4th conference 'Innovations in Fruit Growing-Improving Peach and Apricot Production', Faculty of Agriculture, University of Belgrade, Belgrade.* Poljoprivredni Fakultet, Beograd, Serbia, pp. 75–98.

Iglesias, I. and Echeverría, G. (2022) Current situation, trends and challenges for efficient and sustainable peach production. *Scientia Horticulturae* 296: 110899.

Iglesias, I., Giné-Bordonaba, J., Garanto, X. and Reig, G. (2019) Rootstock affects quality and phytochemical composition of 'Big Top' nectarine fruits grown under hot climatic conditions. *Scientia Horticulturae* 256: 108586.

Indreias, A. (2011) Performance of some peach rootstocks in the Nursery in Southeast Romania. *Acta Horticulturae* 903, 501–506.

Indreias, A. (2013) Breeding program of rootstocks for peach tree at the Research Station for Fruit Growing Constanta, Romania. *Acta Horticulturae* 981, 217–222.

Inglese, P., Caruso, T., Gugliuzza, G. and Pace, L.S. (2002) Crop load and rootstock influence on dry matter partitioning in trees of early and late ripening peach cultivars. *Journal of the American Society for Horticultural Science* 127, 825–830.

Jackson, M.B. (1993) Are plant hormones involved in root to shoot communication? *Advances in Botanical Research* 19, 104–181.

Jiménez, S., Pinochet, J., Abadía, A., Moreno, M.Á. and Gogorcena, Y. (2008) Tolerance response to iron chlorosis of *Prunus* selections as rootstocks. *HortScience* 43, 304–309.

Jiménez, S., Pinochet, J., Romero, J., Gogorcena, Y., Moreno, M.A. and Espada, J.L. (2011) Performance of peach and plum based rootstocks of different vigour on a late peach cultivar in replant and calcareous conditions. *Scientia Horticulturae* 129, 58–63.

Johnson, S., Anderson, R., Autio, W., Beckman, T., Black, B. *et al.* (2011) Performance of the 2002 NC-140 cooperative peach rootstock planting. *Journal of the American Pomological Society* 65, 17–25.

Lang, G. (2019) The cherry industries in the USA: current trends and future perspectives. *Acta Horticulturae* 1235, 119–132.

Layne, R.E.C. (1987) Peach rootstocks. In: Rom, R.C. and Carlson, R.F. (eds) *Rootstocks for Fruit Crops*. Wiley, New York, pp. 185–216.

Layne, R.E.C., Jackson, H.O. and Doud, F.D. (1977) Influence of peach seedling rootstocks on defoliation and cold hardiness of peach cultivars. *Journal of the American Society for Horticultural Science* 102, 89–92.

Legua, P., Pinochet, J., Moreno, M.Á., Martínez, J.J. and Hernández, F. (2012) *Prunus* hybrids rootstocks for flat peach. *Scientia Agricola* 69, 13–18.

Li, K., DiLegge, M.J., Minas, I.S., Hamm, A., Manter, D. and Vivanco, J.M. (2019) Soil sterilization leads to re-colonization of a healthier rhizosphere microbiome. *Rhizosphere* 12: 100176.

Long, L.E., Lang, G.A., Whiting, M.D. and Musacchi, S. (2015) *Cherry Training Systems*. PNW series 667. Oregon State University, Corvallis, Oregon. Available at: https://hdl.handle.net/2376/6087 (accessed 16 December 2022).

Loreti, F. and Massai, R. (1994) Sirio: Nuovo portinnesto ibrido pesco × mandorlo. *L'Informatore Agrario* 28, 47–49.

Loreti, F. and Massai, R. (2002) MiPAF targeted project for evaluation of peach rootstocks in Italy: results of six years of observations. *Acta Horticulturae* 592, 117–124.

Loreti, F. and Massai, R. (2006) 'Castore' and Polluce': two new hybrid rootstocks for peach. *Acta Horticulturae* 713, 275–278.

Manganaris, G.A., Minas, I., Cirilli, M., Torres, R., Bassi, D. and Costa, G. (2022) Peach for the future: a specialty crop revisited. *Scientia Horticulturae* 305: 111390.

Marini, R.P., Sowers, D. and Marini, M.C. (1991) Peach fruit quality is affected by shade during final swell of fruit growth. *Journal of the American Society for Horticultural Science* 116, 383–389.

Martínez-Ballesta, M.C., Alcaraz-López, C., Muries, B., Mota-Cadenas, C. and Carvajal, M. (2010) Physiological aspects of rootstock–scion interactions. *Scientia Horticulturae* 127, 112–118.

Massai, R. and Loreti, F. (2004) Preliminary observations on nine peach rootstocks grown in a replant soil. *Acta Horticulturae* 658, 185–192.

Mestre, L., Reig, G., Betrán, J.A., Pinochet, J. and Moreno, M.Á. (2015) Influence of peach–almond hybrids and plum-based rootstocks on mineral nutrition and yield characteristics of 'Big Top' nectarine in replant and heavy-calcareous soil conditions. *Scientia Horticulturae* 192, 475–481.

Mestre, L., Reig, G., Betrán, J.A. and Moreno, M.Á. (2017) Influence of plum rootstocks on agronomic performance, leaf mineral nutrition and fruit quality of 'Catherina' peach cultivar in heavy-calcareous soil conditions. *Spanish Journal of Agricultural Research* 15: e0901.

Miller, S.T., Otto, K.L., Sterle, D., Minas, I.S. and Stewart, J.E. (2019) Preventive fungicidal control of *Cytospora leucostoma* in peach orchards in Colorado. *Plant Disease* 103, 1138–1147.

Miller, S.T., Sterle, D., Minas, I.S. and Stewart, J.E. (2021) Exploring fungicides and sealants for management of *Cytospora plurivora* infections in western Colorado peach production systems. *Crop Protection* 146: 105654.

Minas I.S. and Sterle D.G. (2020) Differential thermal analysis sheds light on the effect of environment and cultivar in peach floral bud cold hardiness. *Acta Horticulturae* 1281, 385–392.

Minas, I.S., Tanou, G. and Molassiotis, A. (2018) Environmental and orchard bases of peach fruit quality. *Scientia Horticulturae* 235, 307–322.

Minas, I.S., Blanco-Cipollone, F. and Sterle, D. (2020) Near infrared spectroscopy can non-destructively assess the effect of canopy position and crop load on peach fruit maturity and quality. *Acta Horticulturae* 1281, 407–412.

Minas, I.S., Blanco-Cipollone, F. and Sterle, D. (2021) Accurate non-destructive prediction of peach fruit internal quality and physiological maturity with a single scan using near infrared spectroscopy. *Food Chemistry* 335: 127626.

Minas, I.S., Reighard, G.L., Brent Black, B., Cline, J.A., Chavez, D.J. *et al.* (2022) Establishment performance of the 2017 NC-140 semi-dwarf peach rootstock trial across 10 sites in North America. *Acta Horticulturae* 1346, 669–676.

Minas, I.S., Anthony, B.M., Pieper, J.R. and Sterle, D.G. (2023) Large-scale and accurate non-destructive visual to near infrared spectroscopy-based assessment of the effect of rootstock on peach fruit internal quality. *European Journal of Agronomy* 143: 126706.

Moreno, M.Á. (2004) Breeding and selection of *Prunus* rootstocks at the Estación Experimental de Aula Dei, Zaragoza, Spain. *Acta Horticulturae* 658, 519–528.

Moreno, M.Á., Moing, A., Lansac, M., Gaudillère, J.P. and Salesses, G. (1993) Peach/myrobalan plum graft incompatibility in the nursery. *Journal of Horticultural Sciences* 68, 705–714.

Moreno, M.Á., Tabuenca, M.C. and Cambra, R. (1994a) Performance of 'Adafuel' and 'Adarcias' as peach rootstocks. *HortScience* 29, 1271–1273.

Moreno, M.Á., Gaudillère, J.P. and Moing, A. (1994b) Protein and amino acid content in compatible and incompatible peach/plum grafts. *Journal of Horticultural Sciences* 69, 955–962.

Moreno, M.Á., Tabuenca, M.C. and Cambra, R. (1995a) 'Adesoto 101', a plum rootstock for peaches and other stone fruits. *HortScience* 30, 1314–1315.

Moreno, M.Á., Tabuenca, M.C. and Cambra, R. (1995b) Adara, a plum rootstock for cherries and other stone fruit species. *HortScience* 30, 1316–1317.

Moreno, M.Á., Adrada, R., Aparicio, J. and Betrán, J.A. (2001) Performance of 'Sunburst' sweet cherry grafted on different rootstocks. *Journal of Horticultural Sciences* 76, 167–173.

Musacchi, S., Gagliardi, F. and Serra, S. (2015) New training systems for high-density planting of sweet cherry. *HortScience* 50, 59–67.

Nicotra, A. and Moser, L. (1997) Two new plum rootstocks for peach and nectarines: Penta and Tetra. *Acta Horticulturae* 451, 269–271.

Okie, W.R., Reighard, G.L., Beckman, T.G., Nyczepir, A.P., Reilly, C.C. *et al.* (1994a) Field-screening *Prunus* for longevity in the southeastern United States. *HortScience* 29, 673–677.

Okie, W.R., Beckman, T.G., Nyczepir, A.P., Reighard, G.L., Newall, W.C. Jr and Zehr, E.I. (1994b) BY520-9, a peach rootstock for the southeastern United States that increases scion longevity. *HortScience* 29, 705–706.

Orazem, P., Stampar, F. and Hudina, M. (2011) Quality analysis of 'Redhaven' peach fruit grafted on 11 rootstocks of different genetic origin in a replant soil. *Food Chemistry* 124, 1691–1698.

Pascal, T., Pfeiffer, F. and Kervella, J. (2010) Powdery mildew resistance in the peach cultivar Pamirskij 5 is genetically linked with the *Gr* gene for leaf color. *HortScience* 45, 150–152.

Pieper, J.R., Anthony, B.M., Sterle, D.G. and Minas, I.S. (2022) Rootstock vigor and fruit position in the canopy influence peach internal quality. *Acta Horticulturae* 1346, 807–812.

Pina, A., Cookson, S.J., Calatayud, Á., Trinchera, A. and Errea, P. (2017) Physiological and molecular mechanisms underlying graft compatibility. In: Colla, G., Pérez-Alfocea, F. and Schwarz D. (eds) *Vegetable Grafting: Principles and Practices*, pp. 132–154.

Pinochet, J. (2010) 'Replantpac' (Rootpac® R), a plum-almond hybrid rootstock for re-plant situations. *HortScience* 45, 299–301.

Pinochet, J., Calvet, C., Hernández-Dorrego, A., Bonet, A., Felipe, A. and Moreno, M.Á. (1999) Resistance of peach and plum rootstocks from Spain, France, and Italy to rootknot nematode *Meloidogyne javanica*. *HortScience* 34, 1259–1262.

Pinochet, J., Fernandez, C., Cunill, M., Torrents, J., Felipe, A. *et al.* (2002) Response of new interspecific hybrids for peach to root-knot and lesion nematodes, and crown gall. *Acta Horticulturae* 592, 707–716.

Pinochet, J., Cunill, M., Torrents, J., Eremin, G., Eremin, V. *et al.* (2012) Response of low and medium vigour rootstocks for peach to biotic and abiotic stresses. *Acta Horticulturae* 962, 627–632.

Ramming, D.W. and Tanner, O. (1983) 'Nemared' peach rootstock. *HortScience* 18: 376.

Reig, G., Mestre, L., Betrán, J.A., Pinochet, J. and Moreno, M.A. (2016) Agronomic and physicochemical fruit properties of 'Big Top' nectarine budded on peach and plum based rootstocks in Mediterranean conditions. *Scientia Horticulturae* 210, 85–92.

Reig, G., Salazar, A., Zarrouk, O., Font i Forcada, C., Val, J. and Moreno, M.Á. (2019) Long-term graft compatibility study of peach–almond hybrid and plum based rootstocks budded with European and Japanese plums. *Scientia Horticulturae* 243, 392–400.

Reig, G., Garanto, X., Mas, N. and Iglesias, I. (2020) Long-term agronomical perform-ance and iron chlorosis susceptibility of several *Prunus* rootstocks grown under loamy and calcareous soil conditions. *Scientia Horticulturae* 262: 109035.

Reighard, G.L. (2000) Peach rootstocks for the United States: are foreign rootstocks the answer? *HortTechnology* 10, 714–718.

Reighard, G.L. and Loreti, F. (2008) Rootstock development. In: Layne, D.R. and Bassi, D. (eds) *The Peach: Botany, Production and Uses*. CAB International, Wallingford, UK, pp. 193–220.

Reighard, G.L., Newall, W.C. Jr, Zehr, E.I., Okie, W.R., Beckman, T.G. and Nyczepir, A.P. (1997) Field performance of *Prunus* rootstock cultivars and selections on replant soils in South Carolina. *Acta Horticulturae* 451, 243–249.

Reighard, G., Anderson, R., Anderson, J., Autio, W., Beckman, T., *et al.* (2004) Eight-year performance of 19 peach rootstocks at 20 locations in North America. *Journal of the American Pomological Society* 58, 174–202.

Reighard, G.L., Ouellette, D. and Brock, K. (2008) Performance of new *Prunus* rootstocks for peach in South Carolina. *Acta Horticulturae* 772, 237–240.

Reighard, G.L., Beckman, T., Belding, R., Black, B., Byers, P. *et al.* (2011) Six-year performance of 14 *Prunus* rootstocks at 11 sites in the 2001 NC-140 peach trial. *Journal of The American Pomological Society* 65, 26–41.

Reighard, G., Bridges, W. Jr, Archbold, D., Wolfe, D., Atucha, A. *et al.* (2013) NC-140 peach rootstock testing in thirteen US states. *Acta Horticulturae* 1084, 225–232.

Reighard, G.L., Bridges, W., Archbold, D., Atucha, A., Autio, W. *et al.* (2018) Rootstock performance in the 2009 NC-140 peach trial across 11 states. *Acta Horticulturae* 1228, 181–186.

Reighard, G.L., Bridges, W. Jr, Archbold, D., Atucha, A., Autio, W. *et al.* (2020) Nine-year rootstock performance of the NC-140 'Redhaven' peach trial across 13 states. *Journal of The American Pomological Society* 74, 45–56.

Remorini, D., Fei, C., Loreti, F. and Massai, R. (2015) Observations on nine peach rootstocks grown in a replant soil. *Acta Horticulturae* 1085, 131–138.

Renaud, R., Bernhard, R., Grasselly, C. and Dosba, F. (1988) Diploid plum × peach hybrid rootstocks for stone fruit trees. *HortScience* 23, 115–117.

Ritchie, D. and Clayton, C. (1981) Peach tree short life: a complex of interacting factors. *Plant Diease* 65, 462–469.

Robinson, T., Hoying, S., Miranda Sazo, M., DeMarree, A. and Dominguez, L. (2013) A vision for apple orchard systems of the future. *Fruit Quarterly* 21, 11–16.

Salesses, G., Dirlewanger, E., Bonnet, A., Lecouls, A.C. and Esmenjaud, D. (1998) Interspecific hybridization and rootstock breeding for peach. *Acta Horticulturae* 465, 209–217.

Seleznyova, A.N., Thorp, T.G., White, M., Tustin, S. and Costes, E. (2003) Application of architectural analysis and AMAPmod methodology to study dwarfing phenomenon: the branch structure of 'Royal Gala' apple grafted on dwarfing and non-dwarfing rootstock/interstock combinations. *Annals of Botany* 91, 665–672.

Seleznyova, A.N., Tustin, D.S., White, M.D. and Costes, E. (2004) Analysis of the earliest observed expression of dwarfing rootstock effects on young apple trees, using Markovian models. *Acta Horticulturae* 732, 79–84.

Seleznyova, A.N., Tustin, D.S. and Thorp, T.G. (2008) Apple dwarfing rootstocks and interstocks affect the type of growth units produced during the annual growth cycle: precocious transition to flowering affects the composition and vigour of annual shoots. *Annals of Botany* 101, 679–687.

Sherman, W.B., Lyrene, P.M. and Sharpe, R.H. (1991) *Flordaguard Peach Rootstock.* IFAS Circular S-376, University of Florida, Florida.

Stănică, F., Dumitru, L.M. and Ivașcu, A. (2021) Overview of Romanian peach breeding. *Acta Horticulturae* 1304, 1–12.

Szymajda, M., Sitarek, M., Pruski, K. and Żurawicz, E. (2019) A potential of new peach (*Prunus persica* L.) seed tree genotypes for the production of generative rootstocks. *Scientia Horticulturae* 256: 108618.

Tabakov, S.G. and Yordanov, A.I. (2011) Orchard performance of Hungarian apricot cultivar on eleven rootstocks in central south Bulgaria conditions. *Acta Horticulturae* 966, 241–247.

Tabakov, S.G., Yordanov, A.I. and Petrov, M.N. (2021) Study of the influence of five rootstocks on the growth and productivity of three plum cultivars grown in Bulgaria *Acta Horticulturae* 1322, 131–138.

Tombesi, S., Johnson, R.S., Day, K.R. and DeJong, T.M. (2010) Relationships between xylem vessel characteristics, calculated axial hydraulic conductance and size-controlling capacity of peach rootstocks. *Annals of Botany* 105, 327–331.

Tsipouridis, C. and Thomidis, T. (2005) Effect of 14 peach rootstocks on the yield, fruit quality, mortality, girth expansion and resistance to frost damages of May Crest peach variety and their susceptibility on *Phytophthora citrophthora. Scientia Horticulturae* 103, 421–428.

Tustin, D., van Hooijdonk, B. and Breen, K. (2018) The planar cordon – new planting systems concepts to improve light utilisation and physiological function to increase apple orchard yield potential. *Acta Horticulturae* 1128, 1–12.

Tustin, D.S., Breen, K.C. and van Hooijdonk, B.M. (2022) Light utilisation, leaf canopy properties and fruiting responses of narrow-row, planar cordon apple orchard planting systems – a study of the productivity of apple. *Scientia Horticulturae* 294: 110778.

van Hooijdonk, D., Woolley, D.J., Warrington, I.J. and Tustin, D.S. (2010) Initial alteration of scion architecture by dwarfing apple rootstocks may involve shoot–root–shoot signaling by auxin, gibberellin, and cytokinin. *Journal of Horticultural Science and Biotechnology* 85, 59–65.

Webster, A.D. (2004) Vigour mechanisms in dwarfing rootstocks for temperate fruit trees. *Acta Horticulturae* 658, 29–41.

Yadava, U.L. and Doud, S.L. (1989) Rootstock and scion influence growth, productivity, survival, and short life-related performance of peach trees. *Journal of the American Society for Horticultural Science* 114, 875–880.

Zarrouk, O., Gogorcena, Y., Gómez-Aparisi, J., Betrán, J.A. and Moreno, M.Á. (2005) Influence of almond × peach hybrids rootstocks on flower and leaf mineral concentration, yield and vigour of two peach cultivars. *Scientia Horticulturae* 106, 502–514.

Zarrouk, O., Gogorcena, Y., Moreno, M.Á. and Pinochet, J. (2006) Graft compatibility between peach cultivars and *Prunus* rootstocks. *HortScience* 41, 1389–1394.

Zarrouk, O., Testillano, P.S., Risueño, M.C., Moreno, M.Á. and Gogorcena, Y. (2010) Changes in cell/tissue organization and peroxidase activity as markers for early detection of graft incompatibility in peach/plum combinations. *Journal of the American Society for Horticultural Science* 135, 9–17.

Zec, G., Fotirić-Akšić, M., Čolić, S., Vulić, T., Nikolić, D. *et al.* (2013) Influence of vineyard peach selections on vigour and initial yield in peach and nectarine. *Genetika* 45, 11–20.

Zehr, E.I., Miller, R.W. and Smith, F.H. (1976) Soil fumigation and peach rootstocks for protection against peach tree short life. *Phytopathology* 66, 689–694.

Zhang, J., Whiting, M.D. and Zhang, Q. (2015) Diurnal pattern in canopy light interception for tree fruit orchard trained to an upright fruiting offshoots (UFO) architecture. *Biosystems Engineering* 129, 1–10.

Cultivars

Daniele Bassi[1]*, Marco Cirilli[1], Stefano Foschi[2], Ignasi Iglesias[3],
Daniela Giovannini[4], Maria L. Badenes[5†] and Ksenija Gasic[6]
[1]Faculty of Agriculture, University of Milan, Milan, Italy; [2]Rinova, Cesena,
Italy; [3]Agromillora Group, Sant Sadurní d'Anoia, Spain; [4]CREA, Research
Centre for Olive, Fruit and Citrus Crops, Forlì, Italy; [5]Instituto Valenciano de
Investigaciones Agraria, Moncada, Valencia, Spain; [6]Department of Plant
and Environmental Sciences, Clemson University,
Clemson, South Carolina, USA

4.1 BREEDING PROGRAMMES

4.1.1 Qualitative (single) characteristics and their inheritance

The peach (*Prunus persica* (L.) Batsch) is a species with a surprisingly high
number of Mendelian traits, more than 20 subtending to important horticultural characteristics related to tree growth habits, disease resistance and fruit
quality (Salazar *et al.*, 2014). For many of these, linkage and/or association
mapping studies by F_1, F_2 or BC_1 populations or panels of unrelated accessions
have allowed the establishment of their chromosomal positions until the identification of underlined candidate variant(s) and/or gene(s). Monogenic traits
of relevant importance for breeding are those controlling the main commercial
fruit types: peach/nectarine and white/yellow flesh. The recessive glabrous skin
(peach, *G/–* or nectarine, *g/g*) arose from the mutation of a gene controlling
trichome development (*PpeMYB25*) and determining the nectarine phenotype
(Vendramin *et al.*, 2014), while the recessive yellow flesh colour (*y/y*) arose
from different mutation events in the ancestral white-flesh allele (*Y/–*), coding
for a carotenoid biosynthesis gene (*PpCCD4*) (Adami *et al.*, 2013; Falchi *et al.*,
2013). At least two Mendelian 'blood' flesh traits are known: (i) one recessive
locus (*Bf*), mapped to the top of chromosome 4 and present in the 'Harrow
Blood' accession, which induces the early development of anthocyanins in
the fruit flesh beginning at pit hardening, also associated with red midrib leaf
veins (often small trees); and (ii) a second dominant locus (*DBF*), mapped

---

* Email: daniele.bassi@unimi.it

on chromosome 5 and probably controlled by an NAC gene mutation (Zhou *et al.*, 2015), responsible for a fully red mesocarp, 2 weeks before the ripening date, that shows no red on the midrib leaf vein. So far, the blood flesh has been of scarce commercial significance. Among the various shapes of fruits, the 'flat' type (also known as 'saucer' or 'donut') has received increasing breeding interest from the release of 'Stark Saturn' (1985) at Rutgers University, the first flat peach for the fresh market of the modern era. Zhou *et al.* (2020) reported that the trait is dominant (*S/s*) but associated with aborting fruits at the homozygous state (*S/S* or *af/af*).

Fundamental flesh-texture traits at the commercial level are also regulated by Mendelian inheritance: the melting/non-melting texture and flesh adherence to the stone (clingstone/freestone) are both controlled by a copy-number variation of an endopolygalacturonase (endoPG) enzyme involved in cell-wall degradation: one copy (PpendoPGM, H1), two copies (PpendoPGF and PpendoPGM, H2) and none (H3). The accessions with the combinations H1H1, H1H2 or H1H3 have a melting freestone phenotype and those with H2H2 and H2H3 are melting clingstone, while the H3H3 are non-melting clingstone (Gu *et al.*, 2016). The stony hard texture type is characterized by a crunchy texture, caused by a lack of ethylene biosynthesis during fruit ripening due to a mutation in the auxin biosynthesis gene (*YUCCA11*) (Pan *et al.*, 2015; Cirilli *et al.*, 2018). This trait is epistatic to the melting and non-melting traits, and thus difficult to select by on-field evaluation only. Despite a certain commercial potential, breeding interest for this texture type is still minimal due to some (probably pleiotropic) adverse effects on fruit quality.

In terms of fruit quality attributes, the so-called 'low-acid' (LA) trait should be mentioned (*D/d*). The LA phenotype, which is dominant, shows mainly citric and malic acids (~50%) but also quinic (~20%) acid in lower concentrations than standard phenotypes, while total acidity is two- to four-fold lower (0.4 versus 1.4: average values from pooled LA and standard cultivars, respectively). The pH of the LA trait is above 4.0, while in standard phenotypes, the pH is below 3.9. The soluble solids content is comparable to that in the standard phenotype, although LA types show more sucrose and less glucose. For a better taste of LA peaches, soluble solids content above 12% is suggested to overcome the bland feeling of low acidity coupled with too low a sugar content.

Several monogenic traits (often with incomplete dominance) have also been linked to tree growth habits, such as weeping (*Pl*), dwarf (*Dw*), semi-dwarf (*N*), compact (*Ct*), or broomy (*Br*), often well characterized at the molecular level (Hill and Hollender, 2019). Although promising for highly mechanized and super-intensive orchards, some of these forms (e.g. upright and pillar) do not show successful application so far, only holding some interest at the ornamental level. Other traits with simple inheritance have also been reported for resistance to powdery mildew, root-knot nematodes and aphids. However, such resistance traits have often been introgressed from ornamental or wild relatives (e.g. almond, *P. davidiana*), an approach with very poor success in current breeding programmes for the fresh market.

4.1.2   Quantitative characteristics

Characteristics with quantitative inheritance have great importance in selection because they are often related to the horticultural and economic value of the harvested fruit. Traits such as yield, fruit size, colour, texture and taste are under polygenic control and are influenced by environment and orchard management practices and their interactions. Also, many traits associated with tree phenology, such as blooming and the maturity date or chilling/heat requirements, have a complex inheritance pattern, as well as resistance to some important diseases (e.g. sharka, brown rot, *Xanthomonas* spp.). For a breeder, the heritability of a trait under selection is important, as it affects the rapidity and extent of genetic gain through phenotypic-based parental selection. Intrapopulation selection, a recurrent selection method, has been and is still the most common method applied in peach breeding to improve and combine several quantitative traits to increase the frequencies of desirable alleles and maintain a certain degree of heterozygosity (Sansavini *et al.*, 2006). This approach has been the basis for developing varietal ideotypes characterized by improved adaptability to a wide range of environments (from tropical to cold temperate climates) and quality traits (large size, skin coloration and firmness). Several studies have provided heritability estimates of the main agronomic and pomological characteristics, primarily based on progeny mean (Hansche *et al.*, 1972; de Souza *et al.*, 1998a,b; Hernandez Mora *et al.*, 2017; Rawandoozi *et al.*, 2020). The potential of molecular approaches for assisting peach varietal improvement has stimulated wide-scale research to decipher the genetic architecture of the main horticultural traits and the development of breeding tools (Laurens *et al.*, 2018; Iezzoni *et al.*, 2020). Nevertheless, the systematic application of such knowledge and tools is still a long way off in the main peach-breeding programmes.

4.1.3   Ploidy

Peach is a diploid plant with $2n = 16$ chromosomes. Monoploids (or haploids) can sometimes be found in progeny that have received only the basic chromosome number ($n = 8$). This set comes from the female parent, in which a cell of its embryo sac divides without fertilization and creates an embryo. Toyama (1974) searched systematically for monoploids in peach seedlings, trying to isolate them before their germination and obtained pure lines by doubling the chromosome number with colchicine. The seeds of these fruits were able to germinate, and some were triploids, diploids or aneuploids. The triploids were vigorous but with little fertility; the diploids, if they resulted from self-pollination (the monoploid parent was maintained in a greenhouse during flowering but without protection), were pure lines.

4.1.4 First steps in peach breeding

The first peach cultivars came from chance seedlings, eventually propagated by grafting. In this way, each country has constituted a cultivar pool whose richness has continually increased with new discoveries and germplasm exchanges. In the USA, Downing (1866) described 136 cultivars, mentioning hybridization as a new method to obtain cultivars, probably for the first time in fruit culture. Hedrick (1917) described cultivars such as 'Elberta', 'Chinese Cling' and 'J.H. Hale', which were used extensively in the first peach-breeding programmes initiated before the publication of his pomology (Connors, 1917). The contribution of these early breeding programmes has significantly modified the available cultivar assortment worldwide. 'Elberta', a chance seedling from free pollination of 'Chinese Cling' released in 1889, was the first milestone of modern breeding, the progenitor of most of today's yellow-fleshed cultivars. 'J.H. Hale' was the second important introduction (1912), with a higher fruit quality than its probable parent 'Elberta'. This cultivar was widely used in the breeding programmes of the early 1900s, and also in countries outside the USA. The third major US introduction was the peach 'Redhaven' (1940s), which marked a substantial improvement in the aesthetic and commercial quality of peaches, representing the commercial standard in Europe for almost 30 years. The release of 'Le Grand' (1942) marked a milestone in the selection and commercial enhancement of the 'nectarine' type, which began in the early 1900s from seedlings of various origins (such as 'Quetta', 'Goldmine' and 'Lippiatt'). Notably, nectarines appeared in Europe around 1000 AD, such as 'Sbergio', which was probably introduced by the Arabs in Sicily. The name 'nectarine' appeared around 1660, inspired by the German *nektarpfirsich*.

By the 1940s, breeders had intensified efforts to select for season extension and low chilling. Low-chill breeding began in California around 1910, and over time peach cultivation was expanded to wider climatic areas, from Canada to the subtropics. The donor of the low-chill requirement was originally from southern China, imported into the USA from Australia in 1869 under the name of 'Australian Soucer', from which early-used materials such as 'Peento', 'Angel' and 'Jewel' originated. 'Babcock' was the first low-chill cultivar released (1933) and was cultivated until the early 1990s in California. Breeding of other fruit types, such as flat and low acid, was more recent. The variety 'Stark Saturn' (1985), bred as NJF-2 by L.F. Hough at Rutgers University (New Brunswick, New Jersey) and commercialized by Starks Nursery, was probably the first flat peach destined for fresh market consumption. Cold hardiness was the major improvement that allowed expansion of flat peach cultivation in temperate environments.

4.1.5 Breeding goals

The first generation of breeders emphasized improving fruit commercial characteristics such as firmness and attractiveness. A tremendous improvement of fruit overcolor is easily recognizable when comparing the poor red blush of 'Elberta' or the green background colour of the first nectarine cultivars with the almost complete red skin of 'Red Diamond' or 'O'Henry' (1970), prototypes of modern nectarine and peach cultivars, respectively. Furthermore, the progressive increase in flesh firmness, from soft-melting to very firm and (later) slow-softening textures, has been the basis for extending fruit shelf life and harvesting at a more advanced mature stage. The breeding goals of 16 peach (stone fruits) breeders from public institutions collected in a recent worldwide survey are shown in Fig. 4.1.

4.1.6 Diversification for the consumer

To bring diversity, peach breeders draw mainly on Mendelian variability. In the past, white-fleshed peaches dominated the cultivar landscape, being progressively replaced by yellow-fleshed peaches from US programmes. One more Mendelian trait, blood flesh (two loci are known: anthocyanin accumulation in both skin and flesh, regardless of the yellow or white flesh colour), has been introduced in nectarines in France (T. Pascal, personal communication) and the USA. Conversely, anthocyaninless nectarines have been created in the USA.

The first commercial nectarines (fuzzless skin), released in the 1960s, were derived mainly from the stocks of F.W. Anderson and private Californian

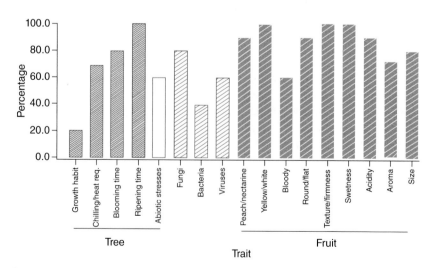

Fig. 4.1. Breeding goals of 16 peach (stone fruit) breeders from public institutions collected in a recent worldwide survey in 2021.

programmes (Bradford Farms and Zaiger Genetics). Releases such as 'Early Sun Grand' (1955), 'Stark RedGold' (1962), 'Armking' (1969) and 'Red Diamond' (1972) marked a rapid improvement in nectarine skin colour, size and taste. In terms of texture, the major trait responsible for firmness, the traditional (soft) melting phenotype, is being progressively replaced in the cultivar market by the slow-softening variant, first introduced worldwide by 'Big Top'. This unique yellow-fleshed nectarine cultivar was first introduced from the USA in 1983 (Zaiger Genetics) and was widely planted in all countries. It quickly became well accepted by growers, markets and consumers because of its solid bright overcolour, good fruit size and firmness (melting but slowly softening compared with the traditional melting phenotype), and LA trait conferring a sweet taste (<6 g malic acid l^{-1}). Two more traits are very popular in terms of firmness: the worldwide popular non-melting (clingstone and anthocyaninless) phenotype (suitable for canning but not exclusively) and the stony hard (differing from the non-melting trait for its crunchy texture and absence of ethylene development). The stony hard type is mainly popular in the Far East. However, there have also been attempts to introduce stony hard peaches in western countries (Liverani *et al.*, 2017; Bassi and Foschi, 2021). Canning (non-melting, clingstone and anthocyaninless) nectarines were also the goal of breeding programmes in Italy and the USA. Introducing the non-melting flesh characteristic in cultivars for the fresh market (with bright red overcolor) allows harvesting the fruits near maturity, at their optimal quality and taste, with an extended postharvest life.

Based on total acidity (TA value; measured as g malic acid l^{-1}), five different peach/nectarine groups have been established: (i) very sweet: TA <3.3; (ii) sweet: TA 3.3–6.0; (iii) equilibrated: TA 6–8; (iv) acidic: TA 8–10; and (v) very acidic: TA >10 (Iglesias and Echeverría, 2021). The 'honey' characteristic (low acid), previously very popular only in Far East countries, is now found in various sources of breeding materials that were only exploited commercially in the 1980s, first in the USA and then in Europe. The private fruit breeder Floyd Zaiger was a pioneer of this trait, which has revolutionized (together with the slow-softening texture) worldwide breeding with important releases such as 'Royal Glory' (1987). Breeding programmes for flat peaches only started at the end of the 1980s, mostly resorting to 'Stark Saturn', particularly in France (noteworthy releases were 'Platina' and 'Mesembrine'), Italy ('UFO' series), China ('Ruipan' series) and the USA (several public or private programmes). Since the early 2000s, Spain has placed special emphasis on breeding flat peaches.

4.1.7 Reduction of production costs

A cultivar has low production costs when it is well adapted to the growing environment. Adaptation to climatic factors, such as the hardiness of dormant buds for northern-limit zones or low-chill requirements for subtropical regions, has been and continues to be thoroughly pursued in breeding programmes. Disease and pest resistance are also desirable traits for lowering production costs and health and environmental hazards. Among the most prominent goals as far as biotic resistance is concerned are bacterial leaf and fruit spot (*Xanthomonas arboricola* pv. *pruni*) in the USA; Brown rot (*Monilinia fructicola* Aderh. et Ruhl. (Honey)) in the USA and Europe; *Cytospora* canker (*Leucostoma persoonii* and *Leucostoma cincta*) in the USA and Italy; green peach aphid (*Myzus persicae*) in France and Italy; leaf curl (*Taphrina deformans*) in the USA and Italy; and powdery mildew (*Podosphaera pannosa*) in France and Italy. However, breeding cultivars combining disease resistance traits and high commercial standards have not yet produced significant results. Additionally, efforts to lower production costs by breeding for modified growth habits, such as 'pillar' (columnar) trees for wall systems (Liverani *et al.*, 2004) or weeping trees for 'vase' systems (open center) training systems, have been explored. Despite these efforts, the spread of these cultivars (particularly with pillar or upright habits) has been minimal.

4.1.8   Low-chill cultivars

A chilling requirement in newly developed peach cultivars has recently received more attention due to climate fluctuations and the need for newly released cultivars to be resilient to climate extreme events to ensure sustainable production (Byrne, 2014; Raseira, 2016). The need for low(er) chilling requirements has been governed by the desire to expand peach and nectarine production to areas with warmer climates as well (Raseira, 2016; Topp *et al.*, 2012). However, recent changes in climate and crop losses due to late spring frosts or insufficient winter chill have added to the urgency in ensuring that newly developed peach and nectarine cultivars exhibit phenotypic plasticity in response to environmental changes. The chilling requirement is one of the traits important for climate resilience; equally important is the heat requirement as they both influence the bloom time and ability to avoid late spring frosts (Okie and Blackburn, 2011). The dynamic relationship between chilling and heat requirements complicates the development of an ideal cultivar with moderate chilling (easy to be satisfied) and a high heat requirement (to delay the bloom time) to avoid low spring temperatures (Bielenberg and Gasic, 2022). Peach and nectarine cultivars patented in the USA in the last three decades have shown a trend towards a medium chilling requirement (450–800 chill hour) (Fig. 4.2). Although the data presented are skewed for those peach (134), and nectarine (112) patents that provided chilling-requirement information, the decrease in chilling requirement in both peach and nectarine cultivars is evident. This trend is going to continue.

4.1.9   Floral biology and selection cycles

In 1806, T. Knight was probably the first botanist to select a peach cultivar by controlled cross-pollination. Although peach is preferentially autogamous,

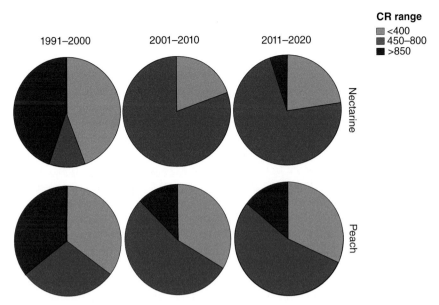

Fig. 4.2. Distribution of chilling requirement (CR) in peach and nectarine cultivars released over the last three decades. Data from the US Patent and Trademark Office (https://patft.uspto.gov/; accessed 16 December 2022).

breeders can include self-pollination and intraspecific crosses in a selection scheme. Seedlings bear their first harvest 2–3 years after obtaining the seeds, depending on climatic and field management. Mutagenesis, polyploidization and interspecific sexual reproduction have been used very little in peach breeding. Results obtained by these methods have been largely unsatisfactory. Alternatively, recombinations obtained by intraspecific crosses have allowed noteworthy improvements of the species and continue to supply most new cultivars. The general breeding schemes are discussed below.

4.2 PARENTAL CHOICE THROUGH THE PHENOTYPIC APPROACH

The straightforward method of crossing two parents with desirable traits and selecting offspring with a new combination of traits for further crosses or direct release has been and still is almost always used in many public and private breeding programmes. This method has led to the improvement of most fruit traits, especially those related to commercial attributes. Self-pollination exploits recombination and recessive characteristics, while intra- and interspecific crossing allow the introduction of new traits within the genetic pool under selection.

4.3 PARENTAL CHOICE THROUGH THE GENOTYPIC APPROACH

The genetic value of a parent can only be assessed through its progeny. This task can easily be achieved in peach because, by self-pollination, it is possible to segregate all Mendelian traits and obtain valuable information on the parental heterozygous state, thus revealing its capacity to transmit important horticultural characteristics. At least six to seven cycles of self-pollination are needed to achieve the homozygosity necessary to examine trait values (Leslie, 1957). Monet (1977) tried to obtain pure lines in peach by genotypic choice of parents, using the following procedure: (i) many cultivars of interest are self-pollinated; the observation of the offspring allows the choice of the best parents; (ii) in each of the progeny of higher value, two of the best individuals are self-pollinated; (iii) comparison of the two offspring allows the choice of the best progeny; (iv) this cycle is repeated. After six self-pollination cycles, the resemblance between individuals is very high. The inbreeding effect is relatively light, probably because peach is an autogamous species. The cross between individuals from offspring after three self-pollination cycles yielded families with good horticultural performances. However, no important heterosis effects were observed after intercrossing pure lines obtained by repeated self-pollination or pure lines obtained by doubling the chromosome number of monoploids, rendering this procedure of little horticultural interest.

4.4 CULTIVAR SELECTION

Successful peach cultivation depends largely on the availability of suitable cultivars (Okie *et al.*, 2008). The key factors for efficient peach production are choosing the best-adapted cultivars for specific soil-climatic conditions, for consumers and the market, in combination with the best rootstock and the optimum technology of production (Iglesias, 2022). Peaches have a short life both on the tree and in the postharvest period; hence, the commercial season of a single cultivar is relatively narrow. As a result, a wide range of cultivars is needed in each peach-producing country to ensure an extended maturity window from April to September (October –March in the southern hemisphere). Depending on the fruit's final destination, a wide variety of fruit types is currently available for growers, sellers and consumers. The fresh market and the processing industry have, in general, different requirements due to the final market to be reached. The main commercial categories are peach and nectarine, yellow and white flesh, round and flat shape, and melting and non-melting flesh.

Although a large number of cultivars released from worldwide breeding programmes over the last decades have increased the options available to growers, it has become increasingly challenging to make the proper selection in terms of adaptability, agronomical performance and fruit quality for a particular location (Iglesias, 2017; Iglesias and Echeverría, 2021). As an example, to illustrate the massive innovation in new cultivars, from 1996 to 2019, of around 1000 new cultivars tested at Institute of Agrifood Research and Technology (IRTA, Catalonia, Spain), 40% were nectarines, 28% peaches, 7% clingstone peaches and 25% flat peaces/nectarines (Iglesias and Echeverría, 2021). However, considering together their field performance, fruit appearance (e.g. colour and size) and eating quality, only 10% were well adapted to the Catalonian environment, and only 6% had a significant impact on the fruit industry, fitting with its requirements, including the postharvest life. The low suitability of many new cultivars to environments different from those they were selected in can partially be explained by the fact that, while high yields, fruit attractiveness and good taste have long been priority goals in peach breeding, adaptability to a wide range of environments has seldom been taken into consideration as a selection criterion (Reig *et al.*, 2015). This highlights the importance of work carried out by new cultivar testing programmes organized in different countries to assess field performance, fruit quality and adaptation to different environments before their commercial scaling.

4.4.1 Examples of testing trials in different countries

Different systems of new cultivar testing programmes have been implemented in the main peach-producing countries. Considering the European Union (EU), in Italy, the 'Liste Varietali' project funded by the Italian Ministry of Agriculture allowed the evaluation in multi-site trials of hundreds of new cultivars, providing for over 20 years competent and, importantly, independent information to the stakeholders (Nencetti *et al.*, 2015). Unfortunately, this valuable project was terminated in 2016. Since then, independent testing of peach in Italy has been carried out in very few cases, mostly on a regional scale (e.g. Agrion in Piedmont and Alsia in Basilicata). In Spain, only the public research institute, IRTA in Catalonia, has developed a continuous and objective evaluation since 1996 (Reig *et al.*, 2015; Iglesias and Echeverría, 2021), obtaining a complete set of data concerning the cultivars most adapted to Ebro Valley climatic conditions (Iglesias, 2014, 2017). In France, objective testing in the south-west of the country is carried out mainly by CTIFL (Balandran), Sudexpe (Marsillargues) and Chambre d'Agriculture des Pyrénées-Orientales (Perpignan). In the USA, the NC-140 peach programme (www.nc140.org; accessed 16 December 2022) allows a common evaluation network of different rootstocks and training systems in 13 producing states, but a corresponding network to test new cultivars has never been implemented. It is left to researchers at State Agricultural Experimental Stations associated with Land Grant Universities or licensed nurseries to conduct regional testing, or in some cases, growers do it themselves. In Brazil, the national institute of research, Embrapa, is responsible for evaluating new cultivars in different states, particularly the low-chilling ones.

4.4.2 The collaborative evaluation system in the EU

Although a number of organizations and research centres in the EU are engaged in assessing cultivar performance based on experimental trials, testing is often carried out with different experimental designs, methods and evaluation protocols. Furthermore, data sharing is poor and infrequent. Such fragmentation limits the completeness of information on the performance of newly introduced cultivars and prevents understanding of the effects of environmental conditions on the field, postharvest and fruit quality performance (Giovannini et al., 2021).

An initiative aiming to implement a collaborative cultivar evaluation system in the EU has recently been embarked upon within the EUFRIN network (https://eufrin.eu; accessed 16 December) by the Apricot and Peach Working Group. A panel of experts in peach cultivar testing from various EU institutes developed a list of 44 descriptors (Table 4.1) to assess the cultivar field value (value for cultivation and use, or VCUS) and related guidelines. Together with traits routinely measured in testing trials, the list also includes traits particularly important in specific environments, such as the chilling requirement (Wert et al., 2007; Ghrab et al., 2014).

The list contains descriptors for a number of traits that are gaining importance in evaluating novel cultivars, as they might negatively impact the pack-out. This is the case for skin speckling, which, in susceptible nectarines, may cause widespread skin blemishes in the stylar end and cheek zones (Topp and Sherman, 2000). Extensive speckling in the skin is more frequent in continental climatic conditions characterized by high temperatures and low relative humidity in the spring–summer period (Iglesias, 2017; Iglesias and Echeverría, 2021). The same pack-out category also includes fruit malformations such as 'fruit doubling', a disorder boosted by severe water stress (e.g. very hot and dry summers) during flower-bud differentiation, affecting up to 50% of the following year's production, depending on the stress intensity and genotype (Handley and Johnson, 2000).

The first EU multi-site peach trial was planted in 2018 at 11 sites, including all the EU leading peach-producing countries (Fig. 4.3). The 18 common cultivars chosen as references are now entering full production. The final goal of this initiative is to improve the dissemination of the newly released cultivars to the benefit of growers; choosing those cultivars evaluated as productive, stable and of excellent quality in their specific environments will increase the competitiveness of the EU peach industry. Finally, this activity

Table 4.1. List of the 44 descriptors selected to assess the field value of the cultivar tested in the EUFRIN evaluation network. Adapted from Giovannini *et al.* (2021).

Category	No.	Trait	Priority
Phenology/vigour	1	Beginning of leaf bud burst	2
	2	Beginning of flowering	1
	3	Full flowering	1
	4	End of flowering	2
	5	Beginning of ripening	1
	6	Additional picking dates	2
	7	Chilling requirement	2
	8	Tree size	1
	9	Tree vigour	1
Productivity	10	Intensity of blooming	1
	11	Fruit set	1
	12	Need for thinning	2
	13	Yield per tree	1
	14	Fruit weight	1
	15	Fruit size distribution	2
Fruit abiotic disorders	16	Split pit (included non-visible)	1
	17	Skin cracking (including stylar cavity closing)	2/1[a]
	18	Double fruits	2
	19	Corky spot	2
	20	Skin speckling (in nectarine)	1
	21	Skin discoloration/bronzing	2
	22	Skin wrinkling	2

Fruit quality	Outer	23	Fruit shape (lateral view)	1
		24	Fruit shape (transversal view)	1
		25	Fruit blush (%)	1
		26	Red over colour pattern	2
	Inner	27	Sugar content	1
		28	Acidity content	1
		29	Fruit taste	1
		30	Flesh firmness	2
		31	Flesh texture	1
Disease susceptibility		32	Brown rot (on flowers and twigs)	2
		33	Leaf curl	2
		34	Powdery mildew	2
Postharvest susceptibility		35	Storability	2
		36	Brown rot susceptibility of fruit	1
		37	Fruit skin inking	2
		38	Flesh browning	2
		39	Mealy or leathery texture	2
Derived parameters		40	Length of blooming period	2
		41	Pack-out	2
		42	Yield efficiency	1
		43	Fruit size uniformity	2
		44	Ripening uniformity	2

[a]Priority 1 for flat peaches and nectarines.

Murcia (1), IMIDA
Murcia (2), IMIDA
Zaragoza, EEAD-CSIC
Lleida, IRTA
Bellegarde, CTIFL
Roma, CREA
Forlì, CREA
Cuneo, AGRION
Tebano di Faenza, RI.NOVA
Bucarest, USAMV
Naoussa, ELGO-DIMITRA

Fig. 4.3. Map of the EU multi-site peach trial locations established in 2018. AGRION, Fondazione per la Ricerca e l'Innovazione; ELGO-DIMITRA, Hellenic Agricultural Organization; CREA, Consiglio per la Ricerca e la Sperimentazione in Agricoltura; CSIC, Consejo Superior de Investigaciones Científicas; EEAD, Estacion Experimental de Aula Dei; CTIFL, Centre Technique Interprofessionnel des Fruits et Légumes; IMIDA, Instituto Murciano de Investigación y Desarrollo Agrario y Medio-ambiental; IRTA, Institute of Agrifood Research and Technology; USAMV, University of Agriculture and Veterinary Medicine; RI.NOVA, Ricerca e Sviluppo.

might facilitate the evolution from a DUS- (distinctness, uniformity and stability) to a VCUS-based evaluation system of new cultivars for the fruit industry. Today, DUS is the only legal requirement for completing the application process for the protection of a new fruit cultivar in Europe. However, given the numerous yearly releases, it seems increasingly important to confirm whether a candidate cultivar is unique and if it has satisfactory value to growers and end-users, which can only be acknowledged by introducing VCUS as a part of the application process (GEVES, 2020). Implementing the VCUS system in the EU could provide an independent guarantee that only cultivars with real improved performance or end-use quality are approved for marketing (Giovannini *et al.*, 2021).

4.5   CULTIVAR DEVELOPMENT

4.5.1   The protection of new cultivars

The nursery sector managed the introduction of new cultivars to the market until the early 1980s, acquiring the new cultivars (or their propagation rights). As tools to counteract unauthorized propagation by growers (or nurseries) are not so easy to implement, the new cultivars could relatively easily be subjected to 'piracy'. Despite this problem, for many years, the nurseries have represented a privileged vehicle that has ensured the renewal of the varietal platform worldwide. A second phase began when public institutions (e.g. the ministries of agriculture in many countries, or the EU) started considering the legal protection of the new cultivars. The approach followed two main lines: (i) the intellectual property linked to the new plant cultivars, by the DUS procedure, to protect the legitimate subjects entitled to the economic exploitation of these novelties (Yang *et al.*, 2021), and (ii) verification of the VCUS (Gilliland and Gensollen, 2010). However, the latter is limited to seed crops, at least in Europe. The regulation of new plant cultivars originates from the so-called UPOV (International Union for the Protection of New Varieties of Plants) Convention, an international agreement signed in Paris on 2 December 1961, initially by seven countries, and amended in 1991.

At the European level, on the model of the UPOV Convention (to which the EU has formally been a party since 2005), EC Regulation No. 2100/94 established the 'community right' (in practice, the patent, although not to be confused with the industrial one) for plant cultivars and the Community Plant Variety Office (CPVO; http://cpvo.europa.eu; accessed 16 December 2022) with headquarters in Angers, France, which manages the entire European system of cultivar rights. The EC Regulation No. 2100/94 therefore allows a single title of protection effective throughout the territory of the EU to be obtained. The new plant cultivars can thus access two levels of protection (one national and the other EU), although these are not interchangeable. The sensitivity towards the protection of new cultivars has been increasing gradually, so much so that, according to CPVO estimates, the number of community rights applications for new cultivars, including fruit crops, has grown in recent years (Fig. 4.4).

Two important clarifications must be made regarding cultivar protection in the EU: (i) the 'variety' right protects from the multiplication of the material and not the cultivar as such, and (ii) the 'variety' right is considered a *sui generis* industrial property right, and more precisely a 'special' patent. In other words, the 'variety' rights, although included in the Industrial Property Code, are placed outside the patent discipline for industrial and biotechnological inventions. In fact, it is precisely the rule of the protection of industrial invention patents that excludes new plant cultivars from patentability.

The distinction between variety rights and industrial invention patents is also confirmed with reference to the different ways in which novelty is usually

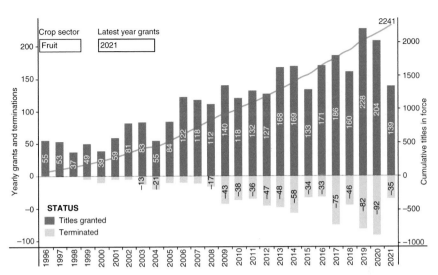

Fig. 4.4. Evolution of fruit protection grants and terminations 1996–2021 in the European Union. Titles granted each year (dark bar), the number of titles terminated (light bar), and the cumulative number of titles in force (grey line) since the inception of the CPVO. Data from CPVO, Angers, France.

assessed, which, as is well known, represents one of the essential protection requirements both for 'variety' rights and for patent invention. Basically, while an industrial invention is new only where it has not been disclosed (i.e. described or exposed) to the public by the same inventor before the patent application is filed, a plant 'variety' is new only at the time of filing the patent application, if the new cultivar has not yet been marketed. The rationale for this significant difference between the two disciplines lies in the fact that, unlike industrial and biotechnological inventions, the breeder of a new plant cultivar must often cultivate the trees outdoors for several cycles to verify the existence of all the other requisites needed to access protection (in particular, homogeneity, distinctiveness and stability) to complete the experimentation process.

Therefore, if the novelty requirement modelled on industrial inventions were to be applied to 'variety' rights, the breeder of new plant 'varieties' would risk seeing the new cultivar disclosed at the experimental phase (pre-market tests), thus compromising the possibility of protection. For this reason, the EU legislation has appropriately linked the novelty requirement for 'variety' rights only to the absence of marketing prior to the filing of the patent application, rather than to the absence of previous disclosures of the 'invention'.

The 'variety' right granted to the breeder, the person who created or discovered and developed the new cultivar, are as follows: (i) production and reproduction; (ii) conditioning for the purpose of reproduction or multiplication; (iii) offer for sale, sale or any other form of marketing; export or import; and

(iv) detention for one of these purposes. The fulfilment of these acts by third parties requires specific authorization from the breeder or the patent owner. In all cases of unauthorized use of the cultivar by third parties, the breeder (i.e. the patent owner) can act to assert their exclusive rights.

The EU regulations also extend the breeder's right to the product of the protected cultivar's harvest (i.e. the fruits). This principle has a considerable impact from an operational point of view in all cases of illegal multiplication of the protected cultivar. In fact, it applies especially when it is impossible to identify the unlawful multiplier, thus allowing the breeders to assert their rights even if only against third parties who have produced and marketed the fruits without authorization from the patent owner. However, it should be noted that the monopoly granted to the breeder of new plant cultivars from the EU regulations is not absolute and is limited by the right of other breeders to do research and innovation. Specifically, the breeder cannot oppose the so-called 'free uses' of the protected cultivar: (i) in the private field; (ii) for non-commercial purposes; (iii) on an experimental basis; or (iv) for breeding purposes. The implementation of these legislative instruments to defend the rights of the breeders has allowed a real flowering (entry on the market) of new breeders (almost all private, considering that the major public agencies, e.g. universities, ministries and other public research institutions, have strongly downsized this activity worldwide), attracted by the prospect of revenues through the collection of royalties. However, it should be noted that, contrary to what one might think, the system of patent rights on new plant cultivars does not hand over almost exclusive control of the genetic asset of the species into the hands of a few breeders. Indeed, like any other industrial property right, the patent right granted to the breeder of new plant cultivars has the sole purpose of ensuring the latter a period of exclusivity to be able to recover the investments made for the obtention of the new cultivar and to generate revenues from the temporary fees (royalties) that third parties are required to pay to exploit the variety. However, at the end of the period of exclusivity, the cultivar becomes free for use by anyone.

In the USA, Plant Variety Protection Office (PVPO) provides intellectual property protection of new cultivars via the Plant Variety Protection Act (PVPA), which grants certificates to protect cultivars for 25 years. PVPA certificates are recognized worldwide and allow the faster filing of Plant Variety Protection (PVP) in other countries. The PVPA certificate provides owners with the rights to exclude others from marketing and selling their cultivars, to manage the use of their cultivars by other breeders, and to enjoy the legal protection of their work (Title 7, Agriculture, Chapter 104, Plant Protection, Sections 7711 and 7721). In the USA, cultivars can be protected via three types of intellectual property protection: (i) PVP certificates to protect seeds, tubers and asexually reproduced plants (issued by the PVPO); (ii) plant patents, to protect asexually reproduced plants (issued by the Patent and Trademark Office (PTO); and

(iii) utility patents for genes, traits, methods, plant parts or cultivars (issued by the PTO) (Shelton and Tracy, 2017). Once a cultivar has been protected as intellectual property, it is typically made available to growers via nurseries with a licence agreement that specifies royalty payment for the use of the cultivar. There are 346 peach and 270 nectarine patents issued under the US PTO (until 2020), with the first nectarine patent issued in 1984 and the first peach patent issued in 1985 (Fig. 4.5, Table 4.2). Patenting of newly released cultivars has increased significantly in the last two decades from 142 (77 peaches and 65 nectarines) registered before 2000 to more than 200 in the last two decades (Table 4.2), reflecting the increase in intellectual property rights awareness in the last two decades.

The PVPO follows the UPOV 1991 agreement and guidance for DUS trials, forms and cooperation between the authorities. Ensuring uniformity, distinctiveness and stability is often done in collaboration with licensed nurseries and/or via testing agreements with growers. The cultivar needs to be new (not sold commercially, or sold for less than 1 year in the USA or less than 4 years internationally), distinct (distinguishable from any other publicly known cultivar), uniform (any variations are describable, predictable and commercially acceptable) and stable (when reproduced it will remain unchanged from the described characteristics). Peach cultivar releases are handled by intellectual property offices at breeders' institutions in the public sector (universities and government, US Department of Agriculture) or by private breeding companies.

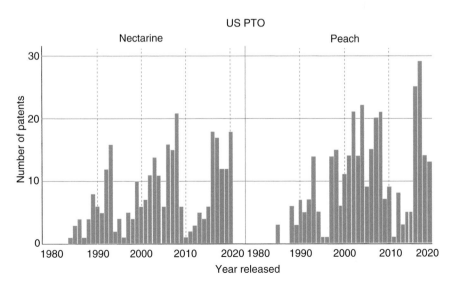

Fig. 4.5. Frequency of patent issue for nectarine and peach cultivars in the USA. Dotted lines designate the decades. Data from the US Patent and Trademark Office (US PTO) (https://patft.uspto.gov/; accessed 16 December 2022).

Table 4.2. Number of patents issued for peach and nectarine cultivars in the USA. Data from the US Patent and Trademark Office (https://patft.uspto.gov/; accessed 16 December 2022).

Patent issued	Peach	Nectarine	Total
1991–2000	77	65	142
2001–2010	152	108	260
2011–2020	117	97	214
Total	346	270	616

4.5.2   Profits of the owner of the patent

Over the years, various methods have been applied to calculate the amount that producers or commercial structures have to pay to be authorized to grow a protected cultivar. Here we describe the three most popular:

1. Royalty per area, calculated on the size of the orchard, independent of tree spacing. This fairly complex management system, introduced in recent times, indirectly tends to favour large growers to the detriment of small farmers, making it more convenient for high-density orchards: the grower pays the amount due directly to the owner of the patent, not to the nursery.
2. Royalties per tree. This is the most popular, where the nurseryman collects the amount of the royalty together with the cost of the scion. It is very easy to manage, does not discourage small farmers and generally disadvantages high-density orchards.
3. Mixed system: royalty per plant plus a royalty per orchard size. This is a rarely applied system because it is expensive and difficult to manage, and is not suitable for very fragmented fruit-farming regions.

Regardless of the collection strategy, royalties are strongly needed for the proper functioning of the production chain to provide financing for cultivar breeding and testing. Some improvements could be proposed to reduce the illegal activities that burden the fruit-growing sectors, particularly in the field of new cultivar protection. Except for special cases, such as the 'club' cultivars, which have well-defined management, the best system would be if the royalty was included in the sale price and collected by the nurseryman, who would pay it to the patent owner. None the less, breeders and cultivar publishers often complain that this way they would lose control of the market and it would become more difficult to fight illegal traffickers. In reality, the first threat of illegality lies in not establishing a consistent and serious marketing strategy with a few trusted licensee nurseries.

REFERENCES

Adami, M., Franceschi, P., Brandi, F., Liverani, A., Giovannini, D. *et al.* (2013) Identifying a carotenoid cleavage dioxygenase (*ccd4*) gene controlling yellow/white fruit flesh color of peach. *Plant Molecular Biology Reporter* 31, 1166–1175.

Bassi, D. and Foschi, S. (2021) A new introduction from the Italian MAS.PES peach breeding program: 'Maissa', a stony hard flat peach. *Acta Horticulturae* 1304, 83–88.

Bielenberg, D.G. and Gasic, K. (2022) Peach [*Prunus persica* (L.) Batsch] cultivars differ in apparent base temperature and growing degree hour requirement for floral bud break. *Frontiers in Plant Science* 13: 801606.

Byrne, D.H. (2014) Progress and potential of low chill peach breeding. *Acta Horticulturae* 1059, 59–66.

Cirilli, M., Giovannini, D., Ciacciulli, A., Chiozzotto, R., Gattolin, S. *et al.* (2018) Integrative genomics approaches validate *PpYUC11*-like as candidate gene for the stony hard trait in peach (*P. persica* L. Batsch). *BMC Plant Biology* 18: 88.

Connors, C.H. (1917) Methods in breeding peaches. *Proceeding of the American Society for Horticultural Science* 14, 126–127.

de Souza, V.A.B., Byrne, D.H. and Taylor, J.F. (1998a) Heritability, genetic and phenotypic correlations, and predicted selection response of quantitative traits in peach: I. An analysis of several reproductive traits. *Journal of American Society for Horticultural Science* 123, 598–603.

de Souza, V.A.B., Byrne, D.H. and Taylor, J.F. (1998b) Heritability, genetic and phenotypic correlations, and predicted selection response of quantitative traits in peach: II. An analysis of several fruit traits. *Journal of American Society for Horticultural Science* 123, 604–611.

Downing, C. (1866) *The Fruits and Fruit Trees of America*. Wiley, New York.

Falchi, R., Vendramin, E., Zanon, L., Scalabrin, S., Cipriani, G. *et al.* (2013) Three distinct mutational mechanisms acting on a single gene underpin the origin of yellow flesh in peach. *Plant Journal* 76, 175–187.

GEVES (2020) DUS and VCUS: the core of variety testing. GEVES, Beaucouzé, France. Available at: www.geves.fr/about-us/variety-study-department/dus-vcus-testing (accessed 5 December 2022).

Ghrab, M., Mimoun, M.B., Masmoudi, M.M. and Mechlia, N.B. (2014) Chilling trends in a warm production area and their impact on flowering and fruiting of peach trees. *Scientia Horticulturae* 178, 87–94.

Gilliland, T.J. and Gensollen, V. (2010) Review of the protocols used for assessment of DUS and VCU in Europe – perspectives. In: Huyghe, C. (ed.) *Sustainable Use of Genetic Diversity in Forage and Turf Breeding*. Springer, Berlin, pp. 261–275.

Giovannini, D., Bassi, D., Cutuli, M., Drogoudi, P., Foschi, S. *et al.* (2021) Evaluation of novel peach cultivars in the European Union: the EUFRIN Peach and Apricot Working Group initiative. *Acta Horticulturae* 1304, 13–20.

Gu, C., Wang, L., Wang, W., Zhou, H., Ma, B. *et al.* (2016) Copy number variation of a gene cluster encoding endopolygalacturonase mediates flesh texture and stone adhesion in peach. *Journal of Experimental Botany* 67, 1993–2005.

Handley, D.F. and Johnson, R.S. (2000) Late summer irrigation of water stressed peach trees reduces fruit doubles and deep sutures. *HortScience* 35, 771.

Hansche, P.E., Hesse, C.O. and Beres, V. (1972) Estimates of genetic and environmental effects on several traits in peach. *Journal of American Society for Horticultural. Science* 97, 76–79.

Hedrick, U.P. (1917) *The Peaches of New York*. J.B. Lyon, Albany, New York.

Hernandez Mora, J.R., Micheletti, D., Bink, M.C.A.M., van de Weg, E., Cantín, C. *et al.* (2017) Integrated QTL detection for key breeding traits in multiple peach progenies. *BMC Genomics* 18: 404.

Hill, J.L. and Hollender, C.A. (2019) Branching out: new insights into the genetic regulation of shoot architecture in trees. *Current Opinion in Plant Biology* 47, 73–80.

Iezzoni, A.F., McFerson, J., Luby, J., Gasic, K., Whitaker, V. *et al.* (2020) RosBREED: bridging the chasm between discovery and application to enable DNA-informed breeding in rosaceous crops. *Horticultural Research* 7: 177.

Iglesias, I. (2014) El melocotón plano en España: 15 años de innovación tecnológica y comercial. *Revista de Fruticultura* 35, 6–31.

Iglesias, I. (2017) Nuevas variedades de melocotón, nectarina y melocotón plano. *Revista de Fruticultura* 58, 12–37.

Iglesias, I. (2022) Situación actual e innovación tecnológica en fruticultura: una apuesta por la eficiencia y la sostenibilidad. *Revista de Fruticultura* 85, 6–45.

Iglesias, I. and Echeverría, G. (2021) Overview of the peach industry in the European Union, with special reference to Spain. *Acta Horticulturae* 1304, 163–176.

Laurens, F., Aranzana, M.J., Arus, P., Bassi, D., Bink, M. *et al.* (2018) An integrated approach for increasing breeding efficiency in apple and peach in Europe. *Horticultural Research* 5: 11.

Leslie, J.W. (1957) A genetic study of inbreeding and crossing inbred lines of peaches. *Proceedings of the American Society for Horticultural Science* 70, 93–103.

Liverani, A., Giovannini, D., Brandi, F. and Merli, M. (2004) Development of new peach varieties with columnar and upright growth habit. *Acta Horticulturae* 663, 381–386.

Liverani, A., Brandi, F., Quacquarelli, I., Sirri, S. and Giovannini, D. (2017) Advanced stony-hard peach and nectarine selections from CREA-FRF breeding program. *Acta Horticulturae* 1172, 2019–2024.

Monet, R. (1977) Amélioration du pêcher par la voie sexuée, exposé d'une mèthode. *Annales de l'Amelioration des Plantes* 27, 223–234.

Monet, R. and Bassi, D. (2008) Classical genetics and breeding. In: Layne, D. R. and Bassi, D. (eds) *The Peach: Botany, Production and Uses.* CAB International, Wallingford, UK, pp. 61–84.

Nencetti, V., Liverani, A. and Sartori, A. (2015) Liste Varietali Pesco 2015. *Terra e Vita* 38, 34–53.

Okie, W.R. and Blackburn, B. (2011) Increasing chilling reduces heat requirement for floral budbreak in Peach. *HortScience* 46, 245–252.

Okie, W.R., Bacon, T. and Bassi, D. (2008) Fresh market cultivar development. In: Layne, D.R. and Bassi, D. (eds) *The Peach: Botany, Production and Uses.* CAB International, Wallingford, UK, pp. 139–174.

Pan, L., Zeng, W., Niu, L., Lu, Z., Liu, H. *et al.* (2015) *PpYUC11*, a strong candidate gene for the stony hard phenotype in peach (*Prunus persica* L. Batsch), participates in IAA biosynthesis during fruit ripening. *Journal of Experimental Botany* 66, 7031–7044.

Raseira, M. (2016) Breeding for Low Chill Areas. Available at: www.researchgate.net/publication/303459946_BREEDING_FOR_LOW_CHILL_AREAS (accessed 5 December 2022).

Rawandoozi, Z., Hartmann, T., Byrne, D. and Carpenedo, S. (2020) Heritability, correlation, and genotype by environment interaction of phenological and fruit quality traits in peach. *Journal of American Society for Horticultural Science* 146, 56–67.

Reig, G., Alegre, S., Gatius, F. and Iglesias, I. (2015) Adaptability of peach cultivars [*Prunus persica* (L.) Batsch] to the climatic conditions of the Ebro Valley, with special focus on fruit quality. *Scientia Horticulturae* 190, 149–160.

Salazar, J.A., Ruiz, D., Campoy, J.A., Sánchez-Pérez, R., Crisosto, C.H. *et al.* (2014) Quantitative trait loci (QTL) and Mendelian trait loci (MTL) analysis in *Prunus*: a breeding perspective and beyond. *Plant Molecular Biology Reporter* 32, 1–18.

Sansavini, S., Gamberini, A. and Bassi, D. (2006) Peach breeding, genetics and new cultivar trends. *Acta Horticulturae* 713, 23–48.

Shelton, A.C. and Tracy, W.F. (2017) Cultivar development in the U.S. public sector. *Crop Science* 57, 1823–1835.

Topp, B.L. and Sherman, W.B. (2000) Nectarine skin speckling is associated with flesh soluble solids content. *Journal of the American Pomological Society* 54, 177–182.

Topp, B.L., Bignell, G.W., Russell, D.M. and Wilk, P. (2012) Breeding low-chill peaches in subtropical Queensland. *Acta Horticulturae* 962, 109–116.

Toyama, T.K. (1974) Haploidy in peach. *HortScience* 9, 187–188.

Vendramin, E., Pea, G., Dondini, L., Pacheco, I., Dettori, M.T. *et al.* (2014) A unique mutation in a *MYB* gene cosegregates with the nectarine phenotype in peach. *PLoS One* 9: e90574.

Wert, T.W., Williamson, J.G., Chaparro, J.X. and Miller, E.P. (2007) The influence of climate on fruit shape of four low-chill peach cultivars. *HortScience* 42, 1589–1591.

Yang, C.J., Russell, J., Ramsay, L., Thomas, W., Powell, W. and Mackay, I. (2021) Overcoming barriers to the registration of new plant varieties under the DUS system. *Communications Biology* 4: 302.

Zhou, H., Lin-Wang, K., Wang, H., Gu, C., Dare, A.P. *et al.* (2015) Molecular genetics of blood-fleshed peach reveals activation of anthocyanin biosynthesis by NAC transcription factors. *Plant Journal* 82, 105–121.

Zhou, H., Ma, R., Gao, L., Zhang, J., Zhang, A. *et al.* (2020) A 1.7-Mb chromosomal inversion downstream of a *PpOFP1* gene is responsible for flat fruit shape in peach. *Plant Biotechnology Journal* 19, 192–205.

IRRIGATION AND WATER MANAGEMENT

Brunella Morandi[1]*, Cristos Xiloyannis[2] and Bartolomeo Dichio[2]

[1]*Department of Agricultural and Food Sciences, University of
Bologna, Bologna, Italy; [2]Dipartimento delle Culture Europee e del
Mediterraneo: Architettura, Ambiente, Patrimoni Culturali – University of
Basilicata, Potenza, Italy*

5.1   WATER RELATIONS AND HYDRAULIC FEATURES

Irrigation is a fundamental practice to guarantee economically sustainable yields and quality of production. In peach, as in other fruit crops, this practice needs to be carefully rationalized in order to avoid both drought and excessive water supplies causing anaerobic stresses. Moreover, due to climate change, several strategies have been developed to reduce irrigation water use, thus improving water-use efficiency (WUE) and often also peach production performance in terms of yield and fruit quality.

Peach trees are characterized by a medium to high xylem hydraulic conductivity with an average vessel diameter around 60 μm and maximum diameter values around 100–130 μm in the 'O'Henry' cultivar (Tombesi *et al.*, 2010). These hydraulic features make this crop particularly vulnerable to embolism, with almost 100% of cavitation with stem-water potential values of −3 MPa (Améglio *et al.*, 1998). In well-watered conditions, peach trees maintain a midday stem-water potential ranging from −0.4 to −1 MPa and varying with several factors such as the environmental conditions, rootstock–scion combination, crop load, phenological stage and root:leaf area ratio (Marsal and Girona, 1997; Naor and Cohen, 2003; Abrisqueta *et al.*, 2015).

In field-grown, drip-irrigated nectarine trees, the roots in the unirrigated inter-row soil produce chemical signals that increase in summer to induce stomatal closure and so increase WUE (Xylogiannis *et al.*, 2020). An abscisic acid root-to-shoot signal issued from the inter-row roots growing in soil that dries out during a Mediterranean summer (hot, with low rainfall) can reduce irrigation requirement (Xylogiannis *et al.*, 2020). High crop loads increase plant water needs, thus generally reducing stem, leaf and fruit water potentials

---

* Email: brunella.morandi@unibo.it

© CAB International 2023. *Peach* (G. Manganaris, G. Costa and C. Crisosto eds)
DOI: 10.1079/9781789248456.0005

(McFadyen *et al.*, 1996; Marsal and Girona, 1997). Similarly, dwarfing rootstocks usually show lower stem-water potentials due to reduced hydraulic conductivity at the grafting point (Basile *et al.*, 2003; Tombesi *et al.*, 2010). Furthermore, the susceptibility of peach to water stress can change, depending on the phenological stage and the root:leaf ratio (Xiloyannis *et al.*, 1993).

During the first cell-division stage, midday stem-water potentials are generally less negative (with stem-water potentials in the range of –0.4 to –0.6 MPa) because of the initial higher soil water content, low evaporative demand, lower canopy development and higher root:leaf ratio. During the following pit-hardening stage, water deficits primarily affect apical and lateral shoot expansion and trunk growth, while having lower impact on fruit growth (Girona *et al.*, 2003). During stage III, the fruit grow at increasing rates due to cell expansion, thus making trees most sensitive to water deficits during this period. Furthermore, water deficits at this stage and during postharvest (for early-maturing cultivars) can induce an increased frequency of anomalous flowers and fruit doubles (Natali *et al.*, 1985) and deep sutures, as well as a decreased fruit set in the next season (Johnson and Phene, 2008). However, it has been shown that the negative effects of water deficit seem to sharply increase with midday stem-water potentials below –2.0 MPa, suggesting this value as a possible threshold for postharvest irrigation scheduling (Naor *et al.*, 2005). In conditions where the water potential is too low, xylem dysfunctionalities begin to occur. For example with stem-water potential reaching values of –2 MPa, Améglio *et al.* (1998) reported 50% embolism, as well as a significant reduction in stomatal conductance, leading to an evident drought stress. Although drought stress negatively impacts yield and fruit size, it has been demonstrated that controlled stress may improve some fruit quality features such as fruit dry matter, soluble solids and polyphenol content without too negative an impact on fruit size (Gelly *et al.*, 2004).

5.2 IRRIGATION SCHEDULING DURING THE SEASON

Peach has been found to be generally quite sensitive to waterlogging and anaerobic conditions at the root level (Schaffer *et al.*, 1992). Therefore, in heavier soils, the correct management of irrigation can be even more critical and also depends on the rootstock and the root's characteristics (Xiloyannis *et al.*, 2012). In peach, as in other crops, traditional irrigation approaches are often based on the classical crop–water balance, which relies on the calculation of the crop potential evapotranspiration (ET_c) according to the classical equation:

$$ET_c = K_c{}^*ET_0$$

where ET_0 represents the potential evapotranspiration determined through a pan evaporimeter and K_c represents the crop coefficient. K_c can be further broken down into K_{bc} representing the crop evapotranspiration and K_e representing the soil evaporation (Allen *et al.*, 1998).

As potential water losses vary with leaf area and canopy light interception, different levels of K_c are foreseen during the growing season: one representing the initial stage ($K_{cInitial}$), mid-season stage (K_{cMid}) and late-season stage (K_{cEnd}), respectively. For peach, K_c values have been estimated as 0.45, 0.90 and 0.65 for $K_{cInitial}$, K_{cMid} and K_{cEnd}, respectively, and as 0.80, 1.15 and 0.85 for $K_{cInitial}$, K_{cMid} and K_{cEnd}, respectively, with ground cover (Allen *et al.*, 1998). However, these values may vary depending on the variety and orchard system. Ayars *et al.* (2003) described a constant increase in peach K_c from 0.2 to 1.1 in 'O'Henry' peaches, with a daily water use ranging from 3 to 7 mm for Californian conditions. These values can bring peach irrigation needs to fairly high values (up to 1000 mm year^{-1}) in dry and hot areas (Ayars *et al.*, 2003). If irrigation volumes (IrrVol) are correctly managed and groundwater is deep, drainage and deep percolation losses, surface runoff losses and groundwater contribution terms are negligible and the water balance equation can thus be simplified as:

$$\text{Irrigation volume}\left(m^3 ha^{-1}\right) = \left(\left(ETc - P_e\right)/D_{eff}\right) \times 10$$

where P_e is the part of the rainfall that is effectively used by the plants and D_{eff} is the distribution efficiency.

In view of the small number of variables to be measured, the simplified water balance can also be applied to irrigation scheduling. D_{eff} expresses the percentage ratio of the amount of water held in the soil and potentially available to the plant to the amount of water applied, and is largely dependent on the irrigation method as well as on the farmer's skill. Determining it properly is thus important for correct irrigation. When managing localized irrigation methods, the simplified water balance has to be referred to the volume of soil wetted by irrigation (container wetted), as that under optimal management conditions, almost all the water used by the plant is taken up from this soil volume where the most absorbing roots are present. The irrigation volume has to replenish the container wetted volume, and it is thus necessary to quantify the amount of water in the container (available water and readily available water) and define the irrigation amount and timing (Xiloyannis *et al.*, 2012).

5.3   DEFICIT IRRIGATION STRATEGIES

As well as traditional irrigation approaches aimed at maintaining trees at full water status, deficit irrigation strategies have also been developed to reduce water supply while obtaining further benefits such as: (i) limited vegetative growth; (ii) optimized yield and fruit quality; and thus (iii) improved WUE.

5.3.1 Regulated deficit irrigation

The most common of these approaches is regulated deficit irrigation (RDI) where deficit irrigation is imposed during the phenological stages that are less susceptible to water stress. In peach, the pit-hardening phase has been identified as an appropriate period for RDI, as fruit growth is minimal and therefore is generally not affected by drought, while shoot growth can be reduced (Chalmers *et al.*, 1981). However, in early-maturing cultivars, this phase is very short, enabling only small water savings. In these cultivars, there is instead a long postharvest period, which is more suitable for reduced irrigation (Johnson *et al.*, 2004; Dichio *et al.*, 2007, 2011). Moderate water deficits applied during stage II improved fruit quality (firmness, colour, improved total soluble solids) without affecting yield (Gelly *et al.*, 2003, 2004). Moderate water stress in stage III also improves fruit quality, but a negative impact on fruit size and yield is likely.

When RDI is applied in the postharvest period, care must be taken to prevent negative effects on flower bud development and thus on fruit set during the next season (Natali *et al.*, 1985; Johnson and Handley, 2000). The regulation of vigour due to moderate water stress possibly reduces the competition for assimilates between reserve tissues and the vegetative apexes, resulting in better light interception, lower water use (Boland *et al.*, 2000; Dichio *et al.*, 2007) and improving bud quality. Water deficit (reduction of 50% of irrigation volume) applied during the postharvest period with early-ripening cultivars and inducing a reduction of midday stem-water potential up to −1.7 MPa can achieve a good yield and a greater accumulation of carbohydrates in the roots and the wood as a result of reduced vegetative growth (Fig. 5.1).

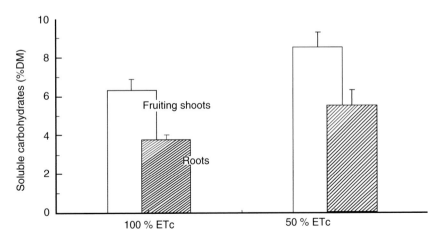

Fig. 5.1. Concentration of soluble carbohydrates (% dry matter (DM) ± standard error) in roots (shaded columns) and fruiting shoots (white columns) in a well-watered peach tree (100% ETc) and subjected to regulated deficit irrigation (RDI) (50% ETc). Measurements were made in November at the end of the second year of RDI application. Adapted from Dichio *et al.* (2007).

Tables 5.1 and 5.2 show the recommended replacement amounts suggested by the Food and Agriculture Organization of the United Nations (FAO) (Steduto *et al.*, 2012), although in most cases, the final effects of water stress on crop performance seem to be very dependent on several orchard, environmental and tree characteristics. A 6-year comparative study between conventional (commercial standard irrigation) and sustainable orchard management practices showed that RDI applied during the postharvest stage, reducing the irrigation to approximately 50% of the plants' requirements, can maximize water productivity without impairing long-term yield (Dichio *et al.*, 2011). This long-term study demonstrated that integrating RDI into wider sustainable fruit-tree orchard management with increased soil carbon inputs resulted in high and stable yields and a high economic water productivity. This information should encourage water-management policy makers to promote wide adoption of RDI in order to reduce agricultural water use to deal with water scarcity in the expected scenario of global change. When applying deficit irrigation protocols, it is important to monitor the level of stress through real-time soil and/or plant sensors in order to provide the right amount of water and avoid excessive drought. Indeed, soil water content changes widely depending

Table 5.1. Suggested limits of midday stem-water potential for late-season cultivars. Data from Steduto *et al.* (2012).

Fruit growth stage	ET_c (%)		Suggested limits of stem-water potential at midday (MPa)
	Shallow soils	Deep soils	
I	100	100	−0.9
II	65	35	−1.8
III	100	100	−1.1
Postharvest	80	50	−1.8

ET_c, crop potential evapotranspiration.

Table 5.2. Suggested limits of midday stem-water potential for early-season cultivars. Data from Steduto *et al.* (2012).

Fruit growth stage	ET_c (%)		Suggested limits of stem-water potential at midday (MPa)
	Shallow soils	Deep soils	
I	100	100	−0.9
III	100	100	−1.0
Early postharvest	50	35	−1.8
Late postharvest	80	80	−1.2[a]

[a]To prevent crop failure the following year.

on the soil type and how dry the environment is. It is advisable to avoid excessive reductions in soil water content as this may lead to long-term stresses in the following season.

5.3.2 Partial rootzone drying

A further deficit irrigation strategy is partial rootzone drying (PRD). This technique allows one part of the root system to be irrigated while the other is kept dry, periodically alternating the two zones (Kriedemann and Goodwin, 2003). PRD is based on the assumption that plant roots in the drying side of the rootzone respond by sending a signal to leaves to reduce stomatal opening and thus water losses (Davies and Zhang, 1991). This reduction should remain quite limited, with little effect on photosynthesis (Jones, 1992). In this way, it is possible to reduce water use by separating the biochemical responses to water deficit from the physical effects of reduced water availability, allowing plant growth to be unaffected. In some crops, previous research has confirmed a water saving of up to 30–50% with no significant yield reduction. However, in peach trees, this technique has not been widely tested, and initial results did not show significant economic advantages from application of PRD under Mediterranean conditions (Vera *et al.*, 2013).

5.4 PRECISE IRRIGATION APPROACHES

Precision irrigation is traditionally defined as the site-specific application of precise amounts of water at precise time across the farm (Smith and Baillie, 2009). Despite some disagreement on the correct definition, precision irrigation can be defined as all innovative technologies aimed at optimizing WUE based on a wide range of information inputs from the farm. The adoption of precision irrigation is thus related to the availability of several new technologies that include the monitoring of environmental, soil and crop conditions.

5.4.1 Online services and models for irrigation scheduling

Precise irrigation approaches include decision support systems (DSSs) implemented on web platforms capable of providing custom-tailored advice to growers on the irrigation needs of their crops. These DSSs are able to estimate crop water requirements and soil water balance based on inputs from local weather stations, as well as information on the orchard features (e.g. soil type, crop, training system, orchard density and age, type and features of the irrigation system) provided directly by the grower. Usually, irrigation is recommended when the estimated crop water requirements (often calculated based on the FAO K_c values) exceed the estimated soil water content. Among these DSSs are the Italian 'Irriframe' (www.irriframe.it/; accessed 16 December 2022) developed by the Emilia Romagna Region and the Spanish 'Riegos Ivia' (https://riegos.ivia.es/; accessed 16 December 2022), developed by the Istituto Valenciano de Investigaciones Agrarias, and many others. These systems are quite easy to use and are offered to growers free of charge. However, their advice is not based on actual data collected at orchard level, and their indications might be inaccurate due to the lack of direct feedbacks for the orchard.

5.4.2　Soil sensors

A wide range of soil sensors are currently available on the market and used in the field to schedule irrigation as single inputs or integrated in more complex DSSs. They include tensiometers, soil psychrometers, time-domain reflectometer and frequency domain reflectometer probes, and several further technologies (Fig. 5.2). However, their values, and consequently the irrigation response

Fig. 5.2. (a–c) Different installation stages for soil humidity sensors (frequency domain reflectometer (FDR); Type SM100 WaterScout Spectrum). (d) Time-domain reflectometer sensors installed at 20, 30 and 60 cm soil depths. (e, f) IRRISENSE system with integrated profile FDR sensors and datalogger systems for data acquisition and remote control.

might change depending on the soil features, and correct installation as well as representative positioning of the probes are essential to obtain reliable measurements. Additionally, optimal soil water content and/or relative humidity might change depending on features of the orchard such as phenological stage and crop load, and therefore a more precise monitoring of the orchard water status should be obtained with the addition of plant sensors, although they are still difficult to apply at the farm level. One major advantage of these new sensors is that they provide continuous measurement of soil moisture and allow automating the readings and data processing through data loggers, which can be connected to the network through automated irrigation management systems. In general, the adopted approaches are based on defining a threshold soil water content beyond which irrigation has to be applied to re-establish optimal moisture values. As this provides continuous soil moisture measurements, irrigation scheduling can be based on the soil water content pattern rather than on an absolute threshold value.

5.4.3 Plant-based sensors

The integration of plant-based sensors in DSSs for irrigation scheduling represents an opportunity to precisely adopt RDI strategies while developing further irrigation-scheduling approaches, aimed at limiting water inputs while maximizing fruit quality and productivity. In peach, as in other crops, plant water status can be assessed through different plant-based parameters such as stem-water potential, sap flow, and stem and/or fruit diameter variations (Fig. 5.3), although the monitoring of these parameters requires the installation of sensors and datalogging systems (Fig. 5.2). A comparative study among different plant indicators found that in peach indices derived from micrometric measurements of trunk diameter, fluctuations were the most sensitive for water stress detection, followed by stem-water potential (Cohen *et al.*, 2001; Conejero *et al.*, 2007). Other indicators such as stomatal conductance, leaf photosynthesis and leaf temperature were less sensitive (Goldhamer *et al.*, 1999). However, to date, plant-based DSSs are rarely used in peach, as in other crops, due to the difficulties in managing the sensors as well as in interpreting the data.

5.4.4 Remote sensing

Remote sensing can be defined as a group of techniques for collecting images or other forms of data from measurements made at a distance. The data are then cross-checked, processed and analysed, and act as inputs for making decisions and providing irrigation recommendations. Remote sensing applications include different technologies for image acquisition such as hyperspectral, multispectral and thermal images. These can be collected remotely by satellites, ultralight aircraft and drones, with precise results (Ballester *et al.*, 2018). In recent years, the

Fig. 5.3. A fruit gauge for the precise and continuous determination of fruit growth.

launch of the European Copernicus programme with its Sentinel satellites has provided open-source and real-time data on a wide range of useful parameters for agriculture. The integration of accurate field data (i.e. from soil and/or plant sensors) with remote sensing information represents one of the most promising approaches towards implementing precision irrigation in orchards, upscaling the accuracy of proximal sensors at the orchard level, while considering information on orchard heterogeneity provided by remote sensing. In particular, the crop water stress index has been introduced as an efficient indicator of crop water status (Jones, 1992). This parameter and its spatial variability within a peach orchard can be estimated remotely based on thermal detection of canopy temperatures from satellites or drones and has been shown to be related to leaf water potential (Bellvert *et al.*, 2016; Park *et al.*, 2017), although this relationship can change depending on the phenological stages. These studies suggest the possibility of mapping the spatial variability of leaf water potential within a peach orchard throughout a growing season, with the possibility of developing DSSs for irrigation scheduling based on remote sensing.

5.5　ORCHARD MANAGEMENT AND WUE

5.5.1　Canopy architecture and WUE

WUE can be defined as the ratio between the amount of fixed carbon dioxide and the water lost by transpiration. About 99.5% of the water absorbed by roots and transferred to the aerial part of the plant is released to the atmosphere through stomatal and cuticular leaf and fruit transpiration. Leaf transpiration is regulated first by the evaporative demand and second by solar radiation, whereas radiation is the major regulating factor for photosynthesis. In peach, leaves receiving an amount of light of approximately 800–1000 μmol PPFD (photosynthetic photon flux density) m^{-2} s^{-1} can achieve maximum photosynthetic values (saturation point). Although leaves exposed to the sun transpire more than shaded leaves (<20% of incoming radiation), they have a WUE value ten times greater than shaded leaves. Therefore, the shadowed part of the canopy is not a source but another sink with a high water use (in some training systems, it can account for as much as 30% of the total water use) and very low WUE. Thus, when choosing a training system, it is important to give due consideration to WUE, which increases with the increase in the ratio between exposed and shaded leaves. WUE can be improved by reduced plant size, correct row orientation, the training system and canopy management (Fig 5.4). In this regard, summer pruning should be performed twice a year, aimed at reducing the leaf area overall by approximately 10 m^2 per plant. Assuming a daily mean leaf transpiration rate of 3 mmol m^{-2} s^{-1}, summer pruning was found to contribute to reducing the transpired water by about 800 m^3 ha^{-1} over approximately 40 days from August (Dichio *et al.*, 2007).

5.5.2   Soil volume explored by roots in irrigation management

The size and characteristics of the soil volume explored by roots should be considered in order to optimize water use in peach orchards. Soil colonization by roots depends mostly on the orchard age, planting density, rootstock, soil type and management. For instance, a vigorous rootstock explores a greater

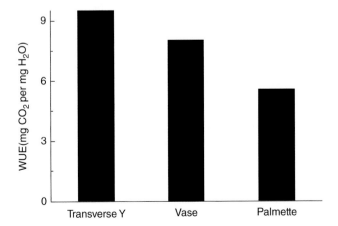

Fig 5.4. Water-use efficiency (WUE) measured in peach trees trained using three different training systems. Adapted from Giugliani *et al.* (1999).

soil volume than a weaker rootstock. For irrigation purposes, the different soil volume explored by roots also corresponds to a different amount of water available to plants. This information is particularly important for calculating the amount of rainfall water stored and potentially usable by the plants. For the purposes of irrigation management, and particularly when localized irrigation methods are used, the total soil volume explored by roots is assumed to be divided into two parts (Fig. 5.5): volume 1 corresponding to the soil portion wetted by irrigation and explored by roots, and volume 2 corresponding to the soil zone explored by roots but not wetted by irrigation. The full use of water reserves, in both compartments, by the plants may be planned to coincide with those growth stages when the crop has little sensibility to water deficit using the RDI technique. If the two compartments are completely emptied at the end of the irrigation season, more rainfall water can be stored again during the rainfall season avoiding an excess of water and damage to the roots.

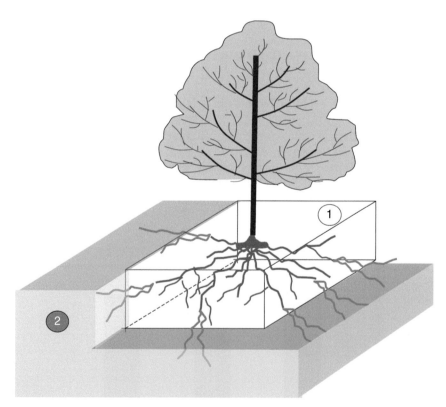

Fig. 5.5. The root system grows some metres from the trunk in mature trees. However, the soil volume wetted by localized irrigation (volume 1) is explored only by part of the total root system (brown roots), while the other roots (grey) explore the non-irrigated soil (volume 2).

5.5.3 Agronomic techniques to increase soil water storage capacity

The water storage capacity of the soil volume explored by roots plays a fundamental role, particularly during the rainfall season when there is no water consumption by trees. Deeper roots allow the plant to explore a larger soil and water volume and thus to improve crop drought tolerance, while reducing the amount of irrigation water. In the absence of chemical, mechanical or any other hindering obstacle, fruit tree roots can develop as deep as 2 m and thus explore a soil reservoir that can store water volumes as high as 4000 m^3 ha^{-1} in soils with high water-holding capacity. To favour the maximum accumulation of water in such a reservoir, sound soil management is necessary, to improve not only the soil's chemical and microbiological fertility and organic matter content but also its structure and hydrological characteristics. The correct choice of rootstock also plays an important role in the soil volume explored by roots, particularly during the first 4–5 years after planting.

5.6 CONCLUSIONS AND PERSPECTIVES

In peach, as in most crops, there is an increasing need to optimize irrigation to better face the many climate change-related challenges, such as increasing evapotranspiration requirements, heat waves and water scarcity. However, besides the need to save water, optimizing irrigation also represents a strong lever towards the improvement of fruit quality. Deficit irrigation is in fact often associated with increased fruit dry-matter content and prolonged shelf life, as well as improved intrinsic quality of the fruit (e.g. increased sweetness and/or nutraceutical components in the fruit tissues) (Falagán *et al.*, 2015).

However, the many precise irrigation technologies already available are currently poorly adopted, as they are perceived as complicated, difficult to apply and with no clear risk:benefit ratios. Advisers and professionals have a key role in helping these technologies to become better known and adopted, while researchers should try to overcome the current constraints, making these approaches more reliable and user-friendly for both growers and advisers.

REFERENCES

Abrisqueta, I., Conejero, W., Valdés-Vela, M., Vera, J., Ortuño, M.F. and Ruiz-Sánchez, M.C. (2015) Stem water potential estimation of drip-irrigated early-maturing peach trees under Mediterranean conditions. *Computers and Electronics in Agriculture* 114, 7–13.

Allen, R.G., Pereira, L.S., Raes, D. and Smith, M. (1998) *Crop Evapotranspiration: Guidelines for Computing Crop Water Requirements.* FAO Irrigation and Drainage Paper 56. Food and Agriculture Organization of the United Nations, Rome. Available at: www.fao.org/3/x0490e/x0490e00.htm#Contents (accessed 5 December 2022).

Améglio, T., Cochard, H., Picon, C. and Cohen, M. (1998) Water relations and hydraulic architecture of peach trees under drought conditions. *Acta Horticulturae* 465, 355–362.

Ayars, J.E., Johnson, R.S., Phene, C.J., Trout, T.J., Clark, D.A. and Mead, R.M. (2003) Water use by drip-irrigated late-season peaches. *Irrigation Science* 22, 187–194.

Ballester, C., Zarco-Tejada, P.J., Nicolás, E., Alarcón, J.J., Fereres, E. *et al.* (2018) Evaluating the performance of xanthophyll, chlorophyll and structure-sensitive spectral indices to detect water stress in five fruit tree species. *Precision Agriculture* 19, 178–193.

Basile, B., Marsal, J. and DeJong, T.M. (2003) Daily shoot extension growth of peach trees growing on rootstocks that reduce scion growth is related to daily dynamics of stem water potential. *Tree Physiology* 23, 695–704.

Bellvert, J., Marsal, J., Girona, J., Gonzalez-Dugo, V., Fereres, E. *et al.* (2016) Airborne thermal imagery to detect the seasonal evolution of crop water status in peach, nectarine and saturn peach orchards. *Remote Sensing* 8: 39.

Boland, A.-M., Jerie, P.H., Mitchell, P.D. and Goodwin I. (2000) Long-term effects of restricted root volume and regulated deficit irrigation on peach: I. Growth and mineral nutrition. *Journal of the American Society for Horticultural Science* 125, 135–142.

Chalmers, D.J., Mitchell, P.D. and van Heek, L. (1981) Control of peach tree growth and productivity by regulated water supply, tree density, and summer pruning. *Journal of the American Society for Horticultural Science* 106, 307–312.

Cohen, M., Goldhamer, D.A., Fereres, E., Girona, J. and Mata M. (2001) Assessment of peach tree responses to irrigation water deficits by continuous monitoring of trunk diameter changes. *Journal of Horticultural Science and Biotechnology* 76, 55–60.

Conejero, W., Alarcón, J.J., García-Orellana, Y., Nicolás, E. and Torrecillas, A. (2007) Evaluation of sap flow and trunk diameter sensors for irrigation scheduling in early maturing peach trees. *Tree Physiology* 27, 1753–1759.

Davies, W.J. and Zhang, J. (1991) Root signals and the regulation of growth and development of plants in drying soil. *Annual Review of Plant Physiology and Plant Molecular Biology* 42, 55–76.

Dichio, B., Xiloyannis C., Sofo, A. and Montanaro, G. (2007) Effects of post-harvest regulated deficit irrigation on carbohydrate and nitrogen partitioning, yield quality and vegetative growth of peach trees. *Plant and Soil* 290, 127–137.

Dichio B., Montanaro G. and Xiloyannis C. (2011) Integration of the regulated deficit irrigation strategy in a sustainable orchard management system. *Acta Horticulturae* 889, 221–226.

Falagán, N., Artés, F., Gómez, P.A., Artés-Hernández, F., Pérez-Pastor, A. *et al.* (2015) Combined effects of deficit irrigation and fresh-cut processing on quality and bioactive compounds of nectarines. *Horticultural Science* 42, 125–131.

Gelly, M., Recasens, I., Mata, M., Arbones, A., Rufat, J. *et al.* (2003) Effects of water deficit during stage II of peach fruit development and postharvest on fruit quality and ethylene production. *Journal of Horticultural Science and Biotechnology* 78, 324–330.

Gelly, M., Recasens, I., Girona, J., Mata, M., Arbones, A. *et al.* (2004) Effects of stage II and postharvest deficit irrigation on peach quality during maturation and after cold storage. *Journal of the Science of Food and Agriculture* 84, 561–568.

Girona, J., Mata, M., Arbonès, A., Alegre, S., Rufat, J. and Marsal, J. (2003) Peach tree response to single and combined regulated deficit irrigation regimes under shallow soils. *Journal of the American Society for Horticultural Science* 128, 432–440.

Giugliani, R., Magnani, E. and Corelli Grappadelli, L. (1999) Relazioni tra scambi gassosi e intercettazione luminosa in chiome di pesco allevate secondo tre forme. *Rivista di Frutticoltura e Ortofloricoltura* 3, 65–69.

Goldhamer, D.A., Fereres, E., Mata, M., Girona, J. and Cohen, M. (1999) Sensitivity of continuous and discrete plant and soil water status monitoring in peach trees subjected to deficit irrigation. *Journal of the American Society for Horticultural Science* 124, 437–444.

Johnson, R.S. and Handley, D.F. (2000) Using water stress to control vegetative growth and productivity of temperate fruit trees. *HortScience* 35, 1048–1050.

Johnson, R.S. and Phene, B.C. (2008) Fruit quality disorders in an early maturing peach cultivar caused by postharvest water stress. *Acta Horticulturae* 792, 385–390.

Johnson, R.S., Ayars, J. and Hxiao, T. (2004) Improving a model for predicting peach tree evapotranspiration. *Acta Horticulturae* 664, 341–346.

Jones, H.G. (1992) *Plants and Microclimate: A Quantitative Approach to Environmental Plant Physiology*, 2nd edn. Cambridge University Press, New York.

Kriedemann, P.E. and Goodwin, I. (2003) *Regulated Deficit Irrigation and Partial Root-zone Drying. An Overview of Principles and Applications*. Irrigation Insights No. 4. Land and Water, Canberra. Available at: www.wineaustralia.com/getmedia/50a58da5-ecc5-4aa1-ad06-8e3c87e38de9/NPI-01-01-RECV-27-10-03-PRD-irrigation-insights&usg (accessed 5 December 2022).

Marsal, J. and Girona, J. (1997) Relationship between leaf water potential and gas exchange activity at different phenological stages and fruit loads in peach trees. *Journal of the American Society for Horticultural Science* 122, 415–421.

McFadyen, L.M., Button, R.J. and Barlow, E.W.R. (1996) Effects of crop load on fruit water relations and fruit growth in peach. *Journal of Horticultural Science* 71, 469–480.

Naor, A. and Cohen, S. (2003) Sensitivity and variability of maximum trunk shrinkage, midday stem water potential, and transpiration rate in response to withholding irrigation from field-grown apple trees. *HortScience* 38, 547–551.

Naor, A., Stern, R., Peres, M., Greenblat, Y., Gal, Y. and Flaishman, M.A. (2005) Timing and severity of postharvest water stress affect following-year productivity and fruit quality of field-grown 'Snow Queen' nectarine. *Journal of the American Society for Horticultural Science* 130, 806–812.

Natali, S., Xiloyannis, C. and Pezzarossa, B. (1985) Relationship between soil water content, leaf water potential and fruit growth during different fruit growing phases of peach trees. *Acta Horticulturae* 171, 167–180.

Park, S., Ryu, D., Fuentes, S., Chung, H., Hernández-Montes, E. and O'Connell, M. (2017) Adaptive estimation of crop water stress in nectarine and peach orchards using high-resolution imagery from an unmanned aerial vehicle (UAV). *Remote Sensing* 9, 828.

Schaffer, B., Andersen, P.C. and Ploetz, R.C. (1992) Responses of fruit crops to flooding. *Horticultural Reviews* 13, 257–313.

Smith, R.J. and Baillie, J.N. (2009) Defining precision irrigation: a new approach to irrigation management. In: *Irrigation Australia 2009: Irrigation Australia Irrigation and Drainage Conference: Irrigation Today - Meeting the Challenge, 18–21 October 2009, Swan Hill, Australia*. Available at: https://eprints.usq.edu.au/19749/ (accessed 5 December 2022).

Steduto, P., Hsiao, T.C., Fereres, E. and Raes, D. (2012) *Crop Yield Response to Water*. FAO Irrigation and Drainage Paper 66. Available at: www.fao.org/3/i2800e/i2800e00. htm (accessed 16 December 2022).

Tombesi, S., Johnson, R.S., Day, K.R. and DeJong, T.M. (2010) Interactions between rootstock, inter-stem and scion xylem vessel characteristics of peach trees growing on rootstocks with contrasting size-controlling characteristics. *AoB PLANTS* 2010: plq013.

Vera, J., Abrisqueta, I., Abrisqueta, J.M. and Ruiz-Sánchez, M.C. (2013) Effect of deficit irrigation on early-maturing peach tree performance. *Irrigation Science* 31, 747–757.

Xiloyannis, C., Massai, R., Piccotino, D., Baroni, G. and Bovo, M. (1993) Method and technique of irrigation in relation to root system characteristics in fruit growing. *Acta Horticulturae* 335, 505–510.

Xiloyannis, C., Dichio, B. and Montanaro, G. (2012) Irrigation in Mediterranean fruit tree orchards. INTECH Open Access Publisher. Available at: https://cdn.intechopen. com/pdfs/34118.pdf (accessed 5 December 2022).

Xylogiannis, E., Sofo, A., Dichio, B., Montanaro, G. and Mininni A.N. (2020) Root-to-shoot signaling and leaf water-use efficiency in peach trees under localized irrigation. *Agronomy* 10: 437.

6

FERTILIZATION AND PLANT NUTRIENT MANAGEMENT

Juan Carlos Melgar[1]*, Bartolomeo Dichio[2] and Cristos Xiloyannis[2]

[1]*Department of Plant and Environmental Sciences, Clemson University, Clemson, South Carolina, USA;* [2]*Dipartimento delle Culture Europee e del Mediterraneo: Architettura, Ambiente, Patrimoni Culturali – University of Basilicata, Potenza, Italy*

6.1   ORCHARD FERTILIZATION

Tree nutrition is an essential part of orchard management, especially under intensive cultivation, and each nutrient plays a fundamental role in the growth, productivity and fruit quality of peach trees. The need to evaluate orchard nutrient status and adjust fertilizer rates has been reported by researchers for decades (Greenham, 1980; Tagliavini *et al.*, 1995). However, maintaining optimal soil fertility levels does not necessarily translate into achieving optimal tree nutrition, as there are numerous biotic and abiotic factors influencing tree nutrient uptake and assimilation, and nutrient losses. Furthermore, synchronizing fertilizer applications with tree demands while minimizing losses to the environment is a complex issue because of the seasonal uptake patterns and nutrient dynamics of deciduous fruit trees, and their ability to store nutrients and reuse them at a later time, as well as climatic conditions and characteristics specific for each orchard. Peach growers are aware of the occurrence of nutrient deficiencies when soil nutrient content is low, or when a nutrient is not available due to physical or chemical characteristics of the soil. Nitrogen and potassium deficiencies are often seen in peach orchards without proper nutritional management, but other macronutrient deficiencies are not common in most peach orchard systems (Johnson, 2008). However, the perception that an annual application of large amounts of fertilizer ensures a good crop (even in orchards established on fertile soils), together with the relatively low cost of

* Email: jmelgar@clemson.edu

© CAB International 2023. *Peach* (G. Manganaris, G. Costa and C. Crisosto eds)   129
DOI: 10.1079/9781789248456.0006

fertilizers in relation to the crop value, and the diversity of orchard variables, often leads to annual, excessive, uncontrolled, calendar-based fertilization rates (Weinbaum *et al.*, 1992; Milosevic and Milosevic, 2009). Overfertilization contributes to pollution of underground waters and air, and may adversely affect productivity and fruit quality (Fig. 6.1) because of: (i) direct effects on the development of physiological disorders and poor quality for storage (Daane *et al.*, 1995); and (ii) indirect effects on flowering, fruit set, and fruit quality, including lower concentration of anthocyanins (Cordts *et al.*, 1987) and reduced antioxidant capacity (Vashisth *et al.*, 2017).

6.2 SOIL AND TISSUE ANALYSES

Soil sampling before planting followed by annual leaf analysis can be used to correct tree nutritional status. The use of soil analysis alone is not sufficient to understand tree nutrient requirements and use, as nutrients may be abundant in the soil but not be available to the trees. However, soil analyses can also provide growers with critical information such as pH level, which can influence nutrient solubility as well as root growth and microbial activity, or cation exchange capacity, which influences soil pH changes. While methods such as

Fig. 6.1. An overfertilized peach orchard with excessive vegetative growth and low yield. Photograph courtesy of J.C. Melgar.

flower analysis have been proposed to allow detection and treatment of deficiencies or low nutrient concentrations at an early stage (Sanz and Montañés, 2008), they are not widely available for growers, and the use of leaf analysis about 90–120 days after full bloom remains the most popular method for diagnosing possible nutritional imbalances and deficiencies.

Interpretations of the leaf analysis are made by comparing the data with sufficiency ranges established for the specific crop, organ and sampling time. Nutritional concentrations associated with these reference values have been associated with optimal tree growth, fruit yield and quality. Nevertheless, interpretation of the leaf analysis must consider many factors that may influence foliar nutrient levels, such as fruit load (or previous yield), as well as other events that remove nutrients from trees (pruning, thinning), tree age and cultivar, seasonal climatic variations, and soil physical and chemical characteristics.

Overall, a nutrient balance is reached when nutrients added to the soil–tree system (from soil/foliar-applied fertilization, atmospheric deposition, mineralization or remobilized from the reserve organs) equal the nutrients removed from the tree (harvested fruit, pruned wood), stored in the reserves, or immobilized by the soil, leached or lost through erosion (runoff) or volatilization (El-Jendoubi *et al.*, 2013). While determining all the components of the balance for specific orchards and conditions is an impractical and laborious task, research studies in the last decade (El-Jendoubi *et al.*, 2013; Zhou and Melgar, 2019) have provided clues on the importance and influence of these factors. Thus, nutrient inputs to achieve profitable and sustainable peach production must be based on the implementation of knowledge-based practices aimed at reducing losses.

6.3 CURRENT AND FUTURE TRENDS IN NUTRITIONAL MANAGEMENT

Current environmental and societal demands require changes that lead the fruit industry to a more sustainable and ecologically friendly production system while increasing yields and adapting to challenges such as soil degradation, and climate variability and change. Fertilizer use is among the orchard production activities with greatest overall impact on the environment (Guo *et al.*, 2018; Medici *et al.*, 2020), and has been estimated to account for 5.6–7.4% of the total footprint of peach/nectarine systems (Cerutti *et al.*, 2010; Ingrao *et al.*, 2015), with urea fertilizer accounting for the highest emissions (Nikkhah *et al.*, 2017). While these estimations may vary largely between production regions, avoiding excessive fertilization and increasing nutrient-use efficiency is a necessary step to reduce greenhouse-gas emissions and improve the sustainability of peach orchards. Other economic and geopolitical factors (e.g. the finite amount of phosphate rocks, which are expected to be depleted within the current century; Gilbert, 2009; de Boer *et al.*, 2018) may also affect the manufacturing, availability, price and use of synthetic fertilizers in the near future, and will demand consideration of the need for annual fertilizer applications at the local level based on orchard characteristics and tree nutritional requirements.

In the last few decades, several fruit-tree systems (e.g. olive, apple) have shifted from extensive, widely spaced orchards to intensive, closely spaced systems with increased use of fertilizers. Although there have been advancements in research on the development of size-controlling peach rootstocks and training systems for high-density peach orchards (Iglesias and Echeverría, 2022), the planting density of peach-production systems has remained largely unchanged. In many peach-producing regions, the open-vase training system limits the options for ground-cover management (including nitrogen-fixing plants) that could improve not only soil nutritional content but also other soil physical, chemical and biological characteristics. This is especially relevant for areas where decades of a production system with the sole focus on productivity has led to depleted soils with low organic matter, low microbial activity, reduced capacity to recycle nutrients, poor water infiltration and increased erosion. Thus, future research studies should address how changes in orchard systems may have an immediate influence on tree nutritional needs and thus on orchard nutritional management.

6.3.1 Orchard characteristics and management

Fertilization programmes are typically planned in spring, before bloom. When interpreting the values of a leaf analysis (from the previous summer), growers and extension specialists should consider the annual nutrient dynamics of perennial, deciduous trees, specifically the mobilization of nutrients to the reserve organs before leaf senescence and dormancy, as well as their remobilization to sink organs in spring. Thus, any specifics of the management or characteristics of the orchard that may have an influence on nutrient dynamics should be taken into account. For instance, the trees shown in Fig. 6.2 were pruned in early autumn (before they were dormant) due to issues related to labour availability at this farm; as leaf senescence had not started by the time these trees were pruned, mobilization of nutrients from the leaves to the storage organs did not occur. Previous research estimates indicate that about 50% of leaf nitrogen content is mobilized to reserves before senescence, which accounts for more than 80% of nitrogen in the reserve organs (Niederholzer *et al.*, 2001). This is especially important for hysteranthous fruit trees such as peach, as leaves unfold following bloom and the nutrients required for growing tissues are not yet supplied by uptake from the soil. As a consequence, the trees in Fig. 6.2 started their next growing season (the following spring) in a poorer nutritional status than if they had been pruned during dormancy. Other orchard

Fig. 6.2. Peach trees pruned in early autumn before leaf senescence. Photograph courtesy of J.C. Melgar.

characteristics such as tree age or ripening season also influence translocation of nutrients to permanent structures and nutrient partitioning to aboveground organs (Zhou and Melgar, 2019, 2020). Furthermore, weather patterns related to climate variability and change are linked to changes in nutrient dynamics, with warm and dry autumn seasons increasing nutrient mobilization to storage organs (Lawrence and Melgar, 2018). These orchard-specific factors should be integrated when determining orchard nutritional needs (e.g. most growers already consider the previous year's yield) to implement holistic fertilization programmes that allow peach growers to achieve optimum productivity while reducing economic and environmental losses.

6.3.2 Organic fertilizers and enhanced-efficiency fertilizers

Organic amendments such as compost have shown an excellent fertilization effect as a replacement for synthetic fertilizers in peach/nectarine orchards, although the nutrients from organic fertilizers do not typically become immediately available for root uptake (Baldi *et al.*, 2016, 2021). Interest in the use of organic amendments in fruit orchards has increased over the last decades, especially due to the increasing availability of organic amendments from material valorization of food waste and the low cost and social benefits related to recycling of municipal waste. However, while organic amendments such as compost can easily be spread with equipment such as manure spreaders at planting or in young orchards, its application can become impractical in mature, traditional, low-density orchards with low scaffolds that do not allow machinery within rows. In these cases, the use of organic fertilizers needs more labour and monetary input compared with the use of chemical fertilizers. Furthermore, peach farmers in many regions use chemical fertilizers instead of organic fertilizers as they fear they may lose income if they switch to organic fertilizers (Xiao *et al.*, 2019).

Incorporation of organic amendments may stimulate nitrous oxide (N_2O) emissions as they provide organic carbon substrates for denitrifying microorganisms (Cheng *et al.*, 2017); however, organic materials with a large carbon:nitrogen ratio can display a strong competition for nitrogen, and consequently inhibit nitrification and denitrification, as reported by Fentabil *et al.* (2016). The use of enhanced-efficiency fertilizers such as nitrification and urease inhibitors have been suggested to decrease nitrogen losses (e.g. nitrate leaching, N_2O emissions) and improve nitrogen-use efficiency (Li *et al.*, 2018). However, they may increase ammonia (NH_3) volatilization, which may also indirectly increase N_2O, and negate or outweigh their benefit. To alleviate this problem, urease inhibitors (with surface-applied urea) could be used alongside nitrification inhibitors. Urease inhibitors limit the breakdown of urea into NH_3 and carbon dioxide, and have been proven to reduce N_2O emissions. Synthetic and organic plant-based nitrification and urease inhibitors have been studied extensively in vegetable crops, but there is little research on fruit trees, and specifically in peach. Villar *et al.* (2018) showed that nitrification inhibitors improved nitrogen-uptake efficiency in peaches, although their performance depended on soil type, climate and field practices, and more studies are needed. In addition, bag-controlled release fertilizer has potential application as a new, environmentally friendly, low-cost and efficient fertilizer option for cleaner production in peach orchards (Xiao *et al.*, 2019).

6.3.3   Advances in breeding, plant physiology and artificial intelligence

As prices of agrochemicals and labour increase, and society and consumers demand enhanced food safety and environmental stewardship, there is a need for efficient precision management, including fine tuning of the timing, amount and location of fertilizer. Foliar fertilization is increasing worldwide to improve tree nutritional status but there is a need for an understanding of tree architecture, leaf morphological and anatomical differences at different parts of the canopy or between different cultivars, surface penetration at different developmental stages, and environmental factors that can increase foliar absorption of fertilizers (Fernández and Bahamonde, 2020). This information, together with the use of laser-guided, variable-rate, intelligent sprayers, will help growers to implement foliar fertilization more effectively than with conventional sprayers, reducing the amount of spray drift and thus costs and potential pollution. The presence of environmental or orchard conditions leading to abiotic or biotic stresses can also influence the response of trees to fertilization. Related to this, the collection of time-series data from unmanned aerial vehicles can also help in setting fertilizer management zones to improve environmental and economic benefits.

Advances in rootstock breeding can also improve tree nutrient efficiency. Recent literature shows that rootstocks influence the seasonal variation in nutrient concentration of peach leaves (Shahkoomahally *et al.*, 2020), and appropriate rootstocks can improve optimal nutrient absorption and translocation, reducing the risk of nutrient leaching without decreasing fruit quality and yield (Padilha Galarça *et al.*, 2015; Jimenes *et al.*, 2018). Thus, it is expected that the adoption of genotypes that are more efficient in the use of nutrients will play a key role in improving orchard nutrient-use efficiency.

6.4 CONCLUDING REMARKS AND FUTURE PERSPECTIVES

Rational fertilization is essential to achieve long-term orchard sustainability (i.e. farm economic viability and stewardship of natural resources), as it considers both practices to obtain high and regular productivity and fruit quality with minimum environmental pollution. Nevertheless, current fertilization practices in most fruit-tree orchards ignore one of the basic principles for rational fertilization of fruit crops: the addition of nutrients only when there is evidence that they are necessary (Fernández-Escobar, 2019).

While smartphones can send alerts to growers about when to irrigate, and drones can help us estimate yield or disease incidence at the orchard level, nutritional management of most fruit orchards continues to be done as in the mid-20th century, i.e. calendar-scheduled, annual applications based on visual judging or grower personal preferences. Thus, there is a need for: (i) integrating orchard-specific factors when determining nutritional needs; and (ii) using rational fertilization principles to adjust current fertilizer recommendations. To achieve these needs, researchers need to: (i) understand the influence of orchard-specific characteristics on tree nutrient dynamics; (ii) determine orchard nutrient restitution needs based on local (orchard-specific) factors such as the previous year's yield, soil type, rootstock, climate, pruning intensity and ripening time; and (iii) develop fertilization models/guidelines that provide holistic fertilization guidelines to the stone-tree fruit industry with accurate, timely, efficient and effective crop nutrient information. These steps will lead the peach industry towards maximum orchard growth, productivity and fruit quality, while maintaining natural resources and improving long-term sustainability and profitability of fruit farms.

REFERENCES

Baldi, E., Marcolini, G., Quartieri, M., Sorrenti, G., Muzzi, E. and Toselli, M. (2016) Organic fertilization in nectarine (*Prunus persica* var. *Nucipersica*) orchard combines nutrient management and pollution impact. *Nutrient Cycling in Agroecosystems* 105, 39–50.

Baldi, E., Cavani, L., Mazzon, M., Marzadori, C., Quartieri, M. and Toselli, M. (2021) Fourteen years of compost application in a commercial nectarine orchard: effect on microelements and potential harmful elements in soil and plants. *Science of the Total Environment* 752: 141894.

Cerutti, A.K., Bagliani, M., Beccaro, G.L. and Bounous, G. (2010) Application of Ecological Footprint Analysis on nectarine production: methodological issues and results form a case study in Italy. *Journal of Cleaner Production* 18, 771–776.

Cheng, Y., Xie, W., Huang, R., Yan, X. and Wang, S. (2017) Extremely high N_2O but unexpectedly low NO emissions from a highly organic and chemical fertilized peach orchard system in China. *Agriculture, Ecosystems & Environment* 246, 202–209.

Cordts, J.M., Scorza, R. and Bell, R. (1987) Effects of carbohydrates and nitrogen on the development of anthocyanins of a red leaf peach (*Prunus persica* (L.) Batsch) *in vitro*. *Plant Cell, Tissue and Organ Culture* 9, 103–110.

Daane, K.M., Johnson, R.S., Michailides, T.J., Crisosto, C.H., Dlott, J.W. *et al.* (1995) Excess nitrogen raises nectarine susceptibility to disease and insects. *California Agriculture* 49, 13–18.

de Boer, M.A., Wolzak, L. and Slootweg, J.C. (2018) Phosphorus: reserves, production, and applications. In: Ohtake, H. and Tsuneda, S. (eds) *Phosphorus Recovery and Recycling*. Springer, Singapore, pp. 75–100.

El-Jendoubi, H., Abadia, J. and Abadia, A. (2013) Assessment of nutrient removal in bearing peach trees (*Prunus persica* L. Batsch) based on whole tree analysis. *Plant and Soil* 369, 421–437.

Fentabil, M.M., Nichol, C.F., Neilsen, G.H., Hannam, K.D., Neilsen, D. *et al.* (2016) Effect of micro-irrigation type, N-source and mulching on nitrous oxide emissions in a semi-arid climate: an assessment across two years in a Merlot grape vineyard. *Agricultural Water Management* 171, 49–62.

Fernández, V. and Bahamonde, H.A. (2020) Advances in foliar fertilizers to optimize crop nutrition. In: Rengel, Z. (ed.) *Achieving Sustainable Crop Nutrition*. Burleigh Dodds Science Publishing, Cambridge, UK, pp. 1–26.

Fernández-Escobar, R. (2019) Olive nutritional status and tolerance to biotic and abiotic stresses. *Frontiers in Plant Science* 10: 1151.

Gilbert, N. (2009) The disappearing nutrient. *Nature* 461, 716–718.

Greenham, D.W.P. (1980) Nutrient cycling: the estimation of orchard nutrient uptake. *Acta Horticulturae* 92, 345–352.

Guo, C., Wang, X., Li, Y., He, X., Zhang, W. *et al.* (2018) Carbon footprint analyses and potential carbon emission reduction in China's major peach orchards. *Sustainability* 10: 2908.

Iglesias, I. and Echeverría, G. (2022) Current situation, trends and challenges for efficient and sustainable peach production. *Scientia Horticulturae* 296: 110899.

Ingrao, C., Matarazzo, A., Tricase, C., Clasadonte, M.T. and Huisingh, D. (2015) Life Cycle Assessment for highlighting environmental hotspots in Sicilian peach production systems. *Journal of Cleaner Production* 92, 109–120.

Jimenes, I.M., Mayer, N.A., dos Santos Dias, C.T., Scarpare Filho, J.A. and da Silva, S.R. (2018) Influence of clonal rootstocks on leaf nutrient content, vigor and productivity of young 'Sunraycer' nectarine trees. *Scientia Horticulturae* 253, 279–285.

Johnson, R.S. (2008) Nutrient and water requirements of peach trees. In: Layne, D.R. and Bassi, D. (eds) *The Peach: Botany, Production and Uses.* CAB International, Wallingford, UK, pp. 303–331.

Lawrence, B.T. and Melgar, J.C. (2018) Variable fall climate influences nutrient resorption and reserve storage in young peach trees. *Frontiers in Plant Science* 9: 1819.

Li, T., Zhang, W., Yin, J., Chadwick, D., Norse, D. *et al.* (2018) Enhanced-efficiency fertilizers are not a panacea for resolving the nitrogen problem. *Global Change Biology* 24, 511–521.

Medici, M., Canavari, M. and Toselli, M. (2020) Interpreting environmental impacts resulting from fruit cultivation in a business innovation perspective. *Sustainability* 12: 9793.

Milosevic, T. and Milosevic, N. (2009) The effect of zeolite, organic and inorganic fertilizers on soil chemical properties, growth and biomass yield of apple trees. *Plant, Soil and Environment* 55, 528–535.

Niederholzer, F.J.A., DeJong, T.M., Saenz, J.-L., Muraoka, T.T. and Weinbaum, S.A. (2001) Effectiveness of fall versus spring soil fertilization of field-grown peach trees. *Journal of the American Society for Horticultural Science* 125, 644–648.

Nikkhah, A., Royan, M., Khojastehpour, M. and Bacenetti, J. (2017) Environmental impacts modeling of Iranian peach production. *Renewable and Sustainable Energy Reviews* 75, 677–682.

Padilha Galarça, S., Lima, C.S.M., Fachinello, J.C., Pretto, A., Vahl, L.C. and Betemps, D.L. (2015) Influence of several rootstocks on foliar nutrition in peach. *Acta Horticulturae* 1084, 75–84.

Sanz, M. and Montañés, L. (2008) Flower analysis as a new approach to diagnosing the nutritional status of the peach tree. *Journal of Plant Nutrition* 18, 1667–1675.

Shahkoomahally, S., Chaparro, J.X., Beckman, T.G. and Sarkhosh, A. (2020) Influence of rootstocks on leaf mineral content in the subtropical peach cv. *UFSun. HortScience* 55, 496–502.

Tagliavini, M., Scudellazi, D., Marangoni, B. and Toselli, M. (1995) Nitrogen fertilization management in orchards to reconcile productivity and environmental aspects. *Fertilizer Research* 43, 93–102.

Vashisth, T., Olmstead, M.A., Olmstead, J. and Colquhoun, T.A. (2017) Effects of nitrogen fertilization on subtropical peach fruit quality: organic acids, phytochemical content, and total antioxidant capacity. *Journal of the American Society for Horticultural Science* 142, 393–404.

Villar, J.M., Pascual, M., Arbonés, A. and Vilarrasa, M. (2018) Using the nitrification inhibitor DMPP to enhance uptake efficiency in a fertigated peach orchard plantation. *Acta Horticulturae* 1217, 201–206.

Weinbaum, S.A., Johnson, R.S. and DeJong, T.M. (1992) Causes and consequences of overfertilization in orchards. *HortTechnology* 2, 112–121.

Xiao, Y., Peng, F., Zhang, Y., Wang, J., Zhuge, Y. *et al.* (2019) Effect of bag-controlled release fertilizer on nitrogen loss, greenhouse gas emissions, and nitrogen applied amount in peach production. *Journal of Cleaner Production* 234, 258–274.

Zhou, Q. and Melgar, J.C. (2019) Ripening season affects tissue mineral concentration and nutrient partitioning in peach trees. *Journal of Plant Nutrition and Soil Science* 182, 203–209.

Zhou, Q. and Melgar, J.C. (2020) Tree age influences nutrient partitioning among annually removed aboveground organs of peach. *HortScience* 55, 560–564.

7

FRUIT THINNING

Guglielmo Costa[1]*, Alessandro Botton[2] and Carlos H. Crisosto[3]
[1]*Department of Agri-Food Sciences and Technologies, University of Bologna, Bologna, Italy; [2]Department of Agronomy, Food, Natural Resources, Animals and Environment, University of Padua, Padua, Italy; [3]Department of Plant Sciences, University of California, Davis, California, USA*

7.1 WHY FRUIT THINNING IS MANDATORY IN PEACH

Most peach genotypes are very fertile and bear an abundance of flowers, and, unless unfavourable conditions occur, fruit set reaches very high percentages. As the physiological, self-regulatory fruit-drop mechanism is insufficient to guarantee fruits of marketable size, vegetation and production should be balanced by thinning to reduce the fruit load (Fig. 7.1). Fruit must be thinned each year not only to improve fruit quality in the season but also to prevent alternate bearing (or biennial bearing) in the following years by balancing the fruit:shoot ratio (Costa *et al.*, 1983). Therefore, fruit thinning is mandatory each year in peach to improve fruit weight and quality, enhance maturity/ripening synchronism and allow optimal flower bud formation (Myers *et al.*, 1993). Taken together, these improvements lead to high prices and high crop value.

Fruit thinning practices, however, are often complicated by interaction with exogenous factors, such as environmental conditions, management practices that affect orchard performance, and endogenous factors determined by orchard genotype and age (Costa *et al.*, 2018), which significantly affect the severity and timing of thinning.

7.2 CONSUMER QUALITY AND THINNING

Fruit quality was defined formerly by fruit size and colour, and it was not known whether these traits were important only for the first buy (Delgado *et al.*, 2013). In the last decade, fruit consumer quality has expanded to encompass sensory properties (appearance, texture, flavour, taste and aroma),

---

* Email: guglielmo.costa@unibo.it

© CAB International 2023. *Peach* (G. Manganaris, G. Costa and C. Crisosto eds)
DOI: 10.1079/9781789248456.0007

Fig. 7.1. Unthinned (a, c) and thinned (b, d) peach trees, and the effects on fruit size at harvest. Photo courtesy of G. Costa.

nutritional value and texture, which together determine fruit quality, economic value and future sells (Crisosto, 2002; Crisosto and Costa, 2008; Delgado *et al.*, 2013). Handlers discovered that consumers rejected fruit with low sugar content, poor aroma, texture and 'off flavour' (chilling injury symptoms), despite a large size and attractive red colour (Lurie and Crisosto, 2005; Manganaris *et al.*, 2019). Consumers encountering peaches with these negative attributes immediately changed their preference towards other fruit types. In the last decades, peach consumption has declined, mainly because of poor understanding and definition of quality, premature harvest, improper handling, chilling injury damage and lack of ripening prior to consumption, resulting in consumer dissatisfaction (Crisosto, 2002). In a detailed sensory study using a large group of peach and nectarine cultivars, cluster analysis revealed two main sensory attributes that were significantly associated with overall liking and consumer acceptance. This and other sensory studies demonstrated that sweetness perception was the main positive driver of approval, and perception of grassy/green fruit aromas and pit aromas ('off flavour') was the main negative driver (Delgado *et al.*, 2013; Bonany *et al.*, 2014). In sound fruit, soluble solids concentration (SSC) was the only instrumental measurement significantly related to consumer approval and preferences (Crisosto and Crisosto, 2005; Crisosto *et al.*, 2006;

Delgado *et al.*, 2013). In California, as an industry guideline, a minimum 10.0% SSC is recommended for mid-acid, yellow-fleshed peaches and nectarines as the quality standard for peaches free of sensory defects such as chilling injury symptoms (Crisosto and Valero, 2008). In the European Union, an SSC of 8.0% and above and firmness of 6.5 kgf or less was established for early- to late-ripening peaches by the European Commission Regulation, which is notably below the current recommendations of SSC of 11.0% or above and fruit firmness of less than 5 kgf (Iglesias and Echeverría, 2009).

Proper fruitlet thinning always increases fruit size and, in most cases, increases colour and SSC, while reducing total production. Thus, a balance between yield and fruit size must be achieved for each orchard condition and cultivar. Generally, maximum profit does not occur at maximum marketable yield, as larger fruit brings a higher market price than flavourful fruit. However, currently, some sales and/or premiums are available for peaches that reach a minimum sweetness perception (consumer quality index). Leaving too many fruits on the tree reduces fruit size and SSC. Therefore, fruit quality can be sacrificed in several ways by incorrect thinning. Grower experience is the best way to understand conditions and determine optimum thinning for each orchard and cultivar. Selectively reducing crop load in the canopy through proper thinning creates better fruit-growing conditions and pesticide exposure, while improving general consumer quality and potential sales.

7.3   CURRENT THINNING METHODS

In general, fruit thinning can be carried out manually, mechanically or chemically, and is usually performed as two or more diverse interventions at different phenological stages. In peach, however, fewer options are available than with other crops, such as apple. Current thinning techniques adopted in peach are described below and their pros and cons discussed briefly.

7.3.1   Hand thinning

Manual/hand thinning is the most widespread method used in peach. It can be very accurate but is expensive and requires many qualified operators within a very short time frame. Manual thinning in peach can be performed either at bloom or at the fruitlet stage. Thinning flowers increases the final fruit size the most, as competition between vegetative and reproductive sinks is relieved earlier (Byers, 1989). However, the economic sustainability of this practice in peach must be assessed carefully, factoring in the economic benefits of the higher prices paid for larger fruits, the desired yield and the risk of spring frosts, which can further reduce fruit set and cause overthinning. For these reasons, flower thinning is mostly performed in early-ripening cultivars with smaller fruits or for experimental purposes. Thinning at the fruitlet stage usually produces less pronounced effects, especially when fruit load is reduced manually at later developmental stages. Peach fruitlet thinning is carried out optimally at around 20 mm and seed length of 9–11 mm (i.e. about 2 weeks before pit hardening or 40–60 days after full bloom with ongoing physiological fruit drop). Performed later, assimilates that have already been diverted to fruitlets would be lost and unavailable for persisting fruits (current season) or differentiating buds (following season).

The amount of fruit thinning (i.e. the number and position of flowers or fruitlets to be removed) must be decided according to practical rules to achieve the best effects. The leaf:fruit ratio, type of shoot and its degree of lignification, and the position of the fruits within the canopy are the main parameters to be considered when performing manual thinning. A standard leaf:fruit ratio value is 40–45 leaves per fruit, with more fruits normally left on the basal part of the shoot, which is more lignified than the apical one. The position of the shoot within the tree is also important, as upper shoots are more exposed to light and have more leaves than lower ones. Consequently, a greater fruit load will be imposed on the upper part of the tree. These general rules are adjusted for genetic background: freestone peaches and nectarines, for example, require stronger thinning than clingstones. Several nectarine genotypes show significant physiological fruit drop during the season, while flat peach genotypes need intensive thinning to reach marketable size.

7.3.2 Mechanical thinning

Both flowers and fruitlets can be removed by machines carrying specialized brushes (Fig. 7.2), rope drags or high-pressure water streams (Byers, 1989). In addition, fruitlets can be thinned by shaking machines attached to either the limb or the trunk, although this technique selectively removes the largest fruits, decreasing fruit yield and value (Berlage and Lanmo, 1982). Mechanical thinning has become particularly attractive over the last 15 years, because there are no reliable chemical thinners in peach and labour for hand thinning is becoming increasingly expensive.

To take full advantage of mechanical thinning techniques and avoid excessive damage, the machine must be adapted to the trees and, in addition, the trees must be adapted to the machine. Training systems may need to be modified, with the best results obtained on hedgerow training systems, where the reproductive structures are fully exposed to the action of the mechanical devices (Damerow and Blanke, 2008; Miller *et al.*, 2011; Schupp and Baugher, 2011; Theron *et al.*, 2015). However, thinning machines unavoidably inflict minor injury to the trees, with consequent formation of scar tissues, which generates some scepticism among growers and sometimes limits their acceptance. Thinning machines represent a significant cost for a single grower, so machine

Fig. 7.2. (a, b) A Darwin string thinner working on peach trees. Note the flowers falling under the action of the thinning brush. Photo courtesy of G. Costa.

thinning solution is more common within cooperatives or other growers' associations. Among the advantages of mechanical thinning are its compatibility with organic orchards, significant time savings, and its independence from both the genotype and the environmental conditions.

7.3.3   Chemical thinning

Chemical thinning is a promising alternative to manual operations for several fruit crops due to its lower cost and fast speed. In drupes like peach, however, chemical thinning is seldom used, because it provides unsatisfactory results, despite numerous thinning agents having been tested over the last 50 years. Nevertheless, much research is currently ongoing to find a chemical thinner that gives reliable results in peach and other stone fruits (Costa and Vizzotto, 2000; Byers *et al.*, 2003; Meland and Birken, 2010; Theron *et al.*, 2017; Costa *et al.*, 2018). Despite the lack of specific chemical thinners for peach, some considerations can be discussed based on several past approaches using diverse formulations at different phenological stages, from dormancy to bloom.

Thinning at dormancy
During bud dormancy, breaking agents can be applied to interrupt dormancy when the chilling requirement is not completely satisfied. Hydrogen cyanamide, applied approximately 40 days before the predicted bloom date, was

found to inhibit flower budburst and resulted in flower bud abscission (Fallahi *et al.*, 1990).

Thinning at bloom

Blossom thinners such as fertilizers, surfactants and desiccants (e.g. endothall, ammonium thiosulfate and long-chained fatty acids) have provided interesting results (Southwick *et al.*, 1996; Fallahi, 1997). Among these chemicals, ammonium thiosulfate is an effective flower thinner for peach and has been used in several countries without specific registration. Taken as a whole, thinners provide contradictory results, with some trials showing commercially unacceptable results, while other trials showed positive effects (Fallahi and Greene, 2010; Meland and Birken, 2010). The best results were achieved with a single application when bloom occurred over a short period. When bloom was prolonged, a single application was insufficient to achieve effective thinning. The economic impact of early thinning depends on several factors, particularly the probability of a subsequent spring frost.

Fruitlet thinning

Fruitlets can potentially be thinned by different bioregulators, few of which work well in peach. Since 1970, many ethephon trials for chemical fruitlet thinning in peach and other stone fruits have taken place in several countries. However, ethephon is not widely used because its efficacy is affected by internal and external factors, such as endogenous metabolic activity, leaf:fruit ratio, fruit load, temperature and humidity (Lavee and Martin, 1974).

Thinning by inhibiting flowering

The potential number of blooming flowers can be reduced by impairing bud break to limit the number of setting flowers, but also by inhibiting vegetative-to-reproductive phase transition during the previous season. Different hormone-like or caustic chemicals can achieve these effects. Among the former, gibberellic acid (GA_3) can inhibit flower bud formation in a wide variety of woody fruit trees, including peach (Zeevaart, 1983). GA_3 can be sprayed from full bloom up to mid-season to inhibit flowering (Byers *et al.*, 1990; Southwick *et al.*, 1996; Costa and Vizzotto, 2000). At the end of the 1990s, GA_3 produced interesting results in some cultivars and growing regions. A formula was registered in the USA under the commercial name of Release LC (Valent BioSciences). However, in regions where winter or spring frosts can damage flower buds, flower bud inhibition is not advisable. The number of flower buds can also be reduced up to 60% by applying some vegetable and petroleum oils during bud dormancy, at concentrations ranging from 75 to 110 l ha^{-1} (Deyton *et al.*, 1992; Myers *et al.*, 1996). These compounds are caustic agents and, under some circumstances, their phytotoxicity may strongly limit their application. These chemical strategies are not used widely due to the risk of cold damage during early fruit development.

7.4   FUTURE PROSPECTS FOR THINNING IN PEACH

Grower expectations of future technological advances in fruit thinning are enormous, especially for peach, which is particularly difficult to thin. Fruit thinning is one of the highest production costs for this crop, affecting its sustainability and potentially determining its future commercial retention or substitution with other crop species. These potential changes can affect the economy of a given production area and the entire supply chain, depriving consumers of a fruit that is beneficial for health. The combination of new peach cultivars, new thinning tools available to growers and ongoing climate changes that are already affecting fruit growing in several areas requires reorganizing fruit-thinning research to meet increasing industry demand by making significant progress in this challenging area.

New thinning tools for peach are under development by exploring all possible approaches. Modern developments include a 'mechatronic' thinning method, in which the precise position of each flower is identified using a computer-imaging program and a device carrying a series of nozzles then directs high-pressure air that removes a programmed number of flowers. However, the high cost of sophisticated robotic tools strongly discourages their commercial development and adoption, limiting them to demonstrations and experimental purposes. Traditional mechanical thinning solutions are undergoing improvements, although no machine, taken alone, can meet all the thinning requirements in peach. The most promising solution for peach would be chemical thinning. New compounds have been proposed as thinning agents: abscisic acid (ABA), a well-known plant growth regulator, and aminocyclopropane-1-carboxylic acid (ACC), the immediate precursor of the gaseous hormone ethylene. ACC has been recently released by Valent BioSciences under the commercial name of ACCEDE. ABA has been tested as a peach thinner at concentrations between 500 and 3000 ppm, applied at petal fall or at 10–15 mm fruit diameter. ACC, the ethylene precursor, has been tested in both peach (Ceccarelli *et al.*, 2016) and plum (Theron *et al.*, 2017). Preliminary results were interesting, but depended on cultivar, time of application and dosage. Further studies are in progress to optimize ACC use and clarify its mechanism of action to optimize the application strategy and control the phytotoxicity occurring with early applications and high dosages.

One possible approach to solve the thinning problem in peach may be to adapt strategies and decision-support systems (DSSs) used in other fruit crops. In apple, for example, fruits can be thinned using 4,6-dinitro-ortho-cresol (DNOC), naphthalene acetic acid (NAA), naphthalene acetic acid amide (NAAm) and 6-benzyladenine (BA) (Costa *et al.*, 2018), which are applied sequentially at specific phenological stages. With the chemicals already available, it may be possible to treat peaches at flower differentiation (gibberellic acid) or during bud dormancy (dormancy-breaking agents), evaluate their efficacy at bloom and still be able to spray surfactants as a blossom thinner, and, if this is still insufficient, thin again with ethephon and other new chemical thinners at the fruitlet stage. Adoption of such thinning strategies, however, requires predictable results at each step by genotype, time of application, dosage and environmental conditions before and after spraying. For this reason, DSSs must be adopted to choose the best application time and optimal concentration. The first studies assumed that time after full bloom could be used to establish the optimal time of application for a chemical thinner, but fruit growth kinetics proved more reliable as a phenological indicator, as it incorporates, to some extent, the influence of climate. Other models consider fruit dry-weight development, especially during stage II of fruit development (14 days after the onset of pit hardening), as it correlates well with the final fruit size and quality (Byers, 1989). Besides adopting the best developmental references, a DSS must also integrate the knowledge available in the specific field and, concurrently, stimulate research to provide innovative knowledge (i.e. new predictors) to enhance its reliability. Establishing research networks to join scientists and technicians in the peach thinning problem is necessary to achieve DSSs that work in the field, significantly improving fruit quality.

REFERENCES

Berlage, A.G. and Lanmo, R.D. (1982) Machine vs hand thinning of peaches. *Transactions of the American Society of Agricultural Engineers* 25, 538–543.

Bonany, J., Carbó, J., Echeverría, G., Hilaire, C., Cottet, V. *et al.* (2014) Eating quality and European consumer acceptance of different peach (*Prunus persica* (L.) Batsch) varieties. *Journal of Food, Agriculture & Environment* 12, 67–72.

Byers, R.E. (1989) Response of peach trees to bloom thinning. *Acta Horticulturae* 254, 125–132.

Byers, R.E., Carbaugh, D.H. and Presley, C.N. (1990) The influence of bloom thinning and GA_3 sprays on flowers bud numbers and distribution in peach trees. *Journal of Horticultural Science* 65, 143–150.

Byers, R.E., Costa, G. and Vizzotto, G. (2003) Flower and fruit thinning of peach and other *Prunus*. *Horticultural Reviews* 28, 351–392.

Ceccarelli, A., Vidoni, S., Rocchi, L., Taioli, M. and Costa, G. (2016) Are ABA and ACC suitable thinning agents for peach and nectarine? *Acta Horticulturae* 1138, 69–74.

Costa, G. and Vizzotto, G. (2000) Fruit thinning of peach trees. *Plant Growth Regulation* 31, 113–119.

Costa, G., Giulivo, C. and Ramina, A. (1983) Effects of the different flower/vegetative buds ratio on the peach fruit abscission and growth. *Acta Horticulturae* 139, 149–160.

Costa, G., Botton, A. and Vizzotto, G. (2018) Fruit thinning: advances and trend. *Horticultural Reviews* 46, 185–226.

Crisosto, C.H. (2002) How do we increase peach consumption? *Acta Horticulturae* 592, 601–605.

Crisosto, C.H. and Costa. G. (2008) Preharvest factors affecting peach quality. In: Layne and Bassi (eds) *The Peach: Botany, Production and Uses.* CAB International, Wallingford, UK, pp. 536–549.

Crisosto, C.H. and Crisosto. G.M. (2005) Relationship between ripe soluble solids concentration (RSSC) and consumer acceptance of high and low acid melting flesh peach and nectarine (*Prunus persica* (L.) Batsch) cultivars. *Postharvest Biology and Technology* 38, 239–246.

Crisosto, C.H. and Valero. D. (2008) Harvesting and postharvest handling of peaches for the fresh market. In: Layne, D.R. and Bassi, D. (eds) *The Peach: Botany, Production and Uses.* CAB International, Wallingford, UK, pp. 575–596.

Crisosto, C.H., Crisosto, G.M., Echeverría, G. and Puy, J. (2006) Segregation of peach and nectarine [*Prunus persica* (L.) Batsch] cultivars according to their organoleptic characteristics. *Postharvest Biology and Technology* 39, 10–18.

Damerow, L. and Blanke, M.M. (2008) A new device for precision and selective flower thinning to regulate fruit set und improve fruit quality. *Acta Horticulturae* 824, 275–281.

Delgado, C., Crisosto, G.M., Heymann, H. and Crisosto, C.H. (2013) Determining the primary drivers of liking to predict consumers' acceptance of fresh nectarines and peaches. *Journal of Food Science* 78, 605–614.

Deyton, D.E., Sams, C.E. and Cummins, J.C. (1992) Application of dormant oil to peach trees modifies bud–twig internal atmosphere. *HortScience* 27, 1304–1305.

Fallahi, E. (1997) Application of endothalic acid, pelargonic acid, and hydrogen cyanamide for blossom thinning in apple and peach. *HortTechnology* 7, 395–399.

Fallahi, E. and Greene, D.W. (2010) The impact of blossom and postbloom thinners on fruit set and fruit quality in apples and stone fruits. *Acta Horticulturae* 884, 179–187.

Fallahi, E., Kilby, M. and Moon, J.W. (1990) Effects of various chemicals on dormancy, maturity and thinning of peaches. In: *Deciduous Fruit and Nut: A College of Agriculture Report, Series P-83.* College of Agriculture, University of Arizona, Tucson, Arizona, pp. 121–128.

Iglesias, I. and Echeverría, G. (2009) Differential effect of cultivar and harvest date on nectarine colour, quality and consumer acceptance. *Scientia Horticulturae* 120, 41–50.

Lavee, S. and Martin, G.C. (1974) Ethephon (1,2-14C(2- chloroethyl)-phosphonic acid) in peach fruits II. Metabolism. *Journal of American Society for Horticultural Science* 99, 100–103.

Lurie, S. and Crisosto. C.H. (2005) Chilling injury in peach and nectarine. *Postharvest Biology and Technology* 37, 195–208.

Manganaris, G.A., Vincente, A.R., Martinez, P. and Crisosto, C.H. (2019) Postharvest physiological disorders in peach and nectarine. In: Tonetto de Freitas, S. and Pareek, S. (eds) *Postharvest Physiological Disorders in Fruits and Vegetable.* CRC Press, Boca Raton, Florida, pp. 253–264.

Meland, M. and Birken, E. (2010) Ethephon as a blossom and fruitlet thinner affects crop load, fruit weight and fruit quality of the European plum cultivar 'Jubileum'. *Acta Horticulturae* 884, 315–321.

Miller, S.S., Schupp, J.R., Baugher, T.A. and Wolford, S.D. (2011) Performance of mechanical thinners for bloom or green fruit thinning in peach. *HortScience* 46, 43–51.

Myers, R E., Dayton, D.E. and Sams, C. (1996) Applying soybean oil to dormant peach trees alters internal atmosphere, reduces respiration, delays bloom and thins flower buds. *Journal of American Society for Horticultural Science* 121, 96–100.

Myers, S.C., King, A. and Savelle, A.T. (1993) Bloom thinning of 'Wimblo' peach and 'Fantasia' nectarine with monocarbamide dihydrogensulfate. *HortScience* 28, 616–617.

Schupp, J.R., and Baugher. T.A. (2011) Peach blossom thinner performance improved with selective pruning. *HortScience* 46, 1486–1492.

Southwick, S.M., Weis, K.G. and Yeager, J.T. (1996) Bloom thinning Loadel cling peach with a surfactant. *Journal of American Society for Horticultural Science* 121, 334–338.

Theron, K.I., de Villiers, M.H.J. and Steyn, W.J. (2015) Is mechanical blossom thinning a viable alternative to hand thinning for stone fruit? *South African Fruit Journal June/July*, 72–73.

Theron, K.I., Human Steenkamp, H. and Steyn, W.J. (2017) Efficacy of ACC (1-aminocyclopropane-1-carboxylic acid) as a chemical thinner alone or combined with mechanical thinning for Japanese plums (*Prunus salicina*). *HortScience* 52, 110–115.

Zeevaart, J.A.D. (1983) Gibberellins and flowering. In: Crozier, A. (ed.) *The Biochemistry and Physiology of Gibberellins*, Vol. 2. Praeger, New York. pp. 333–374.

8

PEACH FRUIT GROWTH, DEVELOPMENT, RIPENING AND POSTHARVEST PHYSIOLOGY

Pietro Tonutti[1]*, Claudio Bonghi[2], Maria F. Drincovich[3] and Livio Trainotti[4]

[1]*Crop Science Research Center, Scuola Superiore Sant'Anna, Pisa, Italy;
[2]Department of Agronomy, Food, Natural resources, Animals and Environment, University of Padua, Agripolis – Legnaro (PD), Italy; [3]Centro de Estudios Fotosintéticos y Bioquímicos, Consejo Nacional de Investigaciones Científicas y Técnicas, Facultad de Ciencias Bioquímicas y Farmacéuticas, Universidad Nacional de Rosario, Rosario, Argentina; [4]Department of Biology, University of Padua, Padua, Italy*

8.1 INTRODUCTION

Peach (*Prunus persica* (L.) Batsch) is a stone fruit of agricultural relevance, not only because of its economic impact but also due to its hedonistic value, unique flavour, and importance as a source of phenolic compounds, cyanogenic glucosides and phyto-oestrogens. In addition, peach has become the reference species for the family *Prunus*, which also encompasses other fruits such as cherries, plums, apricots and almonds (Shulaev *et al.*, 2008). Peach is a fleshy fruit consisting of a single seed surrounded by a pericarp. The pericarp is differentiated into three layers: the endocarp, enclosing the seed, the mesocarp consisting of the soft edible region of the fruit, and the outermost exocarp or skin (Dardick and Callahan, 2014). Peach fruit is classified as a drupe as, during its development, the endocarp undergoes a hardening process with cells differentiating into sclereids with lignified secondary cell walls.

In recent years, there has been an increase in the application of molecular technologies in *Prunus* research: genetic and bioinformatic tools, including annotated whole-genome sequences, transcriptomes, and proteomic and metabolomic information have become essential in basic and applied research, with attention also focused on the dynamics driving fruit development, ripening and postharvest (The International Peach Genome Initiative, 2013; Farinati *et al.*, 2017).

* Email: pietro.tonutti@santannapisa.it

© CAB International 2023. *Peach* (G. Manganaris, G. Costa and C. Crisosto eds)
DOI: 10.1079/9781789248456.0008

This chapter describes the fruit development cycle starting from fruit set until ripening by joining historical information and new insights – the results of molecular and physiological studies carried out with innovative approaches – to provide a detailed picture of the processes characterizing peach fruit growth and maturation. This body of information is essential for understanding peach fruit behaviour at ripening, and for the development of new pre- and postharvest strategies for maintaining global fruit quality after fruit detachment from the tree.

8.2 FRUIT GROWTH AND DEVELOPMENT

8.2.1 Fruit set

The first step in peach fruit development is fruit set; this is established soon after fertilization. Flower fertilization is generally high (up to 90% of flowers) as a consequence of the large presence of self-pollinating cultivars (Szabò and Nyéki, 1999). This reproductive trait is not shared with other important stone-fruit crops such as apricot, sweet cherry and almond, in which the majority of cultivars are self-incompatible through a gametophytic self-incompatibility (GSI) mechanism (Abdallah *et al.*, 2020). The loss of GSI in peach is probably due to an association among mutational events, bottleneck domestication-related gene losses and population expansion (Abdallah *et al.*, 2020). Two ovules are present in the ovary, but generally only one is fertilized. Fertilization is necessary to initiate fruit set and development. In fleshy fruit such as peach, fruit set is usually regulated by the orchestrated action of three hormones, which, depending on the species/genotype, are auxins and/or gibberellins (GAs) and/or cytokinins, and is sustained by the availability of nutrients (calcium, boron, potassium and nitrogen) (Mariotti *et al.*, 2011). In peach, it is difficult to induce parthenocarpy and this method is not used as an agronomic practice, although it can be achieved by GA treatments, which can induce peach ovary development in the absence of fertilization (Crane *et al.*, 1960). With regard to nutrients, a deficiency of boron, particularly at flowering, results in severe pollination problems and poor fruit set. By helping in pollen germination and tube growth, boron can enhance fruit set (Ganie *et al.*, 2013).

8.2.2 The growth pattern

As for any drupe, peach fruit growth follows a double-sigmoid model. This model comes from mathematical analyses of growth rates based on diameter and fresh and dry weight measurements, which define four growth stages, S1–S4 (Fig. 8.1). Fruit development is genetically controlled, and this allowed the selection of cultivars with different harvest dates ranging from 60–70

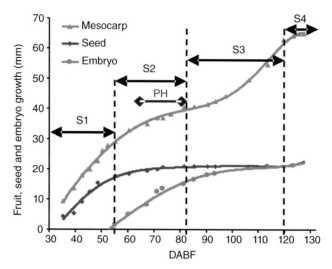

Fig. 8.1. Fruit and seed growth pattern in the cultivar 'Fantasia'. Fruit growth (orange triangles) is expressed as cross-diameter, while length is used for seed (brown diamonds) and embryo (green circles) development. The difference in length between the seed and embryo represents the endosperm, integuments and nucellus as a minimal part of the seed. The fruit developmental cycle is divided into four main stages (S1–S4) according to the first derivative of the fruit growth curve. The purple horizontal line indicates pit hardening (PH). DABF, days after full bloom. Adapted from Bonghi *et al.*, (2011).

(early-ripening) up to 160–170 (late-ripening) days after full bloom (DAFB). Early-ripening cultivars differ from mid-season ones (ripening 110–130 DAFB) in the length of S2 (i.e. the pit hardening stage), in which growth ceases and stone lignification occurs. In early-ripening cultivars, shortening of S2 leads to fruits with immature seeds at ripening, whose embryos need to be rescued with tissue-culture techniques to produce live offspring. Fruit growth rates are fast during S1 and S3, which are sustained mainly by cell duplication and expansion, or cell expansion only, respectively. In late-season cultivars, S3 lasts longer than in mid-season ones. In all genotypes, ripening occurs during S4, in which fruit growth is slow and sustained mainly by the latest event of cell enlargement due to swelling processes (Masia *et al.*, 1992).

Cytofluorimetric analysis has shown that in peach fruit up to four to five rounds of endocycles occur (Farinati *et al.*, 2021). The proportion of clearly endoreduplicated cells (sum of 8C–32C populations, where C represents the amount of DNA in picograms contained within a haploid nucleus) accounted for 39% during S4 (Farinati *et al.*, 2021), a much lower value than that observed in ripening tomato fruit, in which the number of endoreduplicated cells (sum of 8C–256C populations) was more than 85% of total pericarp cells (Cheniclet *et al.*, 2005). On the basis of these observations, the endoreduplication contribution to the final peach fruit size seems to be modest in comparison with that observed in tomato, mainly due to the low number of endoreduplicated cells. This result indicates that endopolyploid cell division would be inhibited in the last phase of peach fruit development. The accumulation in S4 fruit of a *P. persica* homologue of a GUANYLATE-BINDING PROTEIN1 (SlGBP1) (Prupe.4G053400, 70.6% identity), a tomato protein able to maintain endo-polyploid cells in a non-proliferative state (Musseau *et al.*, 2020), supports this view (Nilo-Poyanco *et al.*, 2021). In conclusion, the absolute number of cells seems to be the most important factor impacting peach final fruit size (Scorza *et al.*, 1991), with the influence of cell number on final fruit size accounting for around 60% (Souza *et al.*, 2019).

Most of the information on the control of fruit development and growth is related to hormone actions. Cytokinins and GAs are involved in early phases with ethylene and auxins, which are important also at ripening (Bonghi *et al.*, 2011). The role of the seed, and of the exchange of signals with the mesocarp, is a key aspect of fruit development (Crane *et al.*, 1960), but the nature and the timing of the movements of these signals are far from being understood (Bonghi *et al.*, 2011). Together with hormones, transcription factor proteins and regulatory RNAs, such as microRNAs and long non-coding RNAs, are also possible signalling molecules.

Fruit shape and size are under genetic control (Guo *et al.*, 2020). There are four main fruit shapes: (i) round (globose); (ii) elongated (oval); (iii) flat; and (iv) olecranon, with the latter found almost exclusively in China. The globose and oval shapes are controlled by the same monogenic trait, with elongated dominant over round but both recessive to flat (*S*). This trait in homozygosity (*SS*) induces sterility, as the fruit are aborted a few weeks after fertilization in *SS* individuals (López-Girona *et al.*, 2017; Aranzana *et al.*, 2019). The underlying gene is most likely *Prupe.6G281100*, encoding a leucine-rich-repeat receptor-like kinase. Homologues of this gene are well-known regulators of meristem size and activity (Hu *et al.*, 2018).

Genes involved in the control of fruit size through the regulation of cell pro-liferation have also emerged in quantitative trait locus analyses in several *Prunus* spp. Among such genes, worthy of mention is *PpCNR20* (*Prupe.6G240600*), which is similar to genes controlling cell number (de Franceschi *et al.*, 2013) in the carpel of various crops, such as tomato, aubergine and pepper (Guo and Simmons, 2011). In peaches, the cell number has been claimed to play a major role in establishing drupe sink strength (Yamaguchi *et al.*, 2004). This rela-tionship is crucial as peach is a species in which leaf-out is a late event, and therefore shoot apices can represent competitors for a growing fruit. In par-ticular, during the earlier phases of fruit growth, when cell division primarily occurs, the shoot apex has been demonstrated to be the strongest sink in peach (Génard *et al.*, 1998).

Besides the competition between growing shoots and fruit, *Prunus* spp. are characterized by the presence of an elevated fruit number, which also induces an intensification of competition within the fruit population (Morandi *et al.*, 2008; Costa *et al.*, 2018). The balance among different competing fruit is usually obtained by fruit or flower thinning, a practice carried out annually on peaches and nectarines (Costa *et al.*, 2018). Thinning by reducing the crop load affects the competition for assimilates among the fruit and increases the sink strength of the latter in comparison with the growing shoots (Costa *et al.*, 2018). Thinning strategies may be fine-tuned in terms of fruit position in the canopy or along the 1-year-old shoots, as well as in terms of timing, genotype and environmental conditions (Anthony *et al.*, 2020). Proper thinning avoids carbon starvation to developing fruit, resulting in a more pronounced change in secondary rather than primary metabolites. Fruit phenotypic parameters do not vary in young fruit but are statistically different in fruit at ripening, with fruit from thinned trees being larger and of better quality than those from unthinned ones. At the end of the growing phase, fruit weight is highly variable depending on the genetic background and the agronomic practices (e.g. thinning), and ranges from 80 to more than 500 g. Commercial standards in general require weights from 180 to 230 g. In terms of quality, higher photosynthate availability induced by thinning leads to metabolome changes in young developing fruit that can prime further developmental processes leading to better fruit. Catechin is the molecule that has the best positive correlation with improved fruit quality traits, while the unknown metabolite A244002, induced by stress conditions, also occurring in other plant systems, has the highest negative correlation (Anthony *et al.*, 2020).

Peaches differ from nectarines by their tomentose skin, a dominant trait over glabrous skin. The trait is likely to be controlled by the MYB *Prupe.5G196100* gene, whose correct expression is needed for the proper development of epidermal hairs on peach fruit, and which, in contrast, is impaired due to the insertion of a long-terminal-repeat retroelement in nectarines (Vendramin *et al.*, 2014). MYB transcription factors are important also in controlling the synthesis of anthocyanins, which are important in determining fruit colour in peach fruit skin, and other flavonoids, such as flavonols and catechins. The presence of anthocyanins is independent of the skin yellow or white ground colour. In peach fruit, anthocyanin synthesis, which is high in S1, is controlled at the transcriptional level by the well-known MYB–bHLH–WD40 (MBW) complex (Ravaglia *et al.*, 2013; Rahim *et al.*, 2014; Zhou *et al.*, 2014). In particular, *PpMYB10.1* has a crucial role in regulating the expression of structural genes responsible for the synthesis of the biosynthetic enzymes of the anthocyanin pathway (Ravaglia *et al.*, 2013; Rahim *et al.*, 2014). Anthocyanin content in the peel is maximum at S1, ceases during development and increases again at S4. Expression of the UDP-glucose:flavonoid-3-O-glycosyltransferase (*UFGT*) gene, encoding the key enzyme for their biosynthesis, follows the same pattern (Ravaglia *et al.*, 2013). Flavonol levels are also most abundant in the peel at S1, with flavonol synthase 1 (*FLS1*) expression following the same expression trend. With regard to proanthocyanidins, their levels are maximum at S3 and are abundant both in the peel and mesocarp. Regulation of the synthesis of these compounds is also at the transcriptional level, with anthocyanidin reductase *(ARN)* and leucoanthocyanidin reductase *(LAR)* expression being high in both peel and mesocarp from early S2 up to late S3 (Ravaglia *et al.*, 2013). Although anthocyanin accumulation in peach fruit is usually limited to the skin and the mesocarp portion surrounding the stone, cultivars with completely red flesh are becoming increasingly popular among consumers. Usually, *PpMYB10.1* expression is limited to the epicarp and the inner part of the mesocarp, but, in blood-red-fleshed cultivars it extends to the whole mesocarp, thus switching on anthocyanin synthesis in all cells of the pericarp. The blood-flesh *(BF)* trait that causes the misregulation of *PpMYB10.1* expression is caused by a mutation in the Prupe.5G006200.1 locus, encoding the *BLOOD (BL)* gene (Zhou *et al.*, 2015). *BL* encodes an NAC-domain transcription factor that, interacting with other members of the NAC (NAM, ATAF and CUC) family, suppresses *PpMYB10.1* expression. In the *BF* mutant, which is missing BL, *PpMYB10.1* expression is not suppressed in the mesocarp, thus leading to anthocyanin accumulation in the whole pericarp.

8.3   THE RIPENING SYNDROME

Ripening is the summation of biochemical and physiological changes that occur at the terminal stage of fruit development and render the organ edible and desirable to seed-dispersing organisms. It is a complex physiological syndrome, characterized by chemical and physical processes involving switching off the fruit developmental programme and turning on the initiation of new gene-expression programmes causing the ripening changes (Seymour *et al.* 2013). Peach fruit undergo ripening after reaching the maturation stage (end of S3/beginning of S4) (Tonutti *et al.*, 1991). The main biochemical and structural changes having – directly or indirectly – the most impact on the overall quality and shelf life are:

- increased respiration and ethylene biosynthesis;
- loss of flesh firmness and pulp structural changes due to the depolymerization and solubilization of cell-wall components and the reduced cell turgor;
- changes in taste and overall flavour due to the increase in mono- and disaccharide concentration, the biosynthesis of volatile compounds and the decrease in organic acids;
- turnover of the flesh and skin colour due to the degradation of chlorophyll and accumulation of carotenoid and/or flavonoids; and
- the increase in susceptibility to pathogens to which the fruit responds by accumulating defence proteins, which may be responsible for some allergic reactions in sensitized individuals.

8.3.1   Respiration and ethylene physiology

Peach fruit is classified as climacteric. An increase in respiration and ethylene biosynthesis is observed during ripening, and some of the ripening-related changes are ethylene dependent. Considering respiration, and compared with other species, the average rates are considered low/moderate, ranging from about 15 to 80 ml CO_2 kg^{-1} h^{-1} according to genotype and temperature (Wills *et al.*, 2007). In some genotypes, the respiratory climacteric rise is not always clearly detectable (Brovelli *et al.*, 1999). In general, at their climacteric peak, peaches do not produce high amounts of ethylene when compared with other climacteric fruits, but it must be stressed that ethylene is physiologically active at very low concentrations (0.1–1.0 µl l^{-1}), which are enough to activate ethylene-dependent processes (e.g. pulp softening, synthesis of specific pigments, ethylene autocatalysis) in climacteric fruits that have developed the competence to ripen and have sensitivity to the hormone.

The rate of ethylene biosynthesis at ripening is highly variable in relation to the genotype and phenotype as observed when comparing melting-flesh (MF) and non-melting-flesh (NMF) peaches, with the latter producing, in general, higher levels of ethylene (Brovelli *et al.*, 1999; Ghiani *et al.*, 2011). In contrast to other climacteric fruit species (e.g. tomato), in peach fruit the ethylene biosynthetic peak, concomitant with an increase in respiration rate (Su *et al.*, 2017), occurs at an advanced stage of ripening, which in melting types corresponds to the rapid decrease in flesh firmness (Tonutti *et al.*, 1991, 1997). Thus, peaches are listed among those fruit species in which the ethylene climacteric stage coincides with or follows eating ripeness (roughly corresponding to flesh firmness values of about 25–40 N) and, in melting types, large amounts of ethylene are produced concomitantly with very low flesh firmness values (Tonutti *et al.*, 1996).

The enhancement of ethylene biosynthesis at ripening has been associated with the upregulation of both the ACC synthase (*ACS*) and ACC oxidase (*ACO*) genes, and the increase in the corresponding enzyme activities is responsible for the production of 1-aminocyclopropane-1-carboxylic acid (ACC) and its oxidation to ethylene, respectively. Of the six *PpACS* genes identified in the peach genome, only *PpACS1* shows ripening-related increased expression, thus representing a key player in the determination of ethylene production at the ripe stage (Tatsuki *et al.*, 2006; Zheng *et al.*, 2015). The key role played by *PpACS1* in peach ethylene biosynthesis has been clearly determined following molecular and physiological characterization of the stony-hard (SH) peach type, which exhibits suppressed ethylene production (Haji *et al.* 2001). This behaviour is the result of low *PpACS1* expression during the ripening stage, leading to reduced production of ACC (Tatsuki *et al.*, 2006). Exogenous application of ACC is effective in restoring the ethylene production in ripening SH peaches (Hayama *et al.*, 2008). Interestingly, the suppression of *PpACS1* expression at the late-ripening stage of SH peach fruit has been linked to the low levels of auxins (Tatsuki *et al.*, 2013), and the presence of interdependent mechanisms between auxin and ethylene actions modulating ethylene biosynthesis and peach ripening has been reported (Trainotti *et al.*, 2007; Tadiello *et al.* 2016). Recently, the application of an antibody chip to screen differentially expressed proteins during peach ripening has provided further insights into the interaction between auxin and ethylene attributing to *YUCCA* genes, involved in the indole-3-pyruvic acid pathway, a crucial role in the production of auxin during the climacteric phase (Zeng *et al.*, 2020).

The ripening of peach fruit is strictly associated with the upregulation of *ACO* gene expression. In particular, *PpACO1* mRNA accumulation can be considered a solid molecular marker of the evolution of ripening and the ethylene climacteric rise in peach (Ruperti *et al.*, 2001). Expression of this gene is strongly induced by exogenous ethylene (Rasori *et al.*, 2003), highlighting its role in ethylene autocatalysis in ripening peaches. The responsiveness to ethylene has been related to the presence of ethylene-responsive elements in the promoter region of *PpACO1* (Rasori *et al.*, 2003). The upregulation of *PpACO1* gene transcription is paralleled at ripening by the accumulation of ACO protein (Nilo-Poyanco *et al.*, 2021) and enhancement of enzyme activity (Tonutti *et al.*, 1997).

Ethylene is perceived by specific receptors (ETR and ERS) that, in contrast to other fruit species, have been only partially identified and characterized in peach. Based on transcript accumulation patterns, *PpERS1* and *PpETR2* appear to be more involved than *PpETR1* during peach fruit ripening. In fact, while *PpETR1* shows a constitutive expression pattern and appears to be ethylene independent, *PpERS1* and *PpETR2* transcript accumulation increases during peach fruit ripening, being upregulated by ethylene and downregulated by a continuous application of 1-methylcyclopropene (1-MCP), a competitor of ethylene for its binding sites (Rasori *et al.*, 2002; Dal Cin *et al.*, 2006; Tadiello *et al.*, 2016). Controversial information is reported in the literature concerning the effect of 1-MCP on peach fruit ripening, with effects varying depending on the genotype, concentration of the chemical, time and number of applications. It is, however, generally recognized that the effect of 1-MCP on delaying ripening (flesh softening, in particular) of peaches is much less pronounced compared with other fruit types (e.g. apples), and the chemical is not used in practice as a postharvest treatment on peaches. The observed limited efficacy of 1-MCP on peach fruit (the effect of the ethylene inhibitor rapidly disappears after moving the fruit to air) has been attributed to specific features in terms of ratio, expression pattern and/or turnover of the ethylene receptors (Dal Cin *et al.*, 2006; Ziliotto *et al.*, 2008) and on its effects in stimulating auxin synthesis (Tadiello *et al.*, 2016).

8.3.2 Flesh firmness loss

Flesh firmness loss represents the most limiting quality attribute in the harvest, storage and transportation system of peaches, probably more than in several other perishable fruit species. Gallardo *et al.* (2012) reported that firmness was one of the most important characteristics and priorities for US peach breeders. Several flesh types with specific patterns of firmness loss and flesh texture properties are found within current peach cultivars and breeding (Sandefur *et al.*, 2013).

Besides MF and NMF, SH, slow-melting, slow-ripening and non-softening flesh types have also been identified and are used or available as genetic resources and for breeding programmes. The fresh-market peach industry has been and continues to be dominated by peaches of the MF type in which flesh firmness decreases slowly at the beginning of ripening (softening stage), while rate of the loss of firmness accelerates in the later ripening phase (melting stage). NMF peaches, which maintain higher firmness values at advanced ripening, are characterized by the lack of the melting stage. The loss of firmness, a genetically programmed ripening-related event, is a consequence of the decline in both cell-wall strength and cell-to-cell adhesion that, together with the turgor pressure decrease, leads to flesh softening and an increase in juiciness (Toivonen and Brummell, 2008). Cellulose, hemicellulose and pectins make up 90–95% of the structural components of the cell wall and middle lamella, with the mechanical/elastic properties of the former determined by cellulose and hemicellulose, while the cell-to-cell adhesion is mainly determined by pectins, a major constituent of the middle lamella.

Several biochemical events and processes (e.g. solubilization, depolymerization) occur, and different enzymes, some of them ethylene dependent, are involved in the loss of flesh firmness in peaches. Ripening-related increases have been observed in the gene expression and activities of exo- and endopolygalacturonase (EC3.2.1.67 and EC3.2.1.15), pectin methylesterase (EC3.1.1.11), endo-1,4-β-glucanase (EC3.2.1.4), endo-1,4-β-mannanase (EC3.2.1.78), α-arabinosidase (EC3.2.1.55) and β-galactosidase (EC3.2.1.23) (Trainotti *et al.*, 2003; Brummell *et al.*, 2004).

Considering the complexity of the cell-wall structure and composition, and the number (and activity profiles) of the enzymes, it is unlikely that a single enzyme is responsible for all textural changes. A slight decrease in cellulose content occurs during ripening in most fruits, although this event is often uncoupled from the increase in crystalline cellulose in ripe fruit (Posé *et al.*, 2018). Hemicelluloses (xyloglucans) are the substrate of the hydrolytic endo-1,4-β-glucanase (EG) activity, resulting in a weakening of the cell-wall structure and the enhancement of cell separation phenomena, characterizing ripening. The increase in both EG gene expression (*PpEG1*) and enzyme activity has been observed in the initial phase of peach fruit softening and before the onset of the ethylene climacteric stage (Bonghi *et al.*, 1998). Pei *et al.* (2019) also reported the upregulation of xyloglucan endotransglucosylase/hydrolases (XTHs), responsible for the reduction in mass of wall-bound xyloglucans and consequently the increase the cell-wall extensibility. Worthy of note is the fact that the action of XTHs is induced by xyloglucan oligosaccharides (XGOs) and that, during peach fruit ripening, the downregulation of two GDSL-type esterase/lipase proteins (GELPs) known for their action against XGOs has been reported (Pei *et al.*, 2019). Pectins are the cell-wall components that show the most important changes during ripening, but their role in fruit firmness and softening is still controversial (Paniagua *et al.*, 2014). Depolymerization of pectins during ripening is largely due to a sequential and coordinated action of several pectin-metabolizing enzymes (Morgutti *et al.*, 2006). In this context, endopolygalacturonase (endo-PG) plays a central role in the depolymerization of cell-wall pectins of peach fruit: endo-PGs, which show the highest levels of transcript accumulation and enzyme activity corresponding to advanced ripening (melting stage) and the onset of the ethylene climacteric stage, are essential for the achievement of MF fruit texture due to the loss of cell adhesion, but not for reducing fruit firmness. Cell-wall swelling, occurring as a result of changes to the cell wall during pectin solubilization, most likely represents a main event in MF loss of firmness. In fruits with a crispy texture at ripening (e.g. some apple cultivars), cell-wall swelling is not or is limitedly occurring. This is also the case for NMF peaches in which, due to a mutation in the specific gene, no endo-PG was detected and consequently no loss of cell adhesion was observed (Morgutti *et al.*, 2006). On the basis of these observations, the role of endo-PG activity in the reduction of fruit firmness has been debunked because NMF peaches are able to soften, and, at the same time, a change of symplast/apoplast water status has been suggested as another important mechanism through which peach fruit firmness is regulated. It is noteworthy that monomers (d-galacturonic acid) of pectin depolymerization can enter the biosynthetic pathway of ascorbate (vitamin C) and contribute to the nutritional value of peaches (Canton *et al.*, 2020).

The levels of cell-wall hydrolases change accordingly with the temporal variation in transcription of the corresponding genes of these enzymes, as has been demonstrated in several transcriptomic analyses (Trainotti *et al.*, 2006; Pan *et al.*, 2016; Pei *et al.*, 2019). The different activity patterns of the main hydrolases throughout ripening show the complexity of the loss of firmness process in peaches and the different regulatory roles played by ethylene (Fig. 8.2). Based on a number of studies, it is evident that some cell-wall-related genes are insensitive to ethylene, while others are down- or upregulated. Endo-PGs belong to the latter category, as clearly demonstrated in SH peaches in which the suppression of ethylene production is coupled with a lack of PG mRNA accumulation and the maintenance of high flesh firmness values. By treating SH peaches with exogenous ethylene, PG transcripts are produced and the softening process takes place (Hayama *et al.*, 2006). Together with *PpACO1*, representing a useful marker for the ethylene climacteric rise (Trainotti *et al.*, 2003), the induction of *PpendoPG* expression can be used to detect the transition from the softening to the melting stage of MF ripening peaches (Begheldo *et al.*, 2006).

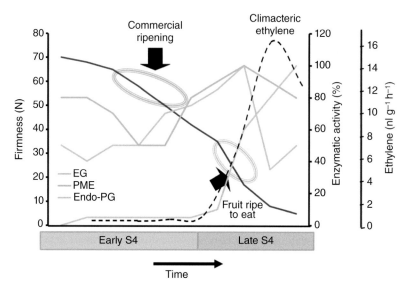

Fig. 8.2. Relationship between the dynamics of firmness loss (red line) of peach fruit and the activities of the main enzymes (endo-β-(1,4)-glucanases (EG), pectin methylesterases (PME) and endopolygalacturonase (endo-PG) involved in de-polymerization of structural polysaccharides of the cell wall. It is noteworthy that the activity of endo-PG increases significantly when the fruit is very soft (<20 N), corresponding to the ethylene production peak (dashed black line). Adapted from Bonghi et al. (2019).

8.3.3 Changes in taste and overall flavour

Sugar and organic acids metabolism

Peach fruit flavour is highly dependent on sugar concentration in the flesh and, more precisely, on the balance between soluble sugars and titratable acidity (Cantin *et al.*, 2009). An increase in sugars, measured by evaluating the soluble solids concentration (SSC), a parameter frequently used to determine harvest time, is also a ripening-related process in peach. There is a general consensus that an SSC reading of at least 12% is required to ensure that peaches are acceptable to most consumers. The total sugar level in the ripe mesocarp of different peach genotypes may vary from about 35.0 to more than 100 g kg⁻¹ of fresh weight, as reported by Cantin *et al.* (2009), who analysed 14 peach and nectarine breeding progenies and 205 seedlings of a breeding population. The sugars in peach flesh are represented by sucrose, glucose, fructose and sorbitol, accounting for up to 60% of fruit dry mass at full maturity (McGlasson *et al.*, 2009), with sucrose being the main carbohydrate accumulated in ripe peaches (Vizzotto *et al.*, 1996). Generally, fructose and glucose are present in equal amounts, and the proportion of these three main sugars is 3:1:1. Sorbitol also contributes to the sugar bulk of ripe peaches but at lower levels than glucose and fructose. Genotype-dependent differences are present in terms of absolute amounts and the relative proportions of these sugars (Walker *et al.*, 2020), with, in general, early-maturing cultivars characterized by lower SSCs than mid- to late-season fruits (McGlasson *et al.*, 2009). The sugar metabolism and accumulation in peaches during ripening is due to the involvement of a number of enzymes responsible for the processes of assimilate import, metabolism and dilution, with relationships among them described in the SUGAR model developed by Génard *et al.* (2003). The differing regulation and intensity of these processes determines the variable levels of sugars accumulating among different crops (grapes, tomato and peaches) and within the same species, with peach cultivars characterized by a large variability in sugar content (Cantin *et al.*, 2009; Dai *et al.*, 2016). This, together with the environmental conditions and management practices effective in modifying the sink strength and overall fruit metabolism, may also explain the marked differences observed in peach genotypes with regard to the fructose:glucose ratio (Desnoues *et al.*, 2014). Sucrose and sorbitol, the two main transportable photosynthates in *Prunus* spp. (Rennie and Turgeon, 2009), are the sugars imported by the fruit. Sucrose accumulates rapidly during the last days of maturation on the tree, and this sugar is hydrolysed into fructose and glucose by different enzymes active in the cytosol (sucrose synthase and neutral invertase), in the cell wall (neutral invertase) and the vacuole (acid invertase). Glucose and fructose accumulate at a nearly constant rate during the last stages of development, but the ratio between these two sugars falls to below 1 at full ripening, indicating that glucose is preferentially used for respiration. Sorbitol could be used to generate fructose by the activity of sorbitol dehydrogenase, and to glucose via sorbitol oxidase. These sorbitol-derived sugars, in turn, can be used to synthesize sucrose (Cirilli *et al.*, 2016). This view is supported by a proteomic analysis in which the enzymes responsible for the conversion of sorbitol into sucrose were induced in the transition from the unripe to the ripe stage (Nilo-Poyanco *et al.*, 2021). Glucose and fructose can be transported/stored in the vacuole and undergo a number of different reactions (phosphorylation, dephosphorylation and interconversion) in the cytosol. Processes such as import, subcellular compartmentation, transport, cleavage and resynthesis, regulated at the level of gene expression (Aslam *et al.*, 2019), are involved in the dynamic modifications of the sugar pool in the mesocarp throughout ripening (Walker *et al.*, 2020). The increase in the content of some other minor soluble sugars (raffinose, fucose and xylose) observed in peach ripe fruit (Monti *et al.*, 2016) is accompanied by an accumulation of the key enzymes involved in their biosynthetic pathways (Nilo-Poyanco *et al.*, 2021). In addition to the effect of genotype, high variability in terms of sugar accumulation and metabolism is also present within the population on the same tree due to inter-fruit competition. One of the most important factors affecting sugar accumulation in ripening peaches is fruit sink demand, activity and utilization (Basile *et al.*, 2007). Fruit-to-fruit competition, fruit position on the tree and fruit metabolism in relation to environmental conditions (sunlight exposure and temperature, also modulated by canopy management practices) are all factors that affect sink strength and overall fruit metabolism, and, consequently, the sugar accumulation pattern at ripening.

Together with soluble sugar content, organic acid concentration is a main driver of the flavour of ripe peaches. Acidity correlates with taste and aroma perception: sweetness is influenced by the sugars:organic acids ratio, while aroma correlates with total acidity and, in particular, malic acid content (Colaric *et al.*, 2005). A major difference exists between normal-acid (NA) and low-acid (LA) cultivars in terms of organic acid accumulation and titratable acidity pattern during fruit growth and development, and the juice pH at ripening, with the LA cultivars characterized by values higher than pH 4, and the NA cultivars with values below pH 3.8 (Dirlewanger *et al.*, 1998). In NA cultivars, after the initial (S1) increase, a marked enhancement of organic acid accumulation occurs in late S3, followed by a decrease characterizing ripening. Compared with NA cultivars, a more marked ripening-related reduced content is observed in LA cultivars (Zheng *et al.*, 2021), in which the S3 increase in organic acid is not present (Moing *et al.*, 2000). Although high diversity for acidity-related traits is present in peach genotypes, ten organic acids (oxalate, *cis*-aconitate, citrate, tartrate, galacturonate, malate, quinate, succinate, shikimate and fumarate) are, at different ratios, commonly present in peach fruit cultivars at the physiological maturity stage, with malate (dicarboxylic acid) and citrate (tricarboxylic acid) together giving the highest contribution and accounting, in a collection of more than 200 accessions, for 62% and 23% of total organic acids, respectively (Baccichet *et al.*, 2021). The pattern of organic acid accumulation changes at maturation, and ripening is highly variable, depending more on the genotype (e.g. NA versus LA cultivars) than agronomic practices and environmental conditions.

In addition, changes and patterns are in relation to the specific organic acid considered and the ratio between the two main organic acids, malate and citrate. In ripening NA cultivars, increases in citric acid have been reported (Lombardo *et al.*, 2011; Monti *et al.* 2016) with more pronounced trends in NA cultivars in which citrate is the main organic acid compared with cultivars with malate as the predominant organic acid (Nowicka *et al.*, 2019; Zheng *et al.*, 2021). In contrast, in LA cultivars, the transition from the pre-climacteric to the climacteric stage is characterized by a reduction in citrate (Zheng *et al.*, 2021). The different patterns of malic acid accumulation and metabolism in peach cultivars appear to be strictly related to the variable capability of the cells to activate proton pumps and hence sequester organic acid in the vacuoles, making them more or less accessible to catabolic enzymes (Etienne *et al.*, 2002). The relationship between malate accumulation and vacuolar functioning has been modelled in peach fruit (Lobit *et al.*, 2006). Zheng *et al.* (2021) report that both storage in the vacuole and metabolism contribute to the malate concentration in peach fruit: in addition to decarboxylation in the cytosol and conversion of tri- and dicarboxylates in the mitochondria or glyoxysomes, malic acid is also depleted due to the activation of gluconeogenesis, as demonstrated by the accumulation of specific proteins involved in this process during ripening (Nilo-Poyanco *et al.*, 2021).

Volatile aroma compounds
Peaches develop a set of volatile organic compounds during the transition from the unripe to the ripe stage, with differences among genotypes, as observed in yellow- versus white-fleshed cultivars, with the latter appreciated for their pronounced and distinct aroma. The observed differences between the aroma of white- and yellow-fleshed cultivars may in part be explained by the differential activity of carotenoid cleavage dioxygenase enzymes, which is higher at late-ripening stages in white-fleshed cultivars, inducing not only changes in pigmentation but also an accumulation of the aroma class of norisoprenoids (Brandi *et al.*, 2011). In unripe peaches, most of the compounds that impart the 'green' aroma are lipid-derived C6 compounds such as *n*-hexanal, (E)-2-hexenal, (E)-2-hexenol, and (Z)-3-hexenol, which decrease during ripening. Correlation analyses have indicated that lipid-derived aldehydes traditionally described as green aromas could be related to immature fruit. For example, the volatile compound 2,4-heptadienal, (E,E) (green/fatty odour) shows the highest correlation with flesh firmness, indicating that it accumulates in immature fruit (Sánchez *et al.*, 2012). The production of esters and lactones, such as (Z)-3-hexenyl acetate, γ-hexalactone, γ-octalactone, γ-decalactone, and δ-decalactone increases with fruit ripening, giving the typical peach aroma. Lactone synthesis shows a clear pattern concomitant with the climacteric rise in ethylene production, with γ-decalactone being the principal volatile compound at the late-ripening stage (Zhang *et al.*, 2010). At least some part of the development of peach aroma is controlled by ethylene. Applications of 1-MCP result in higher levels of C6 compounds (hexenals and hexenols) and lower amounts of esters (Mencarelli *et al.*, 2003), and an altered expression of genes involved in fatty acid metabolism (Ziliotto *et al.*, 2008). Lactones have shown a strong positive correlation with colour and a negative correlation with flesh firmness, suggesting that they are associated with fruit quality and the evolution of ripening (Sánchez *et al.*, 2013). In specific cultivars, the production of terpenoids such as linalool and terpinolene has been associated with a more intense floral and fruity aroma (Visai and Vanoli, 1997).

Flesh and skin colour changes
One of the most important visual quality parameters of peaches is skin colour, which, together with flesh pigmentation, represents a trait that breeders have recently devoted particular attention to. The red coloration is determined by the content and composition of anthocyanins, while the yellow colour is associated with carotenoids (Falchi *et al.*, 2013; Yan *et al.*, 2017). Both skin and flesh colour change during ripening, and these changes are associated with

the degradation of chlorophylls and the consequent unmasking of previously existing carotenoid pigments as well as the new synthesis of carotenoids, responsible for the ground colour, a parameter used for determining ripening evolution and harvest time. An integrated analysis of the transcriptome and metabolome has been carried out in *P. mira*, a wild peach, in three fruit types with various flesh pigmentations (milk-white, yellow and blood) (Ying *et al.*, 2019). The three fruit types displayed diverse transcriptional activities, indicating that differences in fruit flesh pigmentation occurred mainly during the ripening stage.

A general increase in carotenoids, but particularly those characterized by the β configuration, is present in yellow-fleshed ripening peaches, and this is the result of a coordinated upregulation in the transition from S3 to S4 of the phytoene synthase, ζ-carotene desaturase, lycopene β-cyclase and β-carotene hydroxylase genes (Trainotti *et al.*, 2006; Cao *et al.*, 2017), with some showing ethylene-dependent expression (Ziliotto *et al.*, 2008). Together with β-carotene, xantophylls such as zeaxantin and β-cryptoxanthin, synthesized via hydroxylation, represent the main carotenoid pigments in yellow-fleshed peach cultivars, with differences, in terms of total amount and colour intensity, depending on the geno/phenotype, as observed in melting versus non-melting peaches (Karakurt *et al.*, 2000), with the latter, characterized by higher levels of ethylene production, showing more intense accumulation of carotenoids. In addition to the upregulation of the carotenogenic genes, β-carotene and xantophyll accumulation in yellow-fleshed cultivars is related to the downregulation of carotenoid cleavage dioxygenase 4 (CCD4), responsible for the degradation of these pigments. In white-fleshed peaches, this degradative enzyme is active, resulting in reduced accumulation of carotenoids. The white-flesh phenotype (*Y*) is dominant to the yellow-flesh phenotype and has arisen from various ancestral haplotypes by at least three independent and molecularly different mutational events. The *Y* gene encodes CCD4 (PpCCD4; Prupe.1G255500), responsible for carotenoid degradation (Falchi *et al.*, 2013).

Red-skinned peaches have been the objective of the majority of breeding programmes in recent decades. High levels (quantitative and qualitative) of red pigments (anthocyanins) have been sought in cultivars for the fresh market. Studies performed on the factors affecting the red colour development have found that this trait is genetically controlled and that environmental factors in the field such as light and temperature also play an important role in this trait (Andreotti *et al.*, 2008). The main anthocyanins accumulating in the skin of ripening peaches are cyanidin 3-glucoside and cyanidin 3-rutinoside, and their synthesis is the result of the upregulation of specific genes (*LDOX* and *UFGT*) of the flavonoid pathway (Ravaglia *et al.*, 2013). The *UFGT* gene is also highly induced by postharvest treatments with UV and white light, resulting in a marked increase in the coloration of the peel (Ravaglia *et al.*, 2013; Zhao *et al.*, 2017). Genes involved in the early steps of the anthocyanin biosynthetic pathways are highly expressed in the mesocarp of the blood-flesh cultivars.

Transcriptional regulators of genes involved in the control of genes responsible for flavonoid biosynthesis are transcription factors belonging to the MYB, bHLH and WD40 family (called the MBW complex) in which the role of MYB seems to be central for peach fruit (Rahim *et al.*, 2014; Ying *et al.*, 2019).

8.4   POSTHARVEST PHYSIOLOGY IN RELATION TO COLD STORAGE

8.4.1   Metabolic reconfiguration during postharvest refrigeration

Peaches are highly perishable and deteriorate quickly after harvest if stored at ambient temperatures. Thus, prolonging commercial/shelf life and shipping peaches to both internal and distant markets require storage at low temperature. Low temperatures delay ripening and softening, reduce the enzymatic and microbial activity, and slow down the respiration of the fruit. Storage just above the freezing point (0°C) is the most common protocol applied to peaches.

The postharvest ripening process produces several chemical changes in postharvest peaches, which affect their taste and overall flavour. The cold storage after harvest slows down the ripening process, which is restored when the fruit is returned to an ambient temperature. Although cold storage is beneficial for commercialization because of the delay in ripening, softening and decay, it switches on a complex response programme in the fruit, which results in a global reconfiguration of the metabolome, and this may also affect the subsequent ripening process. The biochemical reconfiguration imposed by cold storage impacts the ripening process and may result in changes in the organoleptic properties and overall flavour of the fruit. The reconfiguration of the metabolome by postharvest cold storage has been studied by untargeted metabolomic approaches in several different peach cultivars (Wang *et al.*, 2013; Lauxmann *et al.*, 2014; Bustamante *et al.*, 2016; Brizzolara *et al.*, 2018), revealing the main chemical modifications in the fruit by cold storage (Fig. 8.3).

Sugars, key determinants of the organoleptic quality of the fruit, are drastically modified during peach cold storage. The study of different peach cultivars has revealed that, although several sugar modifications are specific for one or a group of peach cultivars, other sugar changes are common to all the peach cultivars analysed (Bustamante *et al.*, 2016; Brizzolara *et al.*, 2018). One of the sugars that consistently increases due to postharvest cold treatment is raffinose. Raffinose, a trisaccharide composed of galactose, glucose and fructose, is a component of the cold acclimation response in several different plant species (Fürtauer *et al.*, 2019). It has been clearly shown that raffinose protects thylakoid membranes and acts as a potential reactive oxygen species (ROS) scavenger (Nishizawa *et al.*, 2008), but the particular role in the mesocarp of peach deserves further anaylsis. Other sugars such as sucrose, glucose, fructose, isomaltose, trehalose, glucose 6-phosphate, fucose and xylose are also

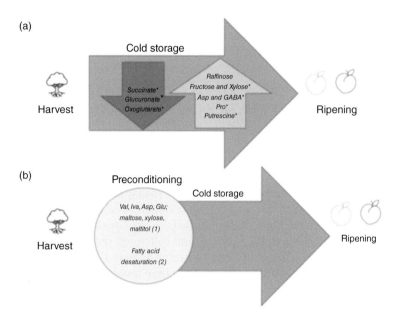

Fig. 8.3. Schematic metabolic rearrangements in peach fruit during cold storage. (a) The most widespread metabolic modifications when peach fruit are stored at low temperatures after harvest are indicated. Compounds marked with an asterisk have been detected in some but not all of the peach cultivars analysed. The biochemical modifications are also dependent on temperature and length of storage. (b) Conditioning treatments, for example 48 h at 0°C (1) or 5 days at 8°C (2) before cold storage, modify the metabolic content to prepare the fruit to cope more effectively with cold storage. Some of the metabolic modifications found in these preconditioning treatments are indicated. All of these metabolic modifications impact the way peach fruit ripen after being transferred to an ambient temperature and on the development (or not) of chilling injury symptoms. Asp, aspartic acid; GABA, γ-aminobutyric acid; Glu, glutamic acid; Iva, isovaline; Pro, proline; Val, valine.

modified as a result of cold treatment of peach fruit. However, the modifications in the levels of these sugars are dependent on the cultivar, the extent of the cold treatment and the temperature of storage. In general, longer cold treatments involve increases in a larger number of sugars and to a greater extent (Bustamante *et al.*, 2016). Sugar modifications are also dependent on the temperature of storage (Brizzolara *et al.*, 2018). For example, fruit stored at 0°C displays higher levels of sucrose than fruit stored at 5°C, which is linked to the higher chilling tolerance at 0°C than at 5°C (Wang *et al.*, 2013). Sugar alcohols such as galactinol or sorbitol are also modified by cold treatment.

The dysfunction of cell membranes at low temperature is one of the major molecular events leading to the development of cell injury. Membrane fluidity, which is essential for the integrity of the cell, is strongly influenced by lipid composition. Thus, it is expected that cold storage of peach fruit will greatly modify the lipid composition of the fruit. The overall changes in the lipidome content of different peach cultivars by cold storage were studied by untargeted approaches (Bustamante *et al.*, 2018). Cold storage induced several different changes in the peach lipidome, which were dependent on both the extent of the cold storage and the peach cultivar. One of the most extended lipid modifications was related to changes in the content of galactolipids, which are linked to plastidic membranes. Adjustments of phospholipids and sphingolipids were also found in peach fruit in response to ethylene (Chen *et al.*, 2021) and other hormones (Chen *et al.*, 2019), when these hormones were applied to induce cold tolerance.

Untargeted metabolomic approaches have revealed that the metabolism of other organic compounds is also altered during postharvest cold storage (Fig. 8.3a). The decrease in 2-oxo-glutarate and succinate levels was linked to postharvest metabolization processes under cold storage. Amino acids such as γ-aminobutyric acid (GABA), proline and aspartic acid were generally increased in the different peach cultivars due to cold storage, while other amino acid modifications were more cultivar-dependent. Putrescine, a polyamine, also increased during cold storage in some peach cultivars and to different extents depending on the temperature of storage. Regarding volatile organic compounds, which are key determinants of the flavour of the fruit, it was found that aldehydes and alcohols generally accumulated more under cold storage (Brizzolara *et al.*, 2018).

8.4.2 Peach fruit ripening after cold storage: chilling injury metabolic alterations

As stated above, cold storage of peach fruit induces an overall metabolic reconfiguration in order to preserve the cell function and integrity of the cell membranes and macromolecules to avoid cell damage. Several different metabolites that are modified during peach cold storage display known roles in the defence against chilling injury (CI). In the case of sugars, they serve as osmoprotectants of biological membranes and macromolecules, as well as having roles in signalling processes (Tarkowski and van den Ende, 2015). Lipid modifications induced by cold are not only linked to the preservation of membrane integrity and fluidity but may also serve as substrates for the generation of signalling lipids, which trigger other metabolic changes and adaptation processes (Hou *et al.*, 2016). Amino acids display a complex pivotal role in plants: they are used for protein biosynthesis, as building blocks for several other biosynthetic pathways, and in signalling processes in relation to stress responses, as in the case of GABA and proline (Bouché and Fromm, 2004; Liang *et al.*, 2013). The metabolic reconfiguration induced by cold is also key in controlling the balance of ROS production, as several of the chemical modifications serve as

ROS scavengers. The imbalance between ROS production and elimination has been highlighted as a causal mechanism of cold injury in peach fruit (Pons Puig *et al.*, 2015; Manganaris *et al.*, 2017; Monti *et al.*, 2019).

Collectively, the diverse array of metabolic modifications after cold storage constitutes the main defence of peach fruit to cope with the stress imposed by cold. When these modifications induced by cold storage are unsuccessful or not sufficient to allow protection against the cold, the low temperature results in damage to the fruit and the development of the disorder known as CI. In peach, CI includes lack of juiciness (mealiness or woolliness), internal browning, flesh bleeding, loss of flavour, loss of ability to ripen and increased incidence of decay. Most CI disorders are observed when the cold-stored fruit are transferred to an ambient temperature, when peach fruit reach their final phase of marketing. At this stage, the fruit display unattractive organoleptic properties, and are rejected and wasted by the consumer.

Tolerance or lower susceptibility to CIs is a multigenic trait because it depends on a complex metabolic rearrangement to build up a robust metabolic barrier to cope with cold. However, although CI tolerance is highly dependent on the capacity of a particular peach cultivar to build up this defence, the results obtained from the metabolic rearrangements induced by cold clearly show that proper postharvest management is also key for CI tolerance. The temperature and length of storage, as well as the application of several different physical (at an appropriate temperature) treatments and/or chemical compounds, may improve the metabolic barrier against cold injury. Different studies have shown that pre-conditioned treatments of peach fruit prior to cold storage are able to increase the tolerance to CI. In one of these studies, a pre-conditioned treatment of 48 h at 20°C prior to low-temperature exposure was able to suppress CI symptoms and induce ethylene biosynthesis when the ripening behaviour was restored after cold treatment. Modified metabolites due to this pre-conditioning treatment included the increase in amino acids (e.g. aspartic acid, glutamic acid, serine, valine, isovaline, phenylalanine), sugars (e.g. maltose, xylose), sugar alcohols (maltitol) and organic acids (glyceric acid and citric acid) (Fig. 8.3b). Several metabolites (e.g. valine and isovaline) have been correlated to CI tolerance during fruit acclimation to a cold environment (Tanou *et al.*, 2017). In another study, the pre-conditioned treatment was performed at 8°C for 5 days before storage at 0°C. This low-temperature conditioning resulted in a higher rate of ethylene production, which allowed proper softening when the fruit were transferred to an ambient temperature for ripening. Moreover, the conditioning treatment triggered lipid rearrangements in the fruit (Fig. 8.3b), which contributed to membrane stability, as well as adjustments to the cell-wall metabolism, which were both linked to increased cold tolerance (Wang *et al.*, 2017).

Overall, although the differential susceptibility to developing CI symptoms is highly dependent on the genotype, proper postharvest management is also able to modify the diverse array of biochemical components produced by the fruit to cope with cold stress and alleviate CI symptoms. This metabolic rearrangement should be sufficient to alleviate CI but should also allow the ripening process to proceed regularly after the cold treatment.

REFERENCES

Abdallah, D., Baraket, G., Perez, V., Salhi Hannachi, A. and Hormaza, J.I. (2020) Self-compatibility in peach [*Prunus persica* (L.) Batsch]: patterns of diversity surrounding the *S*-locus and analysis of SFB alleles. *Horticulture Research* 7: 170.

Andreotti, C., Ravaglia, D., Ragaini, A. and Costa, G. (2008) Phenolic compounds in peach (*Prunus persica*) cultivars at harvest and during fruit maturation. *Annals of Applied Biology* 153, 11–23.

Anthony, B.M., Chaparro, J.M., Prenni, J.E. and Minas, I.S. (2020) Early metabolic priming under differing carbon sufficiency conditions influences peach fruit quality development. *Plant Physiology and Biochemistry* 157, 416–431.

Aranzana, M.J., Decroocq, V., Dirlewanger, E., Iban, E., Zhong Shan, G. *et al.* (2019) *Prunus* genetics and applications after de novo genome sequencing: achievements and prospects. *Horticulture Research* 6, 1–25.

Aslam, M.M., Deng, L., Wang, X., Wang, Y., Pan, L. *et al.* (2019) Expression patterns of genes involved in sugar metabolism and accumulation during peach fruit development and ripening. *Scientia Horticulturae* 257: 108633.

Baccichet, I., Chiozzotto, R., Bassi, D., Gardana, C. and Cirilli, M. (2021) Characterization of fruit quality traits for organic acids content and profile in a large peach germplasm collection. *Scientia Horticulturae* 278: 109865.

Basile, B., Solari, L.I. and Dejong, T.M. (2007) Intra-canopy variability of fruit growth rate in peach trees grafted on rootstocks with different vigour-control capacity. *Journal of Horticultural Science and Biotechnology* 82, 243–256.

Begheldo, M., Manganaris, G.A., Bonghi, C. and Tonutti, P. (2006) Different postharvest conditions modulate ripening and ethylene biosynthetic and signal transduction pathways in Stony Hard peaches. *Postharvest Biology and Technology* 48, 84–91.

Bonghi C., Ferrarese L., Ruperti B., Tonutti P. and Ramina A. (1998) Endo-β-1,4-glucanases are involved in peach fruit growth and ripening and regulated by ethylene. *Physiologia Plantarum* 102, 3546–352.

Bonghi, C., Trainotti, L., Botton, A., Tadiello, A., Rasori, A. *et al.* (2011) A microarray approach to identify genes involved in seed–pericarp cross-talk and development in peach. *BMC Plant Biology* 11: 107.

Bonghi, C., Costa G., Tonutti, P. and Botton A. (2019) Fruit ripening. In: Sansavini, S., Costa, G., Gucci, R., Inglese, P., Ramina, A. *et al.* (eds) *Principles of Modern Fruit Science.* International Society for Horticultural Science, Leuven, Belgium, pp. 137–146.

Bouché, N. and Fromm, H. (2004) GABA in plants: just a metabolite? *Trends in Plant Science* 9, 110–115.

Brandi, F., Bar, E., Mourgues, F., Horváth, G., Turcsi, E. *et al.* (2011) Study of 'Redhaven' peach and its white-fleshed mutant suggests a key role of CCD4 carotenoid dioxygenase in carotenoid and norisoprenoid volatile metabolism. *BMC Plant Biology* 11: 24.

Brizzolara, S., Hertog, M., Tosetti, R., Nicolai, B. and Tonutti, P. (2018) Metabolic responses to low temperature of three peach fruit cultivars differently sensitive to cold storage. *Frontiers in Plant Science* 9: 706.

Brovelli, E.A., Brecht, J.K., Sherman, W.B. and Sims, C.A. (1999) Nonmelting-flesh trait in peaches is not related to low ethylene production rates. *HortScience* 34, 313–315.

Brummell, D.A., Dal Cin, V., Crisosto, C.H. and Labavitch, J.M. (2004) Cell wall metabolism during maturation, ripening and senescence of peach fruit. *Journal of Experimental Botany* 55, 2029–2039.

Bustamante, C.A., Monti, L.L., Gabilondo, J., Scossa, F., Valentini, G. *et al.* (2016) Differential metabolic rearrangements after cold storage are correlated with chilling injury resistance of peach fruits. *Frontiers in Plant Science* 7: 1478.

Bustamante, C.A., Brotman, Y., Monti, L.L., Gabilondo, J., Budde, C.O. *et al.* (2018) Differential lipidome remodeling during postharvest of peach varieties with different susceptibility to chilling injury. *Physiologia Plantarum* 163, 2–17.

Cantin, C.M., Gogorcena, Y. and Moreno, M.A. (2009) Analysis of phenotypic variation of sugar profile in different peach and nectarine [*Prunus persica* (L.) Batsch] breeding progenies. *Journal of the Science of Food and Agriculture* 89, 1909–1917.

Canton, M., Drincovich, M.F., Lara, M.V., Vizzotto, G., Walker, R.P. *et al.* (2020) Metabolism of stone fruits: reciprocal contribution between primary metabolism and cell wall. *Frontiers in Plant Science* 11: 1054.

Cao, S., Liang, M., Shi, L., Shao, J., Song, C. *et al.* (2017) Accumulation of carotenoids and expression of carotenogenic genes in peach fruit. *Food Chemistry* 214, 137–146.

Chen, M., Guo, H., Chen, S., Li, T., Li, M. *et al.* (2019) Methyl jasmonate promotes phospholipid remodeling and jasmonic acid signaling to alleviate chilling injury in peach fruit. *Journal of Agricultural and Food Chemistry* 67, 9958–9966.

Chen, S., Chen, M., Li, Y., Huang, X., Niu, D. *et al.* (2021) Adjustments of both phospholipids and sphingolipids contribute to cold tolerance in stony hard peach fruit by continuous ethylene. *Postharvest Biology and Technology* 171: 111332.

Cheniclet, C., Rong, W.Y., Causse, M., Frangne, N., Bolling, L. *et al.* (2005) Cell expansion and endoreduplication show a large genetic variability in pericarp and contribute strongly to tomato fruit growth. *Plant Physiology* 139, 1984–1994.

Cirilli, M., Bassi, D. and Ciaciulli, A. (2016) Sugars in peach fruit: a breeding perspective. *Horticultural Research* 3: 15067.

Colaric, M., Veberic, R., Stampar, F. and Hudina, M. (2005) Evaluation of peach and nectarine fruit quality and correlations between sensory and chemical attributes. *Journal of the Science of Food and Agriculture* 85, 2611–2616.

Costa, G., Botton, A. and Vizzotto, G. (2018) Fruit thinning: advances and trends. *Horticultural Reviews* 46, 185–226.

Crane, J.C., Primer, P.E. and Campbell, R.C. (1960) Gibberellin induced parthenocarpy in *Prunus*. *Proceedings of the American Society for Horticultural Science* 75, 129–137.

Dai, Z., Wu, H., Baldazzi, V., van Leeuwen, C., Berin, N. *et al.* (2016) Inter-species comparative analysis of components of soluble sugar concentration in fleshy fruits. *Frontiers in Plant Science* 7: 649.

Dal Cin, V., Rizzini, F.M., Botton, A. and Tonutti, P. (2006) The ethylene biosynthetic and signal transduction pathways are differently affected by 1-MCP in apple and peach fruit. *Postharvest Biology and Technology* 42, 125–133.

Dardick, C. and Callahan, A.M. (2014) Evolution of the fruit endocarp: molecular mechanisms underlying adaptations in seed protection and dispersal strategies. *Frontiers in Plant Science* 5: 284.

de Franceschi, P., Stegmeir, T., Cabrera, A., van der Knaap, E., Rosyara, U.R. *et al.* (2013) Cell number regulator genes in *Prunus* provide candidate genes for the control of fruit size in sweet and sour cherry. *Molecular Breeding* 32, 311–326.

Desnoues, E., Gibon, Y., Baldazzi, V., Signoret, V., Gènard, M. and Quilot-Turion, B. (2014) Profiling sugar metabolism during fruit development in a peach progeny with different fructose-to-glucose ratios. *BMC Plant Biology* 14: 336.

Dirlewanger, E., Pronier, V., Parvery, C., Rothan, C., Guy, A. and Monet, R. (1998) Genetic linkage map of peach. *Theoretical and Applied Genetics* 97, 888–895.

Etienne, C., Moing, A., Dirlewanger, E., Raymond, P., Monet, R. and Rothan, C. (2002) Isolation and characterization of six peach cDNAs encoding key proteins in organic acid metabolism and solute accumulation: involvement in regulating peach fruit acidity. *Physiologia Plantarum* 114, 259–270.

Falchi, R., Vendramin, E., Zanon, L., Scalabrin, S., Cipriani, G. *et al.* (2013) Three distinct mutational mechanisms acting on a single gene underpin the origin of yellow flesh in peach. *Plant Journal* 76, 175–187.

Farinati, S., Rasori, A., Varotto, S. and Bonghi, C. (2017) Rosaceae fruit development, ripening and post-harvest: an epigenetic perspective. *Frontiers in Plant Science* 8: 1247.

Farinati, S., Forestan, C., Canton, M., Galla, G., Bonghi, C. and Varotto, S. (2021) Regulation of fruit growth in a peach slow ripening phenotype. *Genes* 12: 482.

Fürtauer, L., Weiszmann, J., Weckwerth, W. and Nägele, T. (2019) Dynamics of plant metabolism during cold acclimation. *International Journal of Molecular Sciences* 20: 5411.

Gallardo, R.K., Nguyen, D., McCracken, V., Yue, C., Luby, J. and McFerson, J.R. (2012) An investigation of trait prioritization in rosaceous fruit breeding programs. *HortScience* 47, 771–776.

Ganie, A.M., Farida, A., Bhat, M.A., Malik, A.R., Junaid, J.M. *et al.* (2013) Boron – a critical nutrient element for plant growth and productivity with reference to temperate fruits. *Current Science* 104, 76–85.

Génard, M., Pagès, L. and Kervella, J. (1998) A carbon balance model of peach tree growth and development for studying the pruning response. *Tree Physiology* 18, 351–362.

Génard, M., Lescourret, F., Gomez, L. and Habib, R. (2003) Changes in fruit sugar concentrations in response to assimilate supply, metabolism and dilution: a modelling approach applied to peach fruit (*Prunus persica*). *Tree Physiology* 23, 373–385.

Ghiani, A., Onelli, E., Aina, R., Cocucci, M. and Citterio, S. (2011) A comparative study of melting and non-melting flesh peach cultivars reveals that during fruit ripening endo-polygalacturonase (endo-PG) is mainly involved in pericarp textural changes, not in firmness reduction. *Journal of Experimental Botany* 62, 4043–4054.

Guo, J., Cao, K., Deng, C., Li, Y., Zhu, G., *et al.* (2020) An integrated peach genome structural variation map uncovers genes associated with fruit traits. *Genome Biology* 21: 258.

Guo, M. and Simmons, C.R. (2011) Cell number counts – the *fw2.2* and *CNR* genes and implications for controlling plant fruit and organ size. *Plant Science* 181, 1–7.

Haji, T., Yaegaki, H. and Yamaguchi, M. (2001) Changes in ethylene production and flesh firmness of melting, nonmelting and stony hard peaches after harvest. *Journal of the Japanese Society for Horticultural Science* 70, 458–459.

Hayama, H., Shimada, T., Fuji, H., Ito, A. and Kashimura, Y. (2006) Ethylene regulation of fruit softening and softening-related genes in peach. *Journal of Experimental Botany* 57, 4071–4077.

Hayama, H., Tatsuki, M., Yoshioka, H. and Nakamura, Y. (2008) Regulation of stony hard peach softening with ACC treatment. *Postharvest Biology and Technology* 50, 231–232.

Hou, Q., Ufer, G. and Bartels, D. (2016) Lipid signalling in plant responses to abiotic stress. *Plant, Cell & Environment* 39, 1029–1048.

Hu, C., Zhu, Y., Cui, Y., Cheng, K., Liang, W. *et al.* (2018) A group of receptor kinases are essential for CLAVATA signalling to maintain stem cell homeostasis. *Nature Plants* 4, 205–211.

Karakurt, Y., Huber, D.J. and Sherman, W.B. (2000) Quality characteristics of melting and non-melting flesh peach genotypes. *Journal of the Science of Food and Agriculture* 80, 1848–1853.

Lauxmann, M.A., Borsani, J., Osorio, S., Lombardo, V.A., Budde, C.O. *et al.* (2014) Deciphering the metabolic pathways influencing heat and cold responses during post-harvest physiology of peach fruit. *Plant, Cell & Environment* 37, 601–616.

Liang, X., Zhang, L., Natarajan, S.K. and Becker, D.F. (2013) Proline mechanisms of stress survival. *Antioxidants and Redox Signaling* 19, 998–1011.

Lobit, P., Génard, M., Soing, P. and Habib, R. (2006) Modelling malic acid accumulation in fruits: relationships with organic acids, potassium, and temperature. *Journal of Experimental Botany* 57, 1471–1483.

Lombardo, V.A., Osorio, S., Borsani, J., Lauxmann, M.A., Bustamante, C.A. *et al.* (2011) Metabolic profiling during peach fruit development and ripening reveals the metabolic networks that underpin each developmental stage. *Plant Physiology* 157, 1696–1710.

López-Girona, E., Zhang, Y., Eduardo, I., Mora, J.R.H., Alexiou, K.G. *et al.* (2017) A deletion affecting an LRR-RLK gene co-segregates with the fruit flat shape trait in peach. *Scientific Reports* 7: 6714.

Manganaris, G.A., Drogoudi, P., Goulas, V., Tanou, G., Georgiadou, E.C. *et al.* (2017) Deciphering the interplay among genotype, maturity stage and low-temperature storage on phytochemical composition and transcript levels of enzymatic antioxidants in *Prunus persica* fruit. *Plant Physiology and Biochemistry* 119, 189–199.

Mariotti, L., Picciarelli, P., Lombardi, L. and Ceccarelli, N. (2011) Fruit-set and early fruit growth in tomato are associated with increases in indoleacetic acid, cytokinin, and bioactive gibberellin contents. *Journal of Plant Growth Regulation* 30, 405–415.

Masia, A., Zanchin, A., Rascio, N. and Ramina, A. (1992) Some biochemical and ultra-structural aspects of peach fruit development. *Journal of the American Society for Horticultural Science* 117, 808–815.

McGlasson, W.B., Golding, J.B. and Holford, P. (2009) Improving the dessert quality of stone fruits. In: Benkeblia, N. (ed.) *Postharvest technologies for horticultural crops, Vol. 2.* Research Signpost, Kerala, India, pp. 49–92.

Mencarelli, F., Cecchi, F., de Santis, D., Aquilani, R., Giuliano, G. and Rotondi, R. (2003) Influence of ethylene on volatiles and carotenoids biosynthesis of peach. In: Vendrell, M., Klee, H., Pech, J.C. and Romojaro, F. (eds) *Biology and Biotechnology of Plant Hormone Ethylene III.* NATO Science Series 348. IOS Press, Amsterdam, pp. 222–226.

Moing, A., Rothan, C., Svanella, L., Just, D., Diakou, P. *et al.* (2000) Role of phosphoenolpyruvate carboxylase in organic acid accumulation during peach fruit development. *Physiologia Plantarum* 108, 1–10.

Monti, L.L., Bustamante, C.A., Osorio, S., Gabilondo, J., Borsani, J. *et al.* (2016) Metabolic profiling of a range of peach fruit varieties reveals high metabolic diversity and commonalities and differences during ripening. *Food Chemistry* 190, 879–888.

Monti, L.L., Bustamante, C.A., Budde, C.O., Gabilondo, J., Müller, G.L. *et al.* (2019) Metabolomic and proteomic profiling of Spring Lady peach fruit with contrasting woolliness phenotype reveals carbon oxidative processes and proteome reconfiguration in chilling-injured fruit. *Postharvest Biology and Technology* 151, 142–151.

Morandi, B., Corelli Grappadelli, L., Rieger, M. and Lo Bianco, R. (2008) Carbohydrate availability affects growth and metabolism in peach fruit. *Physiologia Plantarum* 133, 229–241.

Morgutti, S., Negrini, N., Nocito, F.F., Ghiani, A., Bassi, D. and Cocucci, M. (2006) Changes in endopolygalacturonase levels and characterization of a putative endo-PG gene during fruit softening in peach genotypes with nonmelting and melting flesh fruit phenotypes. *New Phytologist* 171, 315–328.

Musseau, C., Jorly, J., Gadin, S., Sorensen, I., Deborde, C. *et al.* (2020) The tomato guanylate-binding protein SlGBP1 enables fruit tissue differentiation by maintaining endopolyploid cells in a non-proliferative state. *Plant Cell* 32, 3188–3205.

Nilo-Poyanco, R., Moraga, C., Benedetto, G., Orellana, A. and Almeida, AM. (2021) Shotgun proteomics of peach fruit reveals major metabolic pathways associated to ripening. *BMC Genomics* 22: 17.

Nishizawa, A., Yabuta, Y. and Shigeoka, S. (2008) Galactinol and raffinose constitute a novel function to protect plants from oxidative damage. *Plant Physiology* 147, 1251–1263.

Nowicka, P., Wojdyło, A. and Laskowski, P. (2019) Principal component analysis (PCA) of physicochemical compounds' content in different cultivars of peach fruits, including qualification and quantification of sugars and organic acids by HPLC. *European Food Research and Technology* 245, 929–938.

Pan, H.F., Sheng, Y., Gao, Z.H., Chen, H.L., Qi, Y.J. *et al.* (2016) Transcriptome analysis of peach (*Prunus persica* L. Batsch) during the late stage of fruit ripening. *Genetic and Molecular Research* 15: 15049335.

Paniagua, C., Posé, S., Morris, V.J., Kirby, A.R., Quesada, M.A. and Mercado, J.A. (2014) Fruit softening and pectin disassembly: an overview of nanostructural pectin modifications assessed by atomic force microscopy. *Annals of Botany* 114, 1375–1383.

Pei, M., Gu, C. and Zhang, S. (2019) Genome-wide identification and expression analysis of genes associated with peach (*Prunus persica*) fruit ripening. *Scientia Horticulturae* 246, 317–327.

Pons Puig, C., Dagar, A., Marti Ibanez, C., Singh, V., Crisosto, C.H. *et al.* (2015) Pre-symptomatic transcriptome changes during cold storage of chilling sensitive and resistant peach cultivars to elucidate chilling injury mechanisms. *BMC Genomics* 16: 245.

Posé, S., Paniagua, C., Matas, A.J., Gunning, A.P., Morris, V.J. *et al.* (2018) A nanostructural view of the cell wall disassembly process during fruit ripening and postharvest storage by atomic force microscopy. *Trends in Food Science and Technology* 87, 47–58.

Rahim, M.A., Busatto, N. and Trainotti, L. (2014) Regulation of anthocyanin biosynthesis in peach fruits. *Planta* 240, 913–929.

Rasori, A., Ruperti, B., Bonghi, C., Tonutti, P. and Ramina, A. (2002) Characterization of two putative ethylene receptor genes expressed during peach fruit development and abscission. *Journal of Experimental Botany* 53, 2333–2339.

Rasori, A., Bertolasi, B., Furini, A., Bonghi, C., Tonutti, P. and Ramina, A. (2003) Functional analysis of peach ACC oxidase promoters in transgenic tomato and in ripening peach fruit. *Plant Science* 165, 524–530.

Ravaglia, D., Espley, R.V., Henry-Kirk, R.A., Andreotti, C., Ziosi, V. *et al.* (2013) Transcriptional regulation of flavonoid biosynthesis in nectarine (*Prunus persica*) by a set of R2R3 MYB transcription factors. *BMC Plant Biology* 13: 68.

Rennie, E.A. and Turgeon, R. (2009) A comprehensive picture of phloem loading strategies. *Proceedings of the National Academy of Sciences USA* 106, 14162–14167.

Ruperti, B., Bonghi, C., Rasori, A., Ramina, A. and Tonutti, P. (2001) Characterization and expression of two members of the peach 1-aminocyclopropane-1-carboxylate oxidase gene family. *Physiologia Plantarum* 111, 336–344.

Sánchez, G., Besada, C., Badenes, M.L., Monforte, A.J. and Granell, A. (2012) A non-targeted approach unravels the volatile network in peach fruit. *PLoS One* 7: e38992.

Sánchez, G., Venegas-Caleron, M., Salas, J.J., Monforte, A., Badenes, M.L. and Granell A. (2013) An integrative "omics"" approach identifies new candidate genes to impact aroma volatiles in peach fruit. *BMC Genomics* 14: 343.

Sandefur, P., Clark, J.R. and Peace, C. (2013) Peach texture. *Horticultural Reviews* 41, 241–301.

Scorza, R., May, L.G., Purnell, B. and Upchurch, B. (1991) Differences in number and area of mesocarp cells between small- and large-fruited peach cultivars. *Journal of the American Society for Horticultural Science* 116, 861–864.

Seymour, G.B., Pool, M., Giovannoni, J.J. and Tucker G.A. (eds) (2013) *The Molecular Biology and Biochemistry of Fruit Ripening*. Wiley, Ames, Iowa.

Shulaev, V., Korban, S.S., Sosinski, B., Abbott, A.G., Aldwinckle, H.S. *et al.* (2008) Multiple models for Rosaceae genomics. *Plant Physiology* 147, 985–1003.

Souza, F., Alves, E., Pio, R., Castro, E., Reighard, G. *et al.* (2019) Influence of temperature on the development of peach fruit in a subtropical climate region. *Agronomy* 9: 20.

Su, M., Ye, Z., Zhang, B. and Chen, K. (2017) Ripening season, ethylene production and respiration rate are related to fruit non-destructively-analyzed volatiles measured by an electronic nose in 57 peach (*Prunus persica* L.) samples. *Emirates Journal of Food and Agriculture* 29, 807–814.

Szabò, Z. and Nyéki, J. (1999) Self-pollination in peach. *International Journal of Horticultural Science* 5, 76–78.

Tadiello, A., Ziosi, V., Negri, A., Noferini, M., Fiori, G. *et al.* (2016) On the role of ethylene, auxin and a GOLVEN-like peptide hormone in the regulation of peach ripening. *BMC Plant Biology* 16: 44.

Tanou, G., Minas, I.S., Scossa, F., Belghazi, M., Xanthopoulou, A. *et al.* (2017) Exploring priming responses involved in peach fruit acclimation to cold stress. *Scientific Reports* 7: 11358.

Tarkowski, Ł.P. and van den Ende, W. (2015) Cold tolerance triggered by soluble sugars: a multifaceted countermeasure. *Frontiers in Plant Science* 6: 203.

Tatsuki, M., Haji, T. and Yamaguchi, M. (2006) The involvement of 1-aminocyclopropane-1-carboxylic acid synthase isogene, *Pp-ACS1*, in peach fruit softening. *Journal of Experimental Botany* 57, 1281–1289.

Tatsuki, M., Nakajima, N., Fujii, H., Shimada, T., Nakano, M. *et al.* (2013) Increased levels of IAA are required for system 2 ethylene synthesis causing fruit softening in peach (*Prunus persica* L. Batsch). *Journal of Experimental Botany* 64, 1049–1059.

The International Peach Genome Initiative (2013) The high-quality draft genome of peach (*Prunus persica*) identifies unique patterns of genetic diversity, domestication and genome evolution. *Nature Genetics* 45, 487–494.

Toivonen, P.M.A. and Brummell, D.A. (2008) Biochemical bases of appearance and texture changes in fresh-cut fruit and vegetables. *Postharvest Biology and Technology* 48, 1–14.

Tonutti, P., Casson, P. and Ramina, A. (1991) Ethylene biosynthesis during peach fruit development. *Journal of the American Society for Horticultural Science* 116, 274–279.

Tonutti, P., Bonghi, C. and Ramina, A. (1996) Fruit firmness and ethylene biosynthesis in three cultivars of peach (*Prunus persica* L. Batsch). *Journal of Horticultural Science* 71, 141–147.

Tonutti, P., Bonghi, C., Ruperti, B., Tonielli, G.B. and Ramina, A. (1997) Ethylene evolution and 1-aminocyclopropane-1-carboxylate oxidase gene expression during early development and ripening of peach fruit. *Journal of the American Society for Horticultural Science* 122, 642–647.

Trainotti, L., Zanin, D. and Casadoro, G. (2003) A cell-wall oriented genomic approach reveals new and unexpected complexity of the softening in peaches. *Journal of Experimental Botany* 54, 1821–1832.

Trainotti, L., Bonghi, C., Ziliotto, F., Zanin, D., Rasori, A. *et al.* (2006) The use of microarray μPEACH1.0 to investigate transcriptome changes during transition from pre-climacteric to climacteric phase in peach fruit. *Plant Science* 170, 606–613.

Trainotti, L., Tadiello, A. and Casadoro, G. (2007) The involvement of auxin in the ripening of climacteric fruits comes of age: the hormone plays a role of its own and has an intense interplay with ethylene in ripening. *Journal of Experimental Botany* 58, 3299–3308.

Vendramin, E., Pea, G., Dondini, L., *et al.* (2014) A unique mutation in a MYB gene cosegregates with the nectarine phenotype in peach. *PLoS ONE* 9: e90574.

Visai, C. and Vanoli, M. (1997) Volatile compound production during growth and ripening of peaches and nectarines. *Scientia Horticulturae* 70, 15–24.

Vizzotto, G., Pinton, R., Varanini, Z. and Costa, G. (1996) Sucrose accumulation in developing peach fruit. *Physiologia Plantarum* 96, 225–230.

Walker, R.P., Battistelli, A., Bonghi, C., Drincovich, M.F., Falchi, R. *et al.* (2020) Non-structural carbohydrate metabolism in the flesh of stone fruits of the genus *Prunus*. *Frontiers in Plant Science* 7: 649.

Wang, K., Shao, X., Gong, Y., Zhu, Y., Wang, H. *et al.* (2013) The metabolism of soluble carbohydrates related to chilling injury in peach fruit exposed to cold stress. *Postharvest Biology and Technology* 86, 53–61.

Wang, K., Yin, X.R., Zhang, B., Grierson, D., Xu, C.J. and Chen, K.S. (2017) Transcriptomic and metabolic analyses provide new insights into chilling injury in peach fruit. *Plant, Cell & Environment* 40, 1531–1551.

Wills, R., McGlasson, B., Graham, D. and Joyce, D. (2007) Physiology and biochemistry. In: Wills, R., McGlasson, B., Graham, D. and Joyce, D. (eds) *Postharvest: An*

Introduction to the Physiology and Handling of Fruits, Vegetables and Ornamentals, 5th edn. CAB International, Wallingford, UK, pp. 34–62.

Yamaguchi, M., Haji, T. and Yaegaki, H. (2004) Differences in mesocarp cell number, cell length and occurrence of gumming in fruit of Japanese apricot (*Prunus mume* Sieb. Et Zucc.) cultivars during their development. *Journal of the Japanese Society for Horticultural Science* 73, 200–207.

Yan, J., Cai, Z., Shen, Z., Ma, R. and Yu, M. (2017) Proanthocyanidin monomers and cyanidin 3-*O*-glucoside accumulation in blood-flesh peach (*Prunus persica* (L) Batsch) fruit. *Archives in Biological Science* 69, 611–617.

Ying, H., Shi, J., Zhang, S., Pingcuo, G., Wang, S. *et al.* (2019) Transcriptomic and metabolomic profiling provide novel insights into fruit development and flesh coloration in *Prunus mira* Koehne, a special wild peach species. *BMC Plant Biology* 19: 463.

Zeng, W., Niu, L., Wang, Z., Wang, X., Wang, Y. *et al.* (2020) Application of an antibody chip for screening differentially expressed proteins during peach ripening and identification of a metabolon in the SAM cycle to generate a peach ethylene biosynthesis model. *Horticultural Research* 7: 31.

Zhang, B., Shen, J., Wei, W., Xi, W., Xu, C.J. *et al.* (2010) Expression of genes associated with aroma formation derived from the fatty acid pathway during peach fruit ripening. *Journal of Agricultural and Food Chemistry* 58, 6157–6165.

Zhao, Y., Dong, W., Wang, K., Zhang, B., Allan, A.C. *et al.* (2017) Differential sensitivity of fruit pigmentation to ultraviolet light between two peach cultivars. *Frontiers in Plant Science* 8: 1552.

Zheng, B., Zhao, L., Jiang, X., Cherono, S., Liu, J. *et al.* (2021) Assessment of organic acid accumulation and its related genes in peach. *Food Chemistry* 334: 127567.

Zheng, W., Pan, L., Liu, H., Niu, L., Lu, Z. *et al.* (2015) Characterization of 1-aminocyclopropane-1-carboxylic acid synthase (ACS) genes during nectarine fruit development and ripening. *Tree Genetics & Genomes* 11: 18.

Zhou, H., Lin-Wang, K., Wang, H., Gu, C., Dare, A.P. *et al.* (2015) Molecular genetics of blood-fleshed peach reveals activation of anthocyanin biosynthesis by NAC transcription factors. *Plant Journal* 82, 105–121.

Zhou, Y., Zhou, H., Lin-Wang, K., Vimolmangkang, S., Espley, R.V. *et al.* (2014) Transcriptome analysis and transient transformation suggest an ancient duplicated MYB transcription factor as a candidate gene for leaf red coloration in peach. *BMC Plant Biology* 14: 388.

Ziliotto, F., Begheldo, M., Rasori, A., Bonghi, C. and Tonutti, P. (2008) Transcriptome profiling of ripening nectarine (*Prunus persica* L. Batsch) fruit treated with 1-MCP. *Journal of Experimental Botany* 59, 2781–2791.

NON-DESTRUCTIVE PEACH FRUIT MATURITY AND QUALITY ASSESSMENT

Ioannis S. Minas[1]*, Brendon M. Anthony[1], Pavlina Drogoudi[2] and Guglielmo Costa[3]

[1]*Department of Horticulture and Landscape Architecture, Colorado State University, Fort Collins, Colorado, USA; [2]Department of Deciduous Fruit Trees, Institute of Plant Breeding and Genetic Resources, Hellenic Agricultural Organization 'Demeter', Naoussa, Greece; [3]Department of Agri-Food Sciences and Technologies, University of Bologna, Bologna, Italy*

9.1 IMPORTANCE OF PEACH FRUIT MATURITY AND RELATIONSHIP WITH CULTIVAR, HARVEST TIME, POSTHARVEST PERFORMANCE AND EATING QUALITY

Fleshy fruit maturation and ripening is a genetically regulated process that coincides with seed maturation and the advancement of several physicochemical changes (Giovannoni *et al.*, 2017). Most of these physicochemical changes advance taste, nutritional value, appearance and overall consumption experience; however, some might limit the shelf life and postharvest performance of fleshy fruits such as peach. The relatively short harvest window can lead to immature or overripe fruit, in addition to rapid softening and susceptibility to chilling injury (CI), which are factors that limit peach fruit postharvest life (Lurie and Crisosto, 2005). A precise understanding of the processes underlying peach fruit maturation and ripening is key to determining the harvest time during 'on-tree' ripening and the storage duration during postharvest handling ('off-tree'), to reduce postharvest losses and improve consumer acceptance.

Peach is classified as a climacteric fruit, which is characterized by a rapid increase in ethylene emission and respiration at the onset of ripening. Peach 'on-tree' maturity and ripening status determines the magnitude of evolution of traits associated with fruit harvest quality and postharvest performance. Such traits include several sensorial and textural changes resulting from

---

* Email: Ioannis.Minas@colostate.edu

© CAB International 2023. *Peach* (G. Manganaris, G. Costa and C. Crisosto eds)
DOI: 10.1079/9781789248456.0009

increased carbohydrate accumulation (expressed as dry matter content (DMC) and soluble solids concentration (SSC)), ground colour changes from green to yellow/orange due to accumulating pigments (e.g. carotenoids), flesh softening due to cell-wall dismantling, and acidity loss due to increased respiration rates, ethylene biosynthesis and aromatic volatilization (Fig. 9.1). Ethylene coordinates peach ripening (Fig. 9.1) by regulating gene expression responsible for fruit abscission and softening, pigment accumulation and volatile aroma biosynthesis (Trainotti *et al.*, 2007). All of these parameters are related to shipping/storage potential and improved organoleptic characteristics that increase consumer satisfaction (Minas *et al.*, 2018).

Optimum peach harvest time determination is fundamental to increased packouts and consumer acceptance, as well as susceptibility to rapid softening, postharvest diseases and CI that limit storage/shipping or marketing potential (Michailides and Manganaris, 2009; Minas *et al.*, 2018). The goal of fruit harvest handling should be to pick fruit at an optimum maturity stage for commercial use and immediately transport it to the packing facility with minimum mechanical damage (e.g. bruises) and sun exposure. Premature harvesting, which occurs more often in cultivars with a full red overcolour, leads to inferior

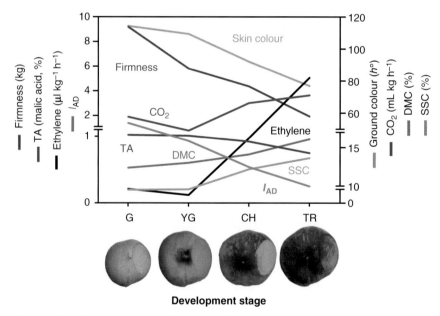

Fig. 9.1. Peach fruit quality and maturity changes during growth, development, maturation and ripening 'on tree' for the stages: G, green (stage S2); YG, yellow green (stage S3); CH, commercial harvest (stage S4I pre-climacteric); TR, tree ripe (stage S4II climacteric). DMC, dry matter content; $h°$, ground skin hue angle; I_{AD}, index of absorbance difference at 670 and 720 nm; SSC, soluble solids concentration; TA, titratable acidity.

consumer quality and has been linked to significant market share loss of major peach-producing countries in Europe (Minas *et al.*, 2018; Manganaris *et al.*, 2022). When peaches are harvested too early (green mature), they might never achieve their full flavour potential (Crisosto, 2002). In contrast, delaying peach harvest may yield improved fruit organoleptic quality, as quality is enhanced with advanced maturity (Fig. 9.1), but this comes at the expense of a shorter storage and shelf life, and higher risk for increased levels of bruising and CI (Bonghi *et al.*, 1999; Lurie and Crisosto, 2005; Manganaris *et al.*, 2017; Tanou *et al.*, 2017).

Peach consumption per capita has been in constant decline in the USA and Europe in recent decades, and this trend has remained unchanged (Minas *et al.*, 2018; Manganaris *et al.*, 2022). Surveys concluded that the main reason consumers decline to repeat fresh peach purchases is due to poor internal fruit quality (Bruhn *et al.*, 1991; Crisosto, 2002; Byrne, 2005). Today, numerous new cultivars are available resulting from breeding programmes, but traits such as improved yields and fruit appearance are prioritized at the expense of eating quality (Iglesias and Echeverría, 2009; Liverani, *et al.*, 2015; Minas, *et al.*, 2018). Thus, the reduced peach consumption rates have been related to immature, tasteless and/or overripe fruit, which results from inappropriate harvest decision making, as well as a variety of textural problems associated with interrupted ripening due to postharvest physiological disorders (e.g. CI) (Bruhn *et al.*, 1991; Crisosto, 2002; Byrne, 2005; Minas *et al.*, 2018). To reverse these trends in peach consumption, the selection of cultivars with improved softening rates, storability, and sensorial and nutritional characteristics is critical. However, optimization of cultural management at the orchard level is necessary to achieve balanced yields and cost efficiency with improved fruit quality. Important orchard and preharvest factors that may also affect peach fruit quality include the fruit maturity stage at harvest, rootstock, training system, crop load management, canopy position and mineral nutrition (Crisosto and Costa, 2008; Minas *et al.*, 2018). Peach internal fruit quality can only be improved preharvest, while optimum postharvest handling, which is also important, can only preserve the final product of optimum orchard factors management. To improve peach eating quality and increase consumption rates, reliable and accurate information on the impact of preharvest factors on fruit maturation and harvested quality is essential.

Preharvest agronomic and environmental factors can severely alter the regulation of the primary and secondary metabolism, which influence the maturation/ripening processes and internal fruit quality (Anthony *et al.*, 2020). More importantly, preharvest factor comparisons performed on fruit of unequal maturity are potentially confounded by the maturation status of the fruit (Minas *et al.*, 2018). Therefore, to truly evaluate the impact of preharvest conditions on fruit quality development, it is critical that comparisons be made on fruit of equal maturity (Minas *et al.*, 2018, 2021; Anthony *et al.*, 2020, 2021). Peach fruit maturity and quality cannot be fully determined by external traits alone, such as size or colour, unless it is for bicoloured cultivars. Ground colour assessment is used to estimate peach maturity and harvest time by using either colorimeters (e.g. Minolta CR-20; Minolta) or colour maps. Colorimeters express colour in numerical terms such as the colour spaces of the hue angle ($h°$) or chroma (C^*) parameter from the Commission Internationale de l'Eclairage. Colour maps use examples of various shades of yellow/orange to match background colours appropriate for optimal maturity and have been demonstrated to be a sufficient index for bicoloured peach cultivars. Due to the early coverage with red blush in fully overcoloured cultivars, ground colour cannot be used as a harvest index.

Precise assessment of maturity and internal quality is essential, but it can be time-consuming and labour-intensive using traditional and standard destructive techniques (Fig. 9.2; Minas *et al.*, 2018). Eating quality and maturity stage at the field and laboratory level can also be evaluated with a sensory assessment of sweetness, acidity, astringency, aroma, etc. In addition, most destructive methodologies are not friendly for large-scale data acquisition or for field use, which can help capture the totality of the maturation process and quality development in the field, in real time. Given the variability of fruit on a tree, or within an orchard, these narrow approaches fail to assess the true influence of various preharvest factors on tree fruit maturity and internal quality (Minas *et al.*, 2021).

Non-destructive techniques were recently developed and can enable rapid and real-time peach maturity and quality assessments in a single scan (Minas *et al.*, 2021). These methods can assist the grower by enabling

Fig. 9.2. Standard destructive maturity and internal quality assessments. (a) Fruit flesh firmness measurement with a motorized probe for user-independent assessment using a Fruit Texture Analyzer (Guss Manufacturing (Pty) Ltd). (b) Determination of soluble solids concentration (°Brix) within the fruit juice using a digital refractometer with automatic temperature compensation (model PR-32α or PAL-1; Atago Co. Ltd). (c) Fruit dry matter content estimation by the difference in mass between the fresh sample of mesocarp tissue and its equivalent dry mass following a 72 h exposure in a forced-air oven at 65°C.

better evaluation of the optimum harvest time, but also the packers/ship-pers so that appropriate sorting is performed and optimum storage duration is decided. At the level of research, these non-destructive technologies can provide large-scale data regarding the true impact of preharvest factors on fruit quality development and can be paired with 'omic' tools to better under-stand the biological characteristics of the fruit, as they pertain to fruit quality metabolism and regulation (Anthony *et al.*, 2020, 2021; Anthony and Minas, 2022). Accurately assessing the impact of preharvest factors on peach ma-turity and quality attributes is imperative for improving consumer satisfaction and postharvest handling (Minas *et al.*, 2020, 2021).

9.2 DESTRUCTIVE METHODS TO ASSESS FRUIT MATURITY AND INTERNAL QUALITY

Peach maturity and quality traits such as flesh firmness (FF), SSC and titrat-able acidity (TA) are the most important indices of harvest maturity, shipping and storage potential, as well as consumer acceptance (Crisosto and Crisosto, 2005). These parameters are easily assessed with simple destructive tools such as penetrometers (FF), refractometers (SSC) and titrators (TA), which can broadly be used by producers, packers, shippers and researchers to determine harvest time and internal quality (Fig. 9.2).

FF in peach is measured with penetrometers that are based on the system developed by Magness and Taylor (1925) using a 8 mm-wide probe that is in-serted into the peeled flesh of a fruit to a depth of 7.9 mm. Penetrometers can be either hand-held gauges for field use (e.g. Effegi Fruit Pressure Tester model FT 327) or benchtop devices with an analogue (e.g. Chatillon) or digital dis-play (e.g. Fruit Firmness Analyzer model 53205; T.R. Turoni Srl) for improved accuracy and consistency in the measurements. There is a perception that the speed and depth of penetration could greatly influence the FF readings. Consequently, an instrument with a motorized probe would provide the most consistent firmness readings, especially when several operators use the same instrument. For this type of user-independent FF assessment, the Fruit Texture Analyzer (Guss Manufacturing (Pty) Ltd) can be utilized for consistent probe penetration depth and speed across fruit samples. This type of instrument is suitable for either laboratory or commercial packinghouse use and provides several data-management features. However, in a comparison among different instruments with the probe being operated either manually (by the user) or automatically (by a motor attached to the device), each instrument was equally accurate with one or several users (Kupferman and Dasgupta, 2001).

FF is used by the peach industry to determine harvest time and shipping/storage potential. The goal of the producers, packers and shippers is to handle peaches with a maturity status that allows for optimum organoleptic charac-teristics within a bruising-safe range. Bruising susceptibility varies among cultivars, but a minimum FF level of 27 N is generally considered safe (Crisosto *et al.*, 2001). Flesh firmness cannot be used alone as an index of minimum maturity, as FF within cultivars varies with flesh typology, fruit size, climate and cultural practices (Serra *et al.*, 2020), although FF thresholds can provide a maximum maturity baseline and are used as the industry's standard bruising risk index. Two main grades of FF are currently used to determine peach harvest time in the USA. The well-mature grade (FF of 45–55 N) is used for fruit intended for 1–2 weeks of storage and long-distance markets (i.e. commercial harvest), while the tree-ripe grade (FF of 30–40 N) is for fruit intended for less than 1 week of storage and shipment to nearby or local markets (Minas *et al.*, 2018).

For practical purposes, peach fruit sweetness is estimated by measuring SSC (°Brix) in a drop of fruit juice using refractometers. SSC is considered to be an acceptable estimate of sugar concentration (Cirilli *et al.*, 2016), and correlates with consumer acceptance (Parker *et al.*, 1991; Ravaglia *et al.*, 1996). A refractometer is used to determine the concentration of soluble solids within the fruit juice. When rays of light pass from the juice, they are bent either towards or away from a normal line. The angle between the normal ray and the refracted ray is called the angle of refraction. The angle of refraction is dependent on the composition of soluble solids, along with the temperature. As the concentration of a particular compound in a solution increases, so does the degree to which the light is bent. For each percentage SSC or °Brix value, there is a corresponding angle of refraction. It is also important to determine the temperature of the testing environment, as temperature affects the angle of refraction (for every 5°C, SSC increases by 0.5%). To make the conversion easier, refractometers are available with scales that are calibrated to read the desired value, in this case, the percentage SSC or °Brix. Available refractometers are either hand-held optic analogue devices with the user reading the measurement by directing the device towards a source of light, or digital refractometers with automatic temperature compensation (e.g. model PR-32α or PAL-1; Atago Co. Ltd) where the device sends the light ray through the glass spot that the fruit juice is dropped on to.

Although SSC correlates with sugar concentration, R^2 values can range from 0.33 to 0.72, depending on the concentration of other molecules such as pectins, salts and organic acids, which contribute to the optical properties of the fruit juice (Cirilli *et al.*, 2016). SSC can vary from day to day and from season to season, depending on recent weather conditions, making it hard to be adopted as a harvest index in wet climates where rain is frequent during the growing season or close to the estimated harvest date. Soluble solids concentration is used mainly as minimum consumer quality index with 10% and 11% SSC being the thresholds for low-acid (TA <0.9%) and high-acid (TA ≥0.9%) peach cultivars, respectively (Kader, 1995; Hilaire, 2003). In addition, 10%, 11% and 12% SSC have been proposed as the minimum thresholds for early, mid- and late-season yellow-fleshed peach cultivars (Testoni, 1995). In general, consumer acceptance correlates with peach sweetness represented by either ripe SSC or the ratio of ripe SSC to ripe TA, and these parameters can be affected by cultivar, maturity status and other orchard factors such as crop load and canopy position (Crisosto and Crisosto, 2005; Minas *et al.*, 2021).

Non-structural carbohydrates (starch and sugars) can be affected by pre-harvest factors, and their concentrations may vary across the different phases of fruit growth and development (Boldingh *et al.*, 2000). It is noteworthy that high correlations were observed between DMC and SSC ($R^2 = 0.96–0.99$) in peach (Minas *et al.*, 2021, 2023). This relationship could be due to the absence of starch in the peach mesocarp. Thus, DMC in peach can provide an accurate estimate of peach sweetness that is represented mainly by SSC. Fruit DMC (%) is estimated by the difference in mass between the fresh sample of mesocarp tissue and the equivalent dry sample. Collectively, peach mesocarp is removed from the side of the fruit using a cork borer and immediately weighed to the nearest 0.001 g with a digital scale (Fig. 9.2). The samples are then dried in a forced-air oven at 65°C until all the water in the mesocarp has been completely dried out (up to 72 h; Minas *et al.*, 2021).

Whichever method is used, standard destructive analysis of peach fruit maturity and internal quality can be performed only on a limited number of samples due to labour and cost restrictions, which may not fully represent the entire lot in the field.

9.3 NON-DESTRUCTIVE METHODS TO ASSESS FRUIT MATURITY AND QUALITY

Judging fruit maturity and quality by size and external colour alone results in varying success in determining appropriate harvest time. Using traditional methods such as FF, DMC, SSC and TA analysis of tree fruit is destructive, time-consuming and requires sample preparation. These labour-intensive limitations make it difficult to assess maturity and internal quality in the field. Such measurements paired with fruit background (i.e. ground) colour are used as harvest and quality indices in peaches and nectarines, and correlate with consumer acceptance and shipping/storage potential (Crisosto and Costa, 2008). In standard bicoloured peach cultivars, harvest time is traditionally de-termined non-destructively with colour maps or colorimeters by comparing skin background colour evolution from green to yellow during 'on-tree' matur-ation and ripening. Peaches with ground colour expressed as $h°$ less than 80° are considered well matured and ready for commercial harvest. However, as mentioned previously, this non-destructive approach cannot be used for newer cultivars that have been selected for improved blush coloration and fully red overcolour coverage prior to maturation (Minas *et al.*, 2018).

The development of non-destructive methods to assess fruit maturity and internal quality offers several advantages. These include accurate evaluation of the orchard maturation, along with a better understanding of quality evolution for real-time decision making about harvest time and postharvest handling considerations (Nicolaï *et al.*, 2014; Minas *et al.*, 2018). Additional advantages include the potential for repeated measurements on the same fruit samples to monitor the physiological evolution of individual fruit in 'on-tree' and 'off-tree' conditions (Minas *et al.*, 2021). When using non-destructive techniques, the user has the possibility of expanding assessments to many or even the whole lot of fruit and has the potential to acquire real-time data on several quality and maturity parameters simultaneously (Minas *et al.*, 2021). The ability to separate fruit into maturity and/or quality classes immediately after harvest is also beneficial with this technology, to sort fruit prior to storage to dramatically optimize postharvest handling and improve consumer quality (Ziosi *et al.*, 2008; Spadoni *et al.*, 2016; Costa *et al.*, 2017). A non-destructive methodology is beneficial for large-scale data acquisition in the field and in the packinghouse to assess and understand the influence of various preharvest/harvest conditions on peach quality and postharvest performance. Real-time assessment of large samples of fruit in a tree canopy or in a sorting line can help optimize orchard, harvest and postharvest management, and can provide critical decision-support information to growers, packers and shippers.

One of the recurring problems that peach researchers have had while trying to evaluate the effects of various preharvest conditions on fruit quality is that fruit maturity, which also affects quality, is affected by many of them. Crop load is an important factor influencing fruit quality and maturity, as a higher crop load will negatively impact fruit quality and delay maturation (Minas *et al.*, 2021). Fruit developing in different canopy positions with differing microclimates will also vary in both maturity and quality (Minas *et al.*, 2018). Therefore, it is difficult to ascertain the true impact of these preharvest factors on internal quality, as differences in quality may simply be a result of variable maturation rather than the preharvest condition itself (Minas *et al.*, 2018, 2021). One way to deal with differences in fruit maturity has been to harvest fruit with a similar ground colour. However, this is only an option for bicoloured peach cultivars and the maturity determination with these methods can be subjective. Non-destructive technology has demonstrated the ability to simultaneously collect data on multiple parameters (e.g. maturity and quality) to help track quality development, while controlling for maturation, throughout the growing season to characterize the true impact of preharvest factors on peach quality (Anthony *et al.*, 2020, 2021; Minas *et al.*, 2021; Anthony and Minas, 2022). Similarly, this technology could allow the development of optimized postharvest protocols to minimize losses and maximize consumer acceptance by segregating fruit based on maturity and/or internal quality (Costa *et al.*, 2017; Minas *et al.*, 2021). Furthermore, by harnessing this technology, peach fruit can be handled or marketed based on non-destructive assessment (e.g. % DMC) at harvest, or on receiving or sorting in the packinghouse.

Among the different technologies that have been used in recent decades, near-infrared (NIR) spectroscopy (NIRS) is a non-destructive option with the most reported applications to determine the peach fruit industry's standard quality and maturity indices, mainly the composition of non-structural carbohydrates and organic acids (Grassi and Alamprese, 2018; Minas *et al.*, 2021). Advanced sensing hardware and multivariate statistics such as partial least squares (PLS) regression analysis have been used to develop meaningful information from analysis of the transmitted NIR by the fruit surface (Nicolaï *et al.*, 2014). However, the necessity of calibrating the instrument for different fruit quality attributes by highly trained personnel makes its adoption by most users (producers, packers, shippers, retailers and researchers) challenging. As a result, non-destructive sensors with accurate models for predicting standard fruit quality (DMC, SSC and TA) and maturity (FF) parameters for a range of fruits (including peach) are not yet commercially available.

9.3.1 NIRS

NIR radiation includes the electromagnetic spectrum between the wavelengths of 780 and 2500 nm. When an object's surface and/or fruit tissues are irradiated with NIR, the reflected or transmitted NIR radiation is subsequently measured by NIRS (Fig. 9.3). Scattering and absorption changes in the spectral properties of NIR radiation that penetrates a specific product surface are wavelength dependent. The chemical composition and microstructure of a product's tissue determines the light-scattering characteristics, which can affect the changes in spectral properties of the penetrating NIR radiation. Using advanced multivariate statistical techniques such as PLS regression analyses, the desired fruit composition information from the convoluted spectra collection can be extracted (Fig. 9.3) (Nicolaï *et al.*, 2014). In peach, NIRS has mainly been used for SSC and DMC prediction; however, different technologies have been combined with NIRS, such as short-wave NIRS, long-wave NIRS or visible light–NIRS (Vis-NIRS) hyperspectral imaging, for early bruising or skin defect detection (Li *et al.*, 2016, 2018).

NIRS is utilized mainly in the form of hand-held 'open-type' devices (e.g. F-750 Produce Quality Meter, Felix Instruments; Fig. 9.4) that can be calibrated for a variety of fruit species/cultivars, as well as a range of quality and maturity traits by the end user (Nicolaï *et al.*, 2014; Minas *et al.*, 2018, 2021). This technology can be combined in modern sorting-line applications for large-scale data collection that can improve both the efficiency and real-time decision making at any point in the fresh-produce supply chain (Slaughter *et al.*, 2003; Nicolaï *et al.*, 2014; Costa *et al.*, 2017). Nevertheless, the need for device calibration for different quality and maturity traits across different cultivars (Zhang *et al.*, 2019) requires highly trained personnel, and this makes its broad adoption by the tree-fruit industry challenging. Again, non-destructive

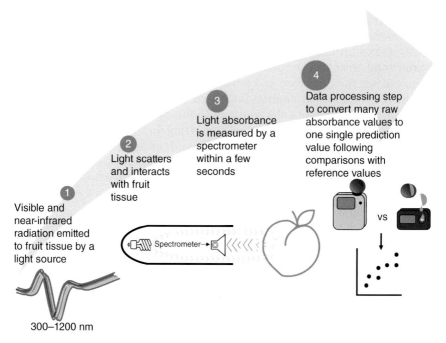

Fig. 9.3. The main steps for assessing fruit quality and maturity non-destructively using 'open-type' visible light–near-infrared sensors.

sensors with accurate prediction models of internal fruit quality (DMC, SSC and TA) and maturity (FF) parameters for peach are not commercially available (Donis-González *et al.*, 2020).

The development of precise and reliable NIRS-based non-destructive tools to assess physicochemical properties of intact fleshy fruit, including peach, has been challenging. The focus for NIRS applications has been the assessment of various internal quality traits in intact fruit at harvest or postharvest (Sánchez *et al.*, 2011; Kumar *et al.*, 2015; Marques *et al.*, 2016; Escribano *et al.*, 2017; Li *et al.*, 2017; Theanjumpol *et al.*, 2019; Zhang *et al.*, 2019). Use of NIRS in the assessment of destructed/processed fruit and vegetable crops and/or peeled fruit could provide an easier way to calibrate accurate and efficient prediction models for many qualitative parameters; however, this approach does not provide a solution for the fresh-produce industry (Lan *et al.*, 2020; Subedi and Walsh, 2020). Non-destructive prediction of SSC or DMC via NIRS in intact peach across many studies has provided a root mean square error of prediction (RMSEP) of more than 1% (or °Brix in the case of SSC) (Minas *et al.*, 2018). It is broadly accepted and scientifically proven that consumers can detect differences in SSC at 1% (or 1°Brix) among fruit samples (Harker *et al.*, 2002), making the existing NIRS technology's effectiveness and adaptation even more challenging.

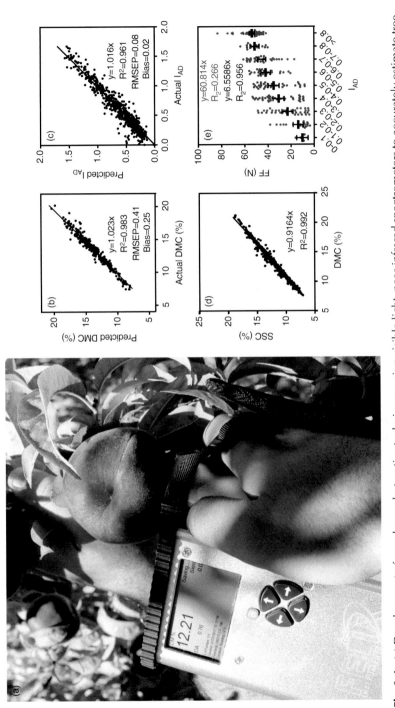

Fig. 9.4. (a) Development of novel non-destructive techniques using visible light–near-infrared spectrometers to accurately estimate tree-fruit quality and maturity with a single scan. (b, c) Accuracy of the created models in terms of linearity (R^2), root mean square error of prediction (RMSEP) and bias is indicated for non-destructive prediction of quality expressed as dry matter content (DMC) (b) and maturity expressed as index of absorbance difference (I_{AD}) (c) in peach fruit. (d) Correlation coefficients of the predicted non-destructive sensor's soluble solids content (SSC) with DMC values. (e) Correlation coefficients of the predicted I_{AD} and actual fruit flesh firmness (FF) values that correspond to individually plotted values of I_{AD} against FF or clustered I_{AD} values against FF. Correlation with FF is significant only when I_{AD} values were clustered at classes of 0.1. Adapted from Minas *et al.* (2021).

The model development process requires the selection of fruit from a wide range of maturity and internal quality for the calibration population, which includes destroying this fruit via destructive quality analysis for validation (Minas *et al.*, 2021). Following the development of novel crop load or rootstock × fruit developmental stage protocols for multivariate NIRS-based prediction model calibration, Minas *et al.* (2021, 2023) were able to assess peach internal quality (DMC and SSC) non-destructively with high accuracy for the first time (Fig. 9.4). While the use of NIRS has demonstrated success in accurately and non-destructively predicting SSC and DMC, there remain other important quality and maturity parameters (TA and FF), as well as textural and storage disorders, that also need to be assessed (Minas *et al.*, 2021). In peach, DMC demonstrates a strong correlation ($R^2 = 0.96–0.99$) with SSC and can be used as a more accurate non-destructive indicator of peach fruit internal quality and sugar content (Fig. 9.4; Minas *et al.*, 2021, 2023).

The DA-Meter
Visible light radiation (Vis) and NIRS have been combined (Vis-NIRS) to create a new non-destructive peach index that has demonstrated a correlation with the onset of endogenous ethylene synthesis and determines physiological maturity/ripening status of peaches and nectarines (Costa *et al.*, 2009). More precisely, this index calculates the absorbance difference (index of absorbance difference, I_{AD}) between two wavelengths (670 and 720 nm) near the absorption peak of chlorophyll *a* ($A_{670nm} – A_{720nm}$; Ziosi *et al.*, 2008). A factory-calibrated ('closed-type') hand-held Vis-NIRS sensor (DA-Meter; T.R. Turoni Srl) can take rapid non-destructive fruit scans (i.e. I_{AD} measurements) that correspond to chlorophyll concentration (i.e. ground colour) a few millimetres below the skin, which provides an estimate of fruit physiological maturity and consumer acceptance (Fig. 9.5) (Costa *et al.*, 2017).

The I_{AD} can be used to determine peach harvest time non-destructively and efficiently, but preliminary work is required to determine the minimum maturity thresholds that are specific for the different cultivars. I_{AD} values range from 0.3 for a tree-ripe (FF 30–40 N) 'Redhaven' (early/mid-season) or 'Suncrest' (mid-season) harvest, while 'Angelus' (late-season) requires a higher I_{AD} value of around 1.0 to achieve tree-ripe maturity (Anthony *et al.*, 2023). Overall variation in FF was limited, while I_{AD}, a more accurate predictor for peach physiological maturity (Costa *et al.*, 2009), demonstrated wider variability among cultivars of distinct harvest windows. For example, 'Bounty' and 'Allstar' had significantly higher I_{AD} values when compared with 'Redhaven' and 'Suncrest' at the tree-ripe stage, while their FF values did not differ statistically. These data support the suggestion that FF and physiological maturity assessed by I_{AD} are not always related, and that different cultivars exhibit different I_{AD} values at tree-ripe maturity (Ziosi *et al.*, 2008). Generally, late-season cultivars tend to exhibit higher I_{AD} values than earlier cultivars at the tree-ripe stage (Anthony *et al.*, 2023).

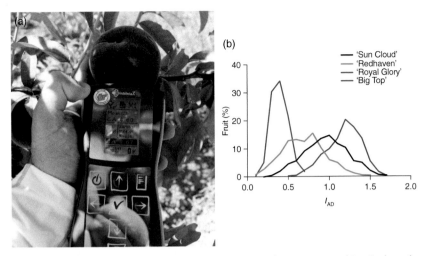

Fig. 9.5. Use of the DA-Meter in the field allows rapid assessment of I_{AD} (index of absorbance difference) that is related to fruit physiological maturity. The distribution of I_{AD} of harvested 'Big Top', Royal Glory', 'Redhaven' and 'Sun Cloud' peach fruit showing that I_{AD} associated physiological maturity is cultivar and flesh typology dependent.

Similarly, a previous study on the mean values and distribution ranges of I_{AD} in a total of 700–1250 fruits from the cultivars 'Big Top,' 'Royal Glory,' 'Redhaven' and 'Sun Cloud' found that they differed significantly (Fig. 9.5). These data suggest that the I_{AD} index is cultivar dependent, and that specific values for each cultivar should be determined in order to be applied commercially. This has been proven to be particularly important in cultivars where external colour cannot define the maturity stage, such as 'Royal Glory' (Manganaris *et al.*, 2017; Minas *et al.*, 2018). Overall, more research data are necessary to establish maturity thresholds for different cultivars and types of postharvest handling for the adoption of this index as a standard commercial practice.

It is important to note, however, that the I_{AD} has been proven to be an accurate, rapid and simple measuring technique to determine peach background colour changes that correlate with physiological maturity (Minas *et al.*, 2018, 2020, 2021). I_{AD} values do not correlate directly with any of the traditional harvest indices (e.g. FF or SSC) when individual values are used (Ziosi *et al.*, 2008; Herrero-Langreo *et al.*, 2011; Shinya *et al.*, 2013; Gonçalves *et al.*, 2016; Minas *et al.*, 2021). Instead, stronger correlations have been reported when clustered values are used (Fig. 9.5) (Ziosi *et al.*, 2008; Bonora *et al.*, 2013; Drogoudi *et al.*, 2017; Minas *et al.*, 2021). However, postharvest changes in the I_{AD} were more closely correlated with changes in FF and less with other traditional maturity/quality indices (e.g. SSC, TA and DMC). Out of 26 peach and nectarine cultivars, changes in the I_{AD} during shelf life were highly correlated with changes in FF in 24 cultivars, with TA in 15 cultivars and with ethylene production rate in 13 cultivars. In all cases, a significant correlation ($r = 0.88$) was determined, suggesting absolute linear correlations (Drogoudi *et al.*, 2016). FF was not significantly correlated with I_{AD} in only two cultivars, the peach 'Opsimo Naoussas' and the nectarine 'Tasty Free'. Both of these cultivars are characterized by slow-softening rates and a late harvest time. Similarly, FF may not relate to I_{AD} in stony-hard cultivars, as they produce little to no ethylene upon ripening and minimally soften at physiological maturity (Serra *et al.*, 2020). The changes in SSC did not correlate with I_{AD} in most cultivars, although for apples, a closer relationship of SSC and I_{AD} has been reported (McGlone *et al.*, 2002; Nyasordzi *et al.*, 2013).

In conclusion, I_{AD} is a particularly important index for bicoloured and more importantly fully red overcoloured cultivars, where harvest time is difficult to estimate due to excess red overcolour on the fruit's skin that obscures the background colour, which is normally used to estimate fruit maturation. I_{AD} identifies physiological changes occurring during ripening, whether or not they lead to modifications in fruit quality determined by means of traditional methodology (Costa *et al.*, 2006, 2009; Ziosi *et al.*, 2008). The effect of crop load and canopy position on fruit physiological maturity at harvest is better described using the non-destructive Vis-NIRS and I_{AD} than the standard destructive FF values (Minas *et al.*, 2021). I_{AD} can be used at harvest, or upon receiving fruit in a packinghouse, to group harvested fruit into homogeneous classes that can ripen uniformly (due to similar maturation status) to optimize postharvest management (Spadoni *et al.*, 2016). In addition, I_{AD} and the DA-Meter can efficiently monitor physiological changes during 'on-tree' and 'off-tree' peach fruit maturation and ripening to determine the impact of optimum harvest time for maximum consumer acceptance (Costa *et al.*, 2006, 2009; Ziosi *et al.*, 2008). The DA-Meter is considered a very valuable research tool and should be used in any peach-related experiment that seeks to control for the confounding variable of maturity to understand the true impact of preharvest factors on peach maturity, internal quality, postharvest performance, molecular biology and ripening physiology (Minas *et al.*, 2018, 2021; Anthony *et al.*, 2020, 2021; Anthony and Minas, 2022).

9.3.2 Fluorescence spectroscopy

Fluorescence spectroscopy is another non-destructive technology that has been used to predict maturity and quality parameters in fleshy fruit. Biological materials fluoresce, emitting energy upon relaxation within the visible electromagnetic range, after being excited by high light energy (e.g. UV light) (Bureau *et al.*, 2009). Prior to fluorescence measurements, the fruit must be adjusted to a completely dark environment for at least 10 min to inhibit interaction with any photosynthetic machinery (Butz *et al.*, 2005). Upon exposure to high light intensity, pigments will absorb and reflect light in particular wavelengths. In particular, chlorophyll *a* and *b* absorb radiation from approximately 250 to 700 nm, with a small portion of this absorbed radiation emitted as red fluorescence. This chlorophyll fluorescence can then be measured in fruit to indirectly evaluate photosynthetic efficiency, along with the physiological maturation status of the fruit. This is understood with fluorescence measurements because, as fruit maturation and ripening advances, photosynthetic rates decline in conjunction with the degradation of chlorophyll content (Butz *et al.*, 2005).

Additional pigments such as anthocyanins, flavanols and carotenoids in the fruit tissues can also react to fluorescence within the 400–800 nm spectral range. In apples, light absorbed in the UV-A range demonstrated a relationship with the total flavanol content, while blue light absorption was related to anthocyanins (Hagen *et al.*, 2006). These secondary metabolites are important quality attributes to monitor and quantify, as they contribute to both the nutraceutical and flavour characteristics of the fruit (e.g. antioxidant activity) (Serra *et al.*, 2020).

In peach, a fluorescence spectrometer prototype was recently used to try and model predictions for maturity (FF and I_{AD}) and quality indices (SSC, and skin and flesh colour) in four peach cultivars (Scalisi *et al.*, 2020). Peach fruit within different maturation classifications as expressed by variable I_{AD} values (immature, harvest-ready and mature) were scanned with the fluorometer and demonstrated unique spectral fluorescence emissions. The predicted values were compared against the observed values to determine the model accuracy as expressed by the coefficient of determination of the cross-validation (R^2_{CV}) and the root mean square error of the cross-validation ($RMSE_{CV}$). PLS analyses revealed that the least accurate prediction models were for the FF and SSC parameters ($R^2_{CV} = 0.5–0.8$; $RMSE_{CV} = 0.7–2.0$), while I_{AD} was the only parameter that was consistently and accurately predicted ($R^2_{CV} > 0.8$; $RMSE_{CV} < 0.25$) across the four cultivars (Scalisi *et al.*, 2020). Overall, these studies demonstrate promising results for the use and application of hand-held fluorometers to predict maturity in the field and packinghouse; however, accuracy must be improved for commercial use.

9.3.3 Hyperspectral and multispectral imaging

Hyperspectral imaging (HSI) and multispectral imaging (MSI) can simultaneously acquire spectral scattering profiles that correlate with fruit structural characteristics and can be used to estimate FF with moderate success (Lu and Peng, 2006; Muhua *et al.*, 2007; Wang *et al.*, 2015). HSI and MSI provide an emerging platform technology that integrates conventional imaging and spectroscopy to attain both spatial and spectral information from an object, which is not possible with NIRS (Gowen *et al.*, 2007). HSI has applications in food safety and quality assessment, such as contaminant detection, defect identification, constituent analysis and quality evaluation, as it enables characterization of complex heterogeneous samples, while the spectral feature allows the identification of a wide range of multi-constituent surface and subsurface features (Table 9.1). HSI and MSI can simultaneously acquire spectral scattering profiles that correlate with structural characteristics and detect common skin defects/bruises on various fruit with singular colour (ElMasry *et al.*, 2008; Nanyam *et al.*, 2012; Lee *et al.*, 2014; Zhang *et al.*, 2020). Peaches are sensitive to bruising from mechanical damage. Analysis of peach fruit bruised regions revealed considerable changes in spectral characteristics over time (2 days) after bruising (Li *et al.*, 2021). Thus, early bruises were able to be identified, as HSI provided a prediction of bumps on peaches. Skin detection in bicoloured peach fruit is also successful with these technologies, as the surface colour is heterogeneous and is generally composed of at least three different colours (e.g. red, green and yellow) (Li *et al.*, 2016, 2018). HSI technology was also used for fungal-disease-related damage where it was possible to quantitatively identify and also visualize the degree of decay for early disease inspection in peach (Sun *et al.*, 2017; Liu *et al.*, 2020). HSI technology was successfully used to distinguish between nectarine cultivars having similar external characters but different physicochemical properties, and performed better than Vis colour imaging or human inspections (Munera *et al.*, 2018a).

Fruit FF can also be estimated by detecting the scattering of light in the flesh using HSI techniques. This approach is feasible as the softening of fruit is due to solubilization of pectins in the intercellular space and cell wall. The pectins contain polar covalent bonds (C–H, O–H and N–H), which can absorb the light of specific wavelengths when the fruit is irradiated by light, revealing complex microstructure information related to the vibration and stretching behaviours of these covalent bonds (McGlone and Kawano, 1998; Wang *et al.*, 2015). An attempt to quantify FF using HSI technology was made in peach with moderate success (Lu and Peng, 2006; Muhua *et al.*, 2007), as well as in persimmon (Munera *et al.*, 2018b) and apple (Peng and Lu, 2008).

In summary, the HSI technology is an innovative method for simultaneously measuring and visualizing physicochemical components in fruits. It has been used successfully for estimating the quality attributes of fruit, including estimation of physicochemical attributes, detection of common contaminants,

Table 9.1. Comparison of RGB (red, green blue) imaging, near-infrared spectroscopy (NIRS), multispectral imaging (MSI) and hyperspectral imaging (HSI). From Gowen *et al.* (2007).

Feature	RGB imaging	NIRS	MSI	HSI
Spatial information	✓		✓	✓
Spectral information		✓	Limited	✓
Multi-constituent information	Limited	✓	Limited	✓
Sensitivity to minor components			Limited	✓

detection of insect and fungal damage, and evaluation of maturity stage. However, existing problems remain such as image focusing, determination of scanning parameters, image processing, and data mining and analysis (Sun *et al.*, 2017). Although it is a powerful technology for fruit and food production monitoring, the current high cost for purchasing this equipment and its processing speed are limiting factors for this technology (Gowen *et al.*, 2007).

9.3.4 X-ray radiation

X-ray radiation is now being pursued for its potential use in fruit quality assessment. X-ray radiation can penetrate material bodies, with the absorbed/ transmitted radiation relating to the density and thickness of the sample (Butz *et al.*, 2005). Water content accounts for the predominant amount of X-ray absorption when fruit is scanned, as water makes up the majority of fruit products. As a result, these predictions are more related to fruit density, rather than chemical composition. It has been suggested that this technology should be used primarily for detecting postharvest disorders such as bitter pit, water core and brown core in apple, along with bruising and CI (i.e. woolliness or mealiness) in other horticultural products such as peach (Butz *et al.*, 2005). Three predominant X-ray methods are used: two-dimensional radiography, line scans and computed tomography (CT) in which a three-dimensional image is produced.

X-ray CT scanners have been used to evaluate internal fruit quality in peaches throughout ripening (Barcelon *et al.*, 1999). X-ray absorption coefficients were used to develop an index known as the 'CT number,' which demonstrated positive relationships with moisture content ($R^2 = 0.97$) and TA ($R^2 = 0.87$), and inverse relationships with pH ($R^2 = 0.89$) and SSC ($R^2 = 0.89$) (Barcelon *et al.*, 1999). Therefore, X-ray CT imaging may be used as an effective non-destructive tool for predicting peach fruit quality. However, the cost, size and access to such equipment may limit its application in the field, although the integration of CT technology may fit well along a sorting line. For field application, recent advancements have enabled the development of hand-held X-ray fluorescence devices. As a result, X-ray fluorescence has been used successfully to evaluate fruit mineral concentration and spatial location (primarily potassium and calcium) non-destructively in apple and pear (Kalcsits, 2016). Although demonstrating success in predicting fruit quality and mineral composition, X-rays, like most non-destructive tools, require optimization and calibration for distinct species, cultivars and tissue types.

9.3.5 Nuclear magnetic resonance (NMR) and magnetic resonance imaging (MRI)

Atoms such as ^{1}H, ^{13}C and ^{31}P have magnetic moments (i.e. magnetic resonance active), which are able to absorb energy when irradiated with specific radiofrequencies (RFs) in a highly magnetic environment (Bureau, 2009). With respect to fruit, ^{1}H is of the most interest, given the large amount of water present in fruit and its high sensitivity to RF. Once irradiated with RF, protons will rotate due to magnetism and, upon signal removal, the atom will return to its relaxed state releasing energy, which can be detected as a subsequent RF signal (Bureau, 2009). MRI can be conducted if magnetic field gradients are applied in three-dimensions around the fruit. NMR detects physical characteristics of fruit, such as the mobility of water and the density of protons (from water molecules). MRI can map the proton density of a fruit, identifying physical characteristics such as defects and injury (Srivastava *et al.*, 2018). MRI can also identify the chemical shifts and relaxation times of magnetic resonance-active nuclei, along with water diffusion constants, which may help provide information to evaluate fruit ripening and maturation (Srivastava *et al.*, 2018), although, to date, NMR and MRI have not demonstrated a correlation with fruit maturity (Hills and Clark, 2003). NMR and MRI technologies may be most applicable along packing/sorting lines, but the technology needs to be ultrafast, efficient in terms of image acquisition under dynamic conditions and within the context of a low-magnetic field (Hills and Clark, 2003).

In peach and nectarine, woolly texture or mealiness (i.e. CI symptoms) is a significant storage disorder, which relates to extracellular water being bound to the cell walls within the fruit (Lurie and Crisosto, 2005). As a result, cell adhesion is reduced, which increases the number and size of the intercellular air spaces. This disorder can be detected using NMR and MRI, as air spaces relate to areas of lower proton density (Hills and Clark, 2003). NMR and CT X-rays were compared for their ability to detect mealiness (Sonego *et al.*, 1995). NMR was effective in identifying mealiness as dark spots in the fruit, which represented low proton density. However, as the total water content in this tissue remains unaffected, NMR alone was not sufficient to fully elicit the physiological basis of mealiness. When coupled with X-ray radiation, CT could identify the presence of intercellular gas in these dark spot locations (Sonego *et al.*, 1995). The coupling of these technologies helped further explain the mechanism behind mealiness. Overall, X-rays are more effective in determining storage disorders as they relate to the movement and binding of water, while NMR and MRI are effective in discriminating areas of higher or lower proton density.

9.3.6 Mechanical properties

Until recently, methods to non-destructively evaluate the textural properties of fruit were unavailable for commercial application. New technologies, such as the Sinclair iQ™ firmness tester (SIQ-FT; Sinclair Systems International), has provided an innovation for the produce supply chain and allows the non-destructive investigation of fruit ripeness and bruising susceptibility, especially in stone fruit (Valero *et al.*, 2007). The SIQ-FT uses a pneumatic impact head, which is equipped with a piezoelectric sensor that returns an FF value (called SFI score) upon impact with the fruit (Valero *et al.*, 2007). The SFI scores range from 0 (soft) to 100 (firm). The SIQ-FT can be integrated into packing lines, being compatible with popular sizing and grading equipment, and works at sorting speeds of ten fruit s^{-1} (Khalifa *et al.*, 2011).

The SIQ-FT has demonstrated high correlation coefficients (R^2) with penetrometers in avocados (0.84), kiwifruit (0.92) and nectarines (0.95) (Khalifa *et al.*, 2011). It was also recently used in a variety of stone fruit species and demonstrated strong relationships with a destructive FF technique, resulting in R^2 values of 0.69, 0.71 and 0.71 for peaches, nectarines, and plums, respectively (Valero *et al.*, 2007). However, these correlation coefficients were deemed too low for commercial adoption, with the authors stating that the techniques were measuring two different properties of the fruit (SIQ-FT = elasticity versus penetrometer = tissue failure). They suggested that the SIQ-FT was more successful in discriminating fruit into different commercial categories, such as ripening stage and/or bruising susceptibility, rather than providing a direct correlation with FF (Valero *et al.*, 2007). Thus, the SIQ-FT might be used to classify stone fruit as 'ready to eat' or 'ready to buy,' and/or 'mature' or 'immature.'

9.3.7 Resonant acoustic vibration method

Resonant acoustic vibration is a promising method for evaluating firmness or texture and contrasts with other non-destructive methods, such as MRI or X-ray CT, because of its relatively low cost and lack of need for lengthy analysis (Ali *et al.*, 2017). The acoustic vibration method, or use of an acoustic firmness sensor (AFS), was initially developed to estimate fruit firmness (Taniwaki *et al.*, 2009) and has recently been commercialized as the AWETA AFS (AWETA International), a system that can be incorporated into commercial sorting/packing lines. The AWETA AFS can also measure fruit mass. The technology has seen increasing use in the evaluation of the maturity of various tree-fruit crops, including persimmon (Taniwaki *et al.*, 2009), pear (Oveisi *et al.*, 2014), apple (de Belie *et al.*, 2000) and kiwifruit (Terasaki *et al.*, 2013).

Upon entry into the AFS, the fruit is gently tapped with a probe and then analysed for its vibration pattern (i.e. resonance-attenuated vibration) by 'listening' to its acoustic spectral signature with a microphone (Feng *et al.*, 2016). This controlled impact provides broadband input energy to excite the fruit, while the microphone allows non-destructive measurement of its structural response based on the elastic mechanical properties of the fruit. The results are then translated into a firmness index, which can be used to sort the fruit (Feng *et al.*, 2016).

When the AWETA AFS was evaluated for accuracy in peach, correlations between destructive and acoustic firmness measurements were significant ($P < 0.01$) but too low, with R^2 values ranging between 0.22 and 0.62 across all cultivars (Yurtlu, 2012). When the AWETA AFS was paired with NIRS (675 and 697 nm), accuracy appeared to improve in assessing peach firmness with acoustic measurements, resulting in R^2 values of 0.76–0.77, depending on the regression method (Lafuente *et al.*, 2015). However, it is important to note that the correlation between fruit elasticity (as measured by the AWETA AFS) and the force requirement for tissue failure (as measured by a penetrometer) weakens as the fruit become too soft and the relationship between these physical properties is reduced.

In most studies, the second and third resonant frequencies (f2 and f3) have been examined as a firmness index because they reflect the internal physical traits of fruit, such as elasticity and viscosity (Cooke, 1972; Drake, 1962). During 'on-tree' peach ripening, a strong positive linear correlation ($R^2 > 0.8$) was observed between penetration forces and f2 or f3 acoustic vibration parameters (Kawai *et al.*, 2018). More importantly, the f3/f2 value could very well predict the occurrence of split pits during 'on-tree' ripening or postharvest (Nakano *et al.*, 2018). Therefore, an AFS could be a powerful tool for: (i) peach sorting according to firmness, maturity class, CI susceptibility and split-pit occurrence in the packinghouse (Hale *et al.*, 2013); and (ii) helping the grower to thin defective fruit efficiently when mobile/hand-held devices are available (Kawai *et al.*, 2018; Nakano *et al.*, 2018).

9.4 SUCCESSFUL APPLICATIONS AND POTENTIAL FUTURE USES OF NON-DESTRUCTIVE TECHNOLOGY

The development of non-destructive methods to measure maturity and internal fruit quality offers several advantages for efficient evaluation of large sample volumes to optimize real-time decision making in the orchard and the packinghouse (Nicolaï *et al.*, 2014). Additionally, the use of these tools is effective in research, targeting a better understanding of the orchard's impact on fruit harvest quality and postharvest performance (Minas *et al.*, 2018). Among the non-destructive methods, use of Vis and NIRS are the most studied technologies to determine standard peach quality (DMC, SSC, TA and overcolour blush) and maturity (ground colour, FF, I_{AD}) traits. The necessity for end-user device calibration, the obscure relationship of results to well-understood standard destructive parameters, and the lack of accuracy at the commercial level for different fruit qualities and maturity attributes render their adoption by producers, packers and shippers challenging, but they remain effective tools in pomological experiments (Crisosto and Costa, 2008; Minas *et al.*, 2021; Anthony and Minas, 2022).

9.4.1 Robust model development and calibration for non-destructive Vis-NIRS technology

Novel crop load or rootstock × fruit developmental stage protocols for calibration of multivariate Vis-NIRS-based prediction have provided accurate non-destructive models to assess peach internal quality and maturity in a single scan (Minas *et al.*, 2021, 2023). Collectively, these novel calibration protocols have provided a broad range of peach internal qualities at similar maturity levels during fruit growth, development and harvest. These experimental approaches provided ideal biological systems for the calibration of non-destructive models to predict peach quality and/or maturity prior to the commercial harvest stage, as well as at harvest (i.e. tree ripe). Selected fruits from the calibration population exemplified a range of internal quality at different maturity levels and sizes. To minimize the effect of temperature on model performance, due to associated changes in the molecular response to light absorbance (second-derivative absorbance spectra), calibration fruit were exposed at three different temperatures (0, 20 and 30°C) prior to scanning with a non-destructive 'open-type' hand-held NIRS sensor to obtain temperature-compensated models (Donis-González *et al.*, 2020; Minas *et al.*, 2021). By scanning at different temperatures, the created models were able to ignore any spectral shifts or changes that are irrelevant to the traits of interest (Fig. 9.4). Such an approach accommodates any need for use under varying temperature conditions, such as the field, laboratory, retail store, cold-storage room, packinghouse or grocery store, which represent the significant domains of the fresh-produce supply chain.

A regression coefficient was calculated with a non-linear iterative PLS regression for each quality trait (DMC or SSC) and for physiological maturity (I_{AD}) within the spectral ranges of 729–935 nm and 600–750 nm, respectively. These coefficients were loaded on to the 'open-type' hand-held device (F-750 Produce Quality Meter; Felix Instruments) and were used to non-destructively predict internal quality and maturity values. Independent validation and regression statistics from the developed prediction models following this approach found that DMC ($R^2 = 0.98$, RMSEP = 0.41%), SSC ($R^2 = 0.96$, RMSEP = 0.58%) and I_{AD} ($R^2 = 0.96$, RMSEP = 0.08) could be measured accurately with a single scan during fruit growth and development (from 70 days after full bloom until the harvest of tree-ripe fruit) using an 'open-type' hand-held Vis-NIRS device (Minas *et al.*, 2021). To the best of our knowledge, this is the best non-destructive prediction performance of Vis-NIRS technology that has been reported in fresh produce and could lead to significant improvements in the commercialization of this technology within numerous other fruit and vegetable crops. As a result, the authors suggested that the development of a pre-calibrated species- and parameter-specific novel concept device could help assess the true impact that preharvest factors have on peach fruit quality and maturity during fruit growth, development and harvest in the field (Minas *et al.*, 2021). This calibration approach could enhance NIRS adaptation across the tree-fruit supply chain, as it can provide a highly acceptable accuracy for commercial use and for real-time decision making in the field/packinghouse (Minas *et al.*, 2021, 2023).

Based on these novel approaches for multivariate Vis-NIRS-based prediction model calibration, three robust individual cultivar models were developed for 'Redhaven' (early/mid-season, bicolour), 'Cresthaven' (late, bicolour) and 'Sierra Rich' (early/mid-season, full red overcolour) to predict physiological maturity (I_{AD}) and quality (DMC) (Minas *et al.*, 2021, 2023). The prediction performance of these cultivar-specific models was validated using independent fruit populations of seven other peach cultivars with variable ripening times that share similar phenotypic characteristics with at least one of the developed models (Anthony *et al.*, 2023). Spectral absorption characteristics highlighted the need for individual or 'cluster-cultivar' models contextualized for similar phenotypic characteristics or ripening time. Individual cultivar models developed were robust, as even for cultivars that had unique spectral profiles, predictions for DMC and I_{AD} retained R^2 values above 0.75 when assessed by each of the three individual cultivar models (Anthony *et al.*, 2023). Prediction accuracy was further improved for almost all model × cultivar combinations across both maturity and quality indices by including a small portion of cultivar-specific fruit in the pre-existing individual cultivar models (Anthony *et al.*, 2023). Overall, linear regressions across most cultivars revealed that R^2 values exceeded 0.90 upon further customization for DMC and I_{AD} predictions (Anthony *et al.*, 2023).

9.4.2 Application of accurate non-destructive Vis-NIRS technology to evaluate the true impact of preharvest factors on peach fruit quality and maturity

Traditional destructive quality and maturity assessments are time-consuming and labour-intensive. This limits the ability to understand the wide variability of fruit quality and maturity on a tree or in an orchard (Minas *et al.*, 2018). By using strong and reliable non-destructive technology, the impact of preharvest factors, such as crop load management, fruit position in the canopy or rootstock, on peach fruit quality and maturity can be accurately and extensively characterized (Minas *et al.*, 2021). Large-scale field validation has shown that heavier crop loads reduced peach quality (DMC and SSC) and delayed maturation (I_{AD}) in the lower canopy, while the fruit canopy position mainly affected fruit quality and maturity only in moderate crop loads (Minas *et al.*, 2021). These data indicate that the true role of preharvest factors on quality development can be captured only when controlling for maturity status as a confounding variable among treatments (Anthony and Minas, 2022).

Through the application of robust NIRS technology, models can be developed and used to predict internal quality (DMC and SSC) and I_{AD} simultaneously from the same piece of fruit. Such an approach can facilitate rapid maturity control to conduct more accurate preharvest factor assessments on quality and enable further biological investigations, as maturity is a regulated process at the molecular level. In particular, for further investigations into the regulatory and biochemical pathways in fruit, such an approach is the only way to ensure that metabolic differences among treatments are associated with preharvest conditions and not with the subsequent effect of fruit maturation.

A crop load study in peach ('Cresthaven') that controlled for maturity (I_{AD}) using robust Vis-NIRS models revealed the true impact of carbon supply on fruit quality development (at three developmental stages: S2, S3 and S4) and metabolism (Anthony *et al.*, 2020). Phenotypes and quality characteristics were minimal early on in S2 but drastically different at harvest (S4) (Anthony *et al.*, 2020). Metabolite profiling (using non-targeted gas chromatography–mass spectrometry) from each carbon supply × developmental stage treatment revealed that vast metabolic shifts occurred between crop loads early in development, but very few differences were present at harvest (Anthony *et al.*, 2020). It was hypothesized that these early metabolic shifts may act as priming events for fruit quality and phenotypic differences observed at harvest. Additionally, given the significance of the primary metabolism in fruit development and maturation, by controlling for maturity between treatments, it was revealed that metabolic differences were minimal at harvest. These metabolic findings were novel to the pomological field and were only obtainable through the use of non-destructive Vis-NIRS technology, which can allow the segregation and sampling of fruit of equal maturity. Therefore, 'omics' studies that investigate the influence of preharvest factors on biological characteristics of the fruit (e.g. metabolism) must control for maturity to yield meaningful results.

In a canopy-position study, two peach cultivars of differing vigour ('Sierra Rich', low vigour; 'Cresthaven', high vigour) were assessed for fruit maturity (I_{AD}) and quality (DMC) at two canopy positions (top and bottom) (Anthony *et al.*, 2021). Prior to maturity control, both cultivars demonstrated significant differences ($P \leq 0.05$) in DMC across canopy positions. The difference in DMC (ΔDMC) between fruit from the top and bottom of the canopy position was 1.5% and 2.0% in the low-vigour and high-vigour cultivars, respectively (Anthony *et al.*, 2021). After maturity control using Vis-NIRS, ΔDMC between positions was reduced in the low-vigour cultivar to 0.7%, but remained significantly different and high at 2.1% in the high-vigour cultivar (Anthony *et al.*, 2021). As a result, it was concluded that quality is more related to the light environment the fruit is experiencing, rather than the canopy position alone (Anthony *et al.*, 2021). This was evidenced by the poor light distribution present in the high-vigour cultivar, which resulted in the bottom-positioned fruit exhibiting inferior quality when compared with the top-positioned fruit, even at uniform maturity levels (Anthony *et al.*, 2021). Uniform light distribution in the low-vigour cultivar demonstrated uniform quality at harvest when maturity was controlled for.

Another preharvest factor that promotes distinct light environments for developing fruits is rootstock vigour classification. Rootstocks with increased vigour reduce light availability in the bottom of the canopy, promoting non-uniform canopies, light distribution and quality at harvest. A recent series of studies was conducted to evaluate the impact of rootstock vigour on peach ('Red Haven') fruit quality (Pieper *et al.*, 2022; Minas *et al.*, 2023). When controlling for maturity using Vis-NIRS models, it was demonstrated that fruit quality (DMC and SSC) improved from vigorous to dwarfing rootstock genotypes due to increased light availability within the canopy.

9.4.3 Current and future uses of non-destructive sensors

Improving the consumer quality of peach fruit by harvesting at the tree-ripe stage usually comes with increased costs due to losses. In industries that are focused on consumer satisfaction (e.g. the Colorado peach industry), every year more than 10% of harvested peach fruit is sacrificed as soft or green due to insufficient harvesting decisions and packinghouse efficiency (Minas *et al.*, 2023). In addition, another 10–15% of packed product is lost because of lack of quality/maturity assessment and mistargeting the right customers (Minas *et al.*, 2023). Thus, improvements in the accuracy of available technological applications for non-destructive assessment of peach fruit maturity and quality are expected to provide significant cost reductions for peach fruit growers and

Fig. 9.6. Commercial application of the AWETA Powervision HS in Talbott Farms packinghouse, Palisade, Colorado. This system is equipped with Vis-NIRS and hyperspectral imaging to inspect external fruit quality such as on-skin defects (mechanical damage, soft spots and russet) and through-skin defects (rot, decay and cracks), as well as shape, size and colour measurements.

packers. Currently, the integration of non-destructive sensors in sorting lines may be one of the most used applications of this technology in modern packinghouse facilities, as the fruit can be separated not only according to different shapes, sizes or colour but also by internal (split pits) or external (bruises and cracks, or potential fungal decay) defects for proper management to minimize postharvest losses and optimize sorting and packing efficiency (Fig. 9.6). More effort should be given to increase accuracy in applications that can detect internal fruit quality (SSC and DMC) and storage disorders such as CI symptoms that are directly related to consumer satisfaction for improved sales. None the less, the large number of cultivars arriving in a packinghouse in a day poses a limitation in the use of these types of applications.

REFERENCES

Ali, M.M., Hashim, N., Bejo, S.K. and Shamsudin, R. (2017) Rapid and nondestructive techniques for internal and external quality evaluation of watermelons: a review. *Scientia Horticulturae* 225, 689–699.

Anthony, B.M., and Minas, I.S. (2021) Optimizing peach tree canopy architecture for efficient light use, increased productivity and improved fruit quality. *Agronomy* 11: 1961.

Anthony, B.M. and Minas, I.S. (2022) Redefining the impact of preharvest factors on peach fruit quality development and metabolism: a review. *Scientia Horticulturae* 297: 110919.

Anthony, B.M., Chaparro, J.M., Prenni, J.E. and Minas, I.S. (2020) Early metabolic priming under differing carbon sufficiency conditions influences peach fruit quality development. *Plant Physiology and Biochemistry* 157, 416–431.

Anthony, B.M., Chaparro, J.M., Sterle, D.G., Prenni, J.E. and Minas, I.S. (2021) Metabolic signatures of the true physiological impact of canopy light environment on peach fruit quality. *Environmental and Experimental Botany* 191: 104630.

Anthony, B.M., Sterle, D.G. and Minas, I.S. (2023) Robust non-destructive individual cultivar models allow for accurate peach fruit quality and maturity assessment following customization in phenotypically similar cultivars. *Postharvest Biology and Technology* 195: 112148.

Barcelon, E.G., Tojo, S. and Watanabe, K. (1999) X-ray computed tomography for internal quality evaluation of peaches. *Journal of Agricultural Engineering Research* 73, 323–330.

Boldingh, H., Smith, G.S. and Klages, K. (2000) Seasonal concentrations of non-structural carbohydrates of five *Actinidia* species in fruit, leaf and fine root tissue. *Annals of Botany* 85, 469–476.

Bonghi, C., Ramina, A., Ruperti, B., Vidrih, R. and Tonutti, P. (1999) Peach fruit ripening and quality in relation to picking time, and hypoxic and high CO_2 short-term postharvest treatments. *Postharvest Biology and Technology* 16, 213–222.

Bonora, E., Noferini, M., Vidoni, S. and Costa, G. (2013) Modeling fruit ripening for improving peach homogeneity *in planta*. *Scientia Horticulturae* 159, 166–171.

Bruhn, C.M., Feldman, N., Garlitz, C., Harwood, J., Ivans, E. *et al.* (1991) Consumer perceptions of quality: apricots, cantaloupes, peaches, pears, strawberries, and tomatoes. *Journal of Food Quality* 14, 187–195.

Bureau, S. (2009) The use of non-destructive methods to analyse fruit quality. *Fresh Produce* 3, 23–34.

Bureau, S., Ruiz, D., Reich, M., Gouble, B., Bertrand, D. *et al.* (2009) Rapid and non-destructive analysis of apricot fruit quality using FT-near-infrared spectroscopy. *Food Chemistry* 113, 1323–1328.

Butz, P., Hofmann, C. and Tauscher, B. (2005) Recent developments in noninvasive techniques for fresh fruit and vegetable internal quality analysis. *Journal of Food Science* 70, 131–141.

Byrne, D.H. (2005) Trends in stone fruit cultivar development. *HortTechnology* 15, 494–500.

Cirilli, M., Bassi, D. and Ciacciulli, A. (2016) Sugars in peach fruit: a breeding perspective. *Horticulture Research* 3: 15067.

Cooke, J.R. (1972) An interpretation of the resonant behavior of intact fruits and vegetables. *Transactions of the American Society of Agricultural Engineers* 15, 1075–1080.

Costa, G., Noferini, M., Fiori, G. and Ziosi, V. (2006) Internal fruit quality: how to influence it, how to define it. *Acta Horticulturae* 712, 339–346.

Costa, G., Noferini, M., Fiori, G. and Torrigiani, P. (2009) Use of Vis/NIR spectroscopy to assess fruit ripening stage and improve management in post-harvest chain. *Fresh Produce* 3, 35–41.

Costa, G., Rocchi, L., Farneti, B., Busatto, N., Spinelli, F. and Vidoni, S. (2017) Use of nondestructive devices to support pre- and postharvest fruit management. *Horticulturae* 3: 12.

Crisosto, C. (2002) How do we increase peach consumption? *Acta Horticulturae* 592, 601–605.

Crisosto C.H. and Costa, G. (2008) Preharvest factors affecting peach quality. In: Layne, D.R. and Bassi, D. (eds) *The Peach: Botany, Production and Uses.* CAB International, Wallingford, UK, pp. 536–549.

Crisosto, C.H. and Crisosto, G.M. (2005) Relationship between ripe soluble solids concentration (RSSC) and consumer acceptance of high and low acid melting flesh peach and nectarine (*Prunus persica* (L.) Batsch) cultivars. *Postharvest Biology and Technology* 38, 239–246.

Crisosto, C.H., Slaughter, D., Garner, D. and Boyd, J. (2001) Stone fruit critical bruising thresholds. *Journal of the American Pomological Society* 55, 76–81.

de Belie, N., Schotte, S., Coucke, P. and de Baerdmaeker, J. (2000) Development of an automated monitoring device to quantify changes in firmness of apples during storage. *Postharvest Biology and Technology* 18, 1–8.

Donis-González, I.R., Valero, C., Momin, M.A., Kaur, A.C. and Slaughter, D. (2020) Performance evaluation of two commercially available portable spectrometers to non-invasively determine table grape and peach quality attributes. *Agronomy* 10: 148.

Drake, B. (1962) Automatic recording of vibrational properties of food stuffs. *Journal of Food Science* 27, 182–188.

Drogoudi P., Pantelidis, G.E., Goulas, V., Manganaris, G.A., Ziogas, V. and Manganaris. A. (2016) The appraisal of qualitative parameters and antioxidant contents during postharvest peach fruit ripening underlines the genotype significance. *Postharvest Biology and Technology* 115, 142–150.

Drogoudi, P., Gerasopoulos, D., Kafkaletou, M. and Tsantili, E. (2017) Phenotypic characterization of qualitative parameters and antioxidant contents in peach and nectarine fruit and changes after jam preparation. *Journal of the Science of Food and Agriculture* 97, 3374–3383.

ElMasry, G., Wang, N., Vigneault, C., Qiao, J. and ElSayed, A. (2008) Early detection of apple bruises on different background colors using hyperspectral imaging. *LWT – Food Science and Technology* 41, 337–345.

Escribano, S., Biasi, W.V., Lerud, R., Slaughter, D.C. and Mitcham, E.J. (2017) Non-destructive prediction of soluble solids and dry matter content using NIR spectroscopy and its relationship with sensory quality in sweet cherries. *Postharvest Biology and Technology* 128, 112–120.

Feng, J., Wohlers, M., Olsson, S.R., White, A., McGlone, V.A. *et al.* (2016) Comparison between an acoustic firmness sensor and a near-infrared spectrometer in segregation of kiwifruit for storage potential. *Acta Horticulturae* 1119, 279–288.

Giovannoni, J., Nguyen, C., Ampofo, B., Zhong, S. and Fei, Z. (2017) The epigenome and transcriptional dynamics of fruit ripening. *Annual Review of Plant Biology* 68, 61–84.

Gonçalves, R.G., Couto, J. and Almeida, D.P. (2016) On-tree maturity control of peach cultivars: comparison between destructive and nondestructive harvest indices. *Scientia Horticulturae* 209, 293–299.

Gowen, A.A., O'Donnell, C.P., Cullen, P.J., Downey, G. and Frias, J.M. (2007) Hyperspectral imaging – an emerging process analytical tool for food quality and safety control. *Trends in Food Science & Technology* 18, 590–598.

Grassi, S. and Alamprese, C. (2018) Advances in NIR spectroscopy applied to process analytical technology in food industries. *Current Opinion in Food Science* 22, 17–21.

Hagen, S.F., Solhaug, K.A., Bengtsson, G.B., Borge, G.I.A. and Bilger, W. (2006) Chlorophyll fluorescence as a tool for non-destructive estimation of anthocyanins and total flavonoids in apples. *Postharvest Biology and Technology* 41, 156–163.

Hale, G., Lopresti, J., Stefanelli, D., Jones, R. and Bonora, E. (2013) Using non-destructive methods to correlate chilling injury in nectarines with fruit maturity. *Acta Horticulturae* 1012, 83–89.

Harker, F.R., Marsh, K.B., Young, H., Murray, S.H., Gunson, F.A. and Walker, S.B. (2002) Sensory interpretation of instrumental measurements 2: sweet and acid taste of apple fruit. *Postharvest Biology and Technology* 24, 241–250.

Herrero-Langreo, A., Lunadei, L., Lleó, L., Diezma, B. and Ruiz-Altisent, M. (2011) Multispectral vision for monitoring peach ripeness. *Journal of Food Science* 76, 178–187.

Hilaire, C. (2003) The peach industry in France: state of art, research and development. In: *Proceedings of the First Mediterranean Peach Symposium, Agrigento, Italy, 10 September 2003*, pp. 27–34.

Hills, B.P. and Clark, C.J. (2003) Quality assessment of horticultural products by NMR. *Annual Reports on NMR Spectroscopy* 50, 76–121.

Iglesias, I. and Echeverría, G. (2009) Differential effect of cultivar and harvest date on nectarine colour, quality and consumer acceptance. *Scientia Horticulturae* 120, 41–50.

Kader, A.A. (1995) Fruit maturity, ripening, and quality relationships. *Perishables Handling Newsletter* 80: 2.

Kalcsits, L.A. (2016) Non-destructive measurement of calcium and potassium in apple and pear using handheld X-ray fluorescence. *Frontiers in Plant Science* 7: 442.

Kawai, T., Matsumori, F., Akimoto, H., Sakurai, N., Hirano, K. *et al.* (2018) Nondestructive detection of split-pit peach fruit on trees with an acoustic vibration method. *Horticulture Journal* 87, 499–507.

Khalifa, S., Komarizadeh, M.H. and Tousi, B. (2011) Usage of fruit response to both force and forced vibration applied to assess fruit firmness – a review. *Australian Journal of Crop Science* 5, 516–522.

Kumar, S., McGlone, A., Whitworth, C. and Volz, R. (2015) Postharvest performance of apple phenotypes predicted by near-infrared (NIR) spectral analysis. *Postharvest Biology and Technology* 100, 16–22.

Kupferman, E. and Dasgupta, N. (2001) *Comparison of Pome Fruit Firmness Testing Instruments*. Fact Sheet 1-12. Postharvest Information Network, Tree Fruit Research and Extension Center, Washington State University, Washington, DC. Available at: www.facchinisrl.eu/ft_review.pdf (accessed 6 December 2022)

Lafuente, V., Herrera, L.J., Pérez, M.D.M., Val, J. and Negueruela, I. (2015) Firmness prediction in *Prunus persica* 'Calrico' peaches by visible/short-wave near infrared spectroscopy and acoustic measurements using optimized linear and non-linear chemometric models. *Journal of the Science of Food and Agriculture* 95, 2033–2040.

Lan, W., Jaillais, B., Leca, A., Renard, C.M. and Bureau, S. (2020) A new application of NIR spectroscopy to describe and predict purees quality from the non-destructive apple measurements. *Food Chemistry* 310: 125944.

Lee, W.-H., Kim, M.S., Lee, H., Delwiche, S.R., Bae, H. *et al.* (2014) Hyperspectral near-infrared imaging for the detection of physical damages of pear. *Journal of Food Engineering* 130, 1–7.

Li, J., Chen, L., Huang, W., Wang, Q., Zhang, B. *et al.* (2016) Multispectral detection of skin defects of bi-colored peaches based on vis–NIR hyperspectral imaging. *Postharvest Biology and Technology* 112, 121–133.

Li, J., Chen, L. and Huang, W. (2018) Detection of early bruises on peaches (*Amygdalus persica* L.) using hyperspectral imaging coupled with improved watershed segmentation algorithm. *Postharvest Biology and Technology* 135, 104–113.

Li, M., Pullanagari, R.R., Pranamornkith, T., Yule, I.J. and East, A.R. (2017) Quantitative prediction of post storage 'Hayward' kiwifruit attributes using at harvest Vis-NIR spectroscopy. *Journal of Food Engineering* 202, 46–55.

Li, X., Liu, Y., Jiang, X. and Wang, G. (2021) Supervised classification of slightly bruised peaches with respect to the time after bruising by using hyperspectral imaging technology. *Infrared Physics & Technology* 113: 103557.

Liu, Q., Zhou, D., Tu, S., Xiao, H., Zhang, B. *et al.* (2020) Quantitative visualization of fungal contamination in peach fruit using hyperspectral imaging. *Food Analytical Methods* 13, 1262–1270.

Liverani, A., Brandi, F., Quacquarelli, I., Sirri, S. and Giovannini, D. (2015) Superior taste and keeping quality are steady goals of the peach breeding activity at CRA-FRF, Italy. *Acta Horticulturae* 1084, 179–185.

Lu, R. and Peng, Y. (2006) Hyperspectral scattering for assessing peach fruit firmness. *Biosystems Engineering* 93, 161–171.

Lurie, S. and Crisosto, C.H. (2005) Chilling injury in peach and nectarine. *Postharvest Biology and Technology* 37, 195–208.

Magness, J.R. and Taylor, G.F. (1925) *An Improved Type of Pressure Tester for the Determination of Fruit Maturity*. Circular No. 350. US Department of Agriculture, Washington, DC.

Manganaris, G.A., Drogoudi, P., Goulas, V., Tanou, G., Georgiadou, E.C. *et al.* (2017) Deciphering the interplay among maturity stage and low-temperature storage on ripening behavior and phytochemical properties on peach and nectarine fruit. *Plant Physiology and Biochemistry* 119, 189–199.

Manganaris, G.A., Minas, I., Cirilli, M., Torres, R., Bassi, D. and Costa, G. (2022) Peach for the future: a specialty crop revisited. *Scientia Horticulturae* 305: 111390.

Marques, E.J.N., de Freitas, S.T., Pimentel, M.F. and Pasquini, C. (2016) Rapid and non-destructive determination of quality parameters in the 'Tommy Atkins' mango using a novel handheld near infrared spectrometer. *Food Chemistry* 197, 1207–1214.

McGlone, V.A. and Kawano, S. (1998) Firmness, dry-matter and soluble-solids assessment of postharvest kiwifruit by NIR spectroscopy. *Postharvest Biology and Technology* 13, 131–141.

McGlone, V.A., Jordan, R.B. and Martinsen, P.J. (2002) Vis/NIR estimation at harvest of pre-and post-storage quality indices for 'Royal Gala' apple. *Postharvest Biology and Technology* 25, 135–144.

Michailides, T.J. and Manganaris, G.A. (2009) Harvesting and handling effects on postharvest decay. *Stewart Postharvest Review* 5, 1–7.

Minas, I.S., Tanou, G. and Molassiotis, A. (2018) Environmental and orchard bases of peach fruit quality. *Scientia Horticulturae* 235, 307–322.

Minas, I.S., Blanco-Cipollone, F. and Sterle, D. (2020) Near infrared spectroscopy can non-destructively assess the effect of canopy position and crop load on peach fruit maturity and quality. *Acta Horticulturae* 1281, 407–412.

Minas, I.S., Blanco-Cipollone, F. and Sterle, D. (2021) Accurate non-destructive prediction of peach fruit internal quality and physiological maturity with a single scan using near infrared spectroscopy. *Food Chemistry* 335: 127626.

Minas, I.S., Anthony, B.M., Pieper, J.R. and Sterle, D.G. (2023) Large-scale and accurate non-destructive visual to near infrared spectroscopy-based assessment of the effect of rootstock on peach fruit internal quality. *European Journal of Agronomy* 143: 126706.

Muhua, L., Peng, F. and Renfa, C. (2007) Non-destructive estimation peach SSC and firmness by mutispectral reflectance imaging. *New Zealand Journal of Agricultural Research* 50, 601–608.

Munera, S., Amigo, J.M., Aleixos, N., Talens, P., Cubero, S. and Blasco, J. (2018a) Potential of VIS-NIR hyperspectral imaging and chemometric methods to identify similar cultivars of nectarine. *Food Control* 86, 1–10.

Munera, S., Besada, C., Blasco, J., Cubero, S., Salvador, A. *et al.* (2018b) Firmness prediction in 'Rojo Brillante' persimmon using hyperspectral imaging technology. *Acta Horticulturae* 1194, 761–768.

Nakano, R., Akimoto, H., Fukuda, F., Kawai, T., Ushijima, K. *et al.* (2018) Nondestructive detection of split pit in peaches using an acoustic vibration method. *Horticulture Journal* 87, 281–287.

Nanyam, Y., Choudhary, R., Gupta, L. and Paliwal, J. (2012) A decision-fusion strategy for fruit quality inspection using hyperspectral imaging. *Biosystems Engineering* 111, 118–125.

Nicolaï, B.M., Bulens, I., de Baerdemaeker, J., de Ketelaere, B., Lammertyn, J. *et al.* (2014) Non-destructive evaluation: detection of external and internal attributes frequently associated with quality and damage. In: Florkowski, W., Banks, N., Shewfelt, R. and Prussia, S. (eds) *Postharvest Handling: A Systems Approach.* Academic Press, Cambridge, Massachusetts, pp. 363–385.

Nyasordzi, J., Friedman, H., Schmilovitch, Z., Ignat, T., Weksler, A. *et al.* (2013) Utilizing the I_{AD} index to determine internal quality attributes of apples at harvest and after storage. *Postharvest Biology and Technology* 77, 80–86.

Oveisi, Z., Minaei, S., Rafiee, S., Eyvani, A. and Borghei, A. (2014) Application of vibration response technique for the firmness evaluation of pear fruit during storage. *Journal of Food Science and Technology* 51, 3261–3268.

Parker, D., Zilberman, D. and Moulton, K. (1991) How quality relates to price in California fresh peaches. *California Agriculture* 45, 14–16.

Peng, Y. and Lu, R. (2008) Analysis of spatially resolved hyperspectral scattering images for assessing apple fruit firmness and soluble solids content. *Postharvest Biology and Technology* 48, 52–62.

Pieper, J.R., Anthony, B.M., Sterle, D.G. and Minas, I.S. (2022) Rootstock vigor and fruit position in the canopy influence peach internal quality. *Acta Horticulturae* 1346, 807–812.

Ravaglia, G., Sansavini, S., Ventura, M. and Tabanelli, D. (1996) Indici di maturazione e miglioramento qualitativo delle pesche. *Frutticoltura* 3, 61–66.

Sánchez, M.T., de la Haba, M.J., Guerrero, J.E., Garrido-Varo, A. and Pérez-Marín, D. (2011) Testing of a local approach for the prediction of quality parameters in intact nectarines using a portable NIRS instrument. *Postharvest Biology and Technology* 60, 130–135.

Scalisi, A., Pelliccia, D. and O'Connell, M.G. (2020) Maturity prediction in yellow peach (*Prunus persica* L.) cultivars using a fluorescence spectrometer. *Sensors* 20: 6555.

Serra, S., Anthony, B., Masia, A., Giovannini, D. and Musacchi, S. (2020) Determination of biochemical composition in peach (*Prunus persica* L. Batsch) accessions characterized by different flesh color and textural typologies. *Foods* 9: 1452.

Shinya, P., Contador, L., Predieri, S., Rubio, P. and Infante, R. (2013) Peach ripening: segregation at harvest and postharvest flesh softening. *Postharvest Biology and Technology* 86, 472–478.

Slaughter, D.C., Thompson, J.F. and Tan, E.S. (2003) Nondestructive determination of total and soluble solids in fresh prune using near infrared spectroscopy. *Postharvest Biology and Technology* 28, 437–444.

Sonego, L., Ben-Arie, R., Raynal, J. and Pech, J.C. (1995) Biochemical and physical evaluation of textural characteristics of nectarines exhibiting woolly breakdown: NMR imaging, X-ray computed tomography and pectin composition. *Postharvest Biology and Technology* 5, 187–198.

Spadoni, A., Cameldi, I., Noferini, M., Bonora, E., Costa, G. and Mari, M. (2016) An innovative use of DA-meter for peach fruit postharvest management. *Scientia Horticulturae* 201, 140–144.

Srivastava, R.K., Talluri, S., Beebi, S.K. and Kumar, B.R. (2018) Magnetic resonance imaging for quality evaluation of fruits: a review. *Food Analytical Methods* 11, 2943–2960.

Subedi, P. and Walsh, K.B. (2020) Assessment of avocado fruit dry matter content using portable near infrared spectroscopy: method and instrumentation optimization. *Postharvest Biology and Technology* 161: 111078.

Sun, Y., Gu, X., Sun, K., Hu, H., Xu, M. *et al.* (2017) Hyperspectral reflectance imaging combined with chemometrics and successive projections algorithm for chilling injury classification in peaches. *LWT – Food Science and Technology* 75, 557–564.

Taniwaki, M., Hanada, T. and Sakurai, N. (2009) Postharvest quality evaluation of "Fuyu" and "Taishuu" persimmons using a nondestructive vibrational method and an acoustic vibration technique. *Postharvest Biology and Technology* 51, 80–85.

Tanou, G., Minas, I.S., Scossa, F., Belghazi, M., Xanthopoulou, A. *et al.* (2017) Exploring priming responses involved in peach fruit acclimation to cold stress. *Scientific Reports* 7: 11358.

Terasaki, S., Sakurai, N., Kuroki, S., Yamamoto, R. and Nevins, D.J. (2013) A new descriptive method for fruit firmness changes with various softening patterns of kiwifruit. *Postharvest Biology and Technology* 62, 85–90.

Testoni, A. (1995) Momento di raccolta, qualit`a, condizionamento e confezionamento delle pesche. In: *Proceedings of the Symposium on La peschicoltura Veronese alle soglie del 2000, Verona, 25 February*, pp. 327–354.

Theanjumpol, P., Wongzeewasakun, K., Muenmanee, N., Wongsaipun, S., Krongchai, C. *et al.* (2019) Non-destructive identification and estimation of granulation in 'Sai Num Pung' tangerine fruit using near infrared spectroscopy and chemometrics. *Postharvest Biology and Technology* 153, 13–20.

Trainotti, L., Tadiello, A. and Casadoro, G. (2007) The involvement of auxin in the ripening of climacteric fruits comes of age: the hormone plays a role of its own and has an intense interplay with ethylene in ripening peaches. *Journal of Experimental Botany* 58, 3299–3308.

Valero, C., Crisosto, C.H. and Slaughter, D. (2007) Relationship between nondestructive firmness measurements and commercially important ripening fruit stages for peaches, nectarines and plums. *Postharvest Biology and Technology* 44, 248–253.

Wang, H., Peng, J., Xie, C., Bao, Y. and He, Y. (2015) Fruit quality evaluation using spectroscopy technology: a review. *Sensors* 15, 11889–11927.

Yurtlu, Y.B. (2012) Comparison of nondestructive impact and acoustic techniques for measuring firmness in peaches. *Journal of Food Agriculture and Environment* 10, 180–185.

Zhang, Y., Nock, J.F., Al Shoffe, Y. and Watkins, C.B. (2019) Non-destructive prediction of soluble solids and dry matter contents in eight apple cultivars using near-infrared spectroscopy. *Postharvest Biology and Technology* 151, 111–118.

Zhang, H., Zhang, S., Dong, W., Luo, W., Huang, Y. *et al.* (2020) Detection of common defects on mandarins by using visible and near infrared hyperspectral imaging. *Infrared Physics and Technology* 108: 103341.

Ziosi, V., Noferini, M., Fiori, G., Tadiello, A., Trainotti, L. *et al.* (2008) A new index based on vis spectroscopy to characterize the progression of ripening in peach fruit. *Postharvest Biology and Technology* 49, 319–329.

10

POSTHARVEST SUPPLY-CHAIN MANAGEMENT PROTOCOLS AND HANDLING OF PHYSIOLOGICAL DISORDERS

George A. Manganaris[1]*, Ariel R. Vicente[2], Gemma Echeverría[3] and Carlos H. Crisosto[4]

[1]*Department of Agricultural Sciences, Biotechnology & Food Science, Cyprus University of Technology, Lemesos, Cyprus; [2]University of La Plata, La Plata, CP, Argentina; [3]Postharvest Program, Institute of Agrifood Research and Technology (IRTA), Lleida, Spain; [4]Department of Plant Sciences, University of California, Davis, California, USA

10.1   HARVEST AND PACKAGING

Fresh-market peaches are produced in the northern hemisphere from April to September and in the southern hemisphere from November to March. However, their availability in stores is rather limited due to their reduced storage potential. Peach harvest has a relatively broad window (firm to fully ripe). The fruit should be harvested with maximum care, regardless of the picking maturity stage. It is crucial to avoid physical damage as this will induce ripening, flavour loss, decay, tissue browning and dehydration. Using clean bags or small containers is recommended to prevent bruising, decay and potential skin inking. Fruit contact with the ground should be avoided to prevent phytosanitary and human disease problems.

Fruit picked at firm stages offers more flexibility regarding postharvest management but sometimes may affect consumer satisfaction. In contrast, fully ripe peaches are highly susceptible to physical damage and decay but have a flavour surplus. The most common practical minimum harvest maturity indexes used are background colour and firmness. As a climacteric fruit, peach background skin-colour changes from green to yellow and/or an even flesh colour are additionally used to ensure that the fruit will ripen properly after harvest during postharvest handling. In highly red-flushed cultivars, the red

* Email: george.manganaris@cut.ac.cy

© CAB International 2023. *Peach* (G. Manganaris, G. Costa and C. Crisosto eds)
DOI: 10.1079/9781789248456.0010

colour may cover background colour development making it difficult to assess minimum maturity, so fruit firmness has been used to assess harvest maturity, including non-destructive assays (Crisosto, 1994; Manganaris *et al.*, 2017). The development of sophisticated non-destructive tools that will allow the determination of fruit quality and the maturity stage in a cost-effective manner is expected to facilitate the logistics regarding storage requirements with minimum risk of chilling injury (CI) symptom development (Manganaris *et al.*, 2022).

The maximum maturity index is defined as the minimum flesh firmness at which fruits can be handled without bruising damage (Crisosto *et al.*, 2001, 2004a; Crisosto and Costa, 2008). Thus, a maximum harvest maturity (critical bruising threshold), based on firmness measured at the weakest fruit spot, is used for fresh commercial cultivars in California, Chile and other countries (Crisosto *et al.*, 2001, 2004a). Maximum maturity indices were developed for different harvesting/packinghouse operations based on their bruising potentials (Table 10.1) and on cultivar critical bruising thresholds that were developed for different stone fruit cultivars (Table 10.2). Impact location on the fruit was an important factor in the determination of critical bruising thresholds, as fruit softening is not even across the fruit surface. In general, yellow-fleshed peaches and nectarines tolerated more physical abuse than white-fleshed peach cultivars. Potential sources of bruising damage during fruit harvesting/packing were determined using an accelerometer (IS-100). A survey of different packinghouse types revealed that bruising potentials varied from 21 to 206 G (Table 10.1). Bruising potential was easily reduced by adding padding material to the packing lines, minimizing height differences at transfer points, synchronizing timing between components and reducing the operating speed. Bruising probabilities for the most-susceptible Californian-grown cultivars at different velocities and G values have been developed (Table 10. 2).

From the consumer point of view, consumers tend to widely accept fruit with force firmness below 3–4 kg cm^{-2} ('ready to transfer or buy') measured on the cheek with an 8 mm tip, while fruit having 1–2 kg cm^{-2} force firmness are considered 'ready to eat', defined as the stage where fruit reached the highest flavour expression for consumers (Crisosto and Mitchell, 2016). Furthermore, it has been validated that non-destructive firmness measurements can be used directly to identify the stage of ripeness ('ready to transfer' and 'ready to eat') and potential susceptibility to bruising during postharvest changes (Crisosto and Valero, 2006; Valero *et al.*, 2007).

An array of protocols for harvesting and packaging have been developed, depending on the destination of the fruit, the desired postharvest life and the available infrastructure (Crisosto and Day, 2012; Crisosto and Mitchell, 2016; Manganaris and Crisosto, 2020).

In most countries, pickers work from the ground or on ladders and the fruit are hand-picked into bags, totes or buckets and dumped into wooden or plastic bins in the field or on top of trailers between rows in the orchard

(Fig. 10.1). Plastic bin liners and padded bin covers have been shown to reduce transport injury in some conditions. Plastic totes are placed directly inside the bins and buckets are placed on modified trailers. Fruit picked at advanced maturity stages, as well as white-fleshed peaches are generally picked and

Table 10.1. Impacts (G force) recorded at transfer points of stone fruit packing lines. From Crisosto *et al.* (2001).

Transfer points	Mean (G)[a]	Standard deviation	Range
Packinghouse A			
Bin dumper	90.7	48.6	24–180
Bin dumper to pony sizer	110.4	12.1	105–131
Pony sizer	70.6	13.3	54–84
To washer/brusher	80.0	16.8	75–98
To sorting tables	102.0	31.6	66–145
To sizers	88.9	9.5	74–97
Sizer cups	67.6	5.3	59–72
Sizer kick out	57	21.3	25–78
Boxing line	71	10.2	55–82
Boxing machine	65	19.8	46–94
Box volume fill	47	24.1	28–89
Box tray pack	60.6	18.5	33–78
Packinghouse B			
Bin dumper	94.3	47.3	38–177
Elevator to pony sizer	121.8	50.3	72–187
Pony sizer to washer/brusher	83.4	10.4	71–98
Brusher to sorting tables	130.9	29.7	58–180
Sorting to sizers	94.2	13.7	72–117
Sizer to sizer cups	61.0	10.3	38–74
Sizer cups kick out	ND	–	–
Drop down to packing belt	94.9	56.9	30–165
Box volume fill	103.8	32.8	70–146
Packinghouse C			
Bin dumper	82.8	16.5	73–107
Dumper to elevator	57.9	26.2	25–114
Conveyor to washer	68.4	21.4	42–106
Washer to waxer	24.5	4.4	19–33
Waxer to sorting tables	25.1	3.5	21–32
Sorting to sizers	90.6	11.6	72–110
Sizers to conveyor	71.6	50.8	23–170
Conveyor to PACKING TABLES	97.5	14.7	83–126
Box tray pack	61.5	31.9	27–117
Box volume fill	143.0	28.1	111–206

[a]Means were calculated using the peak impact measured during each of the ten trips of the instrumented sphere across each transfer point.
ND, not detectable.

Table 10.2. Minimum flesh firmness (measured at the weakest point on the fruit) necessary to avoid commercial bruising at three levels of physical handling. From Crisosto *et al.* (2001).

Cultivar	Drop height[a]			Weakest position
	1 cm (~66 G)	5 cm (~185 G)	10 cm (~246 G)	
Peaches (yellow flesh)				
'Queencrest'	0	4	9	Tip
'Rich May'	0	0	9	Tip
'Kern Sun'	2	6	9	Tip
'Flavorcrest'	3	5	6–9	Tip
'Rich Lady'	6	10	11	Shoulder
'Fancy Lady'	3	7	11	Shoulder
'Diamond Princess'	0	0	9	Shoulder
'Elegant Lady'	3	5	6–9	Shoulder
'Summer Lady'	0	0	8	Shoulder
'O'Henry'	3	5	6–9	Shoulder
'August Sun'	3	4	9	Shoulder
'Ryan Sun'	0	0	10	Shoulder
'September Sun'	0	4	9	Shoulder
Nectarines (yellow flesh)				
'Mayglo'	4	8	11	Tip
'Rose Diamond'	6	7	8	Suture/Shoulder
'Royal Glo'	0	9	11	Shoulder/Tip
'Spring Bright'	6	10	10	Shoulder
'Red Diamond'	6	7	11	Shoulder
'Ruby Diamond'	4	9	9	Shoulder
'Summer Grand'	2	5	6	Shoulder
'Flavortop'	3	6	6	Tip
'Summer Bright'	0	6	8	Shoulder
'Summer Fire'	0	0	9	Shoulder
'August Red'	2	12	12	Shoulder
'September Red'	0	0	10	Shoulder

Fruit firmness was measured with an 8 mm tip.
[a]Dropped on 1/8″ PVC belt. Damaged areas with a diameter greater than or equal to 2.5 mm were measured as bruises. Source: Crisosto *et al.* (2001).

placed into buckets or totes. Depending on the cultivar and specific situation, a worker can usually harvest one-and-a-half to three full-sized bins of fruit per day. Early-season cultivars are usually picked every 2–3 days, but by mid- to late season, the interval can stretch to as much as 7 days between harvests. In general, early-ripening cultivars are harvested twice, while mid- and late-ripening cultivars are harvested three to six times depending on the cultivar, season and prices. Tree heights are commonly 3.7–4.7 m, and workers require ladders to

Fig. 10.1. Peach dumping in plastic bins prior to transfer to the packinghouse.
Photographs courtesy of G. Echeverría.

reach the uppermost fruits. The recent establishment of pedestrian orchards
that include different training/pruning options and the use of size-controlling
rootstocks are reducing the use of ladders as the trees are harvested from the
ground. Ladders are made of aluminium and are 3.7–4.0 m in length. Either
four or six rows are harvested at a time, with an equal number of pickers distributed in each row as conditions warrant. Workers pick an entire tree and
leapfrog one another down the rows. The foreman is responsible for moving
the pickers between rows to maintain uniformity. The bins are then taken to a
centralized area and unloaded from the bin trailers or truck to await loading by
a forklift truck on to flatbed trailers for delivery to the packing facility. Full bins
are typically covered with canvas to prevent heat damage, and loading areas
are usually bordered by large trees for shade, which help reduce fruit exposure
to the sun. In instances where the orchard is close to the packing plant, the fruit
can be conveyed there directly on the bin trailers or truck. The fruit are hauled
for short distances by trailers, but if the distance is greater than 10 km, the
bins or totes are loaded on to a truck for transportation to the packinghouses.
Picking platforms have been tried, but they are not an economically viable way
of reducing reliance on ladders due to their cost and the vast differences in tree
and worker efficiencies.

In a few cases (e.g. Greece), field packing strategies are additionally applied
to freshly consumed freestone and some clingstone cultivars to minimize manipulation and mechanical damage (Fig. 10.2). Such products at an advanced
ripening stage are mainly destined for the domestic market.

The harvested peaches are transported to a packinghouse for cooling,
packaging, storage and distribution. In all situations, at the packinghouse,
some peaches are packed upon arrival from the orchard, while others
are partially cooled and packed the next day. In general, if fruit will not be
packed within 2–3 days, they should be cooled close to 0°C to protect them
from deterioration.

Peach packaging normally includes the following operations: dumping,
washing, rinsing, grading, brushing, fungicide spraying application, sorting
and packing (Fig. 10.3). At the packinghouse, the fruit are dumped and
cleaned using sanitation unit equipment where debris is removed and the fruit

Fig. 10.2. Harvest and in-farm packaging of peach fruit in Greece. Photographs courtesy of G.A. Manganaris.

Fig. 10.3. Operation of a peach packinghouse in Spain. Photographs courtesy of G.A. Manganaris.

are sanitized. Peaches are normally washed and wet brushed to remove the trichomes, or fuzz, which are single-cell extensions of epidermal cells. Water containing chlorine is used to wash and as a first attempt to sanitize peaches and nectarines. Ideally, this area is located outside the packing area. After brushing/washing, the fruit pass through a short drying area in preparation for the waxing/fungicide application where applied and/or allowed. Waxing and approved fungicide treatments follow next in another protected area.

Water-emulsified waxes are normally used, and fungicides may be incorporated into the wax. Waxes are applied cold, and no heated drying is necessary to provide shining and spread, and to hold the applied fungicide. Sorting or grading is done to eliminate fruit with visual defects and sometimes to divert fruit of high surface colour to a high-maturity pack. Attention to detail in sorting-line efficiency is especially important with peaches and nectarines where a range of colours, sizes and shapes of fruit can be encountered. Sizing segregates the fruit by weight or dimension, carried out by operators or electronic computer-controlled systems. Sorting and sizing equipment must be flexible to efficiently handle large volumes of small fruit or smaller volumes of larger fruit. Most yellow-fleshed peaches and nectarines are packed in one- (flat) or two-tray boxes. In some cases, electronic weight sizers are used to automatically fill shipping containers (volume fill packed) by weight. In some cases, mechanical place-packing units use hand-assisted fillers where the operator can control the belt speed to match the flow of fruit into plastic trays. Most white-fleshed and 'tree-ripe' peaches are packed into a one-tray box (flat), punnets or clamshells.

Packinghouses currently offer a wide range of packaging options adapted to the customer's needs. Fruit may be packed in a variety of containers including polyethylene bags, punnets, net bags, and single- or multi-layered boxes. Subject to marketing requests, peaches are packed into single-layer trays, multilayer boxes and netted punnets. First-category fruit is normally exported, whereas second-category fruit is directed to less stringent domestic markets, especially to small stores and organic groceries. Boxes are unitized in a pallet for easy and efficient handling. In most countries, 1.2×1.0 m (UK pallet) is the standard size requested by supermarkets. In the USA, there is maximum weight (36,288 kg) per truck to protect the highway system. Thus, the total number of tiers per pallet will depend on the total weight allowed per container in each country. However, the height of the pallet is limited by the height of the container doors and the container internal-height allowance for proper air distribution (Crisosto and Day, 2012; Crisosto and Mitchell, 2016; Thompson, 2016). In Europe, fruit is commonly packed is a single tray with a net weight of 4.5 kg and a gross weight of 5 kg, while in the USA, most peaches are packed using two trays or layers, to protect ripe peaches, with a box net weight of 9.1 kg and a gross weight of 10 kg. In a very few places, and when less mature peaches are used, the volume fill can be done with a net weight of 11.3 kg and a gross weight of 12.3 kg per box. Under these package conditions, total pallet weight varies from 720 kg (single layer) to 800 kg (two layers) to 994 kg (volume fill) (Crisosto and Mitchell, 2016). All box-pallet loads should be stabilized with netting or strapping and corner boards. The box size and design and pallet placement should include venting areas and perforations (side and/or bottom) to ensure cold air movement throughout the fruit box, and the venting should be designed to provide enough air flow and forced-air path (convective cooling) to reach the fruit inside (Mitchell *et al.*, 1998).

Packages with their own brand, used for high-quality or premium fruit, are exported to both national and international markets in single-layer tray with the dimensions 50 × 30 × 9.5 cm. Peach trays can be made of cardboard, wood or reusable plastic. The requirements for fruit shipped in trays are often more stringent than those for punnet peaches, which also usually include smaller fruit. Once packed and consolidated, the fruit can be forced-air cooled and placed in cold storage (Fig. 10.4).

10.2 COLD STORAGE AND TRANSPORTATION

Peaches are chilling-sensitive fruit, with damage symptoms being more intense between 2 and 8°C. Thus, this temperature range is called the 'killing temperature zone' (Crisosto *et al.*, 1999b; Lurie and Crisosto, 2005; Manganaris *et al.*, 2019; Manganaris and Crisosto, 2020). As keeping the fruit at temperatures higher than 10°C would rapidly result in excessive softening and decay, the fruits should be stored at 0°C or below but above the tissue's freezing point to maximize peach storage potential and shelf-life. Appropriate relative humidity is also crucial to minimize dehydration when condensation does not occur. Maintaining these low pulp temperatures and relative conditions requires knowledge of the freezing point of the fruit, the temperature fluctuations in the storage system, loading techniques and equipment performance.

Peach transportation is commonly conducted by truck when delivery is within a week of the production area. Marine transportation is used for long-distance markets, and the conditions may lead to abnormal ripening due to extended cold storage periods. Air freight is used for transportation to premium markets, particularly during periods with limited availability,

Fig. 10.4. Forced-air cooling facilities after packaging. Photographs courtesy of G.A. Manganaris.

to justify its high cost. Pre-cooling transportation containers at 0°C before loading is crucial, among other recommendations, to ensure that the load is at the desired storage temperature and arrives safely (Thompson, 2016; Thompson and Crisosto, 2016). Stone fruit storage and overseas shipments should be at or below 0°C. The temperature during truck transportation within the USA, Europe, Canada and Mexico should be below 2.2°C. Holding stone fruits at these low temperatures minimizes the losses associated with fungal diseases, excessive softening, water losses and the deterioration resulting from CI in susceptible cultivars (Crisosto *et al.*, 1999b).

CI is the main physiological disorder limiting export and long-distance peach distribution (Lurie and Crisosto, 2005; Manganaris *et al.*, 2006; Falara *et al.*, 2011; Martínez-García *et al.*, 2011; Manganaris and Crisosto, 2020). The different manifestations of CI symptoms in peach are evident as: (i) mealiness or woolliness (perception of a dry and woolly texture upon consumption due to lack of free juice); (ii) leatheriness (hard-textured fruit with no juice); (iii) internal breakdown evident as flesh browning; and (iv) red flesh pigmentation or bleeding (Fig. 10.5). Such symptoms are accompanied by loss of flavour, which is the most frequent complaint by consumers and wholesalers and the main barrier to consumption. 'Off-flavour' development is one of the initial symptoms of CI prior to flesh mealiness and browning development, while susceptibility to CI is largely dependent on genotype and is triggered by a combination of temperature and time of exposure to chilling temperatures. CI represents a major problem because its symptoms remain unnoticed until the peaches reach the customers at a ready-to-eat stage (Crisosto *et al.*, 1999b; Lurie and Crisosto, 2005). At advanced stages, damaged fruit have no obvious abnormal external appearance but lack juiciness and have a highly dry texture that is not related to water loss as both mealy and sound peach and nectarine fruits have a similar water content (Manganaris *et al.*, 2019; Manganaris and Crisosto, 2020).

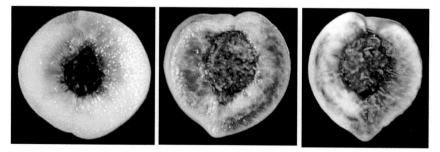

Fig. 10.5. Fruit showing mild severity (left), severe bleeding symptoms (middle) or a combination with a mealy texture (right). Photographs courtesy of C.H. Crisosto.

10.3   STRATEGIES TO ALLEVIATE CI

10.3.1   Selection of chilling-tolerant cultivars

This is the most practical and quickest way to deal with the problem. Peach and nectarine cultivars are characterized by a different degree of CI susceptibility (Fig. 10.6). The susceptibility of cultivars to CI is constantly being evaluated in the most currently planted cultivars from different breeding sources and fruit types (Crisosto *et al.*, 1999b, 2008; Martínez-García *et al.*, 2011; Echeverría *et al.*, 2021). In general, cultivars are segregated into three categories (A, B and C) according to their susceptibility to CI symptoms (mealiness and flesh browning) when exposed to 0 or 5°C storage temperatures. Cultivars in category A do not develop any symptoms of CI after 5 weeks of storage at either temperature. Cultivars in category B develop symptoms only when stored at 5°C within 5 weeks of storage. Cultivars were classified in category C when fruit developed CI symptoms at both storage temperatures within 5 weeks of storage. Most of the yellow- and white-fleshed peach cultivars develop CI symptoms when stored at both storage temperatures (category C).

Based on these data, a market life potential – a concept that can be used for marketing – was developed and is used for different export companies. An application of the market life potential concept was a recent study carried out by the Institute of Agrifood Research and Technology (IRTA), which segregated 29 peach cultivars into five categories according to commercial market life depending on their tolerance to CI (up to 14, 21, 28, 35 or 42 days), as well as logistics information on the transport/marketing period to any port in the world (Echeverría *et al.*, 2021). The results demonstrated the importance of proper genotype selection and temperature management during postharvest handling. Current genotype CI evaluations have revealed that new cultivars are less susceptible to CI due to breeding programme selection (C. Peace, personal communication).

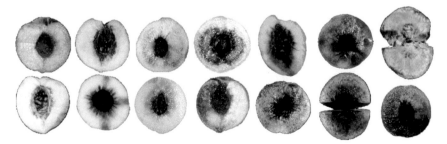

Fig. 10.6. Peach and nectarine cultivars with varying chilling injury manifestations. Photograph courtesy of C. Peace.

10.3.2 Preharvest factors and harvest maturity

To date, few studies have been conducted to dissect a link between fruit susceptibility to CI and preharvest factors, partially due to the large seasonal variations (Campos-Vargas *et al.*, 2006). Low crop loads have been empirically correlated with increased incidence of flesh browning and mealiness. Shaded fruit also had a lower storage potential and were more prone to postharvest disorders (Lurie and Crisosto, 2005; Crisosto and Costa, 2008). Early-harvested fruit are considered to be more susceptible to CI, especially to flesh browning, during storage. However, overripening may also lead to flesh browning, at least in some cultivars, such as 'Big Top'. As a general rule, the maturity stage at harvest appears to have a direct effect on fruit susceptibility to CI, with ripe fruit being less susceptible.

On arrival at the packinghouse, the fruit can be cooled in field bins using forced-air cooling or hydrocooling. Conventional cold storage just above freezing point is the most convenient condition to delay CI manifestations, while avoiding the 'killing temperature zone' (2–8°C) described earlier. The ideal peach storage temperature is −1 to 0°C. The flesh freezing point varies depending on total soluble solids content. Storage-room relative humidity should be maintained at 90–95% and airflow of approximately 0.0236 m^3 s^{-1} t^{-1} is suggested during storage (Mitchell *et al.*, 1998; Crisosto and Mitchell, 2016; Manganaris and Crisosto, 2020).

Application of a controlled atmosphere (6% O_2 + 17% CO_2) has been proven beneficial to delay fruit deterioration (Crisosto *et al.*, 2009; Manganaris and Crisosto, 2020). However, the most evident effect was on controlling flesh browning and softening, with the effects on mealiness and off-flavour development being modest (Crisosto *et al.*, 2009). Modified-atmosphere packaging has been tested in several peach cultivars, mostly without success (Zoffoli *et al.*, 2002; Lurie and Crisosto, 2005). Overall, and regardless of some promising strategies at a laboratory scale, proper temperature management currently remains the most efficient strategy to delay CI.

10.3.3 Conditioning

Conditioning (delayed cold storage) at 20°C and 95% relative humidity, followed by forced-air cooling prior to cold storage can be applied to fruit harvested at the firm-ripe stage to reduce CI susceptibility and ensure even ripening upon removal from cold storage (Crisosto *et al.*, 2004b; Crisosto and Mitchell, 2016). When these treatments are applied properly, market life increased by up to 2 weeks in the cultivars tested (Crisosto *et al.*, 2004b). Careful monitoring of weight loss and firmness during delayed cooling and proper use of fungicides is highly recommended for the success of this strategy (Lurie and Crisosto,

2005). This protocol can also be used to pre-ripen peaches to deliver into the market as a ready-to-eat product (Crisosto *et al.*, 2004b).

Fruit with a greater capacity to produce ethylene after cold storage have been reported to have less severe CI (Zhou *et al.*, 2001a; Giné-Bordonaba *et al.*, 2016). Therefore, 1-methylcyclopropene, an ethylene antagonist that was proven beneficial for shelf-life extension of an array of climacteric type fruits, is considered detrimental for peach fruits destined for cold storage (Dong *et al.*, 2001).

10.3.4 Heat treatments

Heat treatments have shown some benefits on CI prevention, although they have not been applied at a commercial scale (Murray *et al.*, 2007). Their efficacy is also highly dependent on the cultivar, preharvest factors and shipping duration. Intermittent warming has been also reported as a CI-delaying strategy. In this case, fruits are subjected to cold storage with interludes at room temperature. The basis for intermittent warming is to remove the fruit from the stress condition before it gets into the phase at which irreversible damage may occur. When 2 days of intermittent warming at 20°C was applied every 12 days during 0°C storage, mealiness was reduced (Zhou *et al.*, 2001b). This protocol was tested at a commercial scale in South Africa, but it was found to be difficult to apply at this scale, and the benefits are modest.

10.3.5 Chemical treatments

An array of chemical treatments, mainly hormone applications, have been applied to prevent and/or alleviate CI on peach fruit with variable success. Chemical treatments included the application of salicylic acid, methyl jasmonate, oxalic acid, γ-aminobutyric acid, gibberellic acid and glycine betaine (Jin *et al.*, 2009, 2014; Yang *et al.*, 2011; Shan *et al.*, 2016). The most promising results were provided by preharvest gibberellin application, which appeared to induce protection to CI (Pegoraro *et al.*, 2015). This protection has been attributed to the transcriptional changes triggered by gibberellic acid at early stages of fruit development, which could affect subsequent responses to stress after harvest (Pegoraro *et al.*, 2015). Such treatments still need to be validated in commercial settings.

10.4 OTHER PHYSIOLOGICAL DISORDERS

10.4.1 Field skin inking or black staining

This is a type of skin discoloration, causing fruit rejections. The symptoms appear as brown and/or black spots or stripes that are restricted to the skin

Fig. 10.7. Skin inking symptoms in peach fruit. Photograph courtesy of C.H. Crisosto.

(Fig. 10.7). The inked areas are normally small but in extreme cases can reach up to 50% of the fruit surface. Inking symptoms are triggered during harvest and transportation to the packinghouse and normally become evident within 48 h of harvest.

Field inking is believed to be caused by abrasion damage in combination with heavy-metal contamination. The skin cells, which are rich in phenolic compounds, collapse and their contents react with heavy metals turning their colour dark brown/black. Iron, copper and aluminium are the most deleterious heavy-metal compounds that can combine with polyphenols. Trace concentrations of iron (5–10 ppm of iron) may induce inking at a pH of ~3.5. Metal contamination may occur due to dust deposition on the fruit surface or because of preharvest foliar nutrient, fungicide and insecticide sprays used close to harvest that contain the metals. A recent study indicated that bronzing/inking can be induced by multiple factors, including preharvest captan applications and increased rainfall during fruit maturation (Lawton *et al.*, 2022). Orchard management, such as irrigation regime, harvesting method, harvest time and maturity stage, also influenced the incidence of inking; however, further data are needed to fully elucidate the etiology of such symptoms (Reig *et al.*, 2022).

Early studies (Cheng and Crisosto, 1995, 1997; Crisosto *et al.*, 1999a) have proposed the following prevention and mitigation measures to control inking:

- Reduce fruit contamination by keeping picking containers clean and avoid dust contamination on fruits.
- Reduce fruit abrasion damage by treating fruit gently, using air-ride suspension on trailers and avoiding long hauling distances.
- Check water quality for contamination with heavy metals (iron, copper and aluminium) and test pesticides for the presence of heavy metals early in the season.

- Avoid spraying foliar nutrients or preharvest fungicides that contain iron, copper or aluminium within 21 days of the predicted harvest. Chemical manufacturing companies should attempt to identify and remove from their products any potential source of contaminants that may contribute to inking formation, and to develop safe preharvest spray intervals for foliar nutrients, fungicides, miticides and insecticides.
- Growers need to know the composition of the chemicals commonly used on their tree fruit during preharvest and postharvest operations, and to understand how they may affect inking incidence.
- In orchards where inking is a problem, delay packing for around 48 h so that you will be able to remove fruit with field inking before placing the fruit in the box.
- Fine-tune your postharvest fungicide application to ensure that your residues are above the effective minimum recommended but well below the maximum residue limit or tolerance.

10.4.2 Skin burning

Skin burning is another type of skin discoloration that has become a frequent problem on specific susceptible peach and nectarine cultivars (Fig. 10.8) (Cantín *et al.*, 2011). IRTA results from observations over several years have indicated that peach and nectarine skin discolorations (field inking and skin burning) are triggered by a combination of physical damage during harvesting/hauling combined with different postharvest stress factors. However, although field inking and skin burn disorders have similar symptoms, they have different triggers and different biological mechanisms of development, and therefore it is important to understand the differences between these cosmetic skin disorders.

Skin burning symptoms appear as brown and/or black areas that are restricted to the skin. In contrast to field inking, these symptoms are triggered mainly during packing operations, principally at the brushing/washing point, although abrasion that occurs prior to packing may also contribute to its development (Crisosto *et al.*, 2000). Fruit damage is triggered by exposure to high pH and/or dehydration caused by high-velocity, forced-air cooling during packing (Cantín *et al.*, 2011). Symptoms can be observed soon after packing, but the symptoms increase rapidly during cold storage due to dehydration. In fact, it has been observed that most of the intense skin damage in packed fruit occurred on the exposed part of the fruit above the tray receptacle, and no damage occurred under the price look-up sticker (Cantín *et al.*, 2011). Different susceptibility to skin burning has been observed among peach and nectarine cultivars, depending mainly on the specific phenolics in their skin tissues due to co-pigmentation with anthocyanins, resulting in a change in colour of the anthocyanin compound and therefore discoloration of the skin (Cantín *et al.*, 2011).

Fig. 10.8. Skin burning symptoms in peach fruit. Photograph courtesy of C.H. Crisosto.

Some prevention and mitigation measures to control skin burning are as follows:

- Minimize physical damage or abrasion on the fruit surface during pre- and postharvest operations. Handle the fruit gently, use air-ride suspension on trailers, avoid long hauling distances and keep harvest containers free of dirt.
- In a standard packing operation, the pH of washing water in the brushing/ washing or hydrocooling operation should be maintained continuously at around 6.5–7.0. The installation of automated systems using oxidation-reduction potential to monitor and/or adjust active/effective chlorine and pH levels is critical to control disease effectiveness and decrease potential skin burning development.
- Apply dry packing (without brushing or a chlorine rinse) for highly susceptible cultivars.
- Avoid high air velocities during forced-air cooling for skin-burning-susceptible cultivars. For these cultivars, room cooling, without forced air, is suggested.
- As a long-term solution, we suggest screening peach and nectarine breeding parents for their susceptibility to co-pigmentation.

10.4.3   Corky spot

This disorder has been around for long time in California and its intensity varies according to the cultivar and season (Day, 2006). In the Ebro valley, it appeared during the 2006 season in some nectarine cultivars (Fig. 10.9).

Fig. 10.9. Corky spot symptoms in nectarine fruit. Photograph courtesy of G. Echeverría.

Corky spot symptoms appear as dark sunken spots on the surface of the fruit, especially on the sides and blossom end. Internally, the flesh initially has reddish spots that turn brown, corky and dry as the fruit ripen, making them unsuitable for market (Day, 2006; Perís and Alegre, 2012).

This disorder has been attributed to an excess of fertilization and some water-stress conditions during the growing period, which could lead to a nutritional imbalance and a deficiency in calcium fruit content (Day 2006; Perís and Alegre, 2012). A significant decrease in the severity and percentage of fruit affected by corky spot was recorded in calcium-treated fruit (Crisosto *et al.*, 2000; Val *et al.*, 2018). The incidence of the disorder also increased with fruit maturation. The common factors that were monitored in affected orchards were: young trees, vigorous growth, and a dry and hot summer. Seasonal variation of corky spot incidence is thought to be due to hot temperatures prior to harvest. A study in California described calcium and boron deficiency, nutritional imbalance caused by an excess of nitrogen that promotes vigorous tree growth, seasonal cold temperatures and some environmental stress such as water deficits under high evaporative demand conditions as possible causes of this disorder (Day, 2006). Overall, orchard conditions, crop load, cultivar, tree age and summer pruning may affect the incidence of this disorder. Avoiding excess nitrogen and potassium fertilization and water-stress conditions is recommended to prevent corky spot.

10.4.4   Skin bronzing and streaking

Bronzing refers to patches of skin on the fruit, primarily on yellow to light red skin background. Depending on severity, the damage may stretch from a single small patch to covering most of the peach. Although many patches are formed prior to harvest, most of the symptoms only appear after storage.

Peach skin streaking is another form of skin discoloration (Hu *et al.*, 2017; Schmitz and Schnabel, 2019). Streaking refers to symptoms that are consistent with the formation of water droplets formed by dew or rain that slowly run down the fruit. The streaks increase in diameter and end abruptly in a club-shaped fashion. Typically, several streaks of similar form and length are observed on the same fruit in multiple cultivars each season, and streaking incidence may range from non-detectable to over half of the fruit surface. Both bronzing and streaking skin disorders have significant impact on the production of high-quality fruit in the south-eastern USA (G. Schnabel and J.C. Melgar, personal communication).

10.5   CONCLUSION

Peach is a highly perishable product with a limited storage potential. Different handling protocols have been developed throughout the years for proper harvesting, packaging, cooling, storage and distribution of peach fruit. The most relevant issues to consider include the selection of an appropriate firmness, flavour and colour maturity for each distribution setting and avoidance of any type of physical damage. Peach cooling operations most frequently include the use of forced-air cooling and/or hydrocooling equipment and subsequent storage at 0°C, avoiding the 2–8°C killing zone. Controlled and modified atmospheres are used only under specific scenarios because they cause modest benefits and highly variable responses. The fruit is finally delivered in a variety of packaging according to the requirements of the markets and customers. Peach fruit quality can be significantly impaired by different chilling- and non-chilling-related physiological disorders. CI remains a major problem for inappropriate handling and/or long-term peach storage and long-distance markets. Thus, several strategies have been developed for CI alleviation, including the use of cultivars with a better response to low temperatures and the use of proper conditioning treatments. Genetic improvements leading to CI-tolerant cultivars is a priority goal. It is crucially important to pay attention to the steps at harvest and in the packinghouses to ensure the competitiveness and sustainability of the peach industry. Considering the excessive number of available peach and nectarine cultivars, analysis should be redirected towards early- and late-ripening cultivars to increase availability worldwide, offering off-season premium products.

REFERENCES

Campos-Vargas, R., Becerra O., Baeza-Yates R., Cambiazo, V., González, M. *et al.* (2006) Seasonal variation in the development of chilling injury in peach (*Prunus persica* [L.] Batsch) cv. *O'Henry. Scientia Horticulturae* 110, 79–83.

Cantín, C.M., Tian, T., Xiaoqiong, Q. and Crisosto, C.H. (2011) Copigmentation triggers the development of skin burning disorder on peach and nectarine fruit [*Prunus persica* (L.) Batsch]. *Journal of Agricultural and Food Chemistry* 59, 2393–2402.

Cheng, G.W. and Crisosto, C.H. (1995) Browning potential, phenolic composition, and polyphenoloxidase activity of buffer extracts of peach and nectarine skin tissue. *Journal of the American Society of Horticultural Science* 120, 835–838.

Cheng, G.W. and Crisosto, C.H. (1997) Iron–polyphenol complex formation and skin discoloration in peaches and nectarines. *Journal of the American Society for Horticultural Science* 122, 95–99.

Crisosto, C.H. (1994) Stone fruit maturity indices: a descriptive review. *Postharvest News and Information* 5, 65–68.

Crisosto, C.H. and Costa, G. (2008) Preharvest factors affecting peach quality. In: Layne, D.R. and Bassi, D. (eds) *The Peach: Botany, Production and Uses*. CAB International, Wallingford, UK, pp. 536–549.

Crisosto, C.H. and Day, K.R. (2012) Stone fruit. In: Rees, D., Farrell, G. and Orchard, J. (eds) *Crop Post-harvest: Science and Technology: Vol. 3. Perishables*. Wiley, Chichester, UK, pp. 212–225.

Crisosto, C.H. and Mitchell, F.G. (2016) Postharvest handling systems: stone fruits – peach, nectarine, and plum. In: Kader, A.A. and Thompson, J.F. (eds) *Postharvest Technology of Horticultural Crops*, 3rd edn. University of California Agriculture and Natural Resources, Publication No. 3311. University of California, Davis, California, pp. 345–350.

Crisosto, C.H. and Valero, C. (2006) "Ready to eat": maduración controlada de fruta de hueso en cámara. *Horticultura* 29, 32–37.

Crisosto, C.H., Johnson, R.S., Day, K.R., Beede, B. and Andris, H. (1999a) Contaminants and injury induce inking on peaches and nectarines. *California Agriculture* 53, 19–23.

Crisosto, C.H., Mitchell, F.G. and Ju, Z.G. (1999b) Susceptibility to chilling injury of peach, nectarine, and plum cultivars grown in California. *HortScience* 34, 1116–1118.

Crisosto, C.H., Day, K.R., Johnson, R.S. and Garner, D. (2000) Influence of in-season foliar calcium sprays on fruit quality and surface discoloration incidence of peaches and nectarines. *Journal of the American Pomological Society* 54, 118–122.

Crisosto, C.H., Slaughter, D., Garner, D. and Boyd, J. (2001) Stone fruit critical bruising thresholds. *Journal of the American Pomological Society* 55, 76–81.

Crisosto, C.H., Slaughter, D. and Garner, D. (2004a) Developing maximum maturity indices for "Tree ripe" fruit. *Advances in Horticultural Science* 18, 29–32.

Crisosto, C.H., Garner D., Andris H. L. and Day K.R. (2004b) Controlled delayed cooling extends peach market life. *HortTechnology* 14, 99–104.

Crisosto, C.H., Crisosto, G.M. and Day, K.R. (2008) Market life update for peach, nectarine, and plum cultivars grown in California. *Advances in Horticultural Science* 22, 201–204.

Crisosto, C.H., Lurie, S. and Retamales, J. (2009) Stone fruits. In: Yahia, E.M. (ed.) *Modified and Controlled Atmospheres for the Storage, Transportation, and Packaging of Horticultural Commodities*. CRC Press, Boca Raton, Florida, pp. 287–315.

Crisosto, C.H., Echeverría, G. and Manganaris, G.A. (2020) Peach and nectarine postharvest handling. In: Crisosto, C.H. and Crisosto, G.M. (eds) *Mediterranean Tree Fruits and Nuts Postharvest Handling*. CAB International, Wallingford, UK, pp. 53–87.

Day, K.R. (2006) Fruit blemishes (corking and spotting). Central Valley Postharvest Newsletter. *Cooperative Extension University of California* 15, 5–8.

Dong, L., Zhou, H.W., Sonego, L., Lers, A. and Lurie, S. (2001) Ethylene involvement in the cold storage disorder of 'Flavortop' nectarine. *Postharvest Biology and Technology* 23, 105–115.

Echeverría, G., Arrufat, R., Giné-Bordonaba, J., Larrigaudière, C. and Ubach, D. (2021) Capacidad de exportación de 28 variedades de fruta de hueso. Available at: https://issuu.com/horticulturaposcosecha/docs/capacidad_de_exportacion_de_28_variedades_de_fruta (accessed 7 December 2022).

Falara, V., Manganaris, G.A., Ziliotto, F., Manganaris, A., Bonghi, C. *et al.* (2011) A β-d-xylosidase and a PR-4B precursor identified as genes accounting for differences in peach cold storage tolerance. *Functional and Integrative Genomics* 11, 357–368.

Giné-Bordonaba, J., Cantín C.M., Echeverría G., Ubach D. and Larrigaudière C. (2016) The effect of chilling injury-inducing storage conditions on quality and consumer acceptance of different *Prunus persica* cultivars. *Postharvest Biology and Technology* 115, 38–47.

Hu, M.J., Peng, C., Melgar, J.C. and Schnabel, G. (2017) Investigation of potential causes of peach skin streaking. *Plant Disease* 101, 1601–1605.

Jin, P., Zheng, Y., Tang, S., Rui, H. and Wang, C.Y. (2009) A combination of hot air and methyl jasmonate vapor treatment alleviates chilling injury of peach fruit. *Postharvest Biology and Technology* 52, 24–29.

Jin, P., Zhu, H., Wang, L., Shan, T. and Zheng, Y. (2014) Oxalic acid alleviates chilling injury in peach fruit by regulating energy metabolism and fatty acid contents. *Food Chemistry* 161, 87–93.

Lawton, J.M., Melgar, J.C. and Schnabel, G. (2022) Captan-induced bronzing and inking of peach skin and mitigation strategies. *Acta Horticultuarae* 1352, 49–54.

Lurie, S. and Crisosto, C.H. (2005) Chilling injury in peach and nectarine. *Postharvest Biology and Technology* 37, 195–208.

Manganaris, G.A. and Crisosto, C.H. (2020) Stone fruits. In: Gil, M. and Beaudry, R. (eds) *Controlled and Modified Atmosphere for Fresh and Fresh-Cut Produce.* Academic Press, London, pp. 311–322.

Manganaris, G.A., Vasilakakis, M., Diamantidis, G. and Mignani, I. (2006) Cell wall physicochemical aspects of peach fruit related to internal breakdown symptoms. *Postharvest Biology & Technology* 39, 69–74.

Manganaris, G.A., Drogoudi, P., Goulas, V., Tanou, G., Georgiadou, E.C. *et al.* (2017) Deciphering the interplay among genotype, maturity stage and low-temperature storage on phytochemical composition and transcript levels of enzymatic antioxidants on *Prunus persica* fruit. *Plant Physiology and Biochemistry* 119, 189–199.

Manganaris, G.A., Vicente, A.R., Martinez, P. and Crisosto, C.H. (2019) Postharvest physiological disorders in peach and nectarine. In: Tonetto de Freitas, S. and Pareek, S. (eds) *Physiological Disorders in Fruits and Vegetables.* CRC Press, Boca Raton, Florida, pp. 253–264.

Manganaris, G.A., Minas, I., Cirilli, M., Torres, R., Bassi, D. and Costa, G. (2022) Peach for the future: a specialty crop revisited. *Scientia Horticulturae* 305: 111390.

Martínez-García, P.J., Peace, C.P., Parfitt, D.E., Ogundiwin, E.A., Fresnedo-Ramírez, J. *et al.* (2011) Influence of year and genetic factors on chilling injury susceptibility in peach (*Prunus persica* (L.) Batsch). *Euphytica* 185, 267–280.

Mitchell, F.G., Thompson, J.F., Crisosto, C.H. and Kasmire, R.F. (1998) The commodity. In: Thompson, J.F., Mitchell, F.G., Rumsey, T.R., Kasmire, R.F. and Crisosto, C.H. (eds) *Commercial Cooling of Fruits, Vegetables, and Flowers.* DANR Publication #21567. Postharvest Technology Center, UC Davis Department of Plant Sciences, California.

Murray, R., Lucangeli, C., Polenta, G. and Budde, C. (2007) Combined pre-storage heat treatment and controlled atmosphere storage reduced internal breakdown of 'Flavorcrest' peach. *Postharvest Biology and Technology* 44, 116–121.

Pegoraro, C., Tadiello, A., Girardi, C.L., Chaves, F.C., Quecini, V. *et al.* (2015) Transcriptional regulatory networks controlling woolliness in peach in response to preharvest gibberellin application and cold storage. *BMC Plant Biology* 15: 279.

Perís, M. and Alegre, S. (2012) Factors affecting corky spot on nectarine fruits in the Ebro valley in Spain. *Acta Horticulturae* 962, 557–562.

Reig, G., Torguet, S., Sans, L., Sas, M. and Faro, C. (2022) Inking in Spanish peaches: possible causes and solutions. *Acta Horticulturae* 1352, 199–206.

Schmitz, I.T. and Schnabel, G. (2019) Infrequent occurrence of peach skin streaking and the role of rainwater attributes on symptoms development. *Plant Disease* 103, 2606–2611.

Shan, T., Jin, P., Zhang, Y., Huang, Y., Wang, X. and Zheng, Y. (2016) Exogenous glycine betaine treatment enhances chilling tolerance of peach fruit during cold storage. *Postharvest Biology and Technology* 114, 104–110.

Thompson, J. (2016) Transportation. In: Kader, A.A. and Thompson, J.F. (eds) *Postharvest Technology of Horticultural Crops.* University of California Agriculture and Natural Resources, Publication No. 3311. University of California, Davis, California, pp. 259–270.

Thompson, J.F. and Crisosto, C.H. (2016) Handling at destination markets. In: Kader, A.A. and Thompson, J.F. (eds) *Postharvest Technology of Horticultural Crops.* University of California Agriculture and Natural Resources, Publication No. 3311. University of California, Davis, California, pp. 271–278.

Val, J., Del Río, S., Redondo, D. and Díaz, A. (2018) Novel exogenous calcium treatment strategies effectively mitigate physiological disorders in late season peach cultivars. *Acta Horticulturae* 1194, 617–622.

Valero, C., Crisosto, C.H. and Slaughter, D. (2007) Relationship between nondestructive firmness measurements and commercially important ripening fruit stages for peaches, nectarines and plums. *Postharvest Biology and Technology* 44, 248–253.

Yang, A., Cao, S., Yang, Z., Cai, Y. and Zheng, Y. (2011) γ-Aminobutyric acid treatment reduces chilling injury and activates the defense response of peach fruit. *Food Chemistry* 129, 1619–1622.

Zhou, H.W., Dong, L., Ben Arie, R. and Lurie, S. (2001a) The role of ethylene in the prevention of chilling injury in nectarines. *Journal of Plant Physiology* 158, 55–61.

Zhou, H.W., Lurie, S., Ben-Arie, R., Dong, L., Burd, S. *et al.* (2001b) Intermittent warming of peaches reduces chilling injury by enhancing ethylene production and enzymes mediated by ethylene. *Journal of Horticultural Science and Biotechnology* 76, 620–628.

Zoffoli, J.P., Balbontin, S. and Rodríguez, J. (2002) Effect of modified atmosphere packaging and maturity on susceptibility to mealiness and flesh browning of peach cultivars. *Acta Horticulturae* 592, 573–579.

11

PEACH FRUIT QUALITY: COMPONENTS, COMPOSITION, AND NUTRITIONAL AND HEALTH BENEFITS

Gemma Reig[1]*, Luis Cisneros-Zevallos[2], Guglielmo Costa[3] and Carlos H. Crisosto[4]

[1]*Fruit Production Programme, Institute of Agrifood Research and Technology (IRTA), Lleida, Spain; [2]Department of Horticultural Sciences, Texas A&M University, College Station, Texas, USA; [3]Department of Agri-Food Sciences and Technologies, University of Bologna, Bologna, Italy; [4]Department of Plant Sciences, University of California, Davis, California, USA*

11.1 INTRODUCTION

Peach (*Prunus persica* (L.) Batsch) is a fleshy fruit classified as a drupe (Rodríguez *et al.*, 2019), consisting of a single seed surrounded by a pericarp (Bassi and Monet, 2008; Rodríguez *et al.*, 2019). The pericarp is differentiated into three distinct layers: an outer skin or exocarp, a fleshy middle layer or mesocarp, and a hard, woody layer or endocarp (stone) surrounding the seed (kernel) (Fig. 11.1). The exocarp and mesocarp are edible and, depending on consumer preferences, either both or just the mesocarp (flesh) can be eaten fresh.

P. persica is one of the most variable species of fruit. New cultivars are generated each year, and one of the most striking differences among them is the presence/absence of trichomes on the fruit surface. This difference has given rise to the classification of these fruits as peaches, which are pubescent, and nectarines, which are glabrous (Nilo *et al.*, 2012). Peach fruit also present other easily discernible morphological characteristics such as peel colour (greenish, orange, purple, red, yellow or white, with many combinations of these characters), flesh colour (red, yellow, white, yellow-red or white-red), fruit shape (globose, flat or oblong) (Fig. 11.1), fruit size (from 51 mm to >90 mm in diameter), flesh texture (melting, slow melting, non-melting or stony-hard), flesh acidity (subacid (<5 mEq 100 ml⁻¹), sweet (5–9 mEq 100 ml⁻¹), equilibrated (9–12 mEq

---

* Email: reiggemma@gmail.com

© CAB International 2023. *Peach* (G. Manganaris, G. Costa and C. Crisosto eds)
DOI: 10.1079/9781789248456.0011

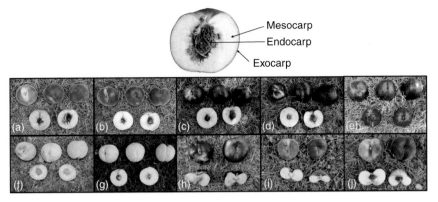

Fig. 11.1. Different peach fruit types and flesh colours: (a) 'Elegant Lady', a yellow-fleshed peach, (b) 'Sweet Regal', a white-fleshed peach, (c) 'Clariss', a yellow-fleshed nectarine, (d) 'Nectarnow', a white-fleshed nectarine, (e) 'Diablotina', a red-fleshed nectarine, (f) 'Ferlot', a yellow-fleshed non-melting peach, (g) 'Ice Peach', a white-fleshed non-melting peach, (h) 'Mistral-30', a yellow-fleshed flat nectarine, (i) 'Osiris', a yellow-fleshed flat peach, and (j) 'SweetCap', a white-fleshed flat peach. Figure courtesy of Gemma Reig, IRTA.

100 ml⁻¹) and acid (12–15 mEq 100 ml⁻¹)) and stone adherence (clingstone, freestone or semi-freestone) (Bassi and Monet, 2008; Iglesias and Echeverría, 2009; UPOV, 2010; Vicente *et al.*, 2011; Reig *et al.*, 2015; Bassi *et al.*, 2016).

As well as fresh consumption, the peach fruit is an interesting material for the food industry (Nowicka and Wojdylo, 2019). The flesh (mesocarp) can be used to manufacture very sweet products such as jams, jellies, juice concentrates, purees and fruit mousse added to yogurts or as an ingredient in smoothie-type products (Nowicka and Wojdylo, 2019). These industrial processes generate large amounts of waste peach flesh, which becomes a by-product. The whole peach fruit can also be processed industrially to produce canning peaches. During this process, the peel and endocarp (kernel included) are usually removed and end up as by-products. In addition, during peach-tree cultivation, large amounts of other by-products are also generated, such as flowers during blooming, stems during harvesting and leaves during winter leaf fall.

11.2 QUALITY DEFINITION

Peach fruit quality, whatever the peach fruit type (peach or nectarine), can be divided into six categories, depending on the trait of interest:

1. External quality: including fruit size and shape, blush or red overcolour coverage (around 50% of blush for traditional or bicolored cultivars, and fully blushed for modern cultivars or those bred based on 'Big Top' characteristics (Font i Forcada *et al.*, 2021)), and absence of external and internal defects.

2. Internal quality: including flesh firmness, soluble solids concentration (SSC), titratable acidity, index of absorbance difference and dry-matter content.

3. Sensorial quality: including flavour, taste, aroma, sweetness, sourness, crispness, firmness, juiciness, fibrousness and ease of breakdown.

4. Nutritional quality: including all soluble sugars, organic acids, minerals, vitamins, fibre, proteins and lipids.

5. Health-promoting quality: including all bioactive compounds and antioxidants.

6. Sanitary quality: including the absence of fungicides, pathogens and heavy metals within the edible parts of the fruit (flesh and peel).

The term fruit quality encompasses many of these properties or traits and, depending on the actor in the supply chain (producer, packer, shipper or consumer), the weight of these quality properties can differ (Crisosto and Costa, 2008; Crisosto and Valero, 2008; Minas *et al.*, 2018; Manganaris *et al.*, 2019). Initially, fruit quality was defined solely by fruit size and colour, and it was uncertain whether these traits were important only for the first buy (Crisosto, 2002; Delgado *et al.*, 2013). In the last decade, fruit consumer quality has expanded to include sensory properties (appearance, texture, flavour, taste and aroma), nutritional value, health properties and safety profile. Together, they determine not only the quality of the fruit but also its economic value and future sales (Crisosto, 2002; Crisosto and Costa, 2008; Delgado *et al.*, 2013; Reig *et al.*, 2013a; Font i Forcada *et al.*, 2019; Manganaris *et al.*, 2019). Handlers discovered that consumers rejected fruit with low sugar content, poor aroma and texture, and 'off-flavour' (chilling injury symptoms), even if they were large-sized and had an attractive red colour (Lurie and Crisosto, 2005; Crisosto and Valero, 2008; Manganaris *et al.*, 2019). Consumers encountering peaches with these negative attributes tend to immediately change their preference to other fruit types.

In recent decades, peach consumption has undergone a continuous decline, mainly because of a poor understanding and definition of fruit quality, premature harvesting, improper handling, chilling injury damage and a lack of ripening prior to consumption, resulting in consumer dissatisfaction (Crisosto, 2002). In a detailed sensory study using a large group of peach and nectarine cultivars, cluster analysis revealed two main sensory attributes that were significantly associated with overall liking and consumer acceptance (Bonany *et al.*, 2014). This and other sensory studies demonstrated that perception of sweetness was the main positive driver of 'liking' and that the perception of grassy/green fruit aromas and pit aromas ('off-flavour') was the main negative driver (Delgado *et al.*, 2013; Bonany *et al.*, 2014). In sound fruit, SSC was the only instrumental measurement significantly related to consumer liking and preferences (Crisosto and Crisosto, 2005; Crisosto *et al.*, 2006; Delgado *et al.*, 2013). In California, as an industry guideline, a minimum 10.0% SSC is recommended for mid-acid yellow-fleshed peaches and nectarines as the quality standard for peaches free of sensory defects such as chilling injury symptoms (Crisosto and Valero, 2008). In the European Union, an SSC value of at least 8.0% and a firmness value of 63.7 N or less have been established for early- to late-ripening peaches, notably below the current recommendations of at least 11.0% SSC and fruit firmness of less than 49 N (Iglesias and Echeverría, 2009).

11.3 COMPOSITION OF PEACH FRUIT: NUTRITIONAL BENEFITS AND HEALTH-PROMOTING POTENTIAL

11.3.1 Exocarp and endocarp layers

Like other tree fruits, the edible parts of peach fruit (flesh and peel) contain many essential components, making it very attractive for fresh consumption. These components are related mainly to internal, sensorial, nutritional and health-promoting qualities. When peach fruit is harvested at its optimal maturity stage, it is an excellent functional food that is rich in macronutrients (carbohydrates, including the soluble sugars sucrose, glucose and fructose, and sugar alcohols (sorbitol), organic acids, fats, proteins and dietary fibre), micronutrients (minerals and vitamins), and volatile and bioactive compounds (Bento *et al.*, 2020). These components, except for the volatile compounds, are described in Tables 11.1 and 11.2.

Peach flesh has a high amount of water (~89 g per 100 g of fruit) (Table 11.1), which aids ingestion and acts as a temperature and pH regulator, maintaining cell and tissue integrity (Sánchez-Moreno *et al.*, 2006). Water content varies among cultivars, with an indirect way of its evaluation being to measure the dry-matter content. Dry matter is everything left in the fruit when all the water is removed; this includes sugars, cell walls, organic acids, fibres and minerals. As it is related to fruit development/maturity, a higher dry-matter content means a higher total carbohydrate level, which translates into better quality and better consumer preference (Musacchi and Serra, 2018). However, to date, no studies have reported on the dry-matter content of the different currently available peach cultivars.

Carbohydrates account for 50–80% of the dry weight, and approximately 9.54 g per 100 g of fruit are carbohydrates (Table 11.1). Sucrose is the major soluble sugar in peach fruit, followed by the reducing sugars (fructose and glucose) and sorbitol (Reig *et al.*, 2013a; Bassi *et al.*, 2016). Sucrose and sorbitol are highly correlated with taste and aroma, and peaches with higher amounts of these sugars, and especially sucrose, have better aromas (Bento *et al.*, 2020). In peach, taste depends largely on water-soluble compounds, such as sugars and organic acids, conferring the sensation of sweetness and/or sourness. Sucrose is also important as a preservative of fruit flavours (Huberlant and Anderson, 2003), while sorbitol additionally affects the texture (Cantín *et al.*, 2009b). Fructose

Table 11.1. Nutritive value per 100 g of raw yellow peach. Data from USDA National Nutrient database (https://data.nal.usda.gov/dataset/usda-national-nutrient-database-standard-reference-legacy-release, accessed 2 January 2023).

Compound	Amount	Compound	Amount
Water	88.87 g	**Phytonutrients**	
Carbohydrates	9.54 g	Carotene-β	0.162 mg
Soluble carbohydrates		Cryptoxanthin-β	0.067 mg
Sucrose	4.76 g	Lutein-zeaxanthin	0.091 mg
Glucose	1.95 g	Tocotrienol-α	0.03 mg
Fructose	1.53 g	Tocopherol-γ	0.02 mg
Lactose	–	**Fatty acids**	
Maltose	0.08 g	Total saturated	0.019 g
Galactose	0.06 g	Hexadecanoic acid	0.017 g
Starch	–	Octadecanoic acid	0.002 g
Dietary fibre	1.5 g	Total monounsaturated	0.067 g
Protein	0.91 g	Hexadecenoic acid	0.002 g
Total lipid (fat)	0.25 g	Octadecenoic acid	0.065 g
Ash	0.43 g	Total polyunsaturated	0.086 g
Cholesterol	–	Octadecadienoic acid	0.084 g
Phystosterols	10 mg	Octadecatrienoic acid	0.002 g
Vitamins		**Amino acids**	
Ascorbic acid (vitamin C)	6.6 mg	Tryptophan	0.01 g
Vitamin A	0.016 mg	Threonine	0.016 g
Pyridoxine (vitamin B6)	0.025 mg	Isoleucine	0.017 g
α-Tocopherol (vitamin E)	0.73 mg	Leucine	0.027 g
Vitamin K	2.6 µg	Lysine	0.03 g
Folate (vitamin B9)	4 µg	Methionine	0.01 g
Niacin (vitamin B3)	0.806 mg	Cystine	0.012 g
Pantothenic acid (vitamin B5)	0.153 mg	Phenylalanine	0.019 g
Riboflavin (vitamin B2)	0.031 mg	Tyrosine	0.014 g
Thiamine (vitamin B1)	0.024 mg	Valine	0.022 g
Choline	6.1 mg	Arginine	0.018 g
Betaine	0.3 mg	Histidine	0.013 g
Minerals		Alanine	0.028 g
Calcium	6 mg	Aspartic acid	0.418 g
Iron	0.25 mg	Glutamic acid	0.056 g
Magnesium	9 mg	Glycine	0.021 g
Phosphorus	20 mg	Proline	0.018 g
Potassium	190 mg	Serine	0.032 g
Sodium	–		
Copper	0.068 mg		
Manganese	0.61 mg		
Zinc	0.17 mg		
Selenium	0.1 µg		
Fluoride	4 µg		

Table 11.2. Classification of bioactive compounds in peach (*Prunus persica*) fruit.

Bioactive compound	Group	Exocarp (peel)	Mesocarp (flesh)	Kernel (seed)	Reference(s)
Phenolic compounds					
Gallic acid	Phenolic acid (hydroxy-benzoic acid)	✓	✓		Manzoor et al. (2012)
Neochlorogenic acid	Phenolic acid (hydroxycinnamic acid)	✓	✓	✓	Stojanovic et al. (2016); Nowicka and Wodjdylo (2019)
Dicaffeoylquinic acid	Phenolic acid (hydroxycinnamic acid)			✓	Senica et al. (2017)
Procyanidin B1	Flavan-3-ols			✓	Nowicka and Wodjdylo (2019)
Cyanidin 3-glucoside	Anthocyanin	✓	✓		Vicente et al. (2011); Reig et al. (2013a)
3-O-p-Coumaroyloquinic acid	Phenolic acid (hydroxycinnamic acid)			✓	Nowicka and Wodjdylo (2019)
Cyanidin 3-rutinoside	Anthocyanin	✓	✓	✓	Reig et al. (2013a); Lara et al. (2020)
(+)-Catechin	Flavan-3-ols	✓	✓	✓	Manzoor et al. (2012); Senica et al. (2017); Nowicka and Wodjdylo (2019); Lara et al. (2020)
Chlorogenic acid	Phenolic acid (hydroxycinnamic acid)	✓	✓	✓	Senica et al. (2017); Nowicka and Wodjdylo (2019); Lara et al. (2020)
Caffeic acid	Phenolic acid (hydroxycinnamic acid)	✓		✓	Vicente et al. (2011); Lara et al. (2020)
Procyanidin B2	Flavan-3-ols			✓	Nowicka and Wodjdylo (2019); Lara et al. (2020)
cis-5-p-Coumaroyloquinic acid	Phenolic acid (hydroxycinnamic acid)			✓	Nowicka and Wodjdylo (2019)
(−)-Epicatechin	Flavan-3-ols			✓	Nowicka and Wodjdylo (2019)
2-O-Caffeoyl-l-malate	Phenolic acid (hydroxycinnamic acid)			✓	Nowicka and Wodjdylo (2019)

Continued

Table 11.2. Continued.

Bioactive compound	Group	Exocarp (peel)	Mesocarp (flesh)	Kernel (seed)	Reference(s)
Ellagic acid	Phenolic acid (hydroxybenzonic acid)			✓	Nowicka and Wodjdylo (2019)
Quercetin 3-O-glucoside	Flavonol	✓		✓	Stojanovic et al. (2016); Nowicka and Wodjdylo (2019)
Luteolin 7-glucoside	Flavonol			✓	Nowicka and Wodjdylo (2019)
(−)-Epicatechin gallate	Flavan-3-ols			✓	Nowicka and Wodjdylo (2019); Lara et al. (2020)
Isorhamnetin 3-O-glucoside	Flavonol			✓	Nowicka and Wodjdylo (2019)
Kaempferol 3-O-glucoside	Flavon			✓	Nowicka and Wodjdylo (2019)
Keaempferol 7-neohesperidoside	Flavon			✓	Nowicka and Wodjdylo (2019)
Kaempferol 3-rutinoside	Flavonol			✓	Senica et al. (2017)
Hesperidin 7-rutinoside	Flavonol			✓	Nowicka and Wodjdylo (2019)
Procyanidin A2	Flavan-3-ols			✓	Nowicka and Wodjdylo (2019)
Tetraterpenoids					
Lutein	Carotenoid		✓		Nowicka and Wodjdylo (2019)
Zeaxanthin	Carotenoid		✓		Nowicka and Wodjdylo (2019)
β-Cryptoxanthin	Carotenoid		✓		Nowicka and Wodjdylo (2019)
β-Carotene	Carotenoid	✓	✓		Liu et al. (2015); Nowicka and Wodjdylo (2019)
β-Cryptoxanthin myristate	Carotenoid		✓		Nowicka and Wodjdylo (2019)
β-Cryptoxanthin palmitate	Carotenoid		✓		Nowicka and Wodjdylo (2019)
Cyanogenic glycoside					
Prunasin	—			✓	Senica et al. (2017); Nowicka and Wodjdylo (2019)
Amygdalin	—			✓	Senica et al. (2017); Nowicka and Wodjdylo (2019)

has a higher level of sweetness than sucrose and glucose, which makes it an important influence in terms of taste (Bento *et al.*, 2020). Different studies conducted on European and US cultivars have reported a wide range of variability in terms of total and individual carbohydrate content (Aubert *et al.*, 2003; Cantín *et al.*, 2009b; Reig *et al.*, 2013a; Font i Forcada *et al.*, 2014b; Kwon *et al.*, 2015; Saidani *et al.*, 2017). Figure 11.2 shows the sugar profile of different peach cultivars, grouped by fruit type. Sugar content is higher in the flesh than in the peel, and sucrose is the main sugar in both tissues (Saidani *et al.*, 2017). However, when comparing fruit types, nectarine fruits generally have a higher glucose content than melting peach fruits, and white-fleshed peaches have higher individual and total carbohydrate content than yellow-fleshed ones. In term of their nutritional value, more than 80% of the total 39 kcal provided by 100 g of raw peach fruit comes from carbohydrates, the main source of energy in the human diet and a provider of important health benefits. By way of example, fructose induces growth of *Bifidobacterium* and *Lactobacillus* spp. in the gastrointestinal tract, while sorbitol can act as a laxative in the body as it promotes the osmotic transfer of water into the bowel, provides dental health benefits and serves as an alternative to glucose for diabetics (Cantín *et al.*, 2009b; Font i Forcada *et al.*, 2014b).

Organic acids represent only 4% of the content of peach nutrients (Bento *et al.*, 2020), accounting for around 0.9–1.6% of fresh weight (Bassi *et al.*, 2016) (Table 11.1). Together, they determine the titratable acidity of the fruit, which is cultivar dependent and mostly under genetic control (Cantín *et al.*, 2010). Malic acid is, in general, the most abundant acid at maturity, followed by quinic, citric and succinic acid, with traces of shikimic acid (Reig *et al.*, 2013a; Bassi *et al.*, 2016; Lyu *et al.*, 2017). Titratable acidity is an important quality indicator and

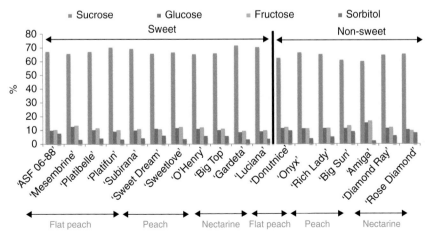

Fig. 11.2. Sugar profile of various peach cultivars. Figure courtesy of Gemma Reig, IRTA.

a major driver of consumer preference. Acidity correlates with the perception of taste and aroma, with sweetness influenced by citric and shikimic acids and the sugars:organic acids ratio, and aroma by total acidity and malic acid content (Colaric *et al.*, 2005). High concentrations of sugar are easily hidden by high concentrations of organic acids that dim the perception of sweetness. This varies according to the predominant organic acid (e.g. citric acid adds more sourness than malic acid), meaning different organic acids require different concentrations to affect sugar perception (Bento *et al.*, 2020). The content and profile of organic acids seem to be quantitatively and qualitatively variable among cultivars (Aubert *et al.*, 2003; Reig *et al.*, 2013a; Saidani *et al.*, 2017; Baccichet *et al.*, 2021). Figure 11.3 shows the organic acid profile of different peach cultivars, grouped by fruit type. In general, acid content is higher in flesh than in peel, with malic acid predominating in both tissues (Saidani *et al.*, 2017). Among fruit types and flesh colours, white nectarines tend to have higher citric acid content, non-melting peaches higher malic acid content and white flat peaches higher shikimic acid content. Yellow-fleshed cultivars tend to have higher malic acid content, whereas those with white flesh have higher citric and shikimic acid content (Reig *et al.*, 2013a).

Peach fruit has a very small amount of fat or fatty acids (Table 11.1). For a 100 g portion of peach, the amount of fat is approximately 0.25 g (Bento *et al.*, 2020), representing a small part of peach fruit composition. Fatty acids can be classified as saturated (butyric, caproic, caprylic, capric, lauric, palmitic, stearic and myristic acids), monounsaturated (oleic and palmitoleic acid) or polyunsaturated (linoleic, linolenic and arachidonic acids). Linoleic and linolenic acids cannot be synthesized in the body and are known as essential fatty acids

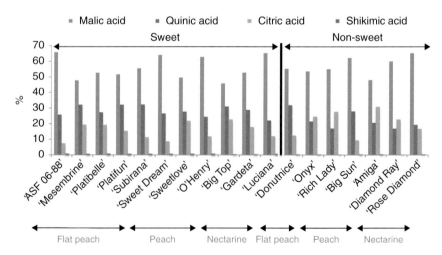

Fig. 11.3. Organic acid profile of several peach cultivars. Figure courtesy of Gemma Reig, IRTA.

(Sánchez-Moreno *et al.*, 2006). Duan *et al.* (2013) reported that the predominant fatty acids in peach fruit are linoleic, palmitic and linolenic acids, with their profiles varying among cultivars. Fatty acids are important because they make up the surface wax, which contributes to the fruit's cosmetic appearance, and the cuticle, which protects the fruit against water loss and pathogens, as well as being constituents of cell membranes, which influence the physiological activities of the fruit (Crisosto and Valero, 2008). As a nutrient, fat serves as a carrier for fat-soluble vitamins, for some fatty acids (linoleic and linolenic acids) that are essential nutrients and can only be ingested with fat, and for some of the bioactive compounds present in fruits such as lipophilic phyto-oestrogens and carotenoids (Sánchez-Moreno *et al.*, 2006). Fatty acids also play an important role in the prevention of cardiovascular diseases and in cell maintenance. Indeed, polyunsaturated fatty acids seem to prevent inflammatory and other chronic diseases and promote low levels of total cholesterol and high-density lipoprotein (Bento *et al.*, 2020). Both linoleic and linolenic acids are needed to build and repair cell walls and tissues in the nervous system and in the formation of prostaglandins, a group of lipids made at sites of tissue damage or infection that are involved in dealing with injury and illness.

Peach fruit has a low protein content (Table 11.1). The small-sized proteins in peach have an important function as enzymes catalysing the various chemical reactions responsible for compositional changes (Crisosto and Valero, 2008). Proteins are made up of long chains of amino acids. The amino acid profile of peach fruit varies among cultivars, and even between the different parts of the fruit, but plays a role in fruit taste. They can be grouped as tasteless (arginine, asparagine, isoleucine, lysine, serine, threonine and valine), bitter (leucine, phenylalanine, tryptophan and tyrosine) and sweet (proline and alanine) amino acids (Bento *et al.*, 2020). However, some proteins can cause food allergies. They are the lipid transfer proteins (LTP). Peach allergy is emerging as a common type of fresh-fruit allergy in Europe, especially in the Mediterranean area, causing stress and anxiety to those affected (Barni *et al.*, 2022). To date, six peach allergens have been recognized (Pru p 1, Pru p 2, Pru p 3, Pru p 4, Pru p 7 and Pru p 9). While all these allergens have been found in both the flesh and peel of the fruit, the concentration of Pru p 3 is seven times higher in the peel than the flesh. Pru p 1 is heat labile and sensitive to gastrointestinal digestion. Thus, only the unprocessed form of the fruit leads to the typical symptoms of oral allergy syndrome (OAS), whereas cooked peach is tolerated. Pru p 3 is resistant to heat and proteolytic digestion. Therefore, the clinical manifestations of Pru p 3 sensitization can range from mild symptoms of OAS to severe systemic allergic reactions (anaphylaxis). Pru p 4 is heat labile and can be destroyed by gastrointestinal digestion. For this reason, the usual clinical manifestation of Pru p 4 sensitization is OAS. Pru p 7 is resistant to heat and proteolytic digestion. Thus, the typical allergic symptom of Pru p 7 sensitization is anaphylaxis. Finally, Pru p 9 is considered a new occupational allergen from peach-tree pollen involved in rhinitis and asthma (Barni *et al.*, 2022).

Peach fruit is a good source of fibre (Table 11.1). Fibre has been called an unavailable carbohydrate and cannot be digested by the secretions in the human gastrointestinal tract (Bento *et al.*, 2020). It can be classified into water-soluble and water-insoluble fibre. The major components of dietary fibre are the polysaccharide celluloses (insoluble), hemicelluloses (soluble), pectins (soluble), gums (soluble) and mucilages (soluble) and the organic polymer lignin (insoluble). Pectin is covalently bound to the cell wall, which is associated with fruit maturity (Lyu *et al.*, 2017). Some studies have reported on the differing fibre profiles of peach cultivars (Lyu *et al.*, 2017). In terms of nutritional benefit, fibre has a regulatory function in the gastrointestinal and circulatory systems, helping to control weight and maintaining low cholesterol levels (Bento *et al.*, 2020). Pectin has been shown to lower blood cholesterol levels and reduce the risk of coronary heart disease (Lyu *et al.*, 2017), and it has been reported that lignin has antiviral and antitumoral effects (Vinardell and Mitjans, 2017).

Fruits, and in particular peach fruit, are a rich source of minerals (Table 11.1). Potassium is by far the most abundant mineral in peach fruit, followed by phosphorus, magnesium, calcium, fluoride, manganese, iron and zinc. Some of these minerals play a key role in metabolism, participating in the synthesis of amino acids and proteins (Bento *et al.*, 2020). Potassium plays an important role in the ionic balance and contributes to the maintenance of cell organization and permeability (Manzoor *et al.*, 2012). Magnesium is an essential factor in nervous-system stability, muscle contraction and as an activator of alkaline phosphatase, and can also be used as an alternative to calcium in the body. Calcium is associated with the cell-wall structure, which is important in fruit softening, whereas calcium in the apoplast is related to senescence (Crisosto and Valero, 2008). Manganese is an important essential element but is only required in trace amounts for the maintenance of proper carbohydrate metabolism, as well as being an antioxidant in superoxide dismutase enzymes. Iron is required for red blood cell formation. Zinc acts as a coenzyme for over 200 enzymes involved in immunity, new cell growth and acid-base regulation. A lack of sufficient amounts of zinc may result in reduced activity of related enzymes (e.g. carbonic anhydrase) (Manzoor *et al.*, 2012). The mineral profile is cultivar dependent (Mihaylova *et al.*, 2021), but little information is available in terms of differences in flesh and peel mineral content. A few studies have reported that peel tissue presents significantly higher amounts of nitrogen, calcium, magnesium, iron and manganese than flesh (Basar, 2006; Manzoor *et al.*, 2012; Saidani *et al.*, 2017), whereas potassium levels are predominant in both flesh and peel (Manzoor *et al.*, 2012; Saidani *et al.*, 2017). Iron is the predominant micronutrient in peel (Saidini *et al.*, 2017). In recent decades, there has been growing interest in incorporating high nutrient levels with sufficient amounts of essential minerals into the regular diet, given their important role as structural components of organs (e.g. bones and teeth) (Maatallah *et al.*, 2020). Potassium is an important component of cell and body fluids that helps regulate heart rate and blood pressure. Fluoride is a component of bones and teeth and is essential for the prevention of dental caries. The importance of calcium is well known in human nutrition for skeletal development and growth, while iron is an important trace element that is a core component of red blood cells and whose deficiency can cause anaemia (Manzoor *et al.*, 2012).

Peach vitamin content can be divided into water soluble (vitamin B12, vitamin C, vitamin B9, vitamin B3, vitamins B1 and B2) and fat soluble (vitamin K, vitamin E and vitamin A) (Table 11.1). The most abundant vitamins in peach are vitamins C, E and B complex (Bento *et al.*, 2020), with vitamin C the most important of them. Although peach fruit is known generally not to accumulate large amounts of vitamin C (Aubert and Chalot, 2020), the vitamin C content in the flesh does vary widely among cultivars (Gil *et al.*, 2002; Font i Forcada *et al.*, 2014b). Regarding flesh colour, Aubert and Chalot (2020) reported that red-fleshed peaches had 40% more ascorbic acid than white ones. Of the edible parts of the fruit, peel tissue has a higher vitamin C content than flesh (Liu *et al.*, 2015; Bassi *et al.*, 2016; Saidani *et al.*, 2017; Bento *et al.*, 2020), helping protect the fruit from oxidative stress caused by light and other environmental and biotic factors (Bassi *et al.*, 2016). As for their health properties, vitamins E and C have both shown preventative effects against cardiovascular disease. Vitamin E acts on low-density lipoproteins preventing their oxidation and the development of atherosclerosis. Vitamin C enhances nitric oxide production, which protects the vascular endothelium against free radicals and the ischaemic heart against apoptosis, and also boosts the immune system, promotes healing and helps prevent cancer. Finally, vitamin A is known to offer protection from lung and oral cancers (Sánchez-Moreno *et al.*, 2006; Bento *et al.*, 2020).

A fruit's aroma is a complex mixture of many volatile organic compounds. It is a central trait that influences consumer fruit quality perception (Lara *et al.*, 2020). The relative contributions of specific aroma volatile compounds to the flavour of peaches and nectarines have been widely examined, and more than 100 compounds have been identified so far (Aubert *et al.*, 2003; Wang *et al.*, 2009; Cano-Salazar *et al.*, 2012). They include aldehydes, alcohols, esters, terpenoids, ketones and lactones (Wang *et al.*, 2009). The most abundant are C_6 compounds, esters, benzaldehyde, linalool, C_{13} norisoprenoids and lactones (Lara *et al.*, 2020). Aldehydes, such as hexanal, benzaldehyde and decanal, among others, are described as being 'grassy' in flavour (Kakiuchi and Ohmiya, 1991). They are known by-products of the enzyme-catalysed breakdown of unsaturated fatty acids (Wang *et al.*, 2009). Alcohols such as (*E*)-3-hexen-1-ol, 2-hexen-1-ol, hexanol and octanol, among others, together with aldehydes, are also known as C_6 compounds (Mohammed *et al.*, 2021). Esters, such as ethyl acetate, (Z)-3-hexenyl acetate, hexyl acetate and 2-hexenyl acetate, among others, are the main contributors to fruity and floral notes (Wang *et al.*, 2009), and often represent the major contribution in peaches. Lactones, particularly γ- and δ-decalactones, have been reported to be the

major contributors to peach aroma, with smaller contributions from other volatiles such as C_6 aldehydes, alcohols and terpenoids (Wang *et al.*, 2009), which
are responsible for the spicy, floral and fruity features in peach. Terpenoids,
such as 2-bornylene, d-limonene, eucalyptol, ocimene, linalool and cam-
phor, among others, contribute fresh floral and fruity characteristics to peach.
Linalool is one of the major compounds providing floral notes to peach fruit
(Wang *et al.*, 2009; Mohammed *et al.*, 2021). The aroma profile is cultivar dependent (Aubert *et al.*, 2003; Eduardo *et al.*, 2010; Cano-Salazar *et al.*, 2012);
however, little information is available on the differences between flesh and
peel volatile compounds, or among fruit types. Jia *et al.* (2005) reported that
total aroma volatile production is higher in the peel than in the flesh. Wang
et al. (2009) reported that flat peaches had significantly higher γ-decalactone
(L6) content than white-fleshed peaches and nectarines, while the linalool
content of nectarines was significantly higher than in white-fleshed peaches.

The bioactive compounds in peach fruit can be categorized into phenolic
compounds and tetraterpenoids (Table 11.2). They play a crucial role in cellular metabolism and physiology (Bvenura *et al.*, 2018), and also participate
in the visual appearance (pigmentation and browning) and taste (astringency)
of the fruit. Phenolic compounds are secondary metabolites, being the main
source of antioxidants, vitamin C and carotenoids in peaches (Gil *et al.*, 2002).
In peach fruit, they are divided into two categories, phenolic acids and flavonoids (Table 11.2). These compounds are more abundant in the peel than in the
flesh (Cevallos-Casals *et al.*, 2006; Manzoor *et al.*, 2012; Tomás-Barberán *et al.*,
2013). This is because phenolic substances generally accumulate in higher
concentrations in the outer tissues of plant parts such as fruit, seeds and bark
due to their prospective function in protection against sun-derived ultraviolet
(UV) radiation. They also act as protectors against pathogen and pest attack
(Toor and Savage, 2005). Among the phenolic acids, chlorogenic and caffeic
acids are the two major ones in the epidermis and subtending cell layers of
peach (Vicente *et al.*, 2011). Flavonoids are a large group of structurally related
compounds that include anthocyanins, flavones, flavonols, flavanones, flavan-
3-ols and isoflavones. Anthocyanins are the most widespread of the flavonoid
pigments. They confer red, blue and purple colours to plant tissue. In peach,
anthocyanins are associated mainly with the peel but can also be found in the
flesh. Cyanidin 3-glucoside has been identified as the main anthocyanin in *P.
persica* along with a smaller amount of cyanidin 3-rutinoside (Orazem *et al.*,
2011; Reig *et al.*, 2013a). Peach and nectarine fruit have lower total phenolics
content compared with other *Prunus* spp. fruits such as red and black plums,
cherries and almonds (Vicente *et al.*, 2011). Despite this, within peach cultivars, wide inter-cultivar differences exist (Font i Forcada *et al.*, 2014a; Kwon
et al., 2015; Stojanovic *et al.*, 2016; Saldani *et al.*, 2017; Lara *et al.*, 2020),
with flavonoids also more abundant in the peel than in the flesh (Cevallos-
Casals *et al.*, 2006; Manzoor *et al.*, 2012; Tomás-Barberán *et al.*, 2013; Kwon
et al., 2015; Stojanovic *et al.*, 2016; Saldani *et al.*, 2017; Serra *et al.*, 2020).

Therefore, consumption of unpeeled peaches can be seen as a valuable source of nutrition with important human health benefits (Serra *et al.*, 2020). In terms of flesh colour, red-fleshed peaches contain higher levels of anthocyanins, cinnamic acids and flavanols when compared with other flesh colours. However, flavan-3-ol levels (in particular catechin) are markedly higher in white-fleshed cultivars (Aubert and Chalot, 2020). Serra *et al.* (2020) also reported that flavan-3-ol content appears to be much higher in white-fleshed cultivars (both in the mesocarp and exocarp).

Carotenoids are a group of pigments responsible for the distinctive yellow and orange colours of the flesh and peel of most *Prunus* spp. (Vicente *et al.*, 2011) (Table 11.2). They are essential structural components of the photosynthetic apparatus, contribute to light absorption in regions of the spectrum where chlorophyll is low and provide protection against photo-oxidation (Vicente *et al.*, 2011). The main carotenoids present in peaches and nectarines are β-carotene, xanthophylls (mono-or dihydroxylated carotenoids), zeaxanthin, β-cryptoxanthin, violaxanthin and lutein (Lara *et al.*, 2020). β-Carotene is a pro-vitamin, which is converted into vitamin A inside the body and is more abundant than β-cryptoxanthin (Gil *et al.*, 2002). The carotenoid profiles of many cultivars have been published in different studies (Giuffrida *et al.*, 2013; Liu *et al.*, 2015; Lara *et al.*, 2020). In all fruit types, peel tissue has more carotenoids than the flesh (Liu *et al.*, 2015), and yellow-fleshed cultivars contain close to ten times more carotenoids than white-fleshed ones (Tomás-Barberán *et al.*, 2013).

In the past few years, studies on peach fruit bioactive compounds have shown that they exert a broad range of health-promoting properties related to the treatment and prevention of cancer and cardiovascular, chronic and neurodegenerative diseases (Maatallah *et al.*, 2020). Eating peaches is believed to reduce the generation of reactive oxygen species in human blood plasma and provides protection from a number of chronic diseases, with phenolic acids, flavonoids and anthocyanins the major sources of their therapeutic effects (Dhingra *et al.*, 2014). In particular, peach fruit polyphenols have shown chemopreventive properties against oestrogen-independent and -dependent breast cancer cells with small or no activity on normal cells (Noratto *et al.*, 2009; Vizzotto *et al.*, 2014), inhibition of growth and induction of differentiation of colon cancer cells (Lea *et al.*, 2008), attenuation of oxidative stress and inflammation in *in vitro* and *ex vivo* studies (Gasparotto *et al.*, 2014a,b), inhibition of tumour growth and metastasis of breast cancer in mice (Noratto *et al.*, 2014), prevention of risk factors for obesity-related metabolic disorders and cardiovascular disease in rats (Noratto *et al.*, 2015) and alteration of intestinal microbiota in rats (Noratto *et al.*, 2014). Peaches also have laxative and antihypertensive properties and are suitable for the prevention of constipation and the treatment of duodenum ulcers. Peach fruit has also been shown to protect rat tissues from nicotine toxicity (Kim *et al.*, 2017).

11.3.2 Kernel

The peach kernel (seed) constitutes around 5–10% of the total fruit. During industrial processing, it is usually removed and becomes a by-product. Peach seed is a good source of nutritional and bioactive compounds, providing seed oil and fatty acids (Lara *et al.*, 2020). However, very few studies evaluating its properties have been performed. Qumar (2016) reported the approximate composition of peach seed (6.3% moisture, 3.0% ash, 36.9% fat, 2.2% proteins, 1.3% fibre and 50.3% carbohydrates). Like peach fruit flesh, the phenolic compounds in peach kernel are divided into the same categories, phenolic acids and flavonoids (Table 11.2). The predominant group is the flavan-3-ols, conferring the typical bitter and tart flavour, followed by phenolic acids, flavonols and flavons (Senica *et al.*, 2017; Nowicka and Wojdylo, 2019; Lara *et al.*, 2020). Peach kernel has a higher content of polymeric procyanidins and phenolic acids compared with the flesh, but a several-fold lower concentration of carotenoids, flavonols and flavons. Importantly, peach kernel has a high content of hydroxycinnamic acids (phenolic acids), identified as the main group of phenolic compounds responsible for *in vitro* anti-cancer (breast, colon and liver), antioxidant and anti-ageing activities (Li *et al.*, 2013). As well as polyphenols, Nowicka and Wojdylo (2019) also found tetraterpenoids, six carotenoids and cyanogenic glycoside in peach kernel. The kernel carotenoid profile is similar to that in peach extracts and other peach products (Giuffrida *et al.*, 2013) and in peach peel and flesh, but the content is totally different.

Most of the peach industry considers peach seed to be a low-value agro-industrial by-product, whereas in Chinese medicine, the peach kernel is one of the nine plant ingredients used in a cocktail for cardiovascular disease (Tu *et al.*, 2003). Such importance is probably related to its fatty acid composition, with a low saturated fatty acid content and high quantities of oleic and linoleic acids. Peach seed is rich in prunasin and diglucoside (R)-amygdalin, which releases hydrocyanic and benzaldehyde acids upon enzymatic hydrolysis (Lara *et al.*, 2020). Prunasin is the predominant cyanogenic glycoside in the kernel (Nowicka and Wojdylo, 2019). The functions and roles of cyanogenic glycosides in plants are numerous: they act as a defence against herbivores, enable the storage of nitrogen required for seedling growth, and either promote or inhibit seed germination (Senica *et al.*, 2017). None the less, despite their positive effects, cyanogenic glycosides are harmful for human consumption. Cyanide poisoning can occur when the seeds are physically processed, crushed or chewed as they release hydrocyanic acid. On this basis, the European Food Safety Authority (EFSA) published a scientific opinion in 2016 on the acute health risks posed by cyanogenic glycoside in which it reported 25-fold lower toxic doses for cyanide at only 20 µg kg^{-1} of body weight. Based on this study, adults could only consume three small apricot kernels and children half of one without risking cyanide poisoning. On the other hand, the World Health Organization (WHO) declared that five apricot kernels can safely be consumed in 1 h and no more than 10 kernels per day by a healthy adult (Senica *et al.*, 2017). However, when the seeds are swallowed whole, they pose no danger as the toxic compounds are not released into the system. In this regard, much attention has been paid to the potential beneficial effects of amygdalin in synthetic forms (leatrile or vitamin B17) in the prevention and even treatment of human diseases (Savic *et al.*, 2016). However, serious adverse effects after amygdalin oral absorption have been reported in several patients, who exhibited typical symptoms of cyanide poisoning (Senica *et al.*, 2017). It is assumed that in lower doses the seeds may have other pro-health properties. The seeds have been found to exhibit high antimicrobial, antioxidant, anti-inflammatory and analgesic activities, and also to inhibit the development of human breast, intestine and liver cancer cells (Nowicka and Wojdylo, 2019). Finally, it has also been demonstrated that peach kernel has a high antioxidant potential and can inhibit enzymes linked to obesity, type 2 diabetes and Alzheimer's disease (Nowicka and Wojdylo, 2019). Elshamy *et al.* (2019) reported anti-inflammatory, antipyretic, and antinoceptive potentials for the nectarine kernel.

11.3.3 By-products

After describing the composition of the fruit with special emphasis on its nutritional and health benefits, it should also be noted that in peach-tree cultivation other materials can be used for animal or human nutrition, as well as for medicinal and pharmaceutical purposes. These materials include flowers, leaves, stems, peel, flesh (remnant or not) and stones (kernel included), and are considered by-products of peach-tree cultivation and peach fruit processing and canning industries. Each year, large amounts of these by-products are generated during blooming, thinning, harvesting, processing and winter leaf fall. In addition to the nutritional and health properties of peach flesh, peel and seed described previously in this chapter, other peach seed by-products can be obtained such as flour to enrich foods and/or animal feed, oil for the cosmetic industry and dietary supplements for human nutrition. Indeed, peach kernel oil, which is rich in unsaturated fatty acids (oleic and linoleic acids) and tocopherols, is highly valued in the cosmetics industry owing to its anti-wrinkle properties and skin protection against UV radiation (Nowicka and Wojdylo, 2019). With regard to peach flowers, several studies have shown that extracts of *P. persica* flowers can reduce UV-induced skin damage and may be useful for protecting against UV-induced DNA damage and carcinogenesis (Kim *et al.*, 2000; Heo *et al.*, 2001). In traditional Chinese folk medicine, *P. persica* flowers have been used as a purgative or diuretic, and have also been used in cosmetology (Han *et al.*, 2015). Peach leaves have pharmaceutical potential and are traditionally used as anthelmintics, laxatives and sedatives (Kazan *et al.*, 2014). Peach leaf extracts were found to have anti-hyperglycaemic effects on postprandial blood glucose levels in glucose-loaded mice (Shirosaki *et al.*, 2012a), with the major active ingredient identified as multiflorin A (Shirosaki *et al.*, 2012b). A decoction of the leaves is used to treat heat rash, skin disease and circulatory troubles (Perry and Metzger, 1980).

11.4   ORCHARD FACTORS AFFECTING PEACH FRUIT QUALITY

It is broadly accepted and scientifically demonstrated that fruit quality is determined in the orchard at the preharvest stage (Crisosto *et al.*, 1997; Minas *et al.*, 2018), with postharvest management only able to maintain the harvested fruit quality (Crisosto and Costa, 2008). While the stage of fruit maturity at harvest has a direct effect on consumer acceptance, there are other factors that determine the final quality of the fruit including, in addition to postharvesting techniques, the various operations (e.g. classification, packaging and transport) that peach fruits undergo before reaching the consumer.

Orchard-related factors often interact in complex ways that depend on the specific cultivar and the particular sensitivities of the fruit growth or development stage (Fig. 11.4). Seasonal characteristics, including those defined in terms of accumulated heat and chill units, and the extent of cool, warm and air-pollution periods during early fruit development, as well as rain intensity and patterns, can increase or decrease the influence of these orchard-related factors (Crisosto *et al.*, 1997). *P. persica* is characterized by the notable variety of peach and nectarine cultivars on the market, making it difficult to study the numerous orchard-related factors in all of them. This can lead, on occasions, to generalizations being applied uniformly to all peach cultivars. Peach consumption has been in decline in recent decades globally (Anthony and Minas,

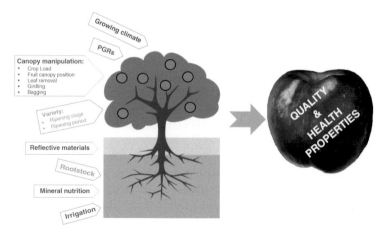

Fig. 11.4. Preharvest factors affecting peach quality and health-promoting properties grouped as genetic (green), environmental (blue), and agronomic (brown) factors. Adapted from Musacchi and Serra (2018) and Minas *et al.* (2018).

2022), due, among other factors, to a flavourless fruit profile, poor texture, inferior quality and inappropriate storage, which can impact consumer acceptance (Crisosto and Labavitch, 2002; Delgado *et al.*, 2013; Tanou *et al.*, 2017; Minas *et al.*, 2018). Therefore, the optimization of preharvest factors (Fig. 11.4) is critical to ensure the successful delivery of consistent, high-quality peaches to the market year after year, and thereby secure consumer trust and repeat sales (Anthony and Minas, 2022).

11.4.1 Cultivar genotype

The cultivar itself is, without doubt, an essential factor that determines the final fruit quality and its acceptance by the consumer. The impact of its adaptation to specific growing regions, the time of maturation within a region or season, and uniformity of ripening within the canopy are critical traits (Reig *et al.* 2017; Minas *et al.*, 2018). Traditionally, breeding programmes for fresh peaches and nectarines have focused on selecting traits associated with fruit size, colour and flesh firmness. These traits are important for growers and retailers, rendering peaches attractive to consumers and resistant to postharvest handling, as well as increasing their shelf life (Cantín *et al.*, 2010). More recently, consumer quality traits (low acidity, high SSC, strong aroma, non-melting flesh) are becoming increasingly important in peach and nectarine breeding programmes (Cirilli *et al.*, 2016). An European consumer study reported that 70% of consumers preferred sweet cultivars (Iglesias and Echeverría, 2009). Breeding programmes around the world are using classic and biotechnological tools to breed cultivars with such traits. In fact, the current trend of most peach-breeding programmes is for sweet or low-acidity cultivars in all fruit types, with almost 100% of the skin surface blushed, except for non-melting peaches. In this latter fruit type, the trend is to create cultivars with 100% yellow skin and equilibrate taste. With regard to the aroma profile, it is not yet considered an objective of the breeding programmes due to its complexity (in terms of the molecular, genetic and physiological mechanisms involved) and the time-consuming evaluation process. However, thanks to new breeding tools, several quantitative trait loci of different volatile organic compounds, such as linalool, nonanal and *p*-menth-1-en-9-al and their candidate genes, have been found (Eduardo *et al.*, 2013; Cao *et al.*, 2021). Despite this, their relevance in peach aroma perception by the consumer is still unclear, especially for linalool, which makes a major contribution to the aroma of peach fruit. Another trait that breeders are currently focusing on is related to genotypes with high antioxidant capacity to improve human health and storage potential (Kader, 1988). Cultivars less susceptible to chilling injury are now being produced commercially and have been well received by consumers. These cultivars have been developed using the marker-assisted selection approach during parent and progeny selection. Finally, another trait that a few peach cultivar breeding programmes are focusing on is related to the bioaccessibility of bioactive compounds, based on the InfoGest consensus method, in the gastro-intestinal tract in order to provide health benefits. The future application of these technologies will depend on successful teamwork among plant breeders, physi-ologists, molecular geneticists and consumer education programmes.

Peach fruit quality, in general, is affected by the ripening stage and harvest time of the cultivar. Flesh firmness decreases as maturity increases, whereas sugars and organic acids increase up to fruit harvesting. Key ripening stages for peach have been defined to assist ripening programmes for consumers in terms of flesh firmness: 'ready to transfer or to buy': 26–35 N; and 'ready to eat': 9–18 N (Crisosto *et al.*, 2001, 2004; Crisosto and Valero, 2006; Valero *et al.*, 2007). Sensory properties (Fig. 11.5) and the composition and content of volatile aroma compounds change during the ripening stages (Cano-Salazar *et al.*, 2012). Generally, green-note volatiles tend to decrease, while the content of fruity-note volatiles accumulates during fruit ripening (Shen *et al.*, 2014). In immature fruits, C_6 compounds are the major contributors, but their levels decrease drastically, whereas lactone, benzaldehyde and linalool levels in-crease significantly during maturation (Wang *et al.*, 2009). Currently, a meas-urement that can be taken without destroying the fruit is gaining importance among peach industry actors, mainly researchers and retailers, to evaluate the ripening stage of the fruit during the last 2 weeks before harvest (stage IV of fruit development). This is the non-destructive index of absorbance difference (I_{AD}), which reflects the chlorophyll content decrease during maturation and ripening of the fruit. I_{AD} ranges from 0.00 to 2.25, with the lowest value repre-senting the complete degradation of chlorophyll of ripe fruit and values above 2 corresponding to very unripe fruit (Musacchi and Serra, 2018). With regard to ripening time, mid- and late-season cultivars are usually higher in flavour and sweetness than early ones, as early ripening corresponds to a lower level of sugars due to the shorter period of fruit development (Bassi *et al.*, 2016). Based on ripening time, another key objective of peach-breeding programmes is to fill the harvest window for each fruit type with cultivars with similar characteris-tics in terms of skin colour, taste, sweetness, texture, aroma and flavour.

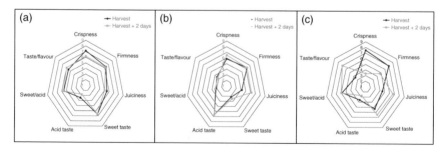

Fig. 11.5. Radar plot of sensory attributes for (a) 'Big Top', (b) 'Crimson Lady' and (c) 'Platibelle'. The values range from 1 (low/dislike) to 9 (high/extremely favourable). Figure courtesy of Gemma Echeverría, IRTA.

11.4.2 Rootstock genotype

It is well known that peach agronomic performance, fruit quality and nutritional composition depend mainly on the scion genotype (Cantín *et al.*, 2009a,b, 2010; Reig *et al.*, 2013a,b, 2015). However, as reported in several studies, the rootstock also plays an important role in determining fruit quality (Orazem *et al.*, 2011; Font i Forcada *et al.*, 2012, 2014b, 2019; Mestre *et al.*, 2015, 2017; Reig *et al.*, 2016, 2020; Iglesias *et al.*, 2019). In fact, peach adaptability to different pedoclimatic conditions and to new training systems that are better adapted to intensive cultivation systems that improve productivity, fruit quality and labour efficiency requires evaluating cultivar and rootstock combination responses to different growing conditions (DeJong, 2005; Reighard and Loreti, 2008).

11.4.3 Growing climate

Peach climatic adaptations, especially the chilling requirement, are very important and one of the top priorities of many peach-breeding programmes. Peaches have a chilling requirement of a certain number of hours of winter temperatures between 0°C and 7.2°C to break dormancy and induce normal bloom and vegetative growth. Therefore, quantification of the chilling requirement of cultivars is a key factor to determine appropriate growth areas for commercial production. For example, insufficient chilling in regions with warm winters can cause aberrant fruit-shape development, among other problems (Kwon *et al.*, 2020). Peach cultivars have chilling requirements that range widely from 8 to 75 chilling portions (Rodríguez *et al.*, 2021), and are accordingly classified as low-, medium- or high-chilling types.

Besides the dormancy period, temperature, solar radiation, photoperiod, precipitation and soil profile during the fruit development period have decisive effects on different attributes of quality. The growing climate and seasonal conditions can affect harvest date, fruit growth patterns and final size at harvest (Minas *et al.*, 2018).

11.4.4 Mineral nutrition

Nutritional status is a well-recognized important factor in quality at harvest and the postharvest life of fruits (Crisosto *et al.*, 1997; Fallahi, *et al.*, 1997; Crisosto and Costa, 2008). Nutrient deficiencies, excesses or imbalances affect fruit growth development and physiology, resulting in disorders that limit the storage life of many fruits (Ferguson *et al.*, 1999; Olivos *et al.*, 2012). Fertilizer application rates vary widely among growers. Generally, they depend on soil type, cropping history and soil test results used to indicate the nitrogen (N), phosphorous (P) and potassium (K) requirements for a given cultivar. To date, fertilization recommendations for fruits have been established primarily for productivity goals, rather than as diagnostics for good flavour quality and optimal postharvest life. The nutrient with the single greatest effect on fruit quality, as well as tree productivity, is N. Research performed over the last 20 years at the Kearney Agricultural Center in Parlier, California, established that, in peach and nectarine orchards grown under Californian conditions, leaf N levels should be kept at between 2.6% and 3.0% for the best fruit quality (Daane *et al.*, 1995). The response of peach and nectarine trees to N fertilization is dramatic, with high N concentrations stimulating vigorous vegetative growth, causing the shading and death of lower fruiting wood. Although trees high in N may look healthy and lush, excess N does not increase fruit size, production or SSC, and can delay fruit maturity, induce poor red colour development, inhibit the ground colour change from green to yellow and result in flavourless fruits. However, N deficiency also leads to small fruit with poor flavour and unproductive trees. Postharvest, fruit water loss from peach and nectarine trees with 3.6% leaf N was greater than that from trees with 2.6% leaf N.

The relationship between fruit N concentration and fruit susceptibility to decay caused by brown rot (*Monilinia fructicola*) has been studied extensively in stored nectarines. Wounded and brown rot-inoculated fruit from 'Fantasia' and 'Flavortop' nectarine trees with leaf N above 2.6% were more susceptible to brown rot than fruit from trees with lower leaf N values. Anatomical observations and cuticle density measurements of the fruit indicated differences in cuticle thickness among 'Fantasia' nectarine fruit from low, intermediate and high N treatments, but this was considered to only partially explain the differences in fruit susceptibility to this disease (Daane *et al.*, 1995; Crisosto *et al.*, 1997). Although calcium (Ca) is classified as a secondary nutrient, it is involved in numerous biochemical and morphological processes in plants, reducing metabolic disorders and decay and maintaining firmness, and has been implicated in many economically important disorders that affect the production and postharvest quality of fruits (Fallahi *et al.*, 1997; Ferguson *et al.*, 1999). Studies have investigated the effects of preharvest foliar Ca sprays on peach fruits. Four to five preharvest foliar sprays of 'Calrico' peaches with Ca propionate (0.5% and 1.0% Ca) reduced SSC. However, application of multiple Ca-containing foliar sprays retained higher peach flesh firmness levels at harvest and following 2 weeks of cold storage at 0°C, with the best results obtained with Ca chloride ($CaCl_2$). When used as adjuvants, Tween-20 and sodium carboxymethyl cellulose increased peach fruit acidity both at harvest and following cold storage. Ca-treated fruits exhibited lower susceptibility to internal browning after cold storage (Val and Fernandez, 2011). Manganaris *et al.* (2005) reported that applying six to ten preharvest sprays using $CaCl_2$ and an ethylenediaminetetraacetic-acid-chelated Ca form increased skin and flesh Ca concentrations, and reduced infection and severity rates of brown rot during cold storage, with the best results obtained on those fruits treated with

CaCl$_2$ spray. In contrast, a screening of several commercial Ca foliar sprays for peaches and nectarines in California (applied every 14 days, starting 2 weeks after full bloom and continuing until 1 week before harvest) showed no effect on the fruit quality of mid- or late-season cultivars (Crisosto and Costa, 2008). These foliar sprays did not affect SSC, firmness, decay incidence or fruit flesh Ca concentration. Fruit flesh Ca concentration measured at harvest varied among cultivars from 200 to 300 µg g^{-1} (dry-weight basis). No decay reduction was seen in 'Jerseyland' peaches grown in Pennsylvania that were treated with ten weekly preharvest sprays of CaCl$_2$ at 0, 34, 67 or 101 kg ha^{-1} (Conway *et al.*, 1987). Even fruit treated at 101 kg ha^{-1}, which had 70% more flesh Ca (490 versus 287 µg g^{-1}, dry-weight basis) than untreated fruit, showed no reduction in decay severity. Such sprays on peaches and nectarines should be treated with caution because their heavy metal (e.g. iron, aluminium, copper) content may contribute to peach and nectarine skin discoloration (Crisosto and Costa, 2008).

Iron deficiency can significantly affect the yield, size, ripening and overall quality of peach fruit. In general, peaches harvested from iron-deficient trees were found to be smaller, with the proportion of commercially acceptable fruits significantly reduced (Minas *et al.*, 2018; Reig *et al.*, 2020), and had more organic acids (especially succinate and quinate), vitamin C and phenolic compounds and lower total sugar:total organic acid ratios (Álvarez-Fernandez *et al.*, 2003).

11.4.5   Irrigation

Climate change projections predict drier and warmer conditions, especially during the summer, and more frequent droughts in major fruit-production regions in the west of the USA and in Mediterranean areas (Minas *et al.*, 2018). With this in mind, research studies have considered the potential impacts of water scarcity (during fruit development or postharvest) on tree productivity and fruit quality.

Regulated deficit irrigation (RDI) has been studied in different peach-production regions as a strategy to reduce water use in dry climates (Minas *et al.*, 2018). In peaches grown in the San Joaquin Valley, California, irrigation regimes of 50%, 100% and 150% evapotranspiration (ET) applied 4 weeks before harvest affected 'O'Henry' peach size and SSC but did not affect the incidence or severity of internal breakdown (Crisosto *et al.*, 1994). In general, fruits from the 50% ET treatment were small but had high SSC levels. Under Mediterranean conditions, in particular under the semi-arid conditions of the Lower Ebro Valley in north-east Spain, during one of the peach fruit growth stages most sensitive to water stress (stage III), López *et al.* (2011) applied four different irrigation regimes (full irrigation, no irrigation, full irrigation for 17 days followed by no irrigation for 33 days, and no irrigation for 34 days followed by full irrigation of 16 days) to the 'Ryan Sun' cultivar. The 'no irrigation' treatment produced a reduction in fruit size, a delay in ripening and an increase in dry-matter content compared with fruits with full irrigation, as well as higher SSC and acidity. A taste panel perceived an increase in firmness, crunchiness and acidity in the 'no irrigation' fruits, but a decrease in sweetness, juiciness and flavour. The application of RDI during a fruit water-stress-tolerant stage (stage II) resulted in improved sensory quality and consumer acceptance of 'Tardibelle' peaches (Vallverdu *et al.*, 2012). In contrast, a study of 'Elberta' peaches grown under different water irrigation regimes during the growing season indicated that drought conditions were detrimental for sensorial quality, as increased fruit sourness and reduced sweetness and juiciness were observed (Rahmati *et al.*, 2015). Finally, RDI strategies may also increase antioxidant activity, vitamin C and total phenolics content (Falagán *et al.*, 2015).

An increase in fruit defects such as deep suture and double-fruit formation was reported for early-season 'Regina' peaches after imposing a postharvest water stress (50% ET) in mid- and late summer during the previous season (Johnson *et al.*, 1992). These defects reduced the final pack-out. A similar regulated water-stress regime applied to other early-season cultivars, such as 'Red Beauty', 'Ambra' and 'Durado' plum cultivars, did not affect the number of double and deep-sutured fruit (Crisosto *et al.*, 1997). Proper water management requires optimization of the balance between yield and postharvest quality (Johnson *et al.*, 1992; Crisosto *et al.*, 1997; Prange and DeEll, 1997).

11.4.6   Canopy manipulation

Crop load
For most fruit crops, thinning during flowering or postbloom (removing fruit/fruitlets) or during stage I of fruit growth (45–60 days after full bloom, DAFB) increases fruit size while reducing total yield. Thus, a balance between yield and fruit size must be achieved. Generally, maximum profit does not occur at maximum marketable yield, as larger fruits bring a higher market price. Thinning 'Contender' peaches at 20 DAFB improved fruit yield and fruit size more than thinning at 0, 10, 30 or 40 DAFB (Njoroge and Reighard, 2008). Leaving too many fruits on a tree reduced fruit size and SSC in early-ripening 'May Glo' nectarine and late-ripening 'O'Henry' peach cultivars (Crisosto *et al.*, 1997). Crop load on 'O'Henry' peach trees affected the incidence of internal breakdown measured after 1, 2 or 3 weeks at 5°C. Despite significant numbers of mealy fruit in all lots, the overall incidence of mealiness and flesh browning was lowest in fruit from the high crop load, intermediate in fruit from the commercial crop load and greatest in fruit grown with the low crop load (Crisosto and Costa, 2008).

Fruit canopy position
Fruit size, ripening time, red overcolour development and storage potential can be affected by canopy position (Minas *et al.*, 2018). Large differences in SSC, acidity and fruit size were also detected between fruits growing on outside versus

inside canopy positions of open vase-trained peach, nectarine and plum trees. In general, fruit size and quality decreased from the top to the lower layers of the tree canopy but with a high variability in the growth of individual fruits (Minas *et al.*, 2018). Regardless of the training system or rootstock, fruit sugar concentration was found to decrease from the canopy top to the bottom (Gullo *et al.*, 2014). Peaches grown under a high light environment (outside the canopy) have a longer storage and market life than peaches grown under a low light environment (inside the canopy), especially with peach cultivars that are more susceptible to chilling injury (Crisosto *et al.*, 1997). The architecture of the canopy can strongly influence light interception, and consequently the fruit sensory and nutritional quality determined as total antioxidant capacity and total phenol concentration, which correlate strongly with red skin overcolour (Gullo *et al.*, 2014). Therefore, more efficient training systems that allow sunlight penetration into the centre and lower canopy areas are recommended to reduce the number of shaded fruits. The KAC-V training system, developed for tree fruits at the University of California, Davis, to increase light interception within the canopy, was found to reduce uneven maturity, increase SSC, promote red colour and reduce the number of pickings (Crisosto and Costa, 2008).

Leaf removal
Summer pruning and leaf pulling around the fruit increases fruit light exposure and, when performed properly, can increase fruit colour without affecting fruit size and SSC (Crisosto and Costa, 2008). Excessive leaf pulling or leaf pulling too close to harvest, however, can reduce both fruit size and SSC in peaches and nectarines.

Girdling
Girdling (a commercial practice in which a cylindrical portion of the phloem of the tree or vine is removed, stopping phloem transport of photosynthates to the roots and the other plant parts until the wound heals) at 4–6 weeks before harvest can increase peach and nectarine fruit size and advance and synchronize maturity (Day and DeJong, 1990). In some reports, girdling increased fruit SSC but also increased fruit acidity and phenolics, resulting in masking of the taste acquired from the additional sugars. Girdling can also cause the stones of peach and nectarine fruits to split, especially if done too early during pit hardening. Fruits with split stones soften more quickly than intact fruit and are more susceptible to decay (Crisosto and Costa, 2008).

Bagging
Peach fruit is easily damaged by insects and birds during the ripening process, which seriously affects the final fruit quality and yield. This problem can be mitigated by the bagging of fruit during peach cultivation (Zu *et al.*, 2020). This a common practice used on the worldwide-known 'Calanda' peaches, grown

in Spain, mainly to avoid damage by *Ceratitis capitata* (Mediterranean fruit fly). However, bagging suppresses the synthesis of anthocyanins at harvest, producing unblemished fruit (Jia *et al.*, 2005) or less redness of the skin depending on the cultivar and the paper bag used (Ma *et al.*, 2021), and inhibits the synthesis and/or degradation of skin chlorophyll (Tombesi *et al.*, 1993). Therefore, peach fruit quality can be affected. Constituents such as sugars, phenols and organic acids have been shown to be sensitive to bagging due to reduced sun exposure (Lima *et al.*, 2012), and anthocyanin accumulation in the pericarp significantly decreased when bagging treatments were applied (Ma *et al.*, 2021). Bagging with orange paper caused no effect on the aroma compounds of whole peach fruit at harvest but significantly increased the (E)-2-hexenal content in the peach skin (Jia *et al.*, 2005). Another report showed that levels of *n*-hexanal and 2-hexenal, as well as C_6-related esters, were significantly lower in peach fruit treated with two-layered paper bags (black inner and brown outer papers) compared with non-bagged fruit (Wang *et al.*, 2010). When bagging, choosing the right colour bag is therefore essential to ensure the quality of the fruit is not diminished. Recently, fruit bagging has been gaining importance in organic peach production to reduce the incidence of disease, with particular attention being paid to *Monilinia* spp. (Balsells-Llauradó *et al.*, 2022).

11.4.7   Reflective materials

Where the season is long and temperatures are very high, it can be difficult to attain the desired red fruit skin colour. In such situations, summer pruning or even the removal of leaves around the fruit may be required to improve the red colour. An alternative that has been proposed to resolve this situation is the use of reflecting material or mulch. Trials carried out with a reflective mulch (Extenday) in Italy on a 'Red Gold' nectarine orchard allowed inducement of higher daily photosynthesis compared with a control treatment. The red colour of the skin was improved by the mulch, as was yield and average fruit weight (Fiori *et al.*, 2002), confirming results obtained in different production areas and orchard situations (Layne *et al.*, 2001). In California, in specific cultivars with high vigour and conditions of poor sunlight canopy penetration, a highly consistent impact on red colour improvement was obtained through leaf removal combined with the use of reflective materials (an aluminium foil laminated to a cloth backing (Namura Co., Japan) and a polyethylene plastic (Specialty Ag., Reedley, California)) (Crisosto and Costa, 2008).

11.4.8   Plant growth regulators

Although the use of plant growth regulators on peaches is not as widespread as in pome fruit (Lurie, 2010), they can be employed to improve stone fruit production and quality at different physiological stages of the tree life. The use of plant growth regulators is regulated and may not be permitted, depending on the country or the production area in question. Likewise, the maximum residue limits of the chemicals to be used must be below the regulatory thresholds. For instance, auxin-like and gibberellin-like substances have been used to increase fruit size and firmness and can delay storage disorders (e.g. flesh browning and woolliness) in peach when applied at the end of pit hardening. Aminoethoxyvinylglycine (AVG) applied to 'Redhaven' peach trees at 62.5–250 mg l^{-1} at 10 days preharvest reduced fruit drop, slightly increased SSC and delayed the onset of endogenous ethylene emission as well as fruit softening (Vizzotto *et al.*, 2002). AVG can also contribute to the optimization of harvest management by delaying picking onset and improving flesh firmness. However, in peach, the effects of AVG are influenced by concentration, cultivar and application time with respect to harvest proximity (Cline, 2006).

Recently introduced formulations, such as abscisic acid and 1-aminocyclo-propane-1-carboxylic acid (under the brand name Accede), have been tested experimentally as potential thinning agents in different stone-fruit-growing locations, cultivars and conditions (Ceccarelli *et al.*, 2016; Costa, 2016; Theron *et al.*, 2017, 2021). The potential benefits in terms of labour reduction, fruit size increase, higher anthocyanin content and improved flavour are still under investigation. Other plant growth regulators that control ethylene production and/or action, including 1-methylcyclopropene, have been tested without showing consistent benefits for peach postharvest life.

REFERENCES

Álvarez-Fernandez, A., Paniagua, P., Abadia, J. and Abadia, A. (2003) Effects of Fe deficiency chlorosis on yield and fruit quality in peach (*Prunus persica* L. Batsch). *Journal of Agricultural and Food Chemistry* 51, 5738–5744.

Anthony, B.M. and Minas, I.S. (2022) Redefining the impact of preharvest factors on peach fruit quality development and metabolism: a review. *Scientia Horticulturae* 297: 110919.

Aubert, C. and Chalot, G. (2020) Physicochemical characteristics, vitamin C, and polyphenolic composition of four European commercial blood-flesh peach cultivars (*Prunus persica* L. Batsch). *Journal of Food Compositions and Analysis* 86: 103337.

Aubert, C., Günata, Z., Christian, A. and Baume, R. (2003) Changes in physicochemical characteristics and volatile constituents of yellow- and white-fleshed nectarines during maturation and artificial ripening. *Journal of Agricultural and Food Chemistry* 51, 3083–3091.

Baccichet, I., Chiozzotto, R., Bassi, D., Gardana, C., Cirilli, M. and Spinardi, A. (2021) Characterization of fruit quality traits for organic acids content and profile in a large peach germplasm collection. *Scientia Horticulturae* 278: 109865.

Balsells-Llauradó, M., Vall-llaura, N., Usall, J., Casals, C., Teixidó, N. and Torres, R. (2022) Impact of fruit bagging and postharvest storage conditions on quality and decay of organic nectarines. *Biological Agriculture & Horticulture in press* [https://doi.org/10.1080/01448765.2022.2113562].

Barni, S., Caimmo, D., Chiera, F., Comberuati, P., Mastorilli, C. *et al.* (2022) Phenotypes and endotypes of peach allergy: what is new? *Nutrients* 14: 998.

Basar, H. (2006) Elemental composition of various peach cultivars. *Scientia Horticulturae* 107, 259–263.

Bassi, D. and Monet, R. (2008) Botany and taxonomy. In: Layne, D.R. and Bassi, D. (eds) *The Peach: Botany, Production and Uses*. CAB International, Wallingford, UK, pp. 1–36.

Bassi, D., Mignani, I., Spinardi, A. and Tura, D. (2016) Peach (*Prunus persica* (L.) Batsch). In: Simmonds, M.S.J. and Preedy, V.R. (eds) *Nutritional Composition of Fruit Cultivars*. Academic Press, San Diego, California, pp. 535–571.

Bento, C., Gonçalves, A.C., Silva, B. and Silva, L.R. (2020) Peach (*Prunus Persica*): phytochemicals and health benefits. *Food Reviews International* 3, 1703–1734.

Bonany, J., Carbó, J., Echeverría, G., Hilaire, C., Cottet, V. *et al.* (2014) Eating quality and European consumer acceptance of different peach (*Prunus persica* (L.) Batsch) varieties. *Journal of Food, Agriculture & Environment* 12, 67–72.

Bvenura, C., Hermaan, N.N.P., Chen, L. and Sivakumar, D. (2018) Nutritional and health benefits of temperature fruits. In: Mir, A., Shah, M.A. and Mir, M.M. (eds) *Postharvest Biology and Technology of Temperate Fruits*. Springer International, Gewebestrasse, Switzerland, pp. 51–75.

Cano-Salazar, J., Echeverría, G., Crisosto, C.H. and López, L. (2012) Cold-storage potential of four yellow-fleshed peach cultivars defined by their volatile compound's emissions, standard quality parameters, and consumer acceptance. *Journal of Agricultural and Food Chemistry* 60, 1266–1282.

Cantín, C.M., Moreno, M.A. and Gogorcena, Y. (2009a) Evaluation of the antioxidant capacity phenolic compounds, and vitamin C content of different peach and nectarine [*Prunus persica* (L.) Batsch]. *Journal of Agricultural Food Chemistry* 57, 4586–4592.

Cantín, C., Gogorcena, Y. and Moreno, M.A. (2009b) Analysis of phenotypic variation of sugar profile in different peach and nectarine [*Prunus persica* (L.) Batsch] breeding progenies. *Journal of Agricultural Food Chemistry* 89, 1909–1917.

Cantín, C.M., Gogorcena, Y. and Moreno, M.A. (2010) Phenotypic diversity and relationships of fruit quality traits in peach and nectarine [*Prunus persica* (L.) Batsch] breeding progenies. *Euphytica* 171, 211–226.

Cao, K., Yang, X., Li, Y., Zhu, G. *et al.* (2021) New high-quality peach (*Prunus persica* L. Batsch) genome assembly to analyze the molecular evolutionary mechanism of volatile compounds in peach fruits. *Plant Journal* 108, 281–295.

Ceccarelli, A., Vidoni, S., Rocchi, L., Taioli, M. and Costa, G. (2016) Are ABA and ACC suitable thinning agents for peach and nectarine? *Acta Horticulturae* 1138, 69–74.

Cevallos-Casals, B.A., Byrne, D., Okie, W.R. and Cisneros-Zevallos, L. (2006) Selecting new peach and plum genotypes rich in phenolic compounds and enhanced functional properties. *Food Chemistry* 96, 273–280.

Cirilli, M., Bassi, D. and Ciacciulli, A. (2016) Sugars in peach fruit: a breeding perspective. *Horticulture Research* 3: 15067.

Cline, J.A. (2006) Effect of aminoethoxyvinylglycine and surfactants on preharvest drop, maturity, and fruit quality of two processing peach cultivars. *HortScience* 41, 377–383.

Colaric, M., Veberic, R., Stampar, F. and Hudina, M. (2005) Evaluation of peach and nectarine fruit quality and correlations between sensory and chemical attributes. *Journal of the Science of Food and Agriculture* 85, 2611–2616.

Conway, W.S., Greene, G.M. and Hickey, K.D. (1987) Effects of preharvest and postharvest calcium treatments of peaches on decay caused by *Monilinia fructicola*. *Plant Disease* 71, 1084–1086.

Costa, G. (2016) Two decades of activity of the "Fruit chemical thinning" working group of the EUFRIN network. *Acta Horticulturae* 1138, 1–8.

Crisosto, C.H. (2002) How do we increase peach consumption? *Acta Horticulturae* 592, 601–605.

Crisosto, C.H. and Costa, G. (2008) Preharvest factors affecting peach quality. In: Layne, D.R. and Bassi, D. (eds) *The Peach: Botany, Production and Uses*. CAB International, Wallingford, UK, pp. 536–549.

Crisosto, C.H. and Crisosto, G.M. (2005) Relationship between ripe soluble solids concentration (RSSC) and consumer acceptance of high and low acid melting flesh peach and nectarine (*Prunus persica* (L.) Batsch) cultivars. *Postharvest Biology and Technology* 38, 239–246.

Crisosto, C.H. and Labavitch, J.M. (2002) Developing a quantitative method to evaluate peach (*Prunus persica*) flesh mealiness. *Postharvest Biology and Technology* 25, 151–158.

Crisosto, C.H. and Valero, C. (2006) "Ready to eat": maduración controlada de fruta de hueso en cámara. *Horticultura* 29, 32–37.

Crisosto, C.H. and Valero, D. (2008) Harvesting and postharvest handling of peaches for the fresh market. In: Layne, D.R. and Bassi, D. (eds) *The Peach: Botany, Production and Uses*. CAB International, Wallingford, UK, pp. 576–596.

Crisosto, C.H., Johnson, R.S., Luza, J.G. and Crisosto, G.M. (1994) Irrigation regimes affect fruit soluble solids content and the rate of water loss of 'O'Henry' peaches. *HortScience* 29, 1169–1171.

Crisosto, C.H., Johnson, R.S., DeJong, T. and Day, K.R. (1997) Orchard factors affecting postharvest stone fruit quality. *HortScience* 32, 820–823.

Crisosto, C.H., Slaughter, D., Garner, D. and Boyd, J. (2001) Stone fruit critical bruising thresholds. *Journal of the American Pomological Society* 55, 76–81.

Crisosto, C.H., Garner, D., Andris, H. and Day, K.R. (2004) Controlled delayed cooling extends peach market life. *HortTechnology* 14, 99–104.

Crisosto, C.H., Crisosto, G.M., Echeverría, G. and Puy, J. (2006) Segregation of peach and nectarine (*Prunus persica* (L.) Batsch) cultivars according to their organoleptic characteristics. *Postharvest Biology and Technology* 39, 10–18.

Daane, K.M., Johnson, R.S., Michailides, T.J., Crisosto, C.H., Dlott, J.W. *et al.* (1995) Excess nitrogen raises nectarine susceptibility to disease and insects. *California Agriculture* 49, 13–17.

Day, K. and DeJong, T. (1990) Girdling of early season 'Mayfire' nectarine trees. *Journal of Horticulture Science* 65, 592–534.

DeJong, T.M. (2005) Using physiological concepts to understand early spring temperature effects on fruit growth and anticipating fruit size problems at harvest. *Summerfruit* 200, 10–13.

Delgado, C., Crisosto, G.M., Heymann, H. and Crisosto, C.H. (2013) Determining the primary drivers of liking to predict consumers' acceptance of fresh nectarines and peaches. *Journal of Food Science* 78, 605–614.

Dhingra, N., Sharma, R. and Kar, A. (2014) Towards further understanding on the antioxidative activities of *Prunus persica* fruit: a comparative study with four different fractions. *Spectrochimica Acta Part A: Molecular and Biomolecular Spectroscopy* 132, 582–587.

Duan, Y., Dong, X., Liu, B. and Li, P. (2013) Relationship of changes in the fatty acid compositions and fruit softening in peach (*Prunus persica* L. Batsch). *Acta Physiologiae Plantarum* 35, 707–713.

Eduardo, I., Chietera, G., Bassi, D., Rossini, L. and Veccietti, A. (2010) Identification of key odor volatile compounds in the essential oil of nine peach accessions. *Journal of the Science of Food and Agriculture* 90, 1146–1154.

Eduardo, I., Chietera, G., Pirona, R., Pacheco, I., Troggio, M. *et al.* (2013) Genetic dissection of aroma volatile compounds from the essential oil of peach fruit: QTL analysis and identification of candidate genes using dense SNP maps. *Tree Genetics & Genomes* 9, 189–204.

Elshamy, A.I, Abdallah, H.M.I., Gendy, A.E.N. G., El-Kashak, W. and Muscatello, B. *et al.* (2019). Evaluation of anti-inflammatory, antinociceptive, and antipyretic activities of Prunus persica var. nucipersica (nectarine) kernel. *Planta Medica*, 85 (11/12), 1016–1023.

Falagán, N., Artés-Hernández, F., Gómez, P.A., Pérez-Pastor, A. and Aguayo, E. (2015) Comparative study on postharvest performance of nectarines grown under regulated deficit irrigation. *Postharvest Biology and Technology* 110, 24–32.

Fallahi, E., Conway, W.S., Hickey, K.D. and Sams, C.E. (1997) The role of calcium and nitrogen in postharvest quality and disease resistance of apples. *HortScience* 32, 831–835.

Ferguson, I., Volz, R. and Woolf, A. (1999) Preharvest factors affecting physiological disorders of fruit. *Postharvest Biology Technology* 15, 255–262.

Fiori, G., Bucchi, F., Corelli Grappadelli, L. and Costa, G. (2002) Effetto dei teli riflettenti a terra sugli scambi gassosi e la qualità delle produzioni in pesco. *Atti VI Giornate Scientifiche SOI*, 157–158.

Font i Forcada, C., Gogorcena, Y. and Moreno, M.A. (2012) Agronomical and fruit quality traits of two peach cultivars on peach–almond hybrid rootstocks growing on Mediterranean conditions. *Scientia Horticulturae* 140, 157–163.

Font i Forcada, C., Gogorcena, Y. and Moreno, M.A. (2014a) Agronomical parameters, sugar profile and antioxidant compounds of "Catherine" peach cultivar influenced by different plum rootstocks. *International Journal of Molecular Sciences* 15, 2237–2254.

Font i Forcada, C., Gradziel, T.M., Gogorcena, Y. and Moreno, M.A. (2014b) Phenotypic diversity among local Spanish and foreign peach and nectarine [*Prunus persica* (L.) Batsch] accessions. *Euphytica* 197, 261–277.

Font i Forcada, C., Reig, G., Giménez, R., Mignard, P., Mestre, L. and Moreno, M.A. (2019) Sugars and organic acids profile and antioxidant compounds of nectarine fruits influenced by different rootstocks. *Scientia Horticulturae* 248, 145–153.

Font i Forcada, C., Reig, G., Fontich, C., Batlle, I., Alegre, S. *et al.* (2021) 'Magna' and 'Blanq' series: two yellow-fleshed and three white-fleshed nectarines. *HortScience* 56, 1130–1131.

Gasparotto, J., Somensi, N., Bortolin, R.C., Moresco, K.S., Girardi, C.S. *et al.* (2014a) Effects of different products of peach (*Prunus persica* L. Batsch) from a variety developed in southern Brazil on oxidative stress and inflammatory parameters *in vitro* and *ex vivo*. *Journal of Clinical Biochemistry and Nutrition* 55, 110–119.

Gasparotto, J., Somensi, N., Bortolin, R.C., Girardi, C.S., Kunzler, A. *et al.* (2014b) Preventive supplementation with fresh and preserved peach attenuates CCl4-induced oxidative stress, inflammation, and tissue damage. *Journal of Nutritional Biochemistry* 25, 1282–1295.

Gil, M.I., Tomás-Barberán, F.A., Hess-Pierce, B. and Kader, A.A. (2002) Antioxidant capacity, phenolic compounds, carotenoids and vitamin C of nectarine, peach and plum cultivars from California. *Journal of Agricultural and Food Chemistry* 50, 4976–4982.

Giuffrida, D., Torre, G., Dugo, P. and Dugo, G. (2013) Determination of the carotenoid profile in peach fruits, juice and jam. *Fruits* 68, 39–44.

Gullo, G., Motisi, A., Zappia, R., Dattola, A., Diamanti, J. and Mezzetti, B. (2014) Rootstock and fruit canopy position affect peach [*Prunus persica* (L.) Batsch] (cv. Rich May) plant productivity and fruit sensorial and nutritional quality. *Food Chemistry* 153, 234–242.

Han, W., Xu, J.D., Wei, F.X., Zheng, Y.D., Ma, J.Z. *et al.* (2015) Prokinetic activity of *Prunus persica* (L.) Batsch flowers extract and its possible mechanism of action in rats. *BioMed Research International* 2015: 569853.

Heo, M.Y., Kim, S.H., Yang, H.E., Lee, S.H., Jo, B.K. and Kim, H.P. (2001) Protection against ultraviolet B- and C-induced DNA damage and skin carcinogenesis by the flowers of *Prunus persica* extract. *Mutation Research: Genetic Toxicology and Environmental Mutagenesis* 496, 47–59.

Huberlant, J. and Anderson, G.H. (2003) Sucrose. In: Caballero, B., Trugo, L.C. and Finglas, P.M. (eds) *Encyclopedia of Food Science, Food Technology and Nutrition*. Academic Press, London, pp. 4431–4439.

Iglesias, I. and Echeverría, G. (2009) Differential effect of cultivar and harvest date on nectarine colour, quality and consumer acceptance. *Scientia Horticulturae* 120, 41–50.

Iglesias, I., Giné-Bordonada, J., Garanto, X. and Reig, G. (2019) Rootstock affects fruit quality and phytochemical composition of "Big Top" nectarine grown under hot climatic conditions. *Scientia Horticulturae* 256: 108586.

Jia, H.J., Araki, A. and Okamoto, G. (2005) Influence of fruit bagging on aroma volatiles and skin coloration of 'Hakuho' peach (*Prunus persica* Batsch). *Postharvest Biology and Technology* 35, 61–68.

Johnson, R.S., Handley, D.F. and DeJong, T. (1992) Long-term response of early maturing peach trees to postharvest water deficit. *Journal of the American Society for Horticultural Science* 69, 1035–1041.

Kader, A.A. (1988) Influence of preharvest and postharvest environment on nutritional composition of fruits and vegetables. In: Quebedeaux, B. and Bliss, F.A. (eds) *Horticulture and Human Health: Contributions of Fruits and Vegetables*. Prentice-Hall, Englewood Cliffs, New Jersey, pp. 18–22.

Kakiuchi, N. and Ohmiya, A. (1991) Changes in the composition and content of volatile constituents in peach fruits in relation to maturity at harvest and artificial ripening. *Journal of the Japanese Society for Horticultural Science* 60, 209–216.

Kazan, A., Koyu, H., Turu, I.C. and Yesil-Celiktas, O. (2014) Supercritical fluid extraction of *Prunus persica* leaves and utilization possibilities as a source of phenolic compounds. *Journal of Supercritical Fluids* 92, 55–59.

Kim, H.J., Park, K.K., Chung, W.Y., Lee, S.K. and Kim, K.R. (2017) Protective effect of white-fleshed peach (*Prunus persica* (L.) Batsch) on chronic nicotine-induced toxicity. *Journal of Cancer Prevention* 22, 22–32.

Kim, Y.H., Yang, H.E., Kim, J.H., Heo, M.Y. and Kim, H.P. (2000) Protection of the flowers of *Prunus persica* extract from ultraviolet B-induced damage of normal human keratinocytes. *Archives of Pharmacal Research* 23, 396–400.

Kwon, J.H., Jun, J.H., Nam, E.Y., Chung, K.H., Hong, S.S. *et al.* (2015) Profiling diversity and comparison of Eastern and Western cultivars of *Prunus persica* based on phenotypic traits. *Euphytica* 206, 401–415.

Kwon, J.H., Nam, E.Y., Yun, S.K., Kim, S.J., Song, S.Y. *et al.* (2020) Chilling and heat require-
    ment of peach cultivars and changes in chilling accumulation spectrums based on
    100-year records in Republic of Korea. *Agricultural and Forest Meteorology* 288–289:
    108009.

Lara, M.V., Bonghi, C., Famiani, F., Vizzotto, G., Walker, R.P. and Drincovich, F. (2020)
    Stone fruit as biofactories of phytochemicals with potential roles in human nutri-
    tion and Health. *Frontiers in Plant Science* 11: 562252.

Layne, D.R., Jiang, Z.W. and Rushing, J.W. (2001) Tree fruit reflective film improves red
    skin coloration and advances maturity in peach. *HortTechnology* 11, 234–242.

Lea, M.A., Ibeh, C., desBordes, C., Vizzotto, M., Cisneros-Zevallos, L. *et al.* (2008)
    Inhibition of growth and induction of differentiation of colon cancer cells by
    peach and plum phenolic compounds. *Anticancer Research* 28, 2067–2076.

Li, F., Li, S., Li, H.B., Deng, G.F., Ling, W.H. *et al.* (2013) Antiproliferative activity of
    peels, pulps and stones of 61 fruits. *Journal of Functional Foods* 5, 1298–1309.

Lima, A.B., Alvarenga, A.A., Malta, M.R., Gebert, D. and Lima, E.B. (2012) Chemical
    evaluation and effect of bagging new peach varieties introduced in southern Minas
    Gerais – Brazil. *Food Science and Technology* 33, 434–440.

Liu, H., Cao, J. and Jiang, W. (2015) Evaluation and comparison of vitamin C,
    phenolic compounds, antioxidant properties and metal chelating activity of
    pulp and peel from selected peach cultivars. *LWT – Food Science and Technology*
    63, 1042–1048.

López, G., Behboudian, M.H., Echeverría, G., Girona, J. and Marsal, J. (2011)
    Instrumental and sensory evaluation of fruit quality for 'Ryan's Sun' peach grown
    under deficit irrigation. *HortTechnology* 21, 712–719.

Lurie, S. (2010) Plant growth regulator for improving postharvest stone fruit quality.
    Acta Horticulturae 884, 189–197.

Lurie, S. and Crisosto, C.H. (2005) Chilling injury in peach and nectarine. *Postharvest
    Biology and Technology* 37, 195–208.

Lyu, J., Liu, X., Bi, J.F., Jiao, Y., Wu, X.Y. and Ruan, W. (2017) Characterization of
    Chinese white-flesh peach cultivars based on principle component and cluster ana-
    lysis. *Journal of Food Science and Technology* 54, 3818–3826.

Ma, Y., Zhao, M., Wu, H., Yuan, C., Li, H. and Zhang, Y. (2021) Effects of fruit bagging
    on anthocyanin accumulation and related gene expression in peach. *Journal of the
    American Society for Horticultural Science* 146, 217–223.

Maatallah, S., Dabbou, S., Castagna, A., Guizani, M., Hajlaoui, H. *et al.* (2020) *Prunus
    persica* by-products: a source of minerals, phenols and volatile compounds. *Scientia
    Horticulturae* 261: 109016.

Manganaris, G.A., Vasilakakis, M., Mignani, I., Diamantidis, G. and Tzavella-Klonari, K.
    (2005) The effect of preharvest calcium sprays on quality attributes, physicochem-
    ical aspects of cell wall components and susceptibility to brown rot of peach fruits
    (*Prunus persica* L. cv. Andross). *Scientia Horticulturae* 107, 43–50.

Manganaris, G.A., Vincente, A.R., Martinez, P. and Crisosto, C.H. (2019) Postharvest
    physiological disorders in peach and nectarine. In: Tonetto de Freitas, S. and
    Pareek, S. (eds) *Physiological Disorders in Fruits and Vegetables*. CRC Press, Boca
    Raton, Florida, pp. 253–264.

Manzoor, M., Anwar, F., Mahmood, Z., Rashid, U. and Ashraf, M. (2012) Variation in
    minerals, phenolics and antioxidant activity of peel and pulp of different varieties
    of peach (*Prunus persica* L.) fruit from Pakistan. *Molecules* 17, 6491–6506.

Mestre, L., Reig, G., Betran, J.A., Pinochet, J. and Moreno, M.A. (2015) Influence of peach almond hybrids and plum-based rootstocks on mineral nutrition and yield characteristics of 'Big Top' nectarine in replant and heavy-calcareous soil conditions. *Scientia Horticulturae* 192, 475–481.

Mestre, L., Reig, G., Betran, J.A. and Moreno, M.A. (2017) Influence of plum rootstocks on agronomic performance, leaf mineral nutrition and fruit quality of 'Catherina' peach cultivar in heavy-calcareous soil conditions. *Spanish Journal of Agricultural Research* 15: e0901.

Mihaylova, D., Popova, A., Desseva, I., Petkova, N., Stoyanova, M., *et al.* (2021) Comparative study of early- and mid-ripening peach (*Prunus persica* L.) varieties: biological activity, macro-, and micro- nutrient profile. *Foods* 10: 164.

Minas, I.S., Tanou, G. and Molassiotis, A. (2018) Environmental and orchard bases of peach fruit quality. *Scientia Horticulturae* 235, 307–322.

Mohammed, J., Belisle, C.E., Wang, S., Itle, R.A., Adhikari, K. and Chavez, D.J. (2021) Volatile profile characterization of commercial peach (*Prunus persica*) cultivars grown in Georgia, USA. *Horticulturae* 7: 516.

Musacchi, S. and Serra, S. (2018) Apple fruit quality: overview on pre-harvest factors. *Scientia Horticulturae* 234, 409–430.

Nilo, R., Campos-Vargas, R. and Orellana, A. (2012) Assessment of *Prunus persica* fruit softening using a proteomics approach. *Journal of Proteomics* 75, 1618–1638.

Njoroge, S.M.C. and Reighard, G.L. (2008) Thinning time during stage I and fruit spacing influences fruit size of "Contender" peach. *Scientia Horticulturae* 115, 352–359.

Noratto, G., Porter, W., Byrne, D. and Cisneros-Zevallos, L. (2009) Identifying peach and plum polyphenols with chemopreventive potential against estrogen-independent breast cancer cells. *Journal of Agricultural and Food Chemistry* 57, 5219–5226.

Noratto, G.D., Garcia-Mazcorro, J.F., Markel, M., Martino, H.S., Minamoto, Y. *et al.* (2014) Carbohydrate-free peach (*Prunus persica*) and plum (*Prunus domestica*) juice affects fecal microbial ecology in an obese animal model. *PLoS One* 9: e101723.

Noratto, G., Martino, H.D., Simbo, S., Byrne, D. and Mertens-Talcott, S.U. (2015) Consumption of polyphenol-rich peach and plum juice prevents risk factors for obesity-related metabolic disorders and cardiovascular disease in Zucker rats. *Journal of Nutritional Biochemistry* 26, 633–641.

Nowicka, P. and Wojdylo, A. (2019) Content of bioactive compounds in the peach kernels and their antioxidant, anti-hyperglycemic, anti-aging properties. *European Food Research and Technology* 245, 1123–1136.

Olivos, A., Johnson, S., Xiaoqiong, Q. and Crisosto, C.H. (2012) Fruit phosphorous and nitrogen deficiencies affect 'Grand Pearl' nectarine flesh browning. *HortScience* 47, 391–394.

Orazem, P., Stampar, F. and Hudina, M. (2011) Fruit quality of Redhaven and Royal Glory peach cultivars on seven different rootstocks. *Journal of Agriculture and Food Chemistry* 59, 9394–9401.

Perry, L.M. and Metzger, J. (1980) *Medicinal Plants of East and Southeast Asia: Attributed Properties and Uses.* MIT Press, Cambridge, Massachusetts.

Prange, R. and DeEll, J.R. (1997) Preharvest factors affecting quality of berry crops. *HortScience* 32, 824–830.

Qumar, N. (2016) Study of nutritional constituents and sensory evaluation of bakery products prepared from seed and bark of *Prunus persica* (peach). *International Journal of Research – Granthaalayah* 4, 12–24.

Rahmati, M., Vercambre, G., Davarynejad, G., Bannayan, M., Azizi, M. and Génard, M. (2015) Water scarcity conditions affect peach fruit size and polyphenol contents more severely than other fruit quality traits. *Journal of the Science of Food and Agriculture* 95, 1055–1065.

Reig, G., Iglesias, I., Gatius, F. and Alegre, S. (2013a) Antioxidant capacity, quality, and anthocyanin and nutrient contents of several peach cultivars [*Prunus persica* (L.) Batsch] grown in Spain. *Journal of Agricultural and Food Chemistry* 61, 6344–6357.

Reig, G., Alegre, S., Gatius, F. and Iglesias, I. (2013b) Agronomical performance under Mediterranean climatic conditions among peach [*Prunus persica* L. (Batsch)] cultivars originated from different breeding programmes. *Scientia Horticulturae* 150, 267–277.

Reig, G., Alegre, S., Gatius, F. and Iglesias, I. (2015) Adaptability of peach cultivars [*Prunus persica* (L.) Batsch] to the climatic conditions of the Ebro Valley, with special focus on fruit quality. *Scientia Horticulturae* 190, 149–160.

Reig, G., Mestre, L., Betran, J.A., Pinochet, J. and Moreno, M.A. (2016) Agronomic and physicochemical fruit properties of 'Big Top' nectarine budded on peach and plum based rootstocks in Mediterranean conditions. *Scientia Horticulturae* 210, 85–92.

Reig, G., Alegre, S., Cantin, C.M., Gatius, F., Puy, J. and Iglesias, I. (2017) Tree ripening and postharvest firmness loss of eleven commercial nectarine cultivars under Mediterranean conditions. *Scientia Horticulturae* 219, 335–343.

Reig, G., Garanto, X., Mas, N. and Iglesias, I. (2020) Long-term agronomical performance and iron chlorosis susceptibility of several *Prunus* rootstocks grown under loamy and calcareous soil conditions. *Scientia Horticulturae* 262: 109035.

Reighard, G.L. and Loreti, F. (2008) Rootstock development. In: Layne, D.R. and Bassi, D. (eds) *The Peach: Botany, Production and Uses*. CAB International, Wallingford, UK, pp 193–220.

Rodríguez, A., Pérez-López, D., Centeno, A. and Ruiz-Ramos, M. (2021) Viability of temperate fruit tree varieties in Spain under climate change according to chilling accumulation. *Agricultural Systems* 186: 102961.

Rodríguez, C.E., Bustamante, C.A., Budde, C.O., Müller, G.L., Drincovich, M.F. and Lara, M.V. (2019) Peach fruit development: a comparative proteomic study between endocarp and mesocarp at very early stages underpins the main differential biochemical processes between these tissues. *Frontiers in Plant Science* 10: 715.

Saidani, F., Giménez, R., Aubert, C., Chalot, G., Betrán, J.A. and Gogorcena, Y. (2017) Phenolic, sugar and acid profiles and the antioxidant composition in the peel and pulp of peach fruits. *Journal of Food Composition and Analysis* 62, 126–133.

Sánchez-Moreno, C., Pascual-Teresa, S., Ancos, B. and Cano, M.P. (2006) Nutritional values of fruits. In: Barta, J., Cano, M.P., Gusek, T., Sidhu, J.S. and Sinha, N.K. (eds) *Handbook of Fruits and Fruit Processing*. Blackwell Publishing, Hoboken, New Jersey, pp. 29–44.

Savic, I.M., Nikolic, V.D., Savic-Gajic, I.M., Kundakovic, T.D., Stanojkovic, T.P. and Najman, S.J. (2016) Chemical composition and biological activity of the plum seed extract. *Advanced Technologies* 5, 38–45.

Senica, M., Stampar, F., Veberic, R. and Mikulic-Petkovsek, M. (2017) Fruit seed of the Rosaceae family: a waste, new life, or a danger to human health? *Journal of Agricultural and Food Chemistry* 65, 10621–10629.

Serra, S., Anthony, B., Masia, A., Giovannini, D. and Musacchi, S. (2020) Determination of biochemical composition in peach (*Prunus persica* L. Batsch) accessions characterized by different flesh color and textural typologies. *Foods* 9: 1452.

Shen, J.Y., Wu, L., Liu, H.R., Zhang, B., Yin, X.R. *et al.* (2014) Bagging treatment influences production of C6 aldehydes and biosynthesis-related gene expression in peach fruit skin. *Molecules* 19, 13461–13472.

Shirosaki, M., Koyama, T. and Yazawa, K. (2012a) Suppressive effect of peach leaf extract on glucose absorption from the small intestine of mice. *Bioscience, Biotechnology and Biochemistry* 76, 89–94.

Shirosaki, M., Goto, Y., Hirooka, S., Masuda, H., Koyama, T. and Yazawa, K. (2012b) Peach leaf contains multiflorin A as a potent inhibitor of glucose absorption in the small intestine in mice. *Biological & Pharmaceutical Bulletin* 35, 1264–1268.

Stojanovic, B.T., Mitic, S.S., Stojanovic, G.S., Mitic, M.N., Kstic, D.A. *et al.* (2016) Phenolic profile and antioxidant activity of pulp and peel from peach and nectarine fruits. *Notulae Botanicae Horti Agrobotanici* 44,175–182.

Tanou, G., Minas, I.S., Scossa, F., Belghazi, M., Xanthopoulou, A. *et al.* (2017) Exploring priming responses involved in peach fruit acclimation to cold stress. *Scientific Reports* 7: 11358.

Theron, K.I., Steenkamp, H., and Steyn, W.J. (2017) Efficacy of ACC (1-aminocyclopropane-1-carboxylic acid) as a chemical thinner alone or combined with mechanical thinning for Japanese plums (*Prunus salacina*). *HortScience* 52, 110–115.

Theron, K.I., Steenkamp, H., Scholtz, A., Lötze, G.F.A., Reynolds, J.S. and Steyn, W.J. (2021) The efficacy of 1-aminocyclopropane-1-carboxylic acid (ACC) in thinning 'Keisie' peaches. *Acta Horticulturae* 1295, 33–40.

Tomás-Barberán, F.A., Ruiz, D., Valero, D., Rivera, D., Obón, C. *et al.* (2013) Heath benefits from pomegrates and stone fruit including plums, peaches, apricots and cherries. In: Skinner, M. and Hunter, D. (eds) *Bioactives in Fruit: Health Benefits and Functional Foods*. Wiley, Chichester, UK, pp. 125–167.

Tombesi, A., Antognozzi, E. and Palliotti, A. (1993) Influence of light exposure on characteristics and storage life of kiwifruit. *New Zealand Journal of Crop and Horticultural Science* 21, 87–92.

Toor, R.K. and Savage, G.P. (2005) Antioxidant activities in different fraction of tomato. *Food Research International* 38, 487–494.

Tu, Z., Han, X., Wang, X., Hou, Y., Shao, B. *et al.* (2003) Protective effects of CVPM on vascular endothelium in rats fed cholesterol diet. *Clinica Chimica Acta* 333, 85–90.

UPOV (2010) *Guidelines for the Conduct of Tests for Distinctness, Uniformity and Stability. Peach (Prunus persica (L.) Batsch)*. TG/53/7. International Union for the Protections of New Varieties of Plants (UPOV), Geneva, Switzerland. Available at: www.upov.int/edocs/tgdocs/en/tg053.pdf (accessed 7 December 2022).

Val, J. and Fernandez, V. (2011) In-season calcium-spray formulations improve calcium balance and fruit quality traits of peach. *Journal of Plant Nutrition and Soil Science* 174, 465–472.

Valero, C., Crisosto, C.H. and Slaughter, D. (2007) Relationship between nondestructive firmness measurement and commercially important ripening fruit stages for peaches, nectarines, and plums. *Postharvest Biology and Technology* 44, 248–253.

Vallverdu, X., Girona, J., Echeverría, G., Marsal, J., Hossein Behboudian, M. and López, G. (2012) Sensory quality and consumer acceptance of "Tardibelle" peach are improved by deficit irrigation applied during stage II of fruit development. *HortScience* 47, 656–659.

Vicente, A.R., Manganaris, G.A., Cisneros-Zevallos, L. and Crisosto, C.H. (2011) Prunus. In: Terry, L.A. (ed.) *Health-Promoting Properties of Fruit and Vegetables*. CAB International, Wallingford, UK, pp. 239–259.

Vinardell, M.P. and Mitjans, M. (2017) Lignins and their derivatives with beneficials effects on human health. *International Journal of Molecular Sciences* 18: 1219.

Vizzotto, M., Porter, W., Byrne, D. and Cisneros-Zevallos, L. (2014) Polyphenols of selected peach and plum genotypes reduce cell viability and inhibit proliferation of breast cancer cells while not affecting normal cells. *Food Chemistry* 164, 363–370.

Vizzotto, G., Casatta, E., Bomben, C., Bregoli, A.M., Sabatini, E. and Costa, G. (2002) Peach ripening as affected by AVG. *Acta Horticulturae* 592, 561–566.

Wang, Y., Yang, C., Li, S., Yang, L., Wang, Y. *et al.* (2009) Volatile characteristics of 50 peaches and nectarines evaluated by HP–SPME with GC–MS. *Food Chemistry* 116, 356–364.

Wang, Y.J., Yang, C.X., Liu, C.Y., Xu, M., Li, S.H. *et al.* (2010) Effects of bagging on volatiles and polyphenols in "Wanmi" peaches during endocarp hardening and final fruit rapid growth stages. *Journal of Food Science* 75, S455–S460.

Zu, M., Fang, W., Chen, C., Wang, L. and Cao, K. (2020) Effects of shading by bagging on carotenoid accumulation in peach fruit flesh. *Journal of Plant Growth Regulation* 40, 1912–1921.

12

COMMON PREHARVEST DISEASES OF PEACH AND NECTARINE CAUSED BY FUNGI AND BACTERIA: BIOLOGY, EPIDEMIOLOGY AND MANAGEMENT

James E. Adaskaveg[1]*, Guido Schnabel[2], David F. Ritchie[3] and Helga Förster[1]

[1]*Department of Microbiology and Plant Pathology, University of California, Riverside, California, USA;* [2]*Department of Plant and Environmental Sciences, Clemson University, Clemson, South Carolina, USA;* [3]*Department of Entomology and Plant Pathology, North Carolina State University, Raleigh, North Carolina, USA*

12.1 INTRODUCTION

There are many, fungal-like and bacterial pathogens that attack the flowers, leaves, fruit, branches, trunks and roots of peach and nectarine trees in production orchards. Many of these organisms also cause important diseases worldwide on other *Prunus* spp. Most cause preharvest diseases and determine the overall productivity and fruit quality during the lifespan of the orchard. Some consistently cause annual production losses, whereas others occur more sporadically and develop under specific climatic conditions or on selected cultivars.

All the diseases discussed are caused by microorganisms that are heterotrophic and require water for growth. The organisms belong to three kingdoms, the Eumycota or True Fungi, the Stramenopila (Chromista) with pathogens in the Oomycota, and the Eubacteria. The True Fungi and Oomycota organisms were previously grouped together as 'Fungi' because of similarities in growth and nutrient assimilation and their reproduction by spores. Members of the True Fungi belong to the phyla Ascomycota, where many of the fungal pathogens are classified, Basidiomycota, and Mucoromycota. These phyla are characterized by the specific ways that sexual spores or meiospores (ascospores, basidiospores and zygospores, respectively) are produced. These spores may be formed in fruiting structures called basidiomes (i.e. Basidiomycota) or ascoma

* Email: jim.adaskaveg@ucr.edu

© CAB International 2023. *Peach* (G. Manganaris, G. Costa and C. Crisosto eds)   261
DOI: 10.1079/9781789248456.0012

(i.e. Ascomycota). Ascoma of the Ascomycota range from cup-shaped apothecia to flask-shaped perithecia to spherical cleistothecia or chasmothecia, or are cavities embedded in fungal tissue called ascostroma. In addition, asexual spores (mitospores) are typically produced. In the Mucoromycota, sporangiospores are produced within sac-like structures called sporangia, whereas in the Ascomycota and sometimes in the Basidiomycota, single- or multi-celled conidia are produced from hyphae or specialized cells. Somatic spores called chlamydospores may be formed by any fungal group as terminal or intercalary thick-walled hyphal cells. Dense survival structures formed by fungal mycelium (i.e. sclerotia) or by mycelium and host tissue (i.e. pseudosclerotia) may also be present. Within the Stramenopila, the phylum Oomycota contains important plant pathogens in the genera *Phytophthora* and *Pythium*. These organisms generally are soil-borne and are adapted to wet environments, and thus are a problem when soil water content is high and not managed. Organisms in the Oomycota produce asexual motile zoospores that are wall-less and flagellate and are capable of swimming towards roots using chemotaxis. Zoospores form in a sac-like cell called a zoosporangium. Oospores are sexually produced spores that are surrounded by a thick wall. These spores, in addition to asexually produced chlamydospores, are survival structures.

Pathogens that infect roots are commonly referred to as soil-borne organisms. Those that occur above ground often have air-borne propagules that have adapted to rain-splash or wind dispersal. Some exclusively reside in root fragments underground and primarily infect trees through root-to-root contact. Bacterial pathogens of peach can be soil-borne and can cause root and crown galls, whereas epiphytic bacteria typically cause flower or bud blast, twig cankers, and leaf, stem and fruit spots. Endophytic bacteria can cause cankers in addition to systemic infections affecting the health of the entire tree.

The purpose of this chapter is to provide the peach and nectarine horticulturalist with an overview of the major fungal and bacterial diseases, including a description of their distribution, the symptoms (description of the disease) and signs (structures of the pathogen) involved with each disease, critical ecological and epidemiological factors, and important management concepts and practices (Table 12.1). Also included are representative disease cycles for the major fungal groups and specific bacterial organisms that visualize host–pathogen relationships over the growing season and illustrate the importance of the different infection processes. More detailed information on each of the diseases discussed can be found in the references provided.

12.2 FLOWER, FOLIAGE AND FRUIT DISEASES

12.2.1 Brown rot

Brown rot is a major fungal disease of all commercially grown *Prunus* spp. in most regions of the world and can result in extensive crop losses (Batra, 1991).

Table 12.1. Common fungal diseases of peach and nectarine and important information on the biology, epidemiology and management of each disease.

Disease	Pathogen	Important reproductive stage(s) for infection[a]	Symptoms of economic significance	Primary management[b]
Blossom, foliage, and fruit diseases				
Brown rot	*Monilinia fructicola, M. laxa, M. fructigena, M. polystroma, M. mumecola, M. yunnanensis*	Conidia (all species), ascospores (*M. fructicola, M. fructigena*)	Fruit decay	Sanitation, fungicides
Jacket rot and green fruit rot	*Botrytis cinerea*	Conidia	Fruit decay	Sanitation, fungicides
	Sclerotinia sclerotiorum	Ascospores	Fruit decay	Orchard floor management, fungicides
Peach leaf curl	*M. fructicola, M. laxa*	Conidia	Fruit decay	Sanitation, fungicides
	Taphrina deformans	Bud conidia	Defoliation, fruit off-grades	Fungicides
Powdery mildew	*Podosphaera pannosa* (= *Sphaerotheca pannosa*), *P. leucotricha, P. clandestina*	Conidia, ascospores	Defoliation, fruit off-grades	Cultivar selection, fungicides
Rust	*Tranzschelia discolor, T. pruni-spinosae*	Urediniospores (aeciospores?)	Defoliation, fruit off-grades	Fungicides (removal of alternative host?)
Scab	*Venturia carpophila*	Conidia	Fruit off-grades	Fungicides
Shot hole	*Wilsonomyces carpophilus*	Conidia	Defoliation, fruit off-grades	Fungicides

Continued

Table 12.1. Continued.

Disease	Pathogen	Important reproductive stage(s) for infection[a]	Symptoms of economic significance	Primary management[b]
Trunk and scaffold diseases				
Leucostoma (Cytospora) canker (syn. perennial canker)	*Leucostoma cincta* (*Cytospora cincta*), *L. persoonii* (*C. leucostoma*), *C. plurivora*	Conidia (ascospores?)	Branch dieback, tree decline and death	Sanitation, pruning practices
Silver leaf disease	*Chondrostereum purpureum*	Basidiospores	Branch dieback, tree decline and death	Sanitation, pruning practices
Fungal gummosis (syn. peach blister canker or ibokawa byo)	*Botryosphaeria dothidea* (*B. obtusa* and *Lasiodiplodia theobromae*)	Ascospores and conidia	Branch dieback, tree decline and death	Sanitation and winter pruning (removal of alternate host?)
Constriction canker	*Diaporthe amygdali*	Conidia	Branch dieback, tree decline and death	Cultivar selection, sanitation and winter pruning, use of low-nitrogen fertilization, and broad-spectrum fungicides
Root and crown diseases				
Armillaria root rot	*Armillaria mellea, A. ostoyae, Desarmillaria tabescens*	Rhizomorphs, mycelium	Root and crown rot, tree decline and death	Site selection (avoidance), sanitation, soil fumigation, root collar excavation, rootstock selection

Disease	Causal agent	Propagule	Symptoms	Management practices
Peach tree short life	*Pseudomonas* spp.	Bacterial cells	Tree decline and death	Site selection (avoidance), soil fumigation, rootstocks
	Cytospora spp.	Conidia (ascospores)	Branch dieback, tree decline and death	Sanitation, pruning practices
Phytophthora root and crown rots	*Phytophthora cactorum, P. cambivora, P. cinnamomi, P. citricola, P. citrophthora, P. cryptogea, P. drechsleri, P. mediterranea, P. niederhauseri, P. parasitica, P. nicotianae, P. syringae and others*	Zoospores, sporangia, oospores, chlamydospores	Root and crown rot, tree decline and death	Irrigation management (e.g. raised beds, sprinkler shields to prevent trunk wetness), rootstocks, fungicides
Verticillium wilt	*Verticillium dahliae*	Microsclerotia	Wilt	Site selection (avoidance), soil fumigation

[a]Propagules with a question mark have a limited or unknown role in the disease cycle.
[b]Management practices with a question mark have a limited or poorly defined role in controlling the disease.

In areas with high rainfall and/or high humidity during the growing season, severe epidemics may occur annually unless intense management practices are employed. In more arid regions, epidemics may still occur when orchard environmental conditions are favourable for the pathogens to cause blossom blight, quiescent infections on developing fruit or brown rot of mature fruit. The disease is caused mostly by three species of the genus *Monilinia*, which can be differentiated by their cultural characteristics (Byrde and Willetts, 1977) and with the use of selective media (Phillips and Harvey, 1975; Amiri *et al.*, 2009). Molecular methods have been developed for genus- and species-specific detection using primers based on random amplified polymorphic DNA sequences (Förster and Adaskaveg, 2000; Côté *et al.*, 2004), internal transcribed spacer (ITS) regions 1 and 2 (Ioos and Frey, 2000; Gell *et al.*, 2007), ribosomal DNA (Snyder and Jones, 1999) and the β-tubulin (Ma *et al.*, 2003; Hu *et al.*, 2011) and cytochrome *b* genes (Miessner and Stammler, 2010; Hily *et al.*, 2011), as well as real-time polymerase chain reaction (PCR) amplifying the ITS region of the nuclear ribosomal RNA gene repeat and using genus- and species-specific fluorogenic-labelled probes (van Brouwershaven *et al.*, 2010).

Monilinia fructicola Winter (Honey), the cause of North American brown rot, is the main pathogen on peaches and other stone fruits in most production regions. This pathogen has been found in North (Canada, Mexico and the USA), Central, and South America (e.g. Argentina, Brazil, Chile), South Africa, Australia and New Zealand (Batra, 1991), but more recently has spread throughout Europe and Asia (e.g. China, India) where it has an established, occasional or restricted distribution (EPPO Global Database, 2021). *Monilinia laxa* (Aderh. & Ruhl.) Honey causes European brown rot on peach and other stone fruits worldwide but is generally less important on peach in areas where *M. fructicola* is also present. The third species, *Monilinia fructigena* Honey ex Whetzel, is not present in North America. In Europe, its primary distribution area, it occurs mainly on pome crops and only occasionally on peach and nectarine.

Another brown rot pathogen, the anamorphic species *Monilinia polystroma* (G. van Leeuwen) L.M. Kohn (syn. *Monilia polystroma* van Leeuwen), was identified based on morphological and molecular differences among isolates from stone and pome fruits in Europe and Japan (van Leeuwen *et al.*, 2002; Poniatowska *et al.*, 2013). This species was previously considered to be *M. fructigena*, and the disease caused by *M. polystroma* is also called Asiatic brown rot. In more recent surveys on peach and plum in China, two additional anamorphic brown rot pathogens were identified: *Monilinia mumecola* (Y. Harada, Y. Sasaki & Sano) Sandoval-Denis & Crous (syn. *Monilia mumecola* Y. Harada, Y. Sasaki & Sano) and *Monilinia yunnanensis* (M.J. Hu & C.X. Luo) Sandoval-Denis & Crous (syn. *Monilia yunnanensis* M.J. Hu & C.X. Luo) (Hu *et al.*, 2011; Yin *et al.*, 2015). *M. mumecola* mostly infects stone fruits, whereas *M. yunnanensis* is primarily found on pome fruits. These two species have not been reported from other continents to date. Based on ITS, glyceraldehyde 3-phosphate dehydrogenase and β-tubulin sequencing, *M. mumecola* is most closely related to *M. laxa*, and these two species also have a similar colony morphology; *M. yunnanensis* was found to be most closely related to *M. fructigena* and *M. polystroma* (Hu *et al.*, 2011; Yin *et al.*, 2015).

The brown rot disease cycle includes blossom and twig blights and, most economically important, pre- and postharvest fruit decay (Fig. 12.1). Primary inoculum sources in the spring are overwintering brown rot fruit mummies on the tree that produce asexual conidia in sporodochia and fruit mummies on the orchard floor that produce sexual fruiting structures (apothecia) and spores (ascospores) (Biggs and Northover, 1985). Twig cankers and fruit peduncles can be additional sources of conidial inoculum (Sutton and Clayton, 1972; Biggs and Northover, 1985). A single blossom may be infected by one or more genotypes of *M. fructicola*, and this diversity is preserved in the corresponding canker (Dowling *et al.*, 2019). Conidia are produced acropetally

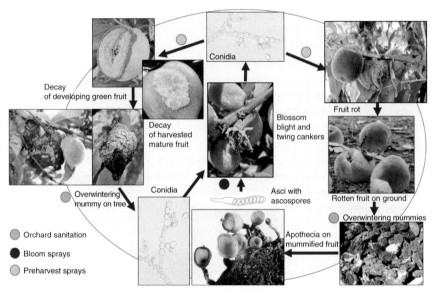

Fig. 12.1. Disease cycle of brown rot of peach caused by *Monilinia fructicola*. The fungal pathogen overwinters in fruit mummies either on the tree or on the orchard floor. In the spring, asexual conidia produced from mummies on the tree and sexual ascospores produced in asci that are formed in fruiting bodies (apothecia) on the ground are the primary inoculum. These spores are wind dispersed to susceptible host tissues (e.g. blossoms) and germinate under favourable wetness and temperature conditions. Infections of blossom tissue can occur within 6–12 h at 15–20°C. After 3–5 days, blossoms become blighted. Diseased blossoms typically remain attached, and the infection spreads into the peduncle and down into the twig. The infection continues with the formation of a twig canker, which often develops a gumdrop as a host response. Conidia form on infected tissue and serve as secondary inoculum for infection of immature (green) and mature fruit. Fruit infections may also result in twig blight during severe outbreaks.

in branched chains. They are hyaline to subhyaline and appear tan to grey en masse. Conidia are mostly single celled, globose to ellipsoid–limoniform and generally have a smooth wall. Blossom blight rarely reduces the crop load, but blighted blossoms and infected shoots provide secondary inoculum for fruit infections later in the season. Spore counts using a microscope have been correlated with use of a species-specific primer pair for *M. fructicola* in a real-time PCR assay, and spores were quantified by air sampling over the season (Luo *et al.*, 2007). Infections can occur over a wide temperature range with an optimum of 22.5–25°C (Biggs and Northover, 1988a; Tamm and Flückiger, 1993). Minimum wetness durations of 3–5 h are required at 20°C, whereas 18 h are needed at 10°C.

The first symptoms of blossom blight are necrosis of the anthers and browning of the filaments that proceeds to the floral tube, ovary and peduncle (Fig. 12.2a). Infections may extend into the twig, which may be girdled. As infected flowers wilt and turn brown, they generally stick firmly to the twig as a gum drop forms where the peduncle attaches to the stem (Fig. 12.2b). In wet weather, infected flowers become covered with greyish to tan sporodochia. Blossom infections by *M. laxa* or by *M. fructicola* may result in twig cankers, which often show extensive gum formation at the advancing margin. The canker may girdle the twig resulting in blight of the distal twig with leaves turning tan to brown and remaining attached. If the branch is not girdled, surrounding healthy tissue will produce callus tissue. On large cankers, sporulation may continue for several years.

Infection of fruit occurs by direct penetration through the cuticle or indirectly through suture cracks or other injuries, or through stomata. Susceptibility of fruit to infection by *Monilinia* spp. is high during the early stages of fruit development, decreases during the green fruit stages, and then

Fig. 12.2. (a, b) Brown rot blossom blight of peach caused by *Monilinia fructicola*. A canker with a gum drop may form at the base of the peduncle.

increases again as the fruit mature and ripen (Biggs and Northover, 1988b; Gradziel, 1994). On mature or ripening fruit, brown rot typically develops as a rapidly spreading, firm, brown decay (Fig. 12.3). Under optimum conditions, decay of ripe peaches infected by *M. fructicola* may be visible within 48 h of infection. Quiescent infections that occur on developing fruitlets and on ripening fruit in less favourable environments (arid and semi-arid climates) may become active when the fruit mature prior to or after harvest. Molecular techniques have been developed for the detection of quiescent infections of stone fruit and for species identification (Förster and Adaskaveg, 2000; Boehm *et al.*, 2001; Ma *et al.*, 2003; Miessner and Stammler, 2010; van Brouwershaven *et al.*, 2010; Hily *et al.*, 2011; Hu *et al.*, 2011). Infections on green fruit that are injured by frost or insects or drop to the orchard floor, especially during late thinning, may lead to sporulating lesions and provide additional inoculum.

Orchard sanitation practices that include removal of mummified fruit and infected twig, as well as twigs with cankers from trees are important components of an integrated management approach. In some peach-production areas, removal of alternative hosts such as wild *Prunus* spp. may also be an effective strategy. No brown rot-resistant peach cultivars are available, but there are considerable differences in susceptibility among peach (cling and freestone) cultivars (Gradziel and Wang, 1993; Bassi *et al.*, 1998). Protective fungicide treatments provide the best control for both blossom blight and fruit rot. The proper use of fungicides with local systemic activity protects flowers and fruit, reduces the amount of sporulation formed on the infected tissue and reduces sources of overwintering inoculum. Blossom applications have to be done as susceptible flower parts are exposed and must be carried out before the occurrence of periods with conducive wetness and temperatures. Fungicides, in particular captan, applied prior to the preharvest period, will significantly reduce brown rot at harvest (Lalancette *et al.*, 2017). However, fungicides need not be

Fig. 12.3. Brown rot fruit rot of peach caused by *Monilinia fructicola*.

applied to immature fruit unless wetness conditions are especially favourable for infection, as is regularly the case in the south-eastern USA, or injury caused by insects or cold has increased the likelihood of disease. Thus, insect control during this period is an important consideration.

Management of brown rot and several other diseases is heavily dependent on fungicides, and thus fungicide resistance management practices are essential to prevent the development of insensitive pathogen populations using the newer single-site mode-of-action materials. Fungicide resistance has developed to methyl benzimidazole carbamates (MBCs) in populations of *M. fructicola* and other *Monilinia* spp. that has resulted in crop losses in many locations, including California, Michigan and South Carolina (Jones and Ehret, 1976; Michailides *et al.*, 1987; Ogawa *et al.*, 1988; Ma *et al.*, 2003; Zhu *et al.*, 2010). Other single-site mode-of-action fungicides have been introduced for brown rot control, including dicarboximides and sterol demethylation inhibitors (DMIs) in the 1980s, quinone outside inhibitors (QoIs) and anilinopyrimidines in the 1990s, and succinate dehydrogenase inhibitors in the 2000s. Of these, DMIs have been the most effective fungicide class for brown rot management and can function both protectively and curatively (Holb and Schnabel, 2007).

Extensive shifts in 50% effective concentrations (EC_{50} values) of the DMI fungicides have been observed in the field (Zehr *et al.*, 1999; Schnabel *et al.*, 2004), and multiple resistance was later detected in *M. fructicola* to both MBC and DMI fungicides in the eastern USA (Chen *et al.*, 2013). Overexpression of *CYP51* genes is a common mechanism conferring resistance to DMI fungicides in fruit-tree fungal pathogens. A 65 bp insertion, termed 'Mona', was located upstream of *MfCYP51* and was associated with the DMI resistance phenotype in isolates from South Carolina, Georgia, Ohio and New York (Luo and Schnabel, 2008). *MfCYP51* overexpression was also found in a DMI-resistant isolate from Brazil; however, the isolate did not possess insertions upstream of *MfCYP51* (Lichtemberg *et al.*, 2017). A large-scale survey of *M. fructicola* isolates from New York and Pennsylvania resulted in the detection of 'Mona' in isolates from only five of nine orchards harbouring DMI resistance or reduced sensitivity, indicating that the 'Mona' element is not universally associated with DMI resistance in *M. fructicola* (Villani and Cox, 2011). Thus, other DMI resistance mechanisms are present in *M. fructicola*. In contrast to the results of Luo and Schnabel (2008), the presence of 'Mona' did not correlate with the DMI-resistant phenotype or overexpression of *MfCYP51* in isolates from Michigan (Lesniak *et al.*, 2021). A second mechanism of resistance, specifically a target gene mutation, was reported in *M. fructicola* isolates from Brazil (Lichtemberg *et al.*, 2017).

12.2.2   Jacket rot and green fruit rot

As flower parts senesce after pollination, they are often colonized by fungi resulting in jacket rot (a disease of the hypanthium) and subsequent green fruit rot. This disease can be a serious problem in wet years or in foggy production areas with prolonged wetness periods during bloom. Jacket rot and green fruit rot can affect all stone-fruit crops; however, peach and nectarine are the least susceptible. One or several fungi including *Monilinia* spp., *Botrytis cinerea* Pers.: Fr. and *Sclerotinia sclerotiorum* (Lib.) de Bary commonly cause jacket rot and green fruit rot, depending on the geographical location and presence of the pathogens. These fungi have a cosmopolitan distribution, and thus this disease presumably occurs throughout temperate fruit-growing regions of the world wherever wetness occurs during and after the bloom period.

The disease begins with the pathogens colonizing the senescent calyx (floral cup or hypanthium known as the 'jacket' or 'shuck' around the developing fruit), petals and other flower parts. As the fruit develop and are in contact with diseased blossom parts, the pathogens can grow into the healthy fruit (Fig. 12.4). A brown lesion develops and spreads quickly into the small or immature fruit. *B. cinerea* is characterized by greyish tufts of conidia on infected senescent or dying plant tissues or on sclerotia on the orchard floor. The oval to elliptical conidia are produced on a swollen cell at the apex of the conidiophore. Germination of conidia occurs with wetness over a wide temperature range with an optimum of 15–20°C. Wetness is not required if relative humidity is above 98%. With *S. sclerotiorum*, white mycelium develops on infected fruit and blossom tissues; no asexual spores are produced. Sexual spores of *S. sclerotiorum* are produced from apothecia that develop from sclerotia on the orchard floor, especially under vegetation of cover crops. The hyaline ascospores are forcibly discharged from the apothecia following changes in relative humidity or by physical disturbance to become air-borne (Fig. 12.5). Infection of floral parts by ascospores requires 48–72 h of wetness with the temperature seldom

Fig. 12.4. Jacket rot of peach caused by *Botrytis cinerea*. The jacket is infected, but the immature fruit is still disease-free.

Fig. 12.5. Apothecia of *Sclerotinia sclerotiorum* are produced on overwintering sclerotia (bottom). Ascospores are forcibly discharged in a spore cloud. Reproduced from Strand (1999) with permission.

being limiting. Sclerotia of this fungus only form after infected structures fall to the ground. *Monilinia* spp. can also cause green fruit rot after partial colonization of blossom tissue, after causing brown rot blossom blight or after insect injury occurs on the developing fruit (Fig. 12.6). Spores of the jacket rot and green fruit rot fungi are wind-borne or are disseminated by splashing rain and germinate to infect the senescent blossom tissues. From infections of flower tissues, the pathogens can move into fruit tissue provided that cool, wet weather occurs that delays the jacket from separating from the developing fruit.

Fig. 12.6. Green fruit rot of peach caused by *Monilinia fructicola*. The decay spreads to neighbouring healthy fruit by contact.

Removal of weeds and senescent plant tissues from the orchard floor is probably beneficial in reducing inoculum levels for jacket rot and green fruit rot caused by *B. cinerea* and *S. sclerotiorum*. Fungicide applications with materials effective against all green fruit rot pathogens during full bloom and after bloom are suggested to prevent losses.

12.2.3 Shot hole

Shot hole disease of stone fruits occurs worldwide but is especially prevalent in the peach-producing areas of the western USA. It is most important on peach, nectarine and apricot. Outbreaks of the disease on peach occur following favourable environments and when no protective fungicide treatments have been applied during the dormant season. Shot hole is caused by the imperfect fungus *Wilsonomyces carpophilus* (Lév.) Adaskaveg, Ogawa & Butler (syn. *Thyrostroma carpophilum* (Lév.) B. Sutton), which was previously classified in the imperfect genera *Coryneum*, *Clasterosporium* and *Stigmina*. *Wilsonomyces* is a monotypic genus. *Helminthosporium carpophilum* was initially described and transferred to different genera until Adaskaveg *et al.* (1990) introduced *Wilsonomyces* to accommodate this species. Sutton regarded *Wilsonomyces* as a synonym of *Thyrostroma* (Sutton, 1997). However, Marin-Felix *et al.* (2017) separated species in these genera based on partial sequencing of the large subunit nuclear ribosomal RNA (LSU rRNA), the ITS and translation elongation factor 1-α (*tef1*) and introduced *Thyrostroma compactum*, confirming that *Wilsonomyces* represents a distinct genus in the Dothidotthiaceae. Ye *et al.* (2020), using morphological and cultural characteristics and multi-locus sequence analysis including the ITS region, LSU rRNA and *tef1* genes, also identified the fungus as *W. carpophilus*.

Symptoms of shot hole occur on twigs, leaves and fruit (Wilson, 1937). The primary damage to peach is the killing of twigs and buds and subsequent infections on the fruit. In the absence of management practices, unsightly superficial fruit infections render the crop unmarketable. On twigs, where the fungus overwinters, lesions first appear as purplish spots 2–3 mm in diameter that enlarge up to 10 mm, turn brown and are slightly sunken (Fig. 12.7). In the light tan centre of the lesions, asexual conidia of the fungus are produced in sporodochia in the spring. The conidia are cylindrical, ellipsoid to fusiform, subhyaline to pale golden brown and have two to 11 dark transverse septa. Conidia are the primary inoculum for new infections; they are rain-splash dispersed to developing flowers and leaves. The conidia may remain viable for several months when kept dry. Infected buds are another overwintering stage of the fungus where primary inoculum is produced in the spring and often throughout the season. Scales of infected buds turn dark and are sometimes covered with gummy exudates. Lesions on leaves and fruit also start out as small purplish spots that turn brown and expand up to 10 mm in diameter (Fig. 12.8). During warm, dry weather, lesions on leaves abscise to produce the typical shot hole symptoms (Shaw *et al.*, 1990). Early dehiscence mostly eliminates the formation of sporodochia on shot hole lesions. In cool, wet environments, however, lesions remain attached to leaves and sporodochia commonly develop in the centre of each lesion (Shaw *et al.*, 1990). Infected leaves may drop, but early defoliation of peach trees is rare. On fruit, lesions turn into raised, corky areas that do not extend into the mesocarp tissues.

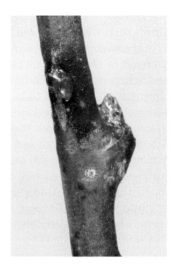

Fig. 12.7. Lesion of the shot hole fungus *Wilsonomyces carpophilus* on peach twig where the pathogen overwinters.

Fig. 12.8. Symptoms of shot hole caused by *Wilsonomyces carpophilus* on peach fruit and leaf.

Conidia of the fungus germinate over a wide temperature range (Smith and Smith, 1942). Optimum conditions for culturing the fungus are the use of β-maltose and asparagine as carbon and nitrogen sources at a pH of 5.5–6 and a temperature of 15°C (Williams and Helton, 1971). Twig infections require at least 24 h of continuous wetness, whereas shorter wetness durations are required for leaf infections. Thus, at 20–25°C, leaf infections occur after 8–12 h of wetness. Epidemics of shot hole occur when high rainfall commonly occurs. For management of the disease in semi-arid climates such as California, dormant fungicide sprays to protect against twig and bud infections are applied in the autumn after leaf drop and before the winter rains begin. Copper treatments, ziram or chlorothalonil are highly effective. In wetter climates, protective fungicides also need to be applied at leaf emergence and throughout fruit set. Ziram, strobilurin (QoI) and DMI fungicides are highly effective (Adaskaveg *et al.*, 2015).

12.2.4 Scab

Scab is an important fungal disease of peach, nectarine and other *Prunus* spp. in warm areas with high rainfall, such as the south-eastern USA, or where orchards are sprinkler irrigated resulting in wet foliage. The disease also has caused epidemics on peach in semi-arid areas such as California in years with unusual high rainfall. Scab is caused by the fungus *Venturia carpophila*

E.E. Fisher. (syn. *Fusicladium carpophilum* (Thüm.) Partridge & Morgan-Jones, *Cladosporium carpophilum* Thüm.). Typically, the species mostly occurs as the anamorph; however, mating-type idiomorphs have been identified, sequenced and compared (Bock *et al.*, 2021). The high genetic diversity of the species suggests that cryptic sexual reproduction is occurring (Bock *et al.*, 2021). Conidia are fusiform, olivaceous-brown, smooth to minutely verruculose with protruding scars on one or both ends, and are mostly single celled. They are produced solitarily or in short chains.

Infections can occur on twigs, leaves and fruit, but symptoms are most noticeable on the fruit. On twigs, where the fungus overwinters, superficial circular to oval lesions that are slightly raised develop on new succulent growth in the spring (Keitt, 1917; Bensaude and Keitt, 1928). These lesions initially appear as water-soaked spots that darken with age and become brown and later purple to dark brown, and have a raised border. At the end of the season, lesions are oval in shape and 3×5–5×8 mm in size. They persist throughout the host dormant period. In the following spring, during periods with relative humidity of 70–100%, the fungus produces abundant primary inoculum in 20–30 h in the form of asexual conidia from olivaceous tufts of mycelium within the lesion (Lawrence and Zehr, 1982; Gottwald, 1983), especially after petal fall until mid-spring. The conidia are wind and rain dispersed. They germinate between 15°C and 30°C with an optimum temperature of 25–30°C (Lawrence and Zehr, 1982). Maximum conidial germination occurs in the presence of liquid water, but conidia can also germinate at 94–100% relative humidity. The fungus does not survive in the twig lesions for more than a second season due to bark formation during twig growth. With unusually severe infections, new shoots of the host may die back. On leaves, symptoms may occur in early summer on the lower surfaces, and are visible as irregular, blotchy lesions, slightly darker in colour than the surrounding healthy tissue. Lesions turn olive green once sporulation of the fungus occurs. Infections of leaves are rarely severe enough to cause serious defoliation of trees. On half- to full-sized fruits approximately 6–8 weeks after petal fall, small circular spots develop mainly on the upper, exposed surfaces. In general, disease symptoms are visible 40–70 days after the conidia are deposited on the fruit. They are 5–10 mm in diameter and are first green and turn olive to black once the fungus sporulates and sometimes have a green to yellow halo (Fig. 12.9). Spots are most common near the stem end and on the outward-facing side of the fruit. The spots are superficial and slowly enlarge. On severely infected fruit, the lesions coalesce and form large, superficial blotches that make the fruit unmarketable. Although scab on peach fruit may develop in a similar way to bacterial spot caused by *Xanthomonas arboricola* pv. *pruni* (Smith) Vauterin, in most early-season infections, bacterial spot is visible as small, irregular, water-soaked, brown spots on fruit that may occur anywhere on the surface and later develop into deep, cavernous-appearing lesions.

Fig. 12.9. Symptoms of scab caused by *Venturia carpophila* (*Fusicladium carpophilum*).

Management of scab is accomplished mainly by the use of fungicidal sprays, which are applied at shuck split and every 2 weeks thereafter through early summer. Scherm *et al.* (2008) reported that conidial production generally commenced before bloom and reached 25% and 90% of the seasonal total by calyx split and 10 weeks after bloom, respectively. Applications as early as petal fall have shown increased disease control where the disease pressure is high. In semi-arid climates, effective management can also be obtained by reducing inoculum dispersal and leaf wetness from foliar sprinkler irrigation. Sanitation by removing infected twigs is impractical because large numbers of over-wintering lesions persist on fruiting wood. Pruning to allow adequate sunlight penetration and unimpeded air movement may improve scab management by facilitating rapid drying and good fungicide coverage.

12.2.5 Rust

Peach rust occurs wherever peach is grown in mild or Mediterranean climates and causes leaf, fruit and stem infections. The incidence of the disease in different years and different production regions is highly variable, and epidemics occur in years with excessive wetness during the growing season (Ogawa and English, 1991; Soto-Estrada and Adaskaveg, 2004). In the south-eastern USA, the disease is occasionally observed in late summer on leaves in orchards that have not been sprayed after harvest. Rust is caused by two species of the biotrophic genus *Tranzschelia*, *T. discolor* (Fuckel) Tranzschel & M.A. Litv. and *T. pruni-spinosae* (Pers.) Dietel, which differ in the morphology of their two-celled teliospores. The isodiametric cells of the teliospores of *T. pruni-spinosae* have uniform, coarsely verrucose walls and are uniformly pigmented. The two cells of the teliospores of *T. discolor* are dissimilar: the apical cell is globose and coarsely verrucose, whereas the basal cell is oblong, lighter in colour and less verrucose. The urediniospores of the two species are morphologically similar. They are rusty-brown, single celled and heavily echinulate, except at the apex, and have two opposing germ pores beneath the apical cap. *T. discolor* is found worldwide, whereas *T. pruni-spinosae* is found mainly in Europe and central and eastern USA (Dunegan, 1938). Different *formae speciales* of *T. discolor* have been described on different *Prunus* spp. (i.e. almond, peach and prune) with strains on peach being designated *T. discolor* f. sp. *persicae* (Bolkan *et al.*, 1985). Although cross-infection of *formae speciales* among host species occurs, virulence is generally reduced.

Tranzschelia spp. are macrocyclic, heteroecious rusts that alternate between species of *Prunus* and genera in the Ranunculaceae (Arthur, 1934). In mild climates, the fungus can survive on *Prunus* spp. without infecting the alternate host (Soto-Estrada and Adaskaveg, 2004). The pathogen can overwinter as teliospores or as mycelium in green shoot infections (Fig. 12.10). The uredinial stage is capable of repeated infections of *Prunus* spp., and windborne urediniospores are disseminated over long distances. Leaf infections can cause defoliation, and fruit infections can cause direct crop loss with serious economic impact.

On peach, in addition to leaves and fruit, the fungus can infect stem tissues, which are important sources of urediniospore primary inoculum in the spring (Ogawa and English, 1991; Soto-Estrada and Adaskaveg, 2004). Urediniospores can germinate within 4 h of wetness, and appressoria form over the stomata 18 h after inoculation. Appressoria do not form over positive replicas of stomata and thus, the urediniospores are non-thigmotropic but may be chemotropically attracted to stomata (Soto-Estrada *et al.*, 2005). Leaf infections develop as angular, yellow lesions (Fig. 12.11) with rusty-brown urediniospore-producing pustules (uredinia) on the lower surfaces (Fig. 12.12) (Goldsworthy and Smith, 1931; Soto-Estrada *et al.*, 2005). Late in the growing season, uredinia develop into telia, which produce dark brown to black teliospores. Heavy leaf infection can result in premature tree defoliation during the autumn and stimulate flowering, which may reduce tree vigour

Fig. 12.10. Sporulating stem lesion of peach rust caused by *Tranzschelia discolor*.

Fig. 12.11. Symptoms of peach rust caused by *Tranzschelia discolor* on peach leaves.

Fig. 12.12. Leaf lesions with uredinia of peach rust caused by *Tranzschelia discolor*.

or productivity in the subsequent season. Symptoms on immature fruit are green, circular lesions, 2–3 mm in diameter. On mature fruit, the lesions are sunken with yellow halos, and the mesocarp below the lesion is discoloured (Fig. 12.13). Current-year peach stems are infected during outbreaks of the disease in autumn, and the fungus survives as mycelium in symptomless stems during the winter. In early spring, infections are first visible as water-soaked lesions. The epidermis then becomes raised and ruptures with the development of uredinia and urediniospores, which function as primary inoculum for leaf infections (Soto-Estrada *et al.*, 2005). Later, as the stems grow in circumference, the lesions split open lengthwise along the stems. The lesions will not continue to produce spores in the following season because infections are then delimited by a wound periderm and sloughed off during secondary growth of the stem (Soto-Estrada *et al.*, 2005). Stem lesions are superficial and are not considered directly damaging to the tree.

Fig. 12.13. Symptoms of peach rust caused by *Tranzschelia discolor* on peach fruit.

On peach, both species of *Tranzschelia* function as asexual fungi. The stages of the fungus that occur on the alternate host have not been observed in many locations, and the aecial stage generally is not considered an important inoculum source. As described above, the fungi overwinter as mycelium in twigs, but may also survive as urediniospores on contaminated twigs or as uredinia on non-abscised leaves. Urediniospores germinate over a wide temperature range (8–38°C, with 13–26°C being optimum) and require wetness or a saturated atmosphere. At 20°C, 18 h of wetness are needed for adequate leaf infection (Soto-Estrada *et al.*, 2005). Thus, for the development of epidemics, the presence of viable spores and conducive wetness periods are required.

For the management of peach rust, preventative applications of fungicides are done before rain events during the spring. The emergence of sporulating stem lesions, as observed during monitoring programmes in the spring, has been used as a starting date for fungicide applications (Soto-Estrada *et al.*, 2003). DMI and QoI fungicides are highly effective, followed by a MBC fungicide and wettable sulfur (Soto-Estrada *et al.*, 2003).

12.2.6   Peach leaf curl

Peach leaf curl is a cosmopolitan disease and occurs wherever peach and nectarine are grown. Historically, economic losses have been reported after winter and spring seasons with high rainfall, but the disease causes negligible crop losses in orchards treated with properly timed applications of fungicides. In mild climates, the disease can occur consistently, but its occurrence is more erratic in most other production areas.

Peach leaf curl is caused by *Taphrina deformans* (Berk.) Tul. Symptoms typically occur on newly developing leaves in the spring (Fig. 12.14). Leaves first develop discoloured areas that thicken and then become wrinkled and puckered, causing the leaves to curl. Infected leaves can range in colour from light green to yellow, red and purple (Fig. 12.15). They may be covered with a subtle white layer of sexually produced spore sacs or asci containing ascospores. Infected leaves eventually turn brown and generally abscise. Defoliated trees will form new leaves, but fruit set will be sparse in the current year and the following season. Fruit infections are less common and are characterized by irregular, raised green and then reddish swellings (Fig. 12.16).

Ascospores are forcibly discharged, and they bud to form binucleate bud conidia (Rossi and Languasco, 2007). The fungus overwinters in the asexual, yeast-like stage that contaminates the twigs and buds of the tree. Emerging leaves are infected when the fungus changes from bud conidia to a parasitic mycelial phase capable of penetrating the intact cuticle (Kramer, 1987). The fungus is homothallic, and the binucleate mycelium grows intercellularly, causing the host tissue to become distorted. *Taphrina* spp. produce the phytohormones indole acetic acid and several cytokinins in culture (Kern and Naef-Roth, 1975), and possibly stimulate the host to produce auxins and cytokinins (Sziraki *et al.*, 1975), which presumably cause hypertrophy and hyperplasia of the palisade mesophyll tissue of the leaf. The binucleate mycelium eventually produces subcuticular chlamydospores on the upper leaf surface that become

Fig. 12.14. Peach leaf curl caused by *Taphrina deformans*.

Fig. 12.15. Advanced symptoms of peach leaf curl caused by *Taphrina deformans* with leaf deformation, discoloration, necrosis and a subtle white layer of sexually produced asci containing ascospores.

the ascogenous cells where karyogamy occurs to form a diploid nucleus. The diploid nucleus divides and two cells are formed: the upper ascogenous cell and the lower stalk cell. Meiosis occurs in the upper cell followed by mitosis, resulting in eight ascospores, whereas the nucleus in the stalk cell degenerates. The naked asci are produced on the leaf surface and the cycle is repeated. Infections occur at temperatures of 10–21°C, but ascospores and bud conidia can survive hot, dry conditions for several months. Periods of cool, wet weather during early bud development favour leaf curl disease. When temperatures at early leaf development are high, infections rarely become established.

Peach leaf curl can be managed with one well-timed preventative fungicide application, either in the late autumn after 90% of the leaves have fallen or in the spring before bud swell. Copper Bordeaux mixtures, fixed copper products, ziram and chlorothalonil have high efficacy against the disease (Adaskaveg *et al.*, 2015). Treatments after infection or symptom development are ineffective. Sanitation and cultural practices do not provide control against this disease. Most peach and nectarine cultivars are susceptible to the disease, but there is a wide range of susceptibilities that is heritable, with peaches

Fig. 12.16. Symptoms of peach leaf curl caused by *Taphrina deformans* on developing peach fruit.

having the cultivar 'Redhaven' in their lineage being less susceptible (Ritchie and Werner, 1981). Vigour of diseased trees should be maintained by irrigation and nitrogen fertilization management, as well as reducing stress from crop load by extra thinning.

12.2.7 Powdery mildew

Powdery mildew of peach and nectarine occurs worldwide but is most damaging in semi-arid growing areas. The disease can be caused by several different species of fungi that commonly occur on rosaceous plants (Yarwood, 1939). Three species have been reported on peach, with *Podosphaera pannosa* (Wallr.:Fr.) Braun & Takamatsu (formerly *Sphaerotheca pannosa* (Wallr.:Fr.) Lév.) being the most important. *Podosphaera leucotricha* (Ellis & Everh.) E.S. Salmon is less common, and *Podosphaera clandestina* (Wallr.:Fr.) Lév. has been reported on peach seedlings in the eastern USA. Fruit infections caused by *P. pannosa* and *P. leucotricha* cause the most economic damage, but leaf infections are important sources of inoculum. In nurseries, powdery mildew leaf infections can cause significant damage to seedlings and small trees.

The susceptibility of peach and other stone-fruit crops varies greatly among cultivars. The eglandular (without glands at the leaf base) peach cultivars (e.g. 'Peak', 'Paloro') are more susceptible than the glandular ones (e.g. 'Johnson', 'Halford', 'Stuart') (Ogawa and Charles, 1956). Furthermore, in some cultivars, tissues also vary in their susceptibility, with fruit being more or less susceptible than leaves, depending on the mildew species involved and the maturity of the host tissue. Leaves, buds, green shoots and fruit are commonly attacked, but flower infections are rare. Powdery mildew resistance has been genetically linked to a monogenic dominant locus named *Vr2* in a linkage group in *Gr* gene (G8) for leaf colour (Pascal *et al.*, 2010). Research continues mapping and identifying candidate genes similar to the *Vr3* gene identified in interspecific almond and peach hybrids for resistance to powdery mildew (Donoso *et al.*, 2016; Marimon *et al.*, 2020b). On twigs, mildews can overwinter as white, dense, felt-like mycelium (Fig. 12.17). The first symptoms on peach leaves and green shoots by *P. pannosa* are small, circular, white, web-like colonies that become powdery once masses of asexual conidia are produced in chains (Fig. 12.18). The leaves may then curl or become stunted. Older infections commonly result in leaf chlorosis and necrosis. Severe infections may result in defoliation.

The fruit are susceptible from the early stages of development until pit hardening on peach, nectarine and plum but not other *Prunus* spp. (Weinhold, 1961). White circular spots may enlarge, coalesce and cover large areas of the fruit (Figs 12.19 and 12.20). Infections usually result in some deformation of

Fig. 12.17. Peach twig with overwintering mycelium and embedded cleistothecia of powdery mildew caused by *Podosphaera pannosa*.

Fig. 12.18. Powdery mildew caused by *Podosphaera pannosa* on peach leaves.

Fig. 12.19. Powdery mildew caused by *Podosphaera pannosa* on developing peach fruit.

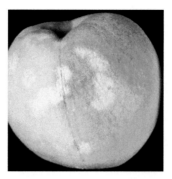

Fig. 12.20. Powdery mildew caused by *Podosphaera pannosa* on mature peach fruit.

Fig. 12.21. Powdery mildew caused by *Podosphaera pannosa* on developing nectarine fruit.

the fruit surface with depressed or slightly raised areas (Fig. 12.21). Infections on peach fruit become necrotic after pit hardening, whereas on nectarine and occasionally also on peach, the tissue remains green. Any fruit with blemishes caused by powdery mildew are generally unmarketable.

Based on indirect evidence, *P. leucotricha* (mainly an apple pathogen) is presumably involved in causing another powdery mildew symptom on peach fruit known as 'rusty spot' (Daines and Trout, 1977; Ries and Royse, 1978). With this disease, small, circular, rusty orange lesions develop on the fruit, which enlarge and may cover the entire fruit. No symptoms occur on leaves and stems. Lesion development has been related to rapid fruit growth. The incidence of disease increases from the shuck-fall stage of fruit development until 60 days after full bloom, and epidemics typically last 17–30 days (Furman *et al.*, 2003a).

P. pannosa overwinters as mycelium in shoots and in the inner bud scales (Weinhold, 1961). In milder climates, young twigs may be covered with dense mats of mycelium. During a winter in California, we found for the first time the sexual fruiting bodies (chasmothecia, formerly referred to as erysiphaceous cleistothecia or perithecia) of the fungus on peach embedded in these mycelial mats (Fig. 12.22). In the spring, newly developing leaves become diseased as they emerge from infected buds. When chasmothecia are present, ascospores are released that also serve as primary inoculum. Because roses are an important host for the pathogen where the disease is not always managed, diseased roses can be major contributors to the development of epidemics of peach powdery mildew. Secondary infections by the wind-disseminated, asexual conidia occur throughout the growing season. Eight ovate to ellipsoid ascospores are produced in one globose to subglobose ascus in a chasmothecium (Fig. 12.22). Conidia are formed apically in chains of septate conidiophores and are hyaline, ellipsoid and have fibrose granules. Conidia germinate at 2–37°C, with an optimum of 21°C. Conidia can germinate in free water and at relative humidities of 43–100%. Excessive durations of wetness, however, will kill conidia of powdery mildew fungi. During periods with warm, humid conditions, the disease can quickly develop into an epidemic. A predictive model for disease progress has been described (Marimon *et al.*, 2020a), which included a threshold based on accumulated degree-days to initiate fungicide programmes at early infection set. Comparative studies between the predictive model and calendar-based treatments resulted in a 33% reduction in the number of fungicide applications (Marimon *et al.*, 2020a).

Management of powdery mildew is by cultural practices and using protective fungicide treatments. Less susceptible cultivars should be planted in areas that commonly have a high incidence of disease. To reduce the relative humidity in the orchard, the frequency of irrigation periods should be minimized, and low-angle sprinklers should be used to keep foliage dry. Fungicide applications are done from full bloom until the pit-hardening stage of fruit development for peach and nectarine. Adequate management of rusty spot was achieved with three to five fungicide applications (Furman *et al.*, 2003b). MBCs, DMIs, succinate dehydrogenase inhibitors and QoI fungicides

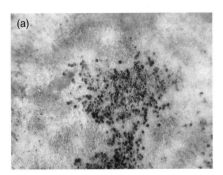
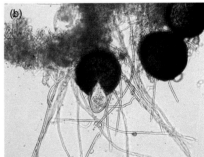

Fig. 12.22. (a) Close-up of chasmothecia of powdery mildew caused by *Podosphaera pannosa* on peach twig. (b) Micrograph of chasmothecium with hyphal-like appendices and a single ascus containing eight ascospores.

are highly effective as preventative, single-site mode-of-action treatments, whereas wettable sulfur, polyoxin D, and bicarbonate-based products can be used as preventative or to reduce inoculum as post-infection treatments. Resistance has been reported to MBCs, DMIs and QoI fungicides. Management must follow rotation practices between different modes of action to prevent the selection of resistant subpopulations of the pathogens.

12.3   TRUNK AND SCAFFOLD DISEASES

12.3.1   Silver leaf disease

Silver leaf disease of peach is caused by the Basidiomycota fungus *Chondrostereum purpureum* (Pers.:Fr.) Pouz. The disease also affects a wide range of other cultivated and non-cultivated hardwood tree species, particularly willow and poplar. Silver leaf disease has been reported from most temperate-zone stonefruit production areas and can also be a problem in nurseries. The fungus can also grow saprophytically on tree logs and prunings.

Leaves of silver leaf-diseased trees become silvery in appearance (Fig. 12.23), which is best noticeable on new growth in the spring when damage by other

Fig. 12.23. Symptoms of silver leaf disease caused by *Chondrostereum purpureum* on peach leaves. Left: diseased leaf. Right: healthy leaf.

diseases and pests does not obscure the symptoms. These symptoms occur soon after the pathogen invades the woody tissues. Silver leaf symptoms result from a toxin produced by the pathogen that causes the upper epidermis to separate from the palisade and mesophyll layers of the leaf. The separated layers then reflect light differently from healthy tissue, giving leaves a silvery or dull metallic appearance. Infected leaves may then become necrotic and abscise. The wood of infected trees becomes discoloured and subsequently a white rot develops. Wood discoloration in branches is often angular to pie-shaped in cross-section, whereas in the trunk, the discoloration occurs in the older secondary xylem (Fig. 12.24). In white rot, the decayed wood becomes mottled to bleached white and eventually spongy soft. Substantial decay of the trunk may occur and extend into the scaffold branches and roots. Leathery fruiting bodies of the fungus may develop in bracket-like clusters on tree trunks and scaffold branches of living trees and on dead wood. They produce sexual basidiospores, which are the only known propagules of the fungus. They are wind disseminated and function as an inoculum in disseminating the organism. On peach, fruiting bodies are not easily found, although they are thought to persist for up to 2 years, producing spores during warm, moist environments at any time of the year.

Like most wood-decaying fungi, *C. purpureum* needs a fresh wound that exposes the wood of the branch or trunk. Wind-disseminated spores of the

Fig. 12.24. Cross-section through a peach trunk infected with *Chondrostereum purpureum*.

fungus are deposited on the moist wood, germinate and quickly invade the xylem tissues of the wood. The fungus grows over a wide temperature range with an optimum of 25°C. Wood-exposing wounds are most susceptible in the first week after injury. Wound healing, which prevents infection by the fungus, progresses at different rates in different climates and at different times of the year.

A wide range of perennial hosts, inoculum production over a long period, the impossibility of protecting all wounded surfaces and the inability to eradicate established infections from tree trunks make silver leaf a difficult disease to control. Management practices should therefore focus on starting with clean nursery stock and preventing establishment of the fungus by minimizing large wood-exposing wounds and using proper pruning practices that promote wound healing. Trees should be maintained in a vigorous growing condition. Sanitation measures to reduce inoculum include removing and burning or burying infected wood and sanitizing pruning equipment before each cut. Chemical protection of pruning wounds is possible (Wicks *et al.*, 1983), but no fungicides are registered. The most effective pruning-wound treatments are biocontrol agents such as *Trichoderma* spp., which act by pre-colonizing the wood and thus excluding the pathogen from the site, and by mycoparasitism.

12.3.2 Perennial canker (*Leucostoma* or *Cytospora* canker)

Perennial canker, also known as *Leucostoma* or *Cytospora* canker, is an important disease of stone-fruit crops including peach and nectarine, especially in colder climates. It is also associated with the peach tree short life syndrome in the south-eastern USA (see section 12.4.4) and is part of the apoplexy disease complex on stone fruits in Europe. The disease also occurs in the Pacific Northwest of the USA, as well as in Canada, South America and Japan. Perennial canker reduces the number of bearing branches, kills twigs and shortens tree life.

Perennial canker of peach in the USA is caused by *Cytospora cincta* Sacc. (teleomorph *Leucostoma cincta* (Sacc.) Höhn.), *Cytospora leucostoma* (synonym: *Leucostoma personii* (Nitschke) Höhn) and *Leucocytospora paraleucostoma* (synonym: *L. parapersonii* Adams, Surve-Iyer and Iezzoni) (Adams *et al.*, 2002). More recently, *Cytospora plurivora* D.P. Lawr. L.A. Holland & Trouillas, sp. nov. and *Cytospora sorbicola* Norphanphoun, Bulgakov, T.C. Wen & K.D. Hyde were described as causal agents of perennial canker of peach (Norphanphoun *et al.*, 2017; Lawrence *et al.*, 2017).

These fungi produce a compound fruiting body that consists of a central pycnidium that produces hyaline asexual conidia. These conidia ooze from the pinhead-sized, black pycnidia to form long orangish tendrils (cirri). Perithecia develop in or under older stroma around the pycnidium. In the perithecia, sexual spores (ascospores) are formed.

The pathogens infect small and large branches of trees (Fig. 12.25), resulting in dieback that continues from apical branches to the main scaffold branches and trunk. On small branches, lesions appear as sunken, zonate, discoloured areas that develop around dead buds or previous years' leaf scars and are observed 2–4 weeks after bud break (Fig. 12.26). Amber gum may ooze from these lesions as they age and darken, unless the branch dies. On large branches, scaffolds and trunks, conspicuous elliptical cankers develop. Copious amounts of gum may exude from the cankers and, with age, the bark dries and cracks form, exposing the blackened tissue beneath (Gauthier *et al.*, 2021). If infections follow winter injury or develop on branches weakened from infections by other pathogens, gumming will not be associated with the cankers. Cankers typically develop concentric rings from alternation between canker extension and callus formation during dormant and growing seasons, respectively. Pycnidia erupt directly through the bark within the branch or trunk canker and give the surface of the bark a pimpled appearance (Fig. 12.27). Thus, pycnidia are not associated with bark lenticels. Discoloured wood, wilting, and chlorotic and dehiscent leaves are other symptoms associated with perennial canker. The disease is sometimes confused with bacterial

Fig. 12.25. Peach tree with *Leucostoma* (*Cytospora*) cankers on scaffold branches. Reproduced from Ogawa *et al.* (1995) with permission from the American Phytopathology Society Press.

Fig. 12.26. *Cytospora* cankers caused by *Leucostoma* (*Cytospora*) *cincta* around the nodes of a peach shoot. Reproduced from Ogawa *et al.* (1995) with permission from the American Phytopathology Society Press.

Fig. 12.27. Pycnidia of *Leucostoma* (*Cytospora*) *cincta* on a cherry twig. Spore masses are exuded by some pycnidia on the upper part of the twig.

canker in the early stages of development. Bacterial canker, however, is irregular in shape and has a non-zonate margin (Ogawa and English, 1991).

Conidia are the primary inoculum for the disease. They are most abundantly produced under the cool, moist conditions of late autumn and early spring but are present throughout the year if rainfall is sufficient. Conidia are dispersed by rain, wind, and possibly birds and wood-boring beetles. Germination requires the presence of free water or 100% relative humidity and a carbon source. Most infections occur on wood injured by sunburn, pruning, insects or rodents. Trees stressed by freezing, nutrient deficiency, and infections by ring nematodes and bacterial canker are predisposed to the disease. Young vigorous trees are less susceptible. On older weakened trees, many new infections may occur at the nodes of 1-year-old shoots.

Because perennial canker develops on weakened trees, requires injuries for infections, and usually follows winter injury in cold climates and sunburn in warm climates, management of perennial canker requires an integrated approach of cultural and pest management practices (Gauthier *et al.*, 2021). In both cold and warm climates, the disease can be kept to a minimum with good cultural practices that ensure tree vigour and hardiness. Painting trunks white with 100% acrylic latex paint prevents sunburn ('south-west') injuries and helps reduce perennial canker. Trees should not be planted in heavy clay or shallow soils where moisture and nutrient stress may occur. Biocontrols and fungicides have not been effective, but spray oils have helped against cold injury and perennial canker in some regions. The cankers may be excised when they are small (less than half the branch width) to prevent or slow the decline and ultimate death of the tree. The application of pruning cut sealants containing fungicides has also been shown to be an effective method to manage the disease (Miller *et al.*, 2019, 2021). In summary, management of perennial canker is based on preventative measures that minimize improper pruning cuts (Biggs, 1989), winter injury, sunburn and insect damage, and promote optimum tree health, facilitate rapid wound healing (Ogawa and English, 1991) and protect wounds from infection (Miller *et al.*, 2019, 2021).

12.3.3 Fungal gummosis

Fungal gummosis was first described in the 1970s almost concurrently in the south-eastern USA (Weaver, 1974; Biggs and Britton, 1988) and in Japan but has since also been found in Australia and China (Wang *et al.*, 2011). The disease has been reported to be caused by three species of *Botryosphaeria*. *Botryosphaeria dothidea* (Moug.: Fr.) Ces. & De Not is considered the primary pathogen because the fungus is able to invade the peach host directly through the bark lenticels (Fig. 12.28a). Two other species, *Botryosphaeria obtusa* (Schwein.) Shoemaker and *Lasiodiplodia theobromae* (Pat.) Griffon & Maubl. (syn. *B. rhodina* (Cooke) Arx), have also been associated with gummosis, but these are primarily wound invaders and are not known to cause infections through bark lenticels (Pusey *et al.*, 1995). In general, *Botryosphaeria* spp. have a broad host range.

Fig. 12.28. Fungal gummosis caused by *Botryosphaeria dothidea*. (a) Blisters develop around lenticels on young bark, which develop in the second or third growing season of the tree. (b) Gumming blisters and necrotic tissue under lenticels, which develop under the bark.

The disease is mainly a problem in orchards that are poorly managed or have water-stressed trees. On tree trunks and large branches, gummosis is characterized by numerous sunken necrotic lesions around lenticels (Fig. 12.28b). Lesions are 5–15 mm in diameter and produce copious amounts of gum. The lesions may coalesce and form large cankers (Fig. 12.29). Severe damage to peach orchards can lead to crop reductions of 25% (Ezra *et al.*, 2017) and up to 40% (Beckman *et al.*, 2011). On young branches, blisters 1–6 mm in diameter develop around lenticels, but there is no gumming. Fungal gummosis significantly depresses tree growth and fruit yield on susceptible peach cultivars (Beckman *et al.*, 2003), and the trees may ultimately die. Lenticel infections by *B. dothidea* occur primarily in the summer months, whereas wound infections by all three species may occur at other times in the growing season during wet weather and when inoculum is available. *B. dothidea* overwinters in diseased bark and woody tissues. Species of *Botryosphaeria* may produce conidia in pycnidia and ascospores in ascostroma that form in branch and trunk cankers. Abundant conidia may be produced that are disseminated by rain splashing.

For management of the disease, the health and vigour of trees should be maintained. Cultural practices include selecting sites to avoid low-lying areas where trees may be prone to cold injury, ensuring adequate air movement in the orchard, minimizing cultivation injuries and preventing irrigation systems from wetting the trunks. Dead trees and infected wood should be removed or destroyed (e.g. by burning, shredding or flail-mowing) to reduce inoculum

Fig. 12.29. Fungal gummosis caused by *Botryosphaeria dothidea*. Lesions may coalesce and form extensive cankers with excessive gumming on lower peach tree trunks.

sources. Winter pruning is encouraged to avoid periods when inoculum is present, but pruning should not be done immediately after rainfall or on trees under water or nutrient stress. Pruning tools should always be disinfested with an approved sanitizer such as rubbing alcohol, quaternary ammonia or sodium hypochlorite. During the summer, inoculum is commonly present, and pruning at this time may result in rapid colonization of wounds by any of the three species of *Botryosphaeria*. Fungicide applications may prevent infections but may not be economically cost-effective in many areas.

Despite cultural practices and activity of some fungicides, management of gummosis remains challenging. There are currently no commercial cultivars available that are resistant to this disease; however, breeding programmes have identified genotypes with resistance to fungal gummosis in interspecific crosses and segregation backcross populations generated using Kansu peach (*Prunus kansuensis* Rehder), almond (*Prunus dulcis* (Mill.) D.A. Webb) and peach (*Prunus persica* (L.) Batsch). Multi-year field trials of hybrids identified a major locus for resistance to *B. dothidea* in *Prunus* spp. that may be useful for marker-assisted breeding programmes (Mancero-Castillo *et al.*, 2018).

12.3.4 Constriction canker and other twig blights

Symptoms of constriction canker, also called *Fusicoccum* canker, caused by *Diaporthe amygdali* (Delacr.) Udayanga, Crous & K.D. Hyde (formerly *Phomopsis amygdali* (Del.) Tuset and Portilla; *Fusicoccum amygdali* Delacr.), appear on leaves but are more damaging on twigs. Lesions on leaves are large, brown, zonate spots that have black asexual pycnidia in the centre. In the spring and summer, cankers develop on twigs around infected buds or nodes of 1-year-old shoots (Fig. 12.30). This is in contrast to cankers caused by brown rot blossom blight that develop from infected blossoms. In addition, lesions of constriction canker exude less gum than those caused by brown rot. As the cankers enlarge, they girdle the twig, causing blighting of the shoots due to the production of a toxin (e.g. fusicoccin) by the pathogen (Fig. 12.31). The fungus may produce numerous conidia in cirri or long tendrils from pycnidia embedded in the twigs (Fig. 12.32). In the autumn, infections occur through leaf scars and during the growing season through buds, scars on bud scales, stipules and fruits, or directly through young shoots. Constriction canker can be managed with the removal of inoculum sources by pruning out infected twigs, but this may not be economically feasible, and the results may be inconsistent (Lalancette and Robison, 2002). Differences in host susceptibility have been found among peach and nectarine cultivars, and planting of these cultivars is an important consideration in establishing an orchard in areas where the disease is a problem. Additionally, fungicides can be applied before bud break and in the autumn (Lalancette and Robison, 2002).

Twig blights can also occur on peach with symptoms ranging from cankers to dieback. *B. obtusa*, *D. amygdali* (syn. *P. amygdali*), *Leucostoma persoonii* (Nitschke) Höhnel and *Cytospora* spp. have been identified as causal agents using colony morphology, conidia size and shape, and ribosomal DNA sequences (Froelich

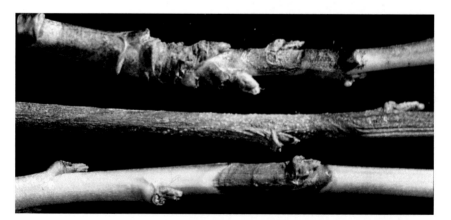

Fig. 12.30. Constriction canker caused by *Diaporthe amygdali*. One-year-old peach twigs with infections around buds. The lower twig shows the zonations around the infection site.

Fig. 12.31. Constriction canker caused by *Diaporthe amygdali*. Flagging and withering of blighted twigs distal to twig cankers.

Fig. 12.32. Cirri (tendrils) exuding pycnidia of *Diaporthe amygdali*. Each cirrus is composed of abundant conidia of the pathogen.

and Schnabel, 2019). Cultivar susceptibility ranges from susceptible to more resistant for the different pathogens. Tolerance to wood-invading pathogens causing twig blights is not necessarily genetically linked to cultivar resistance to bacterial spot or brown rot. *L. persoonii* currently is the most frequently isolated pathogen causing twig cankers followed by *D. amygdali* and *B. obtusa* (Froelich and Schnabel, 2019). All isolates collected of the three species have been found to be sensitive to thiophanate-methyl, pyraclostrobin, and azoxystrobin, whereas all were resistant to boscalid and fluopyram. *D. amygdali* and *B. obtusa* isolates were sensitive to difenoconazole and propiconazole, whereas *L. persoonii* isolates were moderately resistant to these DMI fungicides (Froelich and Schnabel, 2019).

12.4 ROOT AND CROWN DISEASES

12.4.1 Phytophthora root and crown rots

Phytophthora root and crown rots are destructive soil- and water-borne diseases of stone fruits capable of causing enormous economic losses on crops in nurseries and orchards worldwide, as well as environmental damage in natural ecosystems. Peach trees can be affected at all ages, and infections often result in tree death. *Phytophthora* is a genus in the phylum Oomycota, a fungus-like lineage of microscopic eukaryotes in the Kingdom Chromista or Stramenopila. The identification of diseases caused by *Phytophthora* spp. has improved due to better detection methods including the development of selective media, baiting methods and molecular diagnoses. Phytophthora disease incidence has increased because orchards are established in less-than-optimal locations that have poor soil drainage or use improper irrigation practices. Phytophthora root and crown rots can be caused by a number of species in the genus *Phytophthora* (Wilcox and Ellis, 1989; Thomidis, 2003). Species include *P. cactorum* (Lebert & Cohn) J. Schröt., *P. cambivora* (Petri) Buisman, *P. cinnamomi* Rands, *P. citricola* Sawada, *P. citrophthora* (R.E. Sm. & E.H. Sm.) Leonian, *P. cryptogea* Pethybr. & Lafferty, *P. drechsleri* Tucker, *P. megasperma* Drechs., *P. parasitica* Dastur (syn. *P. nicotianae* Breda de Haan) and *P. syringae* (Kleb.) Kleb. More recently, *P. niederhauseri* Z.G. Abad & J.A. Abad was found to attack peach and almond–peach hybrid rootstocks (Browne *et al.*, 2015). In a particular location, several of these species may be present endemically; other species may be introduced by infested irrigation water, soil or plant material. When water is present, all species commonly produce sporangia, which release numerous asexual zoospores, and they may also form resistant chlamydospores and sexual oospores. Ecologically, the different species differ in their geographical distribution, characteristic temperature optima, ability to cause root and/or crown rots, seasonal occurrence and virulence.

Trees infected by *Phytophthora* spp. show poor terminal growth and the leaves are small, chlorotic and sparse (Fig. 12.33). The fruit may be undersized, highly coloured and sunburned. The trees either show dieback and decline progressively over several seasons, or die suddenly in late spring or summer following years with excessive wet weather. These symptoms are often not very characteristic and may be confused with other diseases and disorders. Mild symptoms of root and crown rot are not always noticeable, but the yield will be reduced. Crown rots develop at the root crown and/or base of the trunk (Fig. 12.34). The bark phellogen is killed, and discoloration may extend into the outer layers of the xylem tissues. As the decay expands, a canker develops, often accompanied by the production of copious amounts of amber to brown gum as a host response. Cankers have distinct borders between the diseased and healthy tissues. If the canker encircles the trunk, the tree is girdled and dies. If the lesion ceases expansion or the pathogen dies, callus tissue will grow into the dead areas of the bark as a host response to heal the wound. Cankers

Fig. 12.33. Dieback of peach trees affected by Phytophthora root and crown rot. Reproduced from Strand (1999) with permission.

are usually limited to the crown tissues but may extend up the trunk and occasionally into scaffold branches. Aerial cankers develop when inoculum is disseminated by rain or sprinkler water to the upper parts of the tree, especially the scaffold crotches where water tends to accumulate (Fig. 12.35). Infected roots exhibit decay, sometimes extending to the crown. On feeder roots, portions of the outer cortical tissue disintegrate, leaving only the white, hair-like stele protruding from the decaying outer tissue.

Phytophthora spp. are mainly soil- or water-borne organisms (Fig 12.36). They survive as chlamydospores, hyphae or oospores in root debris in the soil.

Fig. 12.34. Phytophthora crown rot on a peach tree. Reproduced from Strand (1999) with permission.

In the presence of water, the fungi produce abundant sporangia, which release numerous zoospores. Zoospores are motile and are attracted by root exudates to host plant roots where they encyst and infect mainly the fine feeder roots. After each rainfall or irrigation, new generations of sporangia are produced that will restart the disease cycle. Zoospores are considered the main infective propagules, but other structures (i.e. chlamydospores and oospores) may also cause infections. Although *Phytophthora* spp. are pathogenic on their own, in many cases, Phytophthora root rots are increased by root injuries caused by nematodes or other organisms.

Fig. 12.35. A peach tree with an aerial infection of a *Phytophthora* spp. Reproduced from Strand (1999) with permission.

Phytophthora root and crown rots are managed by the use of resistant root-stocks, careful soil water management and fungicides. A considerable degree of resistance among stone-fruit rootstocks is available, but none is immune. In a study, 'Hansen 536' was more susceptible to root and/or crown rot caused by *P. cactorum*, *P. citricola* and *P. megasperma* than 'Lovell', 'Nemaguard', 'Atlas', 'Viking', 'Citation' and 'Marianna 2624' (Browne, 2017). Most plum hybrids were highly and consistently resistant to crown rot caused by *P. niederhauseri*, but only 'Marianna 2624' was highly resistant to both crown and root rot caused by all species evaluated. Some of these rootstocks were also evaluated in a study in Chile, and 'Mr. S 2/5' (*P. cerasifera* × *P. spinosa*) was found to be the most resistant to root rot caused by *P. cryptogea* (Guzman *et al.*, 2007).

Because the zoospores are dependent on water for movement, the disease is greatly affected by soil water management. Ideally, orchards should be es-tablished on well-drained soils that are never water saturated for more than 24 h. In soils with poor drainage, trees should be planted on ridges or raised beds. Flood or furrow irrigation increases the incidence of the disease, whereas irrigating with mini-sprinklers and allowing the surface soil to dry between irrigations will reduce root rot. In addition, sprinklers should never be directed towards the tree trunks and crotches. Control of root and crown rots may also be achieved with the use of systemic fungicides, which are applied to the soil,

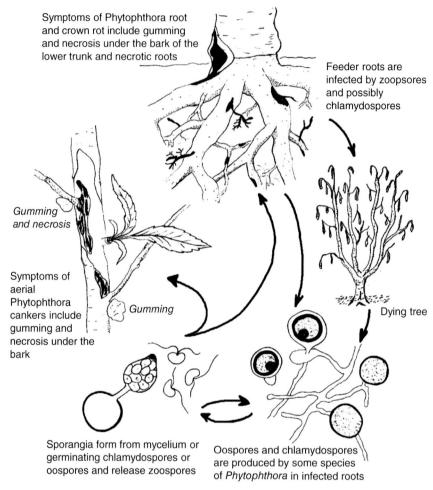

Symptoms of Phytophthora root
and crown rot include gumming
and necrosis under the bark of the
lower trunk and necrotic roots

Feeder roots are
infected by zoopsores
and possibly
chlamydospores

*Gumming
and necrosis*

Symptoms of
aerial
Phytophthora
cankers include
gumming and
necrosis under the
bark

Gumming

Dying tree

Sporangia form from mycelium or
germinating chlamydospores or
oospores and release zoospores

Oospores and chlamydospores
are produced by some species
of *Phytophthora* in infected roots

Fig. 12.36. Disease cycle of Phytophthora root and crown rot of peach. *Phytophthora* spp. are mainly soil-borne organisms that survive as chlamydospores, mycelium (hyphae) or oospores in root debris in the soil. At high soil moisture levels, the fungi produce abundant sporangia with numerous motile zoospores, which are attracted by exudates of host plant roots. Zoospores encyst on root surfaces and infect the fine feeder roots. Tree crowns that are subjected to prolonged wetness from flood irrigation or heavy rain often develop crown rot. After each rainfall or irrigation, new generations of sporangia are produced that will restart the disease cycle. Zoospores are considered the main infective propagules, but other structures (i.e. chlamydospores, oospores) may also cause infections. Propagules of *Phytophthora* spp. are carried to the aerial portions of trees by dust, soil particles and splashing water where they can cause aerial Phytophthora cankers. Trees infected by *Phytophthora* spp. show decreased growth and may ultimately die.

injected, or sprayed or painted on gummosis lesions. In problematic areas, fungicides together with nematicides or a soil fumigant should be used to protect replants during the first 2 years of growth.

12.4.2 Armillaria root rot

Armillaria root rot (also known as shoestring root rot or oak root rot) is an important disease of peach and other stone fruits worldwide, but losses have been highest in production areas in North America. Many native, ornamental and agricultural woody plants are affected by Armillaria root rot. Peach is considered extremely susceptible to the disease (Rhoads, 1950), and no tree stress-reducing management options have been demonstrated to prevent infection or slow disease progression. Trees can be killed at all ages. Severe outbreaks are often observed on land planted successively with peach where inoculum has been allowed to build up gradually (Savage *et al.*, 1953), or where forest trees have recently been removed. In contrast to Phytophthora root rot, Armillaria root rot is also common on well-drained soils.

Early symptoms of Armillaria root rot include poor terminal growth and occasionally undersized, curled leaves on all major limbs (Fig. 12.37). Infected peach trees may collapse suddenly during the summer months with the majority of leaves still attached (Fig. 12.38). In older orchards with widely spaced trees, the disease may expand in a circular fashion, but in high-density orchards, the disease often progresses along tree rows. White mycelial fans that develop in the cambium between the bark and wood on the crown are highly diagnostic signs of the disease (Fig. 12.39). In advanced stages of colonization, white rot wood decay occurs, and the wood is soft, spongy and bleached whitish in colour (Morrison *et al.*, 1991). The decay is a result of the degradation of cellulose, hemicellulose and lignin. The roots may have black, shoestring-like mycelial strands called rhizomorphs attached to the surface (Fig. 12.40). *Desarmillaria tabescens*, however, does not form such rhizomorphs under field conditions (although they have been induced in the laboratory). Typically, in late summer to early autumn, clusters of brown mushrooms develop at the base of infected trees (Fig. 12.41).

Until the late 1970s, *Armillaria mellea* (Vahl:Fr.) Kummer was considered the only causal organism of Armillaria root rot. Based on compatibility assays, it was subdivided into ten biological species (Anderson and Ullrich, 1979), which have since been shown to be well-defined morphologically and genetically distinct species. Three species of *Armillaria* are currently associated with peach. In North America, *A. mellea* sensu stricto (Vahl:Fr.) P. Kumm., *A. ostoyae* (Romagn.) Herink and *D. tabescens* (Scop.) R. A. Koch & Aime comb. nov. (syn. *Armillaria tabescens* and *Clitocybe tabescens*; Koch *et al.*, 2017) have been found in California, Michigan and in the south-eastern USA, respectively. *D. tabescens* was also identified on peach in Japan (Fujii and Hatamoto, 1974).

Fig. 12.37. Leaf symptoms on a peach tree affected by Armillaria root rot.

Fig. 12.38. Peach orchard with an *Armillaria* infection centre, surrounded by healthy-looking trees that are flowering.

Fig. 12.39. Mycelial fans of *Armillaria* sp. under the bark of an infected tree.

Fig. 12.40. A rhizomorph of *Armillaria mellea* (top) and a healthy root (bottom). Reproduced from Strand (1999) with permission.

Fig. 12.41. Basidiomes of *Armillaria mellea* at the base of an infected peach tree.

A. mellea and *A. ostoyae* are the predominant species in Europe. The species can easily be differentiated when fruiting bodies are available. In their absence, labour-intensive compatibility tests (Korhonen, 1978; Anderson and Ullrich, 1979) and, more recently, DNA diagnostics are being used for species identification (Harrington and Wingfield, 1995; White *et al.*, 1998; Sierra *et al.*, 1999; Schnabel *et al.*, 2005).

The disease cycle for Armillaria root rot is shown in Fig. 12.42. *Armillaria* spp. persist in the soil for many years as mycelium in infected plant tissues or as rhizomorphs. The pathogens are most successful in invading healthy tissue if they first become established saprophytically. Thus, very young trees are often killed on replant sites or older trees after they have grown for 3 years or longer. Infections of healthy roots originate from infected root segments in the soil, from root-to-root contact or from contact with rhizomorphs. Spread by root-to-root contact or rhizomorphs was estimated to occur at 0.8–3.2 m year⁻¹ (Kable, 1974). Basidiospores produced on fruiting bodies are not considered major infection propagules in fruit orchards (Rizzo *et al.*, 1998; Termorshuizen, 2000).

Management strategies for *Armillaria* root rot are not very effective. If sites are seriously infected, they are usually abandoned. The most effective strategy is to avoid planting new orchards in infested soil. Options to manage the disease in replant sites are limited and often only marginally effective. Chemical control options with moderate efficacy include pre- and post-plant fumigations to reduce inoculum in the soil (Munnecke *et al.*, 1970; Savage *et al.*, 1974; Adaskaveg *et al.*, 1999). Biological control options have not been effective (Schnabel *et al.*, 2012a). A cultural management option has been developed for peach scheduled to be planted at infested replant

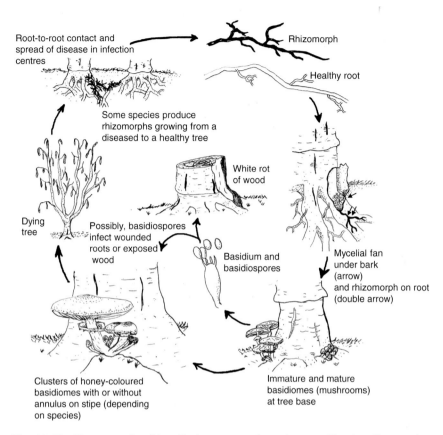

Root-to-root contact and spread of disease in infection centres

Rhizomorph

Healthy root

Some species produce rhizomorphs growing from a diseased to a healthy tree

White rot of wood

Dying tree

Possibly, basidiospores infect wounded roots or exposed wood

Basidium and basidiospores

Mycelial fan under bark (arrow) and rhizomorph on root (double arrow)

Clusters of honey-coloured basidiomes with or without annulus on stipe (depending on species)

Immature and mature basidiomes (mushrooms) at tree base

Fig. 12.42. Disease cycle of Armillaria root rot of peach caused by *Armillaria* and *Desarmillaria* spp. The pathogen persists in the soil as mycelium in infected roots. Some *Armillaria* spp. produce rhizomorphs that are associated with tree roots. New infections of healthy roots originate from infected root segments in the soil, from root-to-root contact, or from contact with rhizomorphs. Basidiospores that are produced on the fruiting bodies are not considered major infection propagules in fruit orchards. Early symptoms of Armillaria root rot include poor tree growth. Infected peach trees may collapse suddenly during summer months with a majority of leaves still attached. White mycelial fans that develop in the cambium between the bark and wood on the crown are highly diagnostic signs of the disease. In advanced stages of colonization, white rot wood decay occurs and the wood is soft, spongy, and bleached whitish in colour. Roots may have black, shoestring-like mycelial strands called rhizomorphs attached to the surface. Typically, in late summer to early fall clusters of brown mushrooms develop at the base of infected trees.

sites. The practice takes advantage of the inability of *D. tabescens* to grow above the soil line in south-eastern USA peach orchards as demonstrated in a proof-of-concept study (Schnabel *et al.*, 2012b). Trees are planted shallow on a raised bed, and the soil is removed from the root crown after 2 years of establishment (Schnabel *et al.*, 2012b; Miller *et al.*, 2020). This practice delayed the onset of Armillaria root rot by about 2 years, thereby substantially reducing tree mortality in two 8-year field trials (Miller *et al.*, 2020). Susceptibility differences in fruit-tree rootstocks to *Armillaria* infection have been reported (Beckman *et al.*, 1998), and the plum–peach hybrid rootstocks 'Sharpe' and 'MP-29' have been identified as having resistance to the disease (Beckman and Pusey, 2001). The latter has been commercialized (Beckman *et al.*, 2012) and is being used by some producers on infested replant sites.

12.4.3   Verticillium wilt

Verticillium wilt of peach and other stone-fruit crops caused by *Verticillium dahliae* Kleb. occurs in many parts of the world and can cause serious economic losses, although the disease has recently been more sporadic. Many plants including tree and annual vegetable crops have been reported to be hosts of this fungal pathogen. Peach and nectarine trees sometimes develop Verticillium wilt when planted in locations where highly susceptible crops such as cotton, tomato or peppers were grown for a number of years. The disease is most serious on young trees, although some peach cultivars may recover with age. Mature trees are most susceptible in cooler climates.

The soil-borne *V. dahliae* produces resistant structures (microsclerotia) that can survive in the soil for many years in the absence of hosts when soil temperatures are between 5°C and 15°C and soil moisture-holding capacity is between 50% and 75%. The conidia are short lived and are considered to have a minor role in infection and dissemination of the fungus from tree to tree. The fungus is found in the highest concentration at a soil depth of 15–30 cm but can be found as deep as 105 cm. The pathogen infects the roots of host plants in the spring and invades the vascular system, moving up through the xylem as mycelium and asexual conidia and interfering with water transport in the xylem. The first symptom of Verticillium wilt is a sudden wilting of leaves on one or more branches during hot weather in the summer (Fig. 12.43). Subsequently, the leaves turn yellow, then brown and curl up. Sometimes, older, lower leaves wither and fall before the younger, upper leaves on the terminal ends of branches. In younger trees, entire tree defoliation may occur, whereas in older trees, only individual branches may be affected. Thus, young trees can be killed, but the primary symptoms on trees that are several years old are poor growth and low productivity. Longitudinally cut branches of diseased trees show brown to black streaks in the sapwood. In cross-sections, a portion of the outer xylem will be stained dark brown to black (Fig. 12.44). In hot weather, the pathogen dies in the upper part of infected trees, and the tree can recover with new growth the following season. The disease cycle, however, can recur each growing season.

Fig. 12.43. A *Verticillium* sp.-infected peach tree with wilting of branches. Reproduced from Strand (1999) with permission.

Fig. 12.44. Cross-section through a branch of a *Verticillium* sp.-infected peach tree with discoloured outer xylem. Reproduced from Strand (1999) with permission.

The most effective management practice for Verticillium wilt is to avoid planting orchards in soils where inoculum of the fungal pathogen has increased on other susceptible crops that have been grown for a number of seasons. The inoculum of the pathogen can be reduced by using chemical fumigation, solarization, flooding fallow fields, growing grass crops for several seasons or any combination of these treatments. Rootstocks resistant to *V. dahliae* are unknown. Minimizing tree stress through maintenance of soil fertility and soil moisture will help trees tolerate the disease and encourage their recovery.

12.4.4 Peach tree short life

Peach tree short life (PTSL) is a disease complex characterized by the sudden wilt and collapse of new growth and death of all aerial portions of the tree (see also section 12.5.1). The disease is part of general replant disorders that refer to problems that diminish tree growth and productivity (Ritchie and Clayton, 1981). PTSL also affects other stone fruits and has been a serious threat to commercial peach growers in the south-eastern USA for more than 100 years (Chandler, 1969). The number of peach trees lost to PTSL in South Carolina roughly tripled when the soil fumigant 1,2-dibromo-3-chloropropane (DBCP) was no longer registered for use. The disease is similar to the bacterial canker complex in California with the exception of cold injury, which kills many trees with PTSL. The bacterial and nematode aspects are similar for the two diseases.

The symptoms defining PTSL syndrome were established in the early 1970s. The canopy of a peach tree suddenly collapses before, during or just after bloom, usually 3–6 years after planting (Fig. 12.45). Tree sap often oozes out of scaffold limbs and the trunk. On many limbs, the tissue just beneath the bark is discoloured (Fig. 12.46) and has a sour odour. Peripheral parts of the tree, although dying, may still reveal healthy tissue underneath the bark. The trees are killed only to the soil line. Later in the season, shoots from the rootstock often emerge (Fig. 12.47) (Ritchie and Clayton, 1981).

PTSL is a complex disease with biotic and abiotic factors that may contribute to the disease directly or indirectly. The immediate causes of death can be cold injury (Nesmith and Dowler, 1976), bacterial canker caused by *Pseudomonas syringae* Van Hall or a combination of the two, and perhaps *Cytospora* canker. Cold injury in the south-eastern USA occurs in late winter after the rest period has been broken and physiological activity resumes. In a normal season, sufficient chill hours have accumulated by late January, and peach trees can lose hardiness after periods of warm weather in the dormant season. Like frost injury, bacterial canker can affect all parts of a peach tree if the tree is stressed. Indirect factors that predispose trees to PTSL include the commonly found ring nematode *Mesocriconema xenoplax* (Raski) Loof and de Grisse (Ritchie and Clayton, 1981), improper rootstock selection (Zehr *et al.*, 1976), time of pruning (Nesmith and Dowler, 1975), root injury, physical characteristics of the orchard site such as pH and soil structure, and planting on land used previously for peach production.

The disease is most common in sandy soils. Losses are more frequent and severe in orchards where peach or other stone fruits have been grown

Fig. 12.45. An orchard with trees affected by peach tree short life disease.

previously than in locations where peach has not been grown before. In clay soils, less difficulty is encountered when peach orchard sites are replanted. Ideally, peach plantings should be established on 'virgin' land with no recent history of growing stone fruit. Site selection in any given region, however, is often compromised by other factors (e.g. urban encroachment, unsuitable microclimate, proximity to packinghouses).

The complexity of PTSL syndrome complicates attempts to effectively manage the disease in commercial orchards. A 10-point programme that emphasizes practices to enhance the health and vigour of peach trees has been developed (Ritchie and Clayton, 1981). The programme consists of the following recommendations: (i) adjust pH to at least 6.5 before planting; (ii) subsoil before planting to break up hardpans and promote root growth; (iii) fumigate on replant sites with sandy soil and other nematode-infested soils; (iv) use rootstocks that are certified to be free of nematodes or have been grown in fumigated soil; (v) select rootstocks tolerant or resistant to nematodes; (vi) apply nutrients and lime as needed; (vii) delay pruning to late winter (February and March); (viii) use recommended herbicides for weed control; (ix) in preplant fumigated sites, postplant fumigate at approximately 2-year intervals or when nematode populations increase; and (x) remove and burn all dead or dying trees. Fumigants such as chloropicrin and methyl-iodide are expensive and only provide benefit for about 2–3 years. Currently, one of the most effective management strategies is the use of rootstocks such as the highly PTSL-tolerant

Fig. 12.46. Discoloured tissue under the bark of a peach tree short life-affected tree.

'Guardian®' (BY520-9 line) (Okie *et al.*, 1994), which has largely replaced other rootstocks in the south-eastern USA. This rootstock is also very resistant to the ring nematode but highly susceptible to *Armillaria* root rot. The BY520-9 lines are genetically highly diverse and research is ongoing to further select lines with horticultural characteristics similar or superior to commercial rootstocks (Wilkins *et al.*, 2002).

12.4.5   Other fungal diseases

Several additional diseases affecting different parts of peach and nectarine trees have been reported. Some of these diseases are limited in their geographical distribution, whereas others are widespread but do not limit production. In favourable environments or due to fungal adaptation, however, many of these diseases can become economically important and limit crop production. Examples of these diseases are listed in Table 12.2.

Fig. 12.47. Rootstock sprouting from a peach tree killed by peach tree short life disease.

Anthracnose caused by *Colletotrichum siamense* Prihastuti, L. Cai & K.D. Hyde and *C. fructicola* Prihastuti, L. Cai & K.D. Hyde of the *C. gloeosporioides* species complex, *C. fioriniae* (Marcelino & Gouli) R.G. Shivas & Y.P. Tan and *C. nymphaeae* (Pass.) Aa of the *C. acutatum* species complex and in rare instances *C. truncatum* have been reported in areas and seasons with high rainfall and warm temperatures during fruit ripening (Bernstein *et al.*, 1995; Grabke *et al.*, 2014; Hu *et al.*, 2015a; Lee *et al.*, 2020; Moreira *et al.*, 2020). The circular lesions are brown or salmon-pink in colour (Fig. 12.48), depending on species. They are firm and slightly sunken, and slowly expand over the fruit surface. Another diagnostic feature is that the lesion in cross-section is cone-shaped, hardened and easily separated from the healthy mesocarp. Asexual fruiting structures of the pathogen (acervuli) are produced in concentric rings on the fruit surface, bearing masses of conidia that are disseminated by splashing rain. Fungicide sprays can be used to manage anthracnose where it is a problem; however, outbreaks of *C. siamense* populations due to fungicide resistance selection have been reported (Hu *et al.*, 2015b).

Other diseases listed in Table 12.2 include black knot caused by *Apiosorina morbosa* (Schwein.:Fr.) Arx, *Botryosphaeria* fruit rot caused by the *Botryosphaeria* spp. discussed above, *Diplodina* fruit rot caused by *Diplodina persicae* Horn & Hawthorne, frosty mildew caused by *Mycosphaerella pruni-persicae* Deighton, *Leucotelium* white rust caused by *Sorataea pruni-persicae*

Table 12.2. Other fungal diseases with variable importance in the production of peach and nectarine, their geographical occurrence, primary symptoms and management practices.

Disease	Pathogen	Geographical distribution	Primary symptoms	Primary management practices[a]
Flower, foliage, and fruit diseases				
Anthracnose	*Colletotrichum fioriniae, C. fructicola, C. nymphaeae, C. siamense, C. truncatum*	Widespread but infrequent	Sunken, firm fruit lesions with concentric rings of sporulation zones	Fungicides (removal of alternate host?)
Diplodina fruit rot	*Diplodina persicae*	South-eastern USA, uncommon	Leaf and fruit spots	Fungicides
Frosty mildew	*Mycosphaerella pruni-persicae*	Widespread but infrequent	White mildew on leaves	None
Leucotelium white rust	*Sorataea pruni-persicae*	Japan, Korea, China	Polygonal to irregular leaf spots	Fungicides (removal of alternate host?)
Target leaf spot	*Phyllosticta persicae*	Italy, India, USA	Leaf spots (target appearance)	None
Trunk and scaffold diseases				
Black knot	*Apiosorina morbosa*	Eastern USA	Twig and branch swellings (knots)	Removal of wild *Prunus* spp., sanitation, and fungicides
Sclerotium stem rot	*Sclerotium rolfsii*	Nursery disease, widespread in warm areas	Stem cankers, wilting	Sanitation, fungicides
Wood-decaying fungi	Species in the Basidiomycota	Widespread in older orchards	Wood rot: white and brown rots	Time of pruning, sanitation, removal of infected branches

Root and crown diseases

Phymatotrichum root rot	*Phymatotrichopsis omnivora*	South-western USA and Mexico	Root rot, tree decline	Integrated approaches: crop rotations, soil fumigation
Rosellinia root rot	*Rosellinia necatrix*	Widespread in warm areas, sporadic in the USA	Root rot, tree decline	Integrated approaches: soil fumigation, fungicides
Violet root rot	*Helicobasidium mompa*	Japan, Korea, China	Root rot, tree decline	Integrated approaches: soil fumigation, fungicides

[a]Management practices with a question mark have a limited or poorly defined role in controlling the disease.

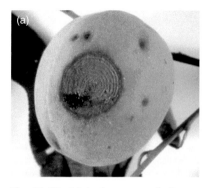

Fig. 12.48. (a) Anthracnose of cling peach caused by a *Colletotrichum* sp. in the *C. acutatum* spp. complex showing concentric pinkish rings where conidia in acervuli (conidioma) of the fungus are produced. (b) Anthracnose of a fresh market peach caused by *C. acutatum* showing a circular lesion with a darker outer ring and a dense salmon-pink centre where the conidia of the fungus are produced.

Tranzschel (formerly *Leucotelium pruni-persicae* (Hori) Tranzschel), target leaf spot caused by *Phyllosticta persicae* Sacc., Phymatotrichum root rot caused by *Phymatotrichopsis omnivora* (Duggar) Hennebert, *Rosellinia* root rot caused by *Rosellinia necatrix* Prill., *Sclerotium* stem rot caused by *Sclerotium rolfsii* Sacc., violet root rot caused by *Helicobasidium mompa* Tanaka, sour pit caused by *Candida inconspicua* (Lodder & Kreger-Van Rij) Meyer & Yarrow, and wood decay caused by a number of fungal species (over 56 species in North America) in the phylum Basidiomycota.

12.5 BACTERIAL DISEASES

12.5.1 Bacterial blast/canker

Bacterial blast and bacterial canker affect all *Prunus* spp. and are a major cause of premature peach tree decline and tree death in many production areas. As indicated in section 12.4.4, bacterial canker can be the immediate cause of tree death in the PTSL complex for trees that are predisposed to biotic and non-biotic factors.

Symptoms first appear in late winter to early spring as trees emerge from dormancy. They include blighted dormant buds, often associated with an elliptical brownish to black canker lesion, as well as cankers on twigs and branches. Branches and scaffold limbs may die, and in severe cases, the entire tree will collapse (Wilson, 1953; Cross, 1966; Hattingh *et al.*, 1989). Upon closer examination of newly affected scaffold limbs, a margin along the limb may be visible, separating asymptomatic from diseased bark and wood. Underneath the bark, non-symptomatic wood is clearly separated from the reddish-brown affected wood (Fig. 12.49). The collapsed bark and diseased wood tissue may extend along the limb all the way to the trunk and may have a sour smell from the infection in early spring, often first noticed during the time of pruning. Young trees (starting 2–3 years after transplanting) are most susceptible to the disease, and this continues until the trees are 6–7 years old. Cold injury can predispose the tree to infection and by itself cause symptoms similar to bacterial canker. With cold injury, however, the bark often splits and separates easily from the wood, which does not happen with bacterial canker. Phytophthora root rot and Armillaria root rot also produce above-ground tree collapse symptoms somewhat like bacterial canker, but the symptoms are not limited to the late-winter and early-spring seasons. In the case of bacterial canker, the roots may remain viable and rootstock suckers may emerge.

Several species of *Pseudomonas* have been associated with tree decline in *Prunus* spp., but thus far, only two pathovars of *Pseudomonas syringae* (i.e. pv. *syringae* and pv. *persicae*) have been described on peach (Wilson, 1953; Roos and Hattingh, 1986; Young, 1988; Menard *et al.*, 2003; Bultreys and Kałuzna, 2010; Kałuzna *et al.*, 2016). Both pathovars affect the scion tissue, but only *P. syringae* pv. *persica* also attacks the roots, preventing sucker formation from the rootstock (Prunier *et al.*, 1970; Young, 1988). Pathogenicity and virulence are associated with the production of phytotoxins and plant hormones. Multiple

Fig. 12.49. A bacterial canker lesion on a scaffold branch of a peach tree caused by *Pseudomonas syringae*.

phytotoxins have been identified in various *Pseudomonas* spp., including syringomycin, syringopeptin, syringolin, phaseolotoxin, mangotoxin and tabtoxin (Patil and Tam, 1972; Turner, 1981; Arrebola *et al.*, 2003; Scholz-Schroeder *et al.*, 2003; Arai and Kino, 2008; Schellenberg *et al.*, 2010). Compounds such as auxin, cytokinins and coronatine produced by the pathogen can interfere with the regulation of plant immune responses (Kosuge *et al.*, 1966; Surico *et al.*, 1985; Glass and Kosuge, 1986; Geng *et al.*, 2014). The fluorescent *P. syringae* pv. *syringae* produces syringomycin, while the non-fluorescent *P. syringae* pv. *persicae* produces persicomycin (Barzic, 1999). The bacteria are ubiquitous in the environment and can be disseminated by wind-blown rain, possibly by pruning tools and by nursery stock. The primary entry into the peach tree is assumed to be through leaf abscission wounds generated during leaf drop in the autumn but possibly also through pruning cuts. Mild temperatures (12–18°C) and wet conditions will favour infection (English *et al.*, 1961). The disease is worse in mild, wet winters than in winters with consistently cold temperatures. The pathogen's tolerance to copper (Popović *et al.*, 2021), which is a widely used bactericide in fruit orchards, makes this control option questionable. In general, however, the focus for managing bacterial canker should be on mitigation of tree stress. Studies conducted over many decades have shown the importance of certain cultural practices on bacterial canker management. These include avoidance of low soil pH (Weaver and Wehunt, 1975) and nutrient deficiency (English *et al.*, 1961), selection of optimal pruning dates (Dowler and Petersen, 1966), disease-tolerant rootstocks and planting sites suitable for peach, as well as management of ring nematodes (Mojtahedi *et al.*, 1975).

12.5.2   Bacterial spot

Bacterial spot of peach and other *Prunus* spp. has been reported from locations worldwide where these crops are grown including Australia, Europe, Japan, South Africa, North America and several countries in South America (Ritchie, 1995; EFSA PLH Panel, 2014). The disease was first described in Michigan on Japanese plum in 1903 (Smith, 1903). Among commercially grown *Prunus* spp., bacterial spot has had the highest economic impact on Japanese plum, peach and nectarine but can also seriously affect apricot, European plum (Stefani, 2010) and almond (Haack *et al.*, 2020). The economic impact of bacterial spot is a result of reduced quality and marketability of fruits, reduced orchard productivity, and increased costs in disease management and nursery production (Stefani, 2010). The disease is most severe in production areas where stone fruits are grown in sandy, light soils, and where environmental conditions are humid, moist and warm during the growing season. Although the causal pathogen is listed as a quarantine organism by the European and Mediterranean Plant Protection Organization

(EPPO, 2006), it is not identified as such by other regional quarantine regulatory agencies (Lamichhane, 2014).

The pathogen was originally described as *Pseudomonas pruni* by E.F. Smith in 1903. The currently accepted name is *Xanthomonas arboricola* pv. *pruni* (Smith) Vauterin, Hoste, Kersters & Swings (*Xap*) (Vauterin *et al.*, 1995). Synonyms include *X. campestris* pv. *pruni* (Smith) Dye and *X. pruni* (Smith) Dowson. *Xap* is a pathogen highly specialized to infect *Prunus* spp. (Stefani, 2010). However, research has shown that pathovars of *X. arboricola* are not confined to nut and stone-fruit trees (Fischer-Le Saux *et al.*, 2015). *Xap* (Gammaproteobacteria, Xanthomonadales, Xanthomonadaceae) is a Gram-negative, motile, rod-shaped (0.2–0.4 × 0.8–1.0 μm), monoflagellated, non-sporulating, aerobic bacterium. The optimum growth range is 24–29°C. Growth on yeast dextrose carbonate or sucrose peptone agar media at 28°C for 48–72 h results in yellow-pigmented, mucoid colonies. Based on identification systems utilizing growth and metabolic reactions on different substrates (e.g. Biolog GN2), strains of this pathovar are able to metabolize a number of sugars, carbohydrates and organic acids.

Multi-locus sequence analysis based on the genes *dnaK*, *fyuA*, *gyrB* and *rpoD* (Young *et al.*, 2008; Garita-Cambronero *et al.*, 2017, 2018) or on partial sequences of *atpD*, *efp* and *glnA* (Boudon *et al.*, 2005; Fischer-Le Saux *et al.*, 2015) have been useful in differentiating the pathovar *pruni* from other pathovars of the species, confirming its phylogenetic proximity to the pathovars *corylina* and *juglandis*. These studies have concluded that these pathovars form monophyletic groups with a low genetic diversity among their strains that must be considered clonal complexes. This is continuing to be confirmed as the genomes of more strains of *Xap* are sequenced (Back *et al.*, 2020).

Genetic variability among strains of *Xap* was assessed by integron gene cassette array analyses, as well as by BOX-PCR and repetitive extragenic palindromic PCR (Barionovi and Scortichini, 2006; Kawaguchi, 2014). Low genetic variability among *Xap* strains was shown, although the study indicated that integrons were involved in configuring the genetic diversity of this species. Despite this, none of the DNA banding patterns observed in *Xap* was associated with any of the host plants or the geographical region from where the strains were isolated.

On *Prunus* spp. hosts, symptoms include numerous small, angular lesions 1–3 mm in diameter on leaves (Fig. 12.50a), fruit and shoots (Roselló *et al.*, 2012). Lesions begin as angular, water-soaked areas surrounded by chlorotic tissue primarily on the underside of leaves and continue to develop into angular brown to purplish-black spots (Fig. 12.50b). Bacterial streaming from a lesion placed into water can be seen with a microscope. Lesion centres may become necrotic and abscise, creating 'shot holes' that can cause the leaf to appear tattered. Lesions are concentrated in areas of the leaf that remain wet for longer periods of time, including the tip, margins and along the mid-rib. Leaves develop the characteristic yellowing and readily defoliate. Leaf symptoms

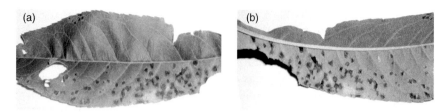

Fig. 12.50. Angular, water-soaked bacterial spot lesions caused by *Xanthomonas arboricola* pv. *pruni* on (a) upper and (b) lower surface of peach leaves.

can appear to be like copper phytotoxicity or shot hole caused by the fungal pathogen *Wilsonomyces carpophilus*. Lesions caused by copper are mostly circular and occur on any part of the leaf and are mostly uniform for trees in the orchard. In contrast, bacterial spot lesions are angular, initially confined by the veinlets of the leaf, have a water-soaked appearance, and mostly occur at the mid-rib, tip and leaf margin, and the disease is spatially irregular within the orchard. Infected leaves may turn completely chlorotic and abscise with severe defoliation on highly susceptible cultivars. Severe defoliation may result in reduced fruit size, sunburn and cracking of fruit, as well as reduced tree vigour and hardiness for overwintering.

The fruit are highly susceptible to infections between the physiological stages of shuck split and pit hardening. These infections usually result in deep lesions extending into the mesocarp (Fig. 12.51). Fruit infections may exude light to dark amber gum. Gumming may accumulate as a distinct drop or continue to form a long tendril on the fruit surface. Lesions that develop from infections after pit hardening usually result in a speckled fruit surface appearance, remaining in and just beneath the fruit skin (Fig. 12.51). Multiple infections can occur on a single fruit, and infected fruit may drop prematurely, especially when infection occurs prior to pit hardening.

Several types of lesion occur on twigs including 'spring' and 'summer' cankers, as well as 'black tip' cankers. Spring cankers develop on shoots of the previous summer's growth from leaf scar infections at the lateral nodes that occurred in late autumn during leaf drop. Spring cankers become visible at or just before the time of leaf emergence and may extend along the twig surface for several centimetres (Fig. 12.52). During the early development of lateral spring cankers, the bark rather than the wood shows symptoms, which can be used to differentiate from bacterial canker caused by *P. syringae* pv. *syringae*. Black tip is a form of spring canker that occurs at the terminal shoot bud, rather than at a lateral bud, and may extend several centimetres down the terminal shoot. Black tip may be visible in early winter but is usually observed in mid- to late winter on the terminal portion of the previous year's shoot. The terminal bud fails to open, the shoot tip turns black and the necrotic tissue extends several centimetres down the shoot (Ritchie, 1995). Summer cankers form on new green shoots and become visible in late spring and summer. The lesions are

Fig. 12.51. Peach fruit with bacterial spot lesions from infections that occurred prior to pit hardening (large, deep lesions) or after pit hardening (small, pinpoint lesions).

initially water soaked and elliptical, and later develop into cankers between the nodes of the shoot. Their role as a source of inoculum during the summer is probably negligible and they are not considered to overwinter the pathogen.

Bacterial spot is widespread in some countries, whereas in others, only local and sporadic outbreaks have been recorded. This pattern is associated with warm, moist environmental conditions during the growing season, particularly during spring. International trade has led to the long-distance dissemination of *Xap*, mainly through latently infected plant material used for propagation (Goodman and Hatting, 1986; López-Soriano *et al.*, 2016). Over short distances, rain, wind and contaminated pruning tools can spread *Xap* among trees and other nearby orchards (Goodman, 1988). Disease spread has also been associated with machinery such as air-blast sprayers in which diseased leaves are collected in the fan area and carried to other orchards and then disseminated as spraying occurs.

The pathogen survives during tree dormancy in symptomless lateral and terminal buds that can develop into twig cankers during late winter and early spring, and epiphytically on symptomless plant surfaces (Feliciano and Daines, 1970; Shepard and Zehr, 1986; Shepard and Zehr, 1994). As temperatures warm up and the trees begin to grow in the spring, the pathogen is carried from between the bud scales with plant growth and from spring cankers from where the pathogen is further disseminated by wind-driven rain and heavy dew. Spring cankers are not observed to produce visible amounts of bacterial

J.E. Adaskaveg et al.

Fig. 12.52 Bacterial spot lesions on a peach twig caused by *Xanthomonas arboricola* pv. *pruni* showing (a) the discoloured bark surface, and (b) internal water-soaked, discoloured internal tissue.

ooze, but at times a 'glistening' appearance can be observed on the cankers. Wetness from dew and fog contributes to bacterial growth on leaf surfaces, and water congestion along leaf margins and tips allows the bacteria to ingress through natural openings such as stomata and hydathodes. Furthermore, the pathogen can egress from stomata of infected leaves prior to symptom development (Miles *et al.*, 1977). Heavy rainfall can drive the bacteria into leaves, fruit and stems, and the disease is then more severe on the windward side of the tree. Secondary spread of the bacterium is from inoculum predominately produced from diseased leaves because bacteria produced from spring cankers decline as tree growth increases and the lateral cankers are walled off. Multiple cycles of infection can occur with each period of precipitation and heavy dew formation. The disease is favoured by warm temperatures (19–28°C) and high humidity, which support multiplication of the bacterium (Morales *et al.*, 2017).

Hot and dry environmental conditions inhibit disease development and spread. In the nursery, budding may spread the disease from infected trees to healthy trees. The pathogenicity of *Xap* relies on a large repertoire of 21 type III effectors (T3Es), which can be delivered directly into the host cells via the type III secretion system (Hajri *et al.*, 2012).

Despite extensive efforts to characterize cultivars of several *Prunus* spp. according to their level of resistance to *Xap* (Bazzi *et al.*, 1990; Garcin and Bresson, 2011; Socquet-Juglard *et al.*, 2012), not much is known about the genetics underlying host defence responses. One major quantitative trait locus (QTL) associated with both fruit and leaf disease incidence in apricot has been identified on linkage group (LG) 5 (Socquet-Juglard *et al.*, 2012). Similarly, for peach, major QTLs for *Xap* resistance have also been identified, two on LG4 for leaf resistance, one on LG5 for both fruit and leaf resistance, and one on each of LG1 and LG6 for fruit resistance (Yang, 2013). In addition, the differential expression of pathogen-related genes in peach was identified by quantitative reverse transcriptase PCR following bacterial spot infection and after methyl jasmonate and ethephon treatments, indicating that the jasmonic acid and ethylene pathways may be involved in disease resistance (Sherif *et al.*, 2011). Host plant defence responses to the pathogen were also investigated using whole-transcriptome sequencing (Socquet-Juglard *et al.*, 2013; Gervasi *et al.*, 2018), and recent studies report transcriptome reprogramming and extensive modulation of gene expression upon *Xap* infection of sensitive and resistant peach genotypes during early leaf infection (Gervasi *et al.*, 2018). Bacterial spot resistance in peach is under polygenic control, and the leaf and fruit responses are regulated by different genes, with alleles conferring susceptibility being more common (Yang, 2013; Gasic *et al.*, 2015).

The success of disease management is based on an integrated approach that includes preventing the introduction and dissemination of the pathogen. It is recommended that where disease is endemic, cultivars having some level of resistance and adapted to the region both climatically and for marketability should be considered for planting. Prevention of pathogen introduction begins with adequate quarantine legislation to protect regional production areas. Because the pathogen can survive in buds during the summer and autumn seasons, bud wood for propagation should only be collected from disease-free trees and orchards. The use of cultivars with reduced susceptibility and suitable agronomic measures, including choosing locations less conducive to disease and avoiding planting near orchards where the disease is present, may limit the development and build-up of the disease in an orchard. Additionally, orchards should be provided with adequate fertilization to avoid excessive foliar growth or weakened trees from poor nutrition, both of which can enhance bacterial spot development. Trees can also be predisposed to severe infection by *Xap*. In sandy soils for example, high levels of ring nematodes (*Criconemella xenoplax*) may induce higher bacterial spot severity (Shepard *et al.*, 1999).

Chemical treatments in orchards to protect leaves, fruit and stems of stone-fruit crops during conducive periods may minimize the occurrence of the disease and the build-up of inoculum. Successful disease management with bactericides is highly dependent on the timing of applications with the objective of reducing the primary inoculum from spring cankers and thus preventing or delaying an epidemic. Pre-dormant or late-autumn (just prior to leaf drop) applications of copper bactericides mixed with lime (i.e. Bordeaux mixtures) or mixed with oil may reduce leaf scar infections, but control success is highly variable. A preventative treatment should also be done as a delayed dormant application just before tree growth resumes in the spring. This is the most important time to start treatment applications because it helps to prevent the spread of inoculum from overwintering cankers to developing leaves and fruit. Copper and copper-based mixtures have generally been reported to have marginal efficacy against the disease. Applications before bloom and during the 3-week period following petal fall are critical for preventing early-season fruit infections and establishment of disease on new foliage, which provides fresh inoculum for subsequent infections (Ritchie and Bennett, 1993). On peach in the eastern USA, three to five applications of copper products are carried out prior to shuck split when limited foliage is present. After shuck split, rates of copper are reduced with each subsequent application to minimize foliar damage, or oxytetracycline is used. This programme is efficacious under low to moderate disease pressure based on environmental conditions (Ritchie, 1999; Ritchie *et al.*, 2008). Once the epidemic is in progress, treatments are at best minimally effective. Materials other than copper and oxytetracycline have generally not provided equivalent control of bacterial spot (Ritchie, 2006, 2008, 2011; McFarland and Lalancette, 2012; Lalancette *et al.*, 2016; Lalancette and Blaus, 2018).

Historically, copper treatments have been limited to autumn leaf drop, dormant and early-season applications. Later spring applications after shuck split of copper-based compounds increase the risk of copper phytotoxicity on *Prunus* spp., with increased risk following repeated applications (Ritchie *et al.*, 2008; Blaauw *et al.*, 2021). Copper ions function in several ways to cause bacterial cell death, including disruption of cell-membrane integrity, induction of reactive oxygen species, DNA damage and inhibition of respiration (Grass *et al.*, 2011; Fones and Preston, 2013; FRAC, 2022), and are dependent on contact for efficacy. Delayed dormant copper applications followed by successively reduced rates with each application of copper (e.g. 10.0, 5.0, 2.5 and 1.0 kg metallic copper equivalents ha^{-1}) during bloom and petal fall have been effective in reducing bacterial spot with minimal phytotoxicity (Blaauw *et al.*, 2021). Other treatments reported to have efficacy with varying results include oxytetracycline, dodine mixed with captan, and zinc-containing fungicides such as ziram and zinc sulfate (Ritchie, 1995). Mancozeb products are currently registered for peach for bacterial spot control in countries other than the USA for managing selected diseases up to 6 weeks before harvest.

12.5.3 Crown gall

Crown gall is a soil-borne bacterial disease caused by *Agrobacterium tumefaciens* that attacks plants in many plant families including species in the family Rosaceae. The pathogen belongs to the Rhizobiaceae, a family of nitrogen-fixing bacteria. Several species in this family induce hyperplastic plant growths. *A. tumefaciens* is also known as Biovar 2 and induces galls in peach. Galls typically develop on the lower stem or trunk and at the crown just below the soil line (Fig. 12.53). Roots are also susceptible, and the disease can occasionally occur on aerial portions of the tree. Galls may develop and enlarge to several centimetres in diameter. The tissue of newly formed galls is soft and spongy, whereas

Fig. 12.53. Galls caused by *Agrobacterium tumefaciens* on peach trees recently dug up from a nursery.

older galls become woody and are often sites of infection for other organisms such as wood-decaying fungi. Young trees with infections may be stunted, and older trees may exhibit symptoms typical of root problems. On mature trees, galls are often discovered upon tree removal. The greatest economic impact of crown gall occurs in the nursery from culling of symptomatic trees.

The bacterium invades through wounds, and infection is favoured by moist conditions and poorly drained soils. The tumourigenic ability is due to genes on a large bacterial plasmid termed the Ti-plasmid. A segment of the Ti-plasmid (T-DNA) is transferred to the plant cell via a conjugation-like process and is then incorporated into the genome (Kado, 2014). Loss of the Ti-plasmid renders strains of *A. tumefaciens* non-tumorigenic, and such strains are classified as *A. radiobacter*. Ti-plasmid transfer is favoured by temperatures just below 20°C, but gall development occurs more rapidly at temperatures higher than 20°C, with galls visible in 2–4 weeks (Fullner *et al.*, 1996).

Crown gall is more common in temperate than in tropical regions. Once cells are transformed, the bacterial presence is no longer needed for tumour development. This is because the T-DNA is now part of the host genome and replicates with the host cell. Gene expression causes the plant to produce excess amounts of indole-3-acetic acid and cytokinin. These growth regulators stimulate hyperplastic growth, which results in gall formation.

Species of *Agrobacterium* are ubiquitous soil bacteria and can be isolated from agricultural and non-agricultural soils (Bouzar and Moore, 1987). Several selective media have been developed to aid in isolation and differentiation of the species (previously known as biovars) (Brisbane and Kerr, 2008). PCR-based DNA primers have also been developed (Haas *et al.*, 1995). The bacterium has the ability to cause latent infections and can exist in soil adhering to root surfaces. This allows the bacterium to be disseminated through nursery stock as well as through movement of soil (Moore, 1976). Species of *Agrobacterium* can survive in the soil for more than 2 years under experimental conditions. Survival, however, is negatively affected by soil temperatures higher than 34°C and by acid conditions.

Management of crown gall relies on prevention of infections. Nursery sites free of the pathogen, inspection of trees at the nursery and planting of disease-free trees are the foundational practices of successfully managing crown gall. The biocontrol *A. radiobacter* can be used for treating seeds used for rootstocks and for treating tree roots at planting time (Jones and Kerr, 1989). Removal of galls on infected trees does not necessarily prevent galls from still developing after planting. Thus, trees with galls should be destroyed. At planting, care should be taken to prevent root injury, and once trees are planted, infections can be minimized by avoiding injuries to the roots, crown and trunk of the tree by equipment used for soil tillage or orchard mowing.

12.5.4   Other bacterial and phytoplasma diseases

Phony peach disease is caused by *Xylella fastidiosa*, a fastidious, xylem-limited, Gram-negative bacterium with an extensive list of hosts, including peach and other *Prunus* spp. (Wells *et al.*, 1983). Periwinkle (*Catharanthus roseus*) is an indicator plant infected with phytoplasmas transmitted by a dodder bridge. Strains of the pathogen are the cause of almond leaf scorch, citrus variegated chlorosis, Pierce's disease of grapevines and plum leaf scald (Mizell *et al.*, 2020). Diseased trees are dwarfed compared with healthy trees and have dark green foliage on branches with shortened internodes that give the tree a compact umbrella appearance (Figs 12.54 and 12.55) (Evert and Smittle, 1989). The disease is spread by sharpshooter leafhoppers that vector the pathogen while feeding. Because the pathogen is most abundant in the roots of infected peach trees, and this is unlike the situation for other hosts that have high concentrations in leaves, peach is considered a dead-end host. The disease is endemic to the south-eastern USA, and sporadic epidemics have occurred in this region as far west as Texas.

Phytoplasmas are phloem-limited, pleomorphic bacteria lacking a cell wall and are transmitted mainly through insect vectors such as leafhoppers or psyllids but also by plant propagation materials and seeds. Peach X-disease was first described in California and is an economic problem in the eastern USA

Fig. 12.54. Stunted peach trees with shortened internodes and dark green foliage affected by phony peach disease caused by *Xylella fastidiosa*.

Fig. 12.55. Phony peach disease caused by *Xylella fastidiosa* showing a shoot with shortened internodes and small fruit (top), compared with an uninfected tree shoot and larger fruit (bottom).

but has not been reported outside North America (Kirkpatrick *et al.*, 1995a). Eastern and western strains of the pathogen are similar, and the term X-disease is currently used to describe a group of yellow-inducing phytoplasmas that are pathogens of *Prunus* spp. Symptoms occur on leaves as irregularly distributed yellow or necrotic spots. The lesions increase in number over the season and may abscise creating a 'shot hole' appearance. The leaves typically roll and have a pale green colour. Abscission of infected leaves at the base of shoots may induce terminal resetting. The symptoms may be unevenly distributed in the tree canopy, and the pathogen may take 2 years to fully colonize a tree. Other diseases caused by the same phytoplasma include peach yellows (Kirkpatrick *et al.*, 1995b), little peach and red suture.

The apple proliferation or AP group of phytoplasmas (group 16Sr X) cause peach yellow leaf roll (PYLR), which is limited to peach trees in several counties in northern California. The disease commonly occurs adjacent to pear orchards that suffer from pear decline. Thus, the PYLR pathogen appears to have evolved as a substrain of the pear decline phytoplasma (Blomquist and Kirkpatrick, 2002).

European stone fruit yellows (ESFY) is a disease of stone fruits including peach that causes tree decline in all Mediterranean countries and north into Germany (Poggi-Pollini *et al.*, 2001). It is the most important phytoplasma-caused disease of stone fruit in Europe. The pathogen also belongs to the AP group 16Sr X of phytoplasmas, and the proposed name is '*Candidatus Phytoplasma prunorum*' (Cieślińska, 2011). The pathogens of these diseases are vectored by psylla insects (Homoptera: Psyllidae) (Carraro *et al.*, 1998). Symptoms of both diseases include leaf rolling (leaves roll upward for PYLR and longitudinally for ESFY) on infected branches, enlarged midribs and veins, and premature leaf drop. Diseased trees have reduced vigour and fruit production, and may die prematurely over 2 or more years.

These diseases are generally managed based on prevention, including pathogen exclusion with the use of phytosanitary regulations such as quarantines to prevent the movement of diseased plants and using certified plant material. Antibacterial treatments are generally not available. Controlling insect vectors, removal and destruction of diseased trees, and eradication of nearby alternative hosts are practices to minimize epidemics.

12.6 CONCLUSION

Numerous fungal and fungal-like organisms, as well as bacteria and phytoplasmas, are pathogens of peach. Many of these organisms have a worldwide distribution, whereas others occur only locally or regionally. Disease incidence is dependent largely on climatic conditions and the distribution of the pathogen in peach-production areas but also on the peach genotype.

Many of the diseases are limiting to crop production if they are not managed effectively. Knowledge about peach diseases has increased over the years. New pathogens have been identified, described and studied at biological, epidemiological and molecular levels. In most cases, they can be effectively managed.

With worldwide trade and increased travel, the potential for pathogens to spread and become limiting factors to peach production has also increased. In the immediate future, peach production will be dependent on the success of programmes that prevent pathogen movement and introduction to new geographical areas. Quarantines that completely restrict plant movement, as well as controlled plant movement (e.g. the federal Inter-Regional No. 2 or IR-2 programme in the USA) and production certification programmes such as the National Clean Plant Network (NCPN) Fruit Trees for breeders and nurseries will prevent or minimize harmful microorganisms from being spread while allowing exchange of diverse genetic plant material. Nevertheless, many of the pathogens and the diseases they cause are endemic to the orchard or region. Therefore, an integrated approach for disease management is necessary. New management practices based on biological and genetic information about the host and pathogen, as well as about novel antimicrobial and host defence chemistries, continue to be developed. There is a high demand for new varieties that are pest and disease resistant, and traditional breeding programmes are in place to produce rootstocks resistant to *Armillaria* root rot and scions that are resistant to brown rot (Fu *et al.*, 2021) and bacterial spot (Vonkreuzhof *et al.*, 2019). Peach may eventually also be genetically modified to resist pests and pathogens. Consumers, however, are demanding non-genetically modified organisms including plants, and thus other approaches such as marker-assisted breeding methods are currently being developed.

REFERENCES

Adams, G.C., Surve-Iyer, R.S. and Iezzoni, A.F. (2002) Ribosomal DNA sequence divergence and group I introns within the *Leucostoma* species *L. cinctum*, *L. persoonii*, and *L. parapersoonii* sp. nov., ascomycetes that cause Cytospora canker of fruit trees. *Mycologia* 94, 947–967.

Adaskaveg, J.E., Ogawa, J.M. and Butler, E.E. (1990) Morphology and ontogeny of conidia in *Wilsonomyces carpophilus*, gen. nov. and comb. nov., causal pathogen of shot hole disease of *Prunus* species. *Mycotaxon* 37, 275–290.

Adaskaveg, J.E., Förster, H., Wade, L., Thompson, D.F. and Connell, J.H. (1999) Efficacy of sodium tetrathiocarbonate and propiconazole in managing *Armillaria* root rot of almond on peach rootstock. *Plant Disease* 83, 240–246.

Adaskaveg, J.E., Duncan, R.A., Hasey, J.K. and Day, K.R. (2015) Peach: UC IPM Pest Management Guidelines. University of California Agriculture and Natural Resources, Publication No. 3454. Available at: https://anrcatalog.ucanr.edu/Details.aspx?itemNo=3454 (accessed 8 December 2022).

Amiri, A., Holb, I.J. and Schnabel, G. (2009) A new selective medium for the recovery and enumeration of *Monilinia fructicola*, *M. fructigena* and *M. laxa* from stone fruits. *Phytopathology* 99, 1199–1208.

Anderson, J.B. and Ullrich, R.C. (1979) Biological species of *Armillaria mellea* in North America. *Mycologia* 71, 402–414.

Arai, T. and Kino, K. (2008) A novel L-amino acid ligase is encoded by a gene in the phaseolotoxin biosynthetic gene cluster from *Pseudomonas syringae* pv. *phaseolicola* 1448A. *Bioscience, Biotechnology and Biochemistry* 72, 3048–3050.

Arrebola, E., Cazorla, F.M., Durán, V.E., Rivera, E., Olea, F. *et al.* (2003) Mangotoxin: a novel antimetabolite toxin produced by *Pseudomonas syringae* inhibiting ornithine/arginine biosynthesis. *Physiological and Molecular Plant Pathology* 63, 117–127.

Arthur, J.C. (1934) *Manual of the Rusts in United States and Canada*. Purdue Research Foundation, Lafayette, Indiana.

Back, C.G., Park, M.J., Park, G.S., Yang, C.Y. and Park J.H. (2020) Complete genome sequence of *Xanthomonas arboricola* pv. *pruni* strain KACC21687 a causal agent for bacterial shot hole on peach. *Korean Journal of Microbiology* 56, 72–73.

Barionovi, D. and Scortichini, M. (2006) Assessment of integron gene cassette arrays in strains of *Xanthomonas fragariae* and *X. arboricola* pvs. *fragariae* and *pruni*. *Journal of Plant Pathology* 88, 279–284.

Barzic, M.R. (1999) Persicomycin production by strains of *Pseudomonas syringae* pv. *persicae*. *Physiological and Molecular Plant Pathology* 55, 243–250.

Bassi, D., Rizzo, M. and Cantoni, L. (1998) Assaying brown rot [*Monilinia laxa* Aderh. et Ruhl. Honey)] susceptibility in peach cultivars and progeny. *Acta Horticulturae* 465, 715–721.

Batra, L.R. (1991) *World Species of Monilinia Fungi: Their Ecology, Biosystematics and Control*. Mycologia Memoir No. 16. J. Cramer, Berlin.

Bazzi, C., Stefani, E., Minardi, P. and Mazzucchi, U. (1990) Suscettibilità comparativa del susino a *Xanthomonas campestris* pv. *pruni*. *L'Informatore Agrario* 46, 71–74.

Beckman, T.G. and Pusey, P.L. (2001) Field testing peach rootstocks for resistance to *Armillaria* root rot. *HortScience* 36, 101–103.

Beckman, T.G., Okie, W.R., Nyczepir, A.P., Pusey, P.L. and Reilly, C.C. (1998) Relative susceptibility of peach and plum germplasm to *Armillaria* root rot. *HortScience* 33, 1062–1065.

Beckman, T.G., Pusey, P.L. and Bertrand, P.F. (2003) Impact of fungal gummosis on peach trees. *Hortscience* 38, 1141–1143.

Beckman, T.G., Reilly, C.C., Pusey, P.L. and Hotchkiss, M. (2011) Progress in the management of peach fungal gummosis (*Botryosphaeria dothidea*) in the Southeastern US peach industry. *Journal of the American Pomological Society* 65, 192–200.

Beckman, T.G., Chaparro, J.X. and Sherman, W.B. (2012) 'MP-29', a clonal interspecific hybrid rootstock for peach. *HortScience* 47, 128–131.

Bensaude, M. and Keitt, G.W. (1928) Comparative studies of certain *Cladosporium* diseases of stone fruits. *Phytopathology* 18, 313–329.

Bernstein, B., Zehr, E.I. and Dean R.A. (1995) Characteristics of *Colletotrichum* from peach, apple, pecan, and other hosts. *Plant Disease* 79, 478–482.

Biggs, A.R. (1989) Effect of pruning technique on *Leucostoma* infection and callus formation over wounds in peach trees. *Plant Disease* 73, 771–773.

Biggs, A. and Britton, K.O. (1988). Presymptom histopathology of peach trees inoculated with *Botryosphaeria obtusa* and *B. dothidea*. *Phytopathology* 78, 1109–1118.

Biggs, A.R. and Northover, J. (1985) Inoculum sources for *Monilinia fructicola* in Ontario peach orchards. *Canadian Journal of Plant Pathology* 7, 302–307.

Biggs, A.R. and Northover, J. (1988a) Influence of temperature and wetness duration on infection of peach and sweet cherry fruits by *Monilinia fructicola*. *Phytopathology* 78, 1352–1356.

Biggs, A.R. and Northover, J. (1988b) Early and late-season susceptibility of peach fruits to *Monilinia fructicola*. *Plant Disease* 72, 1070–1074.

Blaauw, B., Brannen, P., Lockwood, D., Schnabel, G. and Ritchie, D. (2021) *Southeastern Peach, Nectarine and Plum Pest Management and Cultural Guide*. Georgia Extension Bulletin No. 1171. University of Georgia Cooperative Extension Service, Athens, Georgia.

Blomquist, C.L. and Kirkpatrick, B.C. (2002) Identification of phytoplasma taxa and insect vectors of peach yellow leaf roll disease in California. *Plant Disease* 86, 759–763.

Bock, C.H., Young, C.A., Zhang, M., Chen, C., Brannen, P.M. *et al.* (2021) Mating type idiomorphs, heterothallism, and high genetic diversity in *Venturia carpophila*, cause of peach scab. *Phytopathology* 111, 408–424.

Boehm, E.W.A., Ma, Z. and Michailides, T.J. (2001) Species-specific detection of *Monilinia fructicola* from California stone fruits and flowers. *Phytopathology* 91, 428–439.

Bolkan, H.A., Ogawa, J.M., Michailides, T.J. and Kable, P.F. (1985) Physiological specialization in *Tranzschelia discolor*. *Plant Disease* 69, 485–486.

Boudon, S., Manceau, C. and Nottéghem, J.-L. (2005) Structure and origin of *Xanthomonas arboricola* pv. *pruni* populations causing bacterial spot of stone fruit trees in Western Europe. *Phytopathology* 95, 1081–1088.

Bouzar, H. and Moore, L.W. (1987) Isolation of different *Agrobacterium* biovars from a natural oak savanna and tallgrass prairie. *Applied and Environmental Microbiology* 53, 717–721.

Brisbane, P.G. and Kerr, A. (2008) Selective media for three biovars of *Agrobacterium*. *Journal of Applied Microbiology* 54, 425–431.

Browne, G. (2017) Resistance to *Phytophthora* species among rootstocks for cultivated *Prunus* species. *Hortscience* 52, 1471–1476.

Browne, G., Schmidt, L. and Brar, G. (2015) First report of *Phytophthora niederhauserii* causing crown rot of almond (*Prunus dulcis*) in California. *Plant Disease* 99, 1863.

Bultreys, A. and Kałuzna, M. (2010) Bacterial cankers caused by *Pseudomonas syringae* on stone fruit species with special emphasis on the pathovars *syringae* and *morsprunorum* race 1 and race 2. *Journal of Plant Pathology* 92, S21–S33.

Byrde, R.J.W. and Willetts, H.J. (1977) *The Brown Rot Fungi of Fruit: Their Biology and Control*. Pergamon Press, New York.

Carraro, L., Osler, R., Loi, N., Ermacora, P. and Refatti, E. (1998) Transmission of European stone fruit yellows phytoplasma by *Cacopsylla pruni*. *Journal of Plant Pathology* 80, 233–239.

Chandler, W.A. (1969) Reduction in mortality of peach trees following preplant soil fumigation. *Plant Disease Reporter* 53, 49–53.

Chen, F., Liu, X., Chen, S., Schnabel, E. and Schnabel, G. (2013) Characterization of *Monilinia fructicola* strains resistant to both propiconazole and boscalid. *Plant Disease* 97, 645–651.

Cieślińska, M. (2011) European stone fruit yellows disease and its causal agent 'Candidatus Phytoplasma prunorum'. *Journal of Plant Protection Research* 51, 441–447.

Côté, M.J., Prud'homme, M., Meldrum, A.J. and Tardif, M.C. (2004) Variations in sequence and occurrence of SSU rDNA group I introns in *Monilinia fructicola* isolates. *Mycologia* 96, 240–248.

Cross, J.E. (1966) Epidemiological relations of the pseudomonad pathogens of deciduous fruit trees. *Annual Review of Phytopathology* 4, 291–310.

Daines, R.H. and Trout, J.R. (1977) Incidence of rusty spot of peach as influenced by proximity to apple trees. *Plant Disease Reporter* 61, 835–836.

Donoso, J.M., Picañol, R., Serra, O., Howard, W., Alegre, S. *et al.* (2016) Exploring almond genetic variability useful for peach improvement: mapping major genes and QTLs in two interspecific almond × peach populations. *Molecular Breeding* 36: 16.

Dowler, W.M. and Petersen, D.H. (1966) Induction of bacterial canker of peach in the field. *Phytopathology* 56, 989–990.

Dowling, M.E., Bridges, W.C., Cox, B.M., Sroka, T., Wilson, J.R. and Schnabel, G. (2019) Preservation of *Monilinia fructicola* genotype diversity within fungal cankers. *Plant Disease* 103, 526–530.

Dunegan, J.C. (1938) The rust of stone fruits. *Phytopathology* 28, 411–427.

EFSA PLH Panel (2014) Scientific opinion on pest categorization of *Xanthomonas arboricola* pv. *pruni* (Smith, 1903). *EFSA Journal* 12: 3857.

English, H., DeVay, J.E., Lilleland, O. and Davis, J.R. (1961) Effect of certain soil treatments on the development of bacterial canker in peach trees. *Phytopathology* 51: 65.

EPPO (European and Mediterranean Plant Protection Organization) (2006) *Xanthomonas arboricola* pv. *pruni*. *EPPO Bulletin* 36, 129–133.

EPPO Global Database (2021) *Monilinia fructicola*: distribution. European and Mediterranean Plant Protection Organization, Paris. Available at: https://gd.eppo.int/taxon/MONIFC/distribution (accessed 8 December 2022).

Evert, D. and Smittle, D. (1989) Phony disease influences peach leaf characteristics. *HortScience* 24, 1000–1002.

Ezra, D., Hershcovich, M. and Shtienberg, D. (2017) Insights into the etiology of gummosis syndrome of deciduous fruit trees in Israel and its impact on tree productivity. *Plant Disease* 101, 1354–1361.

Feliciano, A. and Daines, R.H. (1970) Factors influencing ingress of *Xanthomonas pruni* through peach leaf scars and subsequent development of spring cankers. *Phytopathology* 60, 1720–1726.

Fischer-Le Saux, M., Bonneau, S., Essakhi, S., Manceau, C. and Jacques, M.A.A. (2015) Aggressive emerging pathovars of *Xanthomonas arboricola* represent widespread epidemic clones distinct from poorly pathogenic strains, as revealed by multilocus sequence typing. *Applied Environmental Microbiology* 81, 4651–4668.

Fones, H. and Preston, G.M. (2013) The impact of transition metals on bacterial plant disease. *FEMS Microbiology Review* 37, 495–519.

Förster, H. and Adaskaveg, J.E. (2000) Early brown rot infections in sweet cherry fruit are detected by *Monilinia*-specific DNA primers. *Phytopathology* 90, 171–178.

FRAC (2022) FRAC Code List 2022: Fungal control agents sorted by cross-resistance pattern and mode of action (including coding for FRAC Groups on product labels)). Fungicide Resistance Action Committee. Available at: www.frac.info/docs/default-source/publications/frac-code-list/frac-code-list-2022--final.pdf (accessed 8 December 2022).

Froelich, M. and Schnabel, G. (2019) Investigation of fungi causing twig blight diseases on peach trees in South Carolina. *Plant Disease* 103, 705–710.

Fu, W., da Silva Linge, C. and Gasic, K. (2021) Genome-wide association study of brown rot (*Monilinia* spp.) tolerance in peach. *Frontiers in Plant Science* 12: 354.

Fujii, S. and Hatamoto, M. (1974) Peach withering disease caused by *Armillariella tabescens*. *Shokubutsu Boeki* 28, 1–4.

Fullner, K.J., Cano Lara, J. and Nester, E.W. (1996) Pilus assembly by *Agrobacterium* T-DNA transfer genes. *Science* 273, 1107–1109.

Furman, L.A., Lalancette, N. and White, J.F. (2003a) Peach rusty spot epidemics: Temporal analysis and relationship to fruit growth. *Plant Disease* 87, 366–374.

Furman, L.A., Lalancette, N. and White, J.F. (2003b) Peach rusty spot epidemics: Management with fungicide, effect on fruit growth, and the incidence–lesion density relationship. *Plant Disease* 87, 1477–1486.

Garcin, A. and Bresson, J. (2011) Sensibilité des arbres ànoyau au *Xanthomonas* – bilan de huit ans d'expérimentation. *L'Arboriculture Fruitiére* 653, 30–33.

Garita-Cambronero, J., Palacio-Bielsa, A., López, M.M. and Cubero, J. (2017) Pangenomic analysis permits differentiation of virulent and non-virulent strains of *Xanthomonas arboricola* that cohabit *Prunus* spp. and elucidate bacterial virulence factors. *Frontiers in Microbiology* 8: 573.

Garita-Cambronero, J., Palacio-Bielsa, A. and Cubero, J. (2018). *Xanthomonas arboricola* pv. *pruni*, causal agent of bacterial spot of stone fruits and almond: its genomic and phenotypic characteristics in the *X. arboricola* species context. *Molecular Plant Pathology* 19, 2053–2065.

Gasic, K., Reighard, G., Okie, W., Clark, J., Gradziel, T. *et al.* (2015) Bacterial spot resistance in peach: functional allele distribution in breeding germplasm. *Acta Horticulturae* 1084, 69–74.

Gauthier, N., Smigell, C. and Lyons, K. (2021) *Bacterial Canker and Perennial Canker of Stone Fruit.* Plant Pathology Fact Sheet PPFS-FR-T-08. University of Kentucky, Lexington, Kentucky.

Gell, I., Cubero, J. and Melgajero, P. (2007) Two different PCR approaches for universal diagnosis of brown rot and identification of *Monilinia* spp. in stone fruit trees. *Journal of Applied Microbiology* 103, 2629–2637.

Geng, X., Jin, L., Shimada, M., Kim, M.G. and Mackey, D. (2014) The phytotoxin coronatine is a multifunctional component of the virulence armament of *Pseudomonas syringae*. *Planta* 240, 1149–1165.

Gervasi, F., Ferrante, P., Dettori, M.T., Scortichini, M. and Verde, I. (2018) Transcriptome reprogramming of resistant and susceptible peach genotypes during *Xanthomonas arboricola* pv. *pruni* early leaf infection. *PLoS One* 13: e0196590.

Glass, N.L. and Kosuge, T. (1986) Cloning of the gene for indoleacetic acid-lysine synthetase from *Pseudomonas syringae* subsp. *savastanoi*. *Journal of Bacteriology* 166, 598–603.

Goldsworthy, M.C. and Smith, R.E. (1931) Studies on a rust of clingstone peaches in California. *Phytopathology* 21, 133–168.

Goodman, C.A. (1988) Mechanical transmission of *Xanthomonas campestris* pv. *pruni* in plum nursery trees. *Plant Disease* 72: 643.

Goodman, C.A. and Hatting, M.J. (1986) Transmission of *Xanthomonas campestris* pv. *pruni* in plum and apricot nursery trees by budding. *HortScience* 21, 995–996.

Gottwald, T.R. (1983) Factors affecting spore liberation by *Cladosporium carpophilum*. *Phytopathology* 73, 1500–1505.

Grabke, A., Williamson, M., Henderson, G.W. and Schnabel, G. (2014) First report of anthracnose on peach fruit caused by *Colletotrichum truncatum* in South Carolina. *Plant Disease* 98: 1154.

Gradziel, T.M. (1994) Changes in susceptibility to brown rot with ripening in three cling-stone peach genotypes. *Journal of the American Horticultural Society* 119, 101–105.

Gradziel, T.M. and Wang, D. (1993) Evaluation of brown rot resistance and its relation to enzymatic browning in clingstone peach germplasm. *Journal of the American Horticultural Society* 118, 675–679.

Grass, G., Rensing, C. and Solioz, M. (2011) Metallic copper as an antimicrobial surface. *Applied Environmental Microbiology* 77, 1541–1547.

Guzman, G., Latorre, B.A., Torres, R. and Wilcox, W.F. (2007) Relative susceptibility of peach rootstocks to crown gall and *Phytophthora* root and crown rot in Chile. *Ciencia e Investigación Agraria* 34, 31–40.

Haack, S.E., Wade, L., Förster, H. and Adaskaveg, J.E. (2020) Epidemiology and management of bacterial spot of almond caused by *Xanthomonas arboricola* pv. *pruni*, a new disease in California. *Plant Disease* 104, 1685–1693.

Haas, J.H., Moore, L.W., Ream, W. and Manulis, S. (1995) Universal PCR primers for detection of phytopathogenic *Agrobacterium* strains. *Applied and Environmental Microbiology* 61, 2879–2884.

Hajri, A., Pothier, J.F., Fischer-Le Saux, M., Bonneau, S. and Poussier, S. (2012) Type three effector gene distribution and sequence analysis provide new insights into the pathogenicity of plant-pathogenic *Xanthomonas arboricola*. *Applied Environmental Microbiology* 78, 371–384.

Harrington, T.C. and Wingfield, B.D. (1995) A PCR based identification method for species of *Armillaria*. *Mycologia* 87, 280–288.

Hattingh, M., Roos, I. and Mansvelt, E. (1989) Infection and systemic invasion of deciduous fruit trees by *Pseudomonas syringae* in South Africa. *Plant Disease* 73, 784–789.

Hily, J.M., Singer, S.D., Villani, S.M. and Cox, K.D. (2011) Characterization of the *cytochrome b* (*cyt b*) gene from *Monilinia* species causing brown rot of stone and pome fruit and its significance in the development of QoI resistance. *Pest Management Science* 67, 385–396.

Holb, I.J. and Schnabel, G. (2007) Differential effects of triazoles on mycelial growth and disease measurements of *Monilinia fructicola* isolates with reduced sensitivity to DMI fungicides. *Crop Protection* 26, 753–759.

Hu, M.J., Cox, K.D., Schnabel, G. and Luo, C.X. (2011) *Monilinia* species causing brown rot of peach in China. *PLoS One* 6: e24990.

Hu, M.J., Grabke, A. and Schnabel, G. (2015a) Investigation of the *Colletotrichum* species complex causing peach anthracnose in South Carolina. *Plant Disease* 99, 797–805.

Hu, M.J., Grabke, A., Dowling, M.E., Holstein, H. and Schnabel, G. (2015b) Resistance in *Colletotrichum siamense* from peach and blueberry to thiophanate-methyl and azoxystrobin. *Plant Disease* 99, 806–814.

Ioos, R. and Frey, P. (2000) Genomic variation within *Monilinia laxa*, *M. fructigena* and *M. fructicola*, and application to species identification by PCR. *European Journal Plant Pathology* 106, 373–378.

Jones, A.L. and Ehret, G.R. (1976) Isolation and characterization of benomyl-tolerant strains of *Monilinia fructicola*. *Plant Disease Reporter* 60, 765–769.

Jones, D.A. and Kerr, A. (1989) *Agrobacterium radiobacter* strain K1026, a genetically engineered derivative of strain K84, for biological control of crown gall. *Plant Disease* 73, 15–18.

Kable, P.F. (1974) Spread of *Armillariella* sp. in a peach orchard. *Transaction of the British Mycology Society* 62, 89–98.

Kado, C.I. (2014) Historical account on gaining insights on the mechanism of crown gall tumorigenesis induced by *Agrobacterium tumefaciens*. *Frontiers in Microbiology* 5: 340.

Kałuzna, M., Willems, A., Pothier, J.F., Ruinelli, M., Sobiczewski, P. and Puławska, J. (2016) *Pseudomonas cerasi* sp. nov. (non Griffin, 1911) isolated from diseased tissue of cherry. *Systematic and Applied Microbiology* 39, 370–377.

Kawaguchi, A. (2014) Genetic diversity of *Xanthomonas arboricola* pv. *pruni* strains in Japan revealed by DNA fingerprinting. *Journal of General Plant Pathology* 80, 366–369.

Keitt, G.W. (1917) *Peach Scab and its Control*. Bulletin 395. US Department of Agriculture, Washington, DC.

Kern, H. and Naef-Roth, S. (1975) Zur Bildung von Auxinen und Cytokininen durch Taphrina-Arten. *Journal of Phytopathology* 83, 193–222.

Kirkpatrick, B.C., Uyemoto, J.K. and Purcell, A.H. (1995a) X-disease. In: Ogawa, J.M., Zehr, E.I., Bird, G.W., Ritchie, D.F., Uriu, K. and Uyemoto, J.K. (eds) *Compendium of Stone Fruit Diseases*. American Phytopathology Society Press, St. Paul, MN, USA, pp. 57–59.

Kirkpatrick, B.C., Uyemoto, J.K. and Purcell, A.H. (1995b) Peach yellows. In: Ogawa, J.M., Zehr, E.I., Bird, G.W., Ritchie, D.F., Uriu, K. and Uyemoto, J.K. (eds) *Compendium of Stone Fruit Diseases*. American Phytopathology Society Press, St. Paul, Minnesota, p. 57.

Koch, R.A., Wilson, A.W., Sene, O., Henkel, T.W. and Aime, M.C. (2017) Resolved phylogeny and biogeography of the root pathogen *Armillaria* and its gasteroid relative, *Guyanagaster*. *BMC Evolutionary Biology* 17: 33.

Korhonen, K. (1978) Interfertility and clonal size in the *Armillaria mellea* complex. *Karstenia* 18, 31–42.

Kosuge, T., Heskett, M.G. and Wilson, E.E. (1966) Microbial synthesis and degradation of indole-3-acetic acid. I. The conversion of L-tryptophan to indole-3-acetamide by an enzyme system from *Pseudomonas savastanoi*. *Journal of Biological Chemistry* 241, 3738–3744.

Kramer, C.L. (1987) The Taphriniales. *Studies in Mycology* 30, 151–166.

Lalancette, N. and Blaus, L. (2018) Management of peach bacterial spot: comparison of antibiotics, copper and SAR materials, 2017. *Plant Disease Management Reports* 12: PF030.

Lalancette, N. and Robison, D.M. (2002) Effect of fungicides, application timing, and canker removal on incidence and severity of constriction canker of peach. *Plant Disease* 86, 721–728.

Lalancette, N., Blaus, L. and Rossi, S. (2016) Efficacy of kasugamycin for management of peach bacterial spot, 2015. *Plant Disease Management Reports* 10: ST006.

Lalancette, N., Blaus, L.L., Gager, J.D. and McFarland, K.A. (2017) Contribution of mid-season cover sprays to management of peach brown rot at harvest. *Plant Disease* 101, 794–799.

Lamichhane, J.R. (2014) *Xanthomonas arboricola* diseases of stone fruit, almond, and walnut trees: progress toward understanding and management. *Plant Disease* 98, 1600–1610.

Lawrence, D.P., Travadon, R., Pouzoulet J., Rolshousen, P.E., Wilcox, W.F. and Baumgartner, K. (2017) Characterization of *Cytospora* isolates from wood cankers of declining grapevine in North America, with the descriptions of two new *Cytospora* species. *Plant Pathology* 66, 713–725.

Lawrence, E.G. and Zehr, E.I. (1982) Environmental effects on the development and dissemination of *Cladosporium carpophilum* on peach. *Phytopathology* 72, 773–776.

Lee, D.M., Hassan, O. and Chang, T. (2020) Identification, characterization, and pathogenicity of *Colletotrichum* species causing anthracnose of peach in Korea. *Mycobiology* 48, 210–218.

Lesniak, K.E., Peng, J., Proffer, T.J., Outwater, C.A., Eldred, L.I. *et al.* (2021) Survey and genetic analysis of demethylation inhibitor fungicide resistance in *Monilinia fructicola* from Michigan orchards. *Plant Disease* 105, 958–964.

Lichtemberg, P.S.F., Luo, Y., Morales, R.G., Muehlmann-Fischer, J.M., Michailides, T.J. and May de Mio, L.L. (2017) The point mutation G461S in the *MfCYP51* gene is associated with tebuconazole resistance in *Monilinia fructicola* populations in Brazil. *Phytopathology* 107, 1507–1514.

López-Soriano, P., Boyer, K., Cesbron, S., Morente, M.C., Peñalver, J. *et al.* (2016) Multilocus variable number of tandem repeat analysis reveals multiple introductions in Spain of *Xanthomonas arboricola* pv. *pruni*, the causal agent of bacterial spot disease of stone fruits and almond. *PLoS One* 11: e0163729.

Luo, C.-X. and Schnabel, G. (2008) The cytochrome P450 lanosterol 14α-demethylase gene is a demethylation inhibitor fungicide resistance determinant in *Monilinia fructicola* field isolates from Georgia. *Applied and Environmental Microbiology* 74, 359–366.

Luo, Y., Ma, Z., Reyes, H.C., Morgan, D. and Michailides, T.J. (2007) Quantification of airborne spores of *Monilinia fructicola* in stone fruit orchards of California using real-time PCR. *European Journal of Plant Pathology* 118, 145–154.

Ma, Z.H., Yoshimura, M.A. and Michailides, T.J. (2003) Identification and characterization of benzimidazole resistance in *Monilinia fructicola* from stone fruit orchards in California. *Applied and Environmental Microbiology* 69, 7145–7152.

Mancero-Castillo, D., Beckman, T.G., Harmon, P.F. and Chaparro, J.X. (2018) A major locus for resistance to *Botryosphaeria dothidea* in *Prunus*. *Tree Genetics & Genomes* 14: 26.

Marimon, N., Eduardo, I., Martínez-Minaya, J., Vicent, A. and Luque, J. (2020a) A decision support system based on degree-days to initiate fungicide spray programs for peach powdery mildew in Catalonia, Spain. *Plant Disease* 104, 2418–2425.

Marimon, N., Luque, J., Arús, P. and Eduardo, I. (2020b) Fine mapping and identification of candidate genes for the peach powdery mildew resistance gene *Vr3*. *Horticultural Research* 7: 175.

Marin-Felix, Y., Groenewald, J.Z., Cai, L., Chen, Q. and Crous, P.W. (2017) Genera of phytopathogenic fungi: GOPGY 1. *Studies in Mycology* 86, 99–216.

McFarland, K.A. and Lalancette, N. (2012) Management of bacterial spot on peach with novel bactericides, 2011. *Plant Disease Management Reports* 6: ST009.

Menard, M., Sutra, L., Luisetti, J., Prunier, J. and Gardan, L. (2003) *Pseudomonas syringae* pv. *avii* (pv. nov.), the causal agent of bacterial canker of wild cherries (*Prunus avium*) in France. *European Journal of Plant Pathology* 109, 565–576.

Michailides, T.J., Ogawa, J.M. and Opgenorth, D.C. (1987) Shift of *Monilinia* spp. and distribution of isolates sensitive and resistant to benomyl in California prune and apricot orchards. *Plant Disease* 71, 893–896.

Miessner, S. and Stammler, G. (2010) *Monilinia laxa*, *M. fructigena* and *M. fructicola*: risk estimation of resistance to QoI fungicides and identification of species with cytochrome *b* gene sequences. *Journal of Plant Disease Protection* 117, 162–167.

Miles, W.G., Daines, R.H. and Due, J.W. (1977). Presymptomatic egress of *Xanthomonas pruni* from infected peach leaves. *Phytopathology* 67, 895–897.

Miller, S.B., Gasic, K., Reighard, G.R., Henderson, W.G., Rollins, P.A. *et al.* (2020) Preventative root collar excavation reduces peach tree mortality caused by *Armillaria* root rot on replant sites. *Plant Disease* 104, 1274–1279.

Miller, S.T., Otto, K. L., Sterle, D., Mins, I.S. and Steward, J.E. (2019) Preventive fungicidal control of *Cytospora leucostoma* in peach orchards in Colorado. *Plant Disease* 103, 1138–1147.

Miller, S.T., Sterle, D., Minas, I.S. and Stewart, J.E. (2021) Exploring fungicides and sealants for management of *Cytospora plurivora* infections in western Colorado peach production systems. *Crop Protection* 146: 105654.

Mizell, R.F., Anderson, P.C., Tipping, C. and Brodbeck, B.V. (2020) *Xylella fastidiosa Diseases and Their Leafhopper Vectors*. ENY-683, University of Florida, University Cooperative Extension Program. Available at: https://edis.ifas.ufl.edu/pdf/IN/IN17400.pdf (accessed 8 December 2022).

Mojtahedi, H., Lownsbery, B.F. and Moody, E.H. (1975) Ring nematodes increase development of bacterial canker in plums. *Phytopathology* 65, 556–559.

Moore, L.W. (1976) Latent infections and seasonal variability of crown gall development in seedlings of three *Prunus* species. *Phytopathology* 66, 1097–1101.

Morales, G., Llorente, I., Montesinos, E. and Moragrega, C. (2017) A model for predicting *Xanthomonas arboricola* pv. *pruni* growth as a function of temperature. *PLoS One* 12: e0177583.

Moreira, R.R., Silva, G.A. and de Mio, L.L.M. (2020) *Colletotrichum acutatum* complex causing anthracnose on peach in Brazil. *Australasian Plant Pathology* 49, 179–189.

Morrison, D.E., Williams, R.E. and Whitney, R.D. (1991) Infection, disease development, diagnosis, and detection. In: Shaw, C.G. and Kile, G. (eds) *Armillaria Root Disease*. Agricultural Handbook 691, Forest Service, US Department of Agriculture, Washington, DC, pp. 62–75.

Munnecke, D.E., Wilbur, W.D. and Kolbezen, M.J. (1970) Dosage response of *Armillaria mellea* to methyl bromide. *Phytopathology* 60, 992–993.

Nesmith, W.C. and Dowler, W.M. (1975) Soil fumigation and fall pruning related to peach tree short life. *Phytopathology* 65, 277–280.

Nesmith, W.C. and Dowler, W.M. (1976) Cultural practices affect cold hardiness and peach tree short life. *Journal of the American Horticulture Society* 101, 116–119.

Norphanphoun, C., Doilom, M., Daranagama, D.A., Phookamsak, R., Wen, T.C. *et al.* (2017) Revisiting the genus *Cytospora* and allied species. *Mycosphere* 8, 51–97.

Ogawa, J.M. and Charles, F.M. (1956) Powdery mildew of peach trees. *California Agriculture* 10: 7.

Ogawa, J.M. and English, H. (1991) *Diseases of Temperature Zone Tree Fruit and Nut Crops.* Division of Agriculture and Natural Resources, Publication No. 3345. University of California, Oakland, CA.

Ogawa, J.M., Manji, B.T., Adaskaveg, J.E. and Michailides, T.J. (1988) Population dynamics of benzimidazole-resistant *Monilinia* species on stone fruit trees in California. In: Delp, C.J. (ed.) *Fungicide Resistance in North America.* American Phytopathology Society Press, St. Paul, Minnesota, pp. 36–39.

Ogawa, J.M., Zehr, E.I., Bird, G.W., Ritchie, D.F., Uriu, K. and Uyemoto, J.K. (1995) *Compendium of Stone Fruit Diseases.* American Phytopathology Society Press, St. Paul, Minnesota.

Okie, W.R., Beckman, T.G., Nyczepir, A.P., Reighard, G.L., Newall, W.C. and Zehr, E.I. (1994) BY5209, a peach rootstock for the southeastern United States that increases scion longevity. *HortScience* 29, 705–706.

Pascal, T., Pfeiffer, F. and Kervella, J. (2010) Powdery mildew resistance in the peach cultivar Pamirskij 5 is genetically linked with the *Gr* gene for leaf color. *HortScience* 45, 150–152.

Patil, S.S. and Tam, L.Q. (1972) Mode of action of the toxin from *Pseudomonas phaseolicola*: I. Toxin specificity, chlorosis, and ornithine accumulation. *Plant Physiology* 49, 803–807.

Phillips, D.J. and Harvey, J.M. (1975). Selective medium for detection of inoculum of *Monilinia* spp. on stone fruits. *Phytopathology* 65, 1233–1236.

Poggi-Pollini, C., Bissani, R. and Ginchedi, L. (2001) Occurrence of European stone fruit phytoplasma (ESFYP) infection in peach orchards in northern-central Italy. *Journal of Phytopathology* 149, 725–730.

Poniatowska, A., Michalecka, M. and Bielenin, A. (2013) Characteristic of *Monilinia* spp. fungi causing brown rot of pome and stone fruits in Poland. *European Journal of Plant Pathology* 135, 855–865.

Popović, T., Menković, J., Prokić, A., Zlatković, N. and Obradović, A. (2021) Isolation and characterization of *Pseudomonas syringae* isolates affecting stone fruits and almonds in Montenegro. *Journal of Plant Diseases and Protection* 128, 391–405.

Prunier, J.P., Luisetti, J. and Gardan, L. (1970) Etudes sur les bacterioses des arbres fruitiers. II. Caractérisation d'un *Pseudomonas* non fluorescent, agent d'une bacteriose nouvelle chez le pêcher. *Annales de Phytopathologie* 2, 168–197.

Pusey, P.L., Kitajima, H. and Wu, Y. (1995) Fungal gummosis. In: Ogawa, J.M., Zehr, E.I., Bird, G.W., Ritchie, D.F., Uriu, K. and Uyemoto, J.K. (eds) *Compendium of Stone Fruit Diseases.* American Phytopathology Society Press, St. Paul, Minnesota, pp. 33–34.

Rhoads, A.S. (1950) *Clitocybe Root Rot of Woody Plants in the Southeastern United States.* Circular 853. US Department of Agriculture, Washington, DC.

Ries, S.M. and Royse, D.J. (1978) Peach rusty spot epidemiology: incidence as affected by distance from a powdery mildew-infected apple orchard. *Phytopathology* 68, 896–899.

Ritchie, D.F. (1995) Bacterial spot. In: Ogawa, J.M., Zehr, E.I., Bird, G.W., Ritchie, D.F., Uriu, K. and Uyemoto, J.K. (eds) *Compendium of Stone Fruit Diseases.* American Phytopathological Society, St. Paul, Minnesota, pp. 50–52.

Ritchie, D.F. (1999) Sprays for control of bacterial spot of peach cultivars having different levels of disease susceptibility, 1998. *Fungicide and Nematicide Tests* 54, 63–64.

Ritchie, D.F. (2006) Copper material, FlameOut, Prophyt, and Serenade ASO for bacterial spot management on peaches, 2005. *Fungicide and Nematicide Tests* 61: STF007.

Ritchie, D.F. (2008) Evaluation of materials and application schedules for bacterial spot management on peaches, 2007. *Plant Disease Management Reports* 2: STF010.

Ritchie, D.F. (2011) Actigard applied as peach soil drench and drip compared to standard foliar sprays – 2009, 2010. *Plant Disease Management Reports* 5: ST013.

Ritchie, D.F. and Bennett, M.H. (1993) Use of early season sprays to manage bacterial spot on fruit of peaches. *Fungicide and Nematicide Tests* 48: 62.

Ritchie, D.F. and Clayton, C.N. (1981) Peach tree short life: a complex of interacting factors. *Plant Disease* 65, 462–469.

Ritchie, D.F. and Werner, D.J. (1981) Susceptibility and inheritance of susceptibility to peach leaf curl in peach and nectarine cultivars. *Plant Disease* 65, 731–734.

Ritchie, D.F., Barba, M. and Pagani, M.C. (2008) Diseases caused by prokaryotes – bacteria and phytoplasmas. In: Layne, D.R. and Bassi, D. (eds) *The Peach: Botany, Production and Uses*. CAB International, Wallingford, UK, pp. 416–418.

Rizzo, D.M., Whiting, E.C. and Elkins, R.B. (1998) Spatial distribution of *Armillaria mellea* in pear orchards. *Plant Disease* 82, 1226–1231.

Roos, I.M. and Hattingh, M. (1986) Resident populations of *Pseudomonas syringae* on stone fruit tree leaves in South Africa. *Phytophylactica* 18, 55–58.

Roselló, M., Santiago, R., Palacio-Bielsa, A., García-Figueres, F., Montón, C., Cambra, M. and López, M. (2012) Current status of bacterial spot of stone fruits and almond caused by *Xanthomonas arboricola* pv. *pruni* in Spain. *Journal of Plant Pathology* 94, 15–21.

Rossi, V. and Languasco, L. (2007) Influence of environmental conditions on spore production and budding in *Taphrina deformans*, the causal agent of peach leaf curl. *Phytopathology* 97, 359–365.

Savage, E.F., Weinberger, J.H., Luttrell, E.S. and Rhoads, A.S. (1953) Clitocybe root rot – a disease of economic importance in Georgia peach orchards. *Plant Disease Reporter* 37, 269–270.

Savage, E.F., Hayden, R.A. and Futral, J.G. (1974) Effect of soil fumigants on growth, yield, and longevity of Dixired peach trees. *University of Georgia, Research Bulletin* 148, 3–22.

Schellenberg, B., Ramel, C. and Dudler, R. (2010) *Pseudomonas syringae* virulence factor syringolin A counteracts stomatal immunity by proteasome inhibition. *Molecular Plant–Microbe Interactions* 23, 1287–1293.

Scherm, H., Savelle, A.T., Boozer, R.T. and Foshee, W.G. (2008) Seasonal dynamics of conidial production potential of *Fusicladium carpophilum* on twig lesions in southeastern peach orchards. *Plant Disease* 92, 47–50.

Schnabel, G., Bryson, P.K., Bridges, W.C. and Brannen, P.M. (2004) Reduced sensitivity in *Monilinia fructicola* to propiconazole in Georgia and implications for disease management. *Plant Disease* 88, 1000–1004.

Schnabel, G., Ash, J.S. and Bryson, P.K. (2005) Identification and characterization of *Armillaria tabescens* from the southeastern United States. *Mycological Research* 109, 1208–1222.

Schnabel, G., Rollins, P.A. and Henderson, G.W. (2012a) Field evaluation of *Trichoderma* spp. for control of *Armillaria* root rot. *Plant Health Progress* 12, doi:10.1094/PHP-2011-1129-01-RS.

Schnabel, G., Agudelo, P., Henderson, G.W. and Rollins, P.A. (2012b) Above-ground root collar excavation of peach trees for *Armillaria* root rot management. *Plant Disease* 96, 681–686.

Scholz-Schroeder, B.K., Soule, J.D. and Gross, D.C. (2003) The *sypA*, *sypB*, and *sypC* synthetase genes encode twenty-two modules involved in the nonribosomal peptide synthesis of syringopeptin by *Pseudomonas syringae* pv. *syringae* B301D. *Molecular Plant Microbe Interactions* 16, 271–280.

Shaw, D.A., Adaskaveg, J.E. and Ogawa, J.M. (1990) Influence of wetness period and temperature on infection and development of shot-hole disease of almond caused by *Wilsonomyces carpophilus*. *Phytopathology* 80, 749–756.

Shepard, D.P. and Zehr, E.I. (1986) Epiphytic behavior of *Xanthomonas campestris* pv. *pruni* on peach and plum. *Phytopathology* 76: 1070.

Shepard, D.P. and Zehr, E.I. (1994) Epiphytic persistence of *Xanthomonas campestris* pv. *pruni* on peach and plum. *Plant Disease* 78, 627–629.

Shepard, D.P., Zehr, E.I. and Bridges, W.C. (1999) Increased susceptibility to bacterial spot of peach trees growing in soil infested with *Criconemella xenoplax*. *Plant Disease* 83, 961–963.

Sherif, S., Paliyath, G. and Jayasankar, S. (2011) Molecular characterization of peach PR genes and their induction kinetics in response to bacterial infection and signaling molecules. *Plant Cell Reporter* 4, 697–711.

Sierra, A.P., Whitehead, D.S. and Whitehead, M.P. (1999) Investigation of a PCR-based method for the routine identification of British *Armillaria* species. *Mycological Research* 103, 1631–1636.

Smith, C.O. and Smith, D.J. (1942) Host range and growth-temperature relations of *Coryneum beijerinckii*. *Phytopathology* 32, 221–225.

Smith, E.F. (1903) Observations on a hitherto unreported bacterial disease, the cause of which enters the plant through ordinary stomata. *Science* 17, 456–457.

Snyder, C.L. and Jones, A.L. (1999) Genetic variation between strains of *Monilinia fructicola* and *Monilinia laxa* isolated from cherries in Michigan. *Canadian Journal of Plant Pathology* 21, 70–77.

Socquet-Juglard, D., Duffy, B., Pothier, J.F., Christen, D. and Gessler, C. (2012) Identification of a major QTL for *Xanthomonas arboricola* pv. *pruni* resistance in apricot. *Tree Genetics and Genomes* 9, 409–421.

Socquet-Juglard, D., Kamber, T., Pothier, J.F., Christen, D., Gessler, C. *et al.* (2013) Comparative RNA-Seq analysis of early-infected peach leaves by the invasive phytopathogen *Xanthomonas arboricola* pv. *pruni*. *PLoS One* 8: e54196.

Soto-Estrada, A. and Adaskaveg, J.E. (2004) Temporal and quantitative analysis of stem lesion development and foliar disease progression of peach rust in California. *Phytopathology* 94, 52–60.

Soto-Estrada, A., Förster, H., Hasey, J. and Adaskaveg, J.E. (2003) New fungicides and application strategies based on inoculum and precipitation for managing stone fruit rut on peach in California. *Plant Disease* 87, 1094–1101.

Soto-Estrada, A., Förster, H., DeMason, D.A. and Adaskaveg, J.E. (2005) Initial infection and colonization of leaves and stems of cling peach by *Tranzschelia discolor*. *Phytopathology* 95, 942–950.

Stefani, E. (2010) Economic significance and control of bacterial spot/canker of stone fruits caused by *Xanthomonas arboricola* pv. *pruni*. *Journal of Plant Pathology* 92, 99–104.

Strand, L. (1999) Integrated Pest Management for Stone Fruits. University of California Statewide Integrated Pest Management Program. Agriculture and Natural Resource Publication No. 3389. University of California, Oakland, California.

Surico, G., Iacobellis, N.S. and Sisto, A. (1985) Studies on the role of indole-3-acetic acid and cytokinins in the formation of knots on olive and oleander plants by *Pseudomonas syringae* pv. *savastanoi*. *Physiological Plant Pathology* 26, 309–320.

Sutton, B.C. (1997) On *Stigmina*, *Wilsonomyces*, and *Thyrostroma* (Hyphomycetes). *Arnoldia* 14, 33–35.

Sutton, T.B. and Clayton, C.N. (1972) Role and survival of *Monilinia fructicola* in blighted peach branches. *Phytopathology* 62, 1369–1373.

Sziraki, I., Balaazs, E. and Kirali, Z. (1975) Increase levels of cytokinin and indoleacetic acid in peach leaves infected with *Taphrina deformans*. *Physiological Plant Pathology* 5, 45–50.

Tamm, L. and Flückiger, W. (1993) Influence of temperature and moisture on growth, spore production and conidial germination of *Monilinia laxa*. *Phytopathology* 83, 1321–1326.

Termorshuizen, A.J. (2000) Ecology and epidemiology of *Armillaria*. In: Fox, R.T.V. (ed.) *Armillaria Root Rot: Biology and Control of Honey Fungus*. Intercept, Andover, UK, pp. 45–64.

Thomidis, T. (2003) The *Phytophthora* diseases of stone fruit trees in Greece. *Archives of Phytopathology and Plant Protection* 36, 69–80.

Turner, J.G. (1981) Tabtoxin, produced by *Pseudomonas tabaci*, decreases *Nicotiana tabacum* glutamine synthetase *in vivo* and causes accumulation of ammonia. *Physiological Plant Pathology* 19, 57–67.

van Brouwershaven, I.R., Bruil, M.L., van Leeuwen, G.C.M. and Kox, L.F.F. (2010) A real-time (TaqMan) PCR assay to differentiate *Monilinia fructicola* from other brown rot fungi of fruit crops. *Plant Pathology* 59, 548–555.

van Leeuwen, G.C.M., Baayen, R.P., Holb, I.J. and Jeger, M.J. (2002) Distinction of the Asiatic brown rot fungus *Monilia polystroma* sp. nov. from *M. fructigena*. *Mycological Research* 106, 444–451.

Vauterin, L., Hoste, B., Kersters, K. and Swings, J. (1995) Reclassification of *Xanthomonas*. *International Journal of Systematic and Evolutionary Bacteriology* 45, 472–489.

Villani, S.M. and Cox, K.D. (2011) Characterizing fenbuconazole and propiconazole sensitivity and prevalence of 'Mona' in isolates of *Monilinia fructicola* from New York. *Plant Disease* 95, 828–834.

Vonkreuzhof, M.W., Worthington, M.L., Clark, J.R., Gasic, K., da Silva Linge, C. and Fu, W. (2019) Mapping QTLs for peach (*Prunus persica*) resistance to bacterial spot (*Xanthomonas arboricola*) pv. *pruni* (*Xap*). *HortScience* 54, S370–S371.

Wang, F., Zhao, L., Li, G., Huang, J. and Hsiang, T. (2011) Identification and characterization of *Botryosphaeria* spp. causing gummosis of peach trees in Hubei Province, central China. *Plant Disease* 95, 1378–1384.

Weaver, D.J.A. (1974) Gummosis disease of peach trees caused by *Botryosphaeria dothidea*. *Phytopathology* 64, 1429–1432.

Weaver, D.J. and Wehunt, E.J. (1975) Effect of soil pH on susceptibility of peach to *Pseudomonas syringae*. *Phytopathology* 65, 984–989.

Weinhold, A.R. (1961) The orchard development of peach powdery mildew. *Phytopathology* 51, 478–481.

Wells, J.M., Raju, B.C. and Nyland, G. (1983) Isolation, culture, and pathogenicity of the bacterium causing phony disease of peach. *Phytopathology* 73, 859–862.

White, E.E., Dubertz, C.P., Cruickshank, M.G. and Morrison, D.J. (1998) DNA diagnostic for *Armillaria* species in British Columbia: within and between species variation in the IGS-1 and IGS-2 regions. *Mycologia* 90, 25–131.

Wicks, T.J., Volle, D. and Lee, T.C. (1983) Effect of fungicides on infection of apricot and cherry pruning wounds inoculated with *Chondrostereum purpureum*. *Animal Production Science* 23, 91–94.

Wilcox, W.F. and Ellis, M.A. (1989) *Phytophthora* root and crown rot of peach in the eastern Great Lakes region. *Plant Disease* 73, 794–498.

Wilkins, B.S., Ebel, R.C., Dozier, W.A., Pitts, J., Eakes, D.J. *et al.* (2002) Field performance of Guardian™ peach rootstock selections. *HortScience* 37, 1049–1052.

Williams, R.E. and Helton, A.W. (1971) An optimum environment for the culturing of *Coryneum carpophilum*. *Phytopathology* 61, 829–830.

Wilson, E. (1953) Bacterial canker of stone fruits. In: *The Yearbook of Agriculture*. US Department of Agriculture, Washington, DC, pp. 722–729.

Wilson, E.E. (1937) The shot-hole disease of stone-fruit trees. *California University Agriculture Experiment Station Bulletin* 608, 3–40.

Yang, N. (2013) Mapping quantitative trait loci associated with resistance to bacterial spot (Xanthomonas arboricola pv. pruni) in peach. PhD thesis, Clemson University. Clemson, South Carolina.

Yarwood, C.E. (1939) Powdery mildews of peach and rose. *Phytopathology* 29, 282–284.

Ye, S., Jia, H., Cai, G., Tian, C. and Ma, R. (2020) Morphology, DNA phylogeny, and pathogenicity of *Wilsonomyces carpophilus* isolate causing shot-hole disease of *Prunus divaricata* and *Prunus armeniaca* in wild-fruit forest of western Tianshan mountains, *China. Forests* 11: 319.

Yin, L.-F., Chen, S.-N., Chen, G.-K., Schnabel, G., Du, S.-F. *et al.* (2015) Identification and characterization of three *Monilinia* species from plum in China. *Plant Disease* 99, 1775–1783.

Young J.M. (1988) *Pseudomonas syringae* pv. *persicae* from nectarine, peach, and Japanese plum in New Zealand. *OEPP/EPPO Bulletin* 18, 141–151.

Young, J.M., Park, D.-C., Shearman, H.M. and Fargier, E. (2008) A multilocus sequence analysis of the genus *Xanthomonas*. *Systematic and Applied Microbiology* 31, 366–377.

Zehr, E.I., Miller, R.W. and Smith, F.H. (1976) Soil fumigation and peach rootstocks for protection against peach tree short life. *Phytopathology* 66, 689–694.

Zehr, E.I., Luszcz, L.A., Olien, W.C., Newall, W.C. and Toler, J.E. (1999) Reduced sensitivity in *Monilinia fructicola* to propiconazole following prolonged exposure in peach orchards. *Plant Disease* 83, 913–916.

Zhu, F.X., Bryson, P.K., Amiri, A. and Schnabel, G. (2010) First report of the β-tubulin E198A allele for fungicide resistance in *Monilinia fructicola* from South Carolina. *Plant Disease* 94, 1511.

13

MAJOR POSTHARVEST DISEASES OF PEACH AND NECTARINE WITH A REVIEW OF THEIR MANAGEMENT

James E. Adaskaveg* and Helga Förster
Department of Microbiology and Plant Pathology, University of California, Riverside, California, USA

13.1   INTRODUCTION

Peaches and nectarines are grown commercially in most countries with temperate or Mediterranean climates. The largest peach and nectarine producers worldwide are China, Spain, Italy, Turkey, Greece, Iran and the USA with yields of 15.0, 1.4, 1.0, 0.9, 0.9, 0.7 and 0.6 million t in 2020 (FAOSTAT, 2020). The most important postharvest decays of peach fruit are caused by fungi, and they occur worldwide wherever the crop is grown (Table 13.1). These organisms are ubiquitous, heterotrophic and require water for growth. The major pathogens belong to the Eumycota in the phyla Ascomycota and Mucoromycota, which are characterized by the specific ways that sexual spores or meiospores are produced. In the Ascomycota, sexual meiospores (result of meiosis) are called ascospores and are produced within an ascus that may or may not be formed inside a fruiting structure called an ascoma. Asexual mitospores (a result of mitosis) called conidia are produced from septate hyphae and are typically dispersed by wind and wind-driven rain. The Ascomycota fungi that cause postharvest fruit decays are generally also flower and foliar pathogens. In the Mucoromycota, the sexual zygospore is always produced without a fruiting body. Karyogamy and meiosis occur in the zygosporangium and zygospore, respectively. The zygospore produces the germ sporangium and releases sporangiospores. Abundant asexual mitospores are produced inside a sac-like structure, the sporangium, which develops from coenocytic mycelium. The Mucoromycota are soil-borne organisms and can complete their life cycle in the soil. Most fungi in both phyla are necrotrophs, killing the host tissues in advance of colonization.

* Email: jim.adaskaveg@ucr.edu

© CAB International 2023. *Peach* (G. Manganaris, G. Costa and C. Crisosto eds)
DOI: 10.1079/9781789248456.0013

Table 13.1. Postharvest fungal diseases of peach and nectarine and important information on their biology, epidemiology and management.

Disease	Pathogen	Important reproductive stage(s) for infection	Symptoms of economic significance	Primary management
Common				
Brown rot	*Monilinia fructicola, M. laxa, M. fructigena, M. mumecola, M. polystroma, M. yunnanensis*	Conidia	Fruit decay	Preharvest management of blossom blight, green fruit rot and fruit decays. Harvest and postharvest practices include:
Grey mould	*Botrytis cinerea*	Conidia	Fruit decay	Minimize injuries, temperature management, sanitation, pre- and postharvest fungicides
Rhizopus rot	*Rhizopus stolonifer*	Sporangiospores	Fruit decay	Minimize injuries, temperature management, sanitation, postharvest fungicides
Gilbertella decay	*Gilbertella persicaria*	Sporangiospores	Fruit decay	Minimize injuries, temperature management, sanitation, postharvest fungicides
Mucor decay	*Mucor piriformis*, other *Mucor* spp.	Sporangiospores	Fruit decay	Minimize injuries, temperature management, sanitation, postharvest fungicides
Sour rot	*Geotrichum candidum*	Arthroconidia	Fruit decay	Temperature management, sanitation, pre- and postharvest fungicides

Less frequent				
Alternaria rot	*Alternaria alternata, other Alternata* spp.	Conidia	Decay of old fruit	Minimize injuries, temperature management, sanitation, fungicides
Black mould	*Aspergillus niger*, species in *A. niger* complex	Conidia	Decay of old fruit	Minimize injuries, temperature management, sanitation, fungicides
Blue mould	*Penicillium expansum*, other *Penicillium* spp.	Conidia	Decay of old fruit	Minimize injuries, temperature management, sanitation, fungicides
Botryosphaeria fruit rot	*Botryosphaeria dothidea, B. obtusa, Lasiodiplodia theobromae*	Conidia	Fruit decay (sunken soft lesions, no sporulation)	Preharvest disease management with fungicides, sanitation
Cladosporium rot	*Cladosporium herbarum*	Conidia	Decay of old fruit	Minimize injuries, temperature management, sanitation, postharvest fungicides
Sour pit	*Candida inconsipicua*	Conidia	Fruit decay	None available

Ecologically, fungi have different strategies of survival. Some, such as members of the Mucoromycota, grow, reproduce and colonize suitable substrates rapidly in disruptive environments and are known as ruderals or R-strategists. Other fungi, such as the postharvest decay-causing Ascomycota, can grow and reproduce extensively in a stable environment with a large carrying capacity, and are known as K-strategists. *Monilinia* and *Botrytis* spp. can cause quiescent infections (i.e. tissue penetration but delayed colonization until physiological changes of the host occur) during early fruit development (Adaskaveg *et al.*, 1999) or active infections during favourable field or postharvest environments, and can exploit injuries with massive amounts of air-borne spores that contaminate fruit surfaces.

Management strategies for postharvest decay of peach and other stone fruits employ a wide range of practices that begin with appropriate harvest and handling practices to minimize fruit injuries and end with suitable storage conditions of the commodity (Crisosto and Mitchell, 2002; Adaskaveg *et al.*, 2002, 2022, 2023; Adaskaveg and Förster, 2022). Grading of fruit in the packinghouse eliminates most fruit with visible injuries and of low quality, and this represents a first step in preparing and packing the commodity for marketing. Oxidation treatments predominate sanitation practices used for handling peaches and other stone fruits. These treatments are essential for removing and inactivating spores of decay-causing fungi, as well as minimizing potential contamination with human bacterial pathogens (Bartz *et al.*, 2022). Low temperatures that slow the fruit respiration rate and high relative humidity that prevents water loss are generally ideal for storing commodities over an extended period and distributing them over long distances. Fungicide and some biological treatments provide a sustainable, decay-free supply of stone-fruit commodities and allow marketing of fruit locally and over long distances (Eckert and Ogawa, 1988). Currently, temperature management that is carefully monitored throughout postharvest handling, transportation and marketing and postharvest fungicide treatments are considered the most effective strategies to prevent losses from decay and to maintain postharvest crop life. A review of postharvest decays as well as management strategies and future goals is presented in this chapter. More thorough reviews of postharvest decays of stone fruit and other crops and practices for their management are available (Snowdon, 1990; Mari *et al.*, 2020; Adaskaveg and Förster, 2022; Adaskaveg *et al.*, 2022, 2023). The biology and epidemiology of many decay-causing fungi has been described in detail previously (Ogawa and English, 1991; Ogawa *et al.*, 1995).

13.2   POSTHARVEST DECAY OF PEACH

Peach is affected by few postharvest decays; however, the major decays of brown rot, grey mould and Rhizopus rot develop worldwide, occur every year and can cause extensive crop losses. Most of the minor decays, including those caused by *Aspergillus, Penicillium* and *Cladosporium* spp., generally only affect overripe fruit or fruit that have been stored for prolonged time periods, and therefore can be minimized using appropriate marketing strategies (Table 13.1).

13.2.1 Brown rot and grey mould

Brown rot caused by *Monilinia fructicola* Winter (Honey), *M. laxa* (Aderh. & Ruhl.) Honey, *M. fructigena* Honey in Whetzel, and other species such as *M. polystroma* (G. van Leeuwen) L.M. Kohn (anamorphic syn. *Monilia polystroma* van Leeuwen), *M. mumecola* (Y. Harada, Y. Sasaki & Sano) Sandoval-Denis & Crous (anamorphic syn. *Monilia mumecola* Y. Harada, Y. Sasaki & Sano) and *M. yunnanensis* (M.J. Hu & C.X. Luo) Sandoval-Denis & Crous is the most important postharvest decay of peach and other stone fruits (Mari *et al.*, 2020; Adaskaveg *et al.*, 2022). The causal species are difficult to differentiate morphologically but can be identified using molecular methods (Gell *et al.*, 2007; Wang *et al.*, 2018). Their distribution varies by geographical production region. Losses from brown rot occur every year but are especially high in areas where the more aggressive *M. fructicola* is present, and without management, large amounts of the crop may be destroyed (Villarino *et al.*, 2016). With worldwide trade of fruit commodities, this pathogen has been introduced into areas where it previously was not known to occur such as Chile (Latorre *et al.*, 2014) and Europe (Abate *et al.*, 2018). Grey mould, caused by *Botrytis cinerea* Pers.: Fr., also occurs every year, but damage is generally less serious than for brown rot. In South Africa, however, *B. cinerea* is considered the most important pathogen causing blossom blight and postharvest decay of early and mid-season cultivars of nectarine and plum (Fourie *et al.*, 2002). Preharvest aspects of the brown rot and grey mould pathogens were discussed in Chapter 12 (this volume).

  Fungal inoculum starts increasing in the orchard when blossom blight and jacket rot or green fruit rot are not managed (Ogawa and English, 1991). At harvest time when air-borne, surface-deposited conidia gain entry through small wounds on the fruit that commonly occur during picking, bulk handling, and transport to a packinghouse, postharvest decay can be initiated. Conidia germinate and start growing into the fruit within 4–6 h at 20–25°C. Some of the postharvest brown rot and grey mould decay originates from quiescent infections. These infections occur during early fruit development, and pathogen growth is delayed until fruit maturity when the fungus can actively cause decay and colonize the fruit. A model has been developed to predict pre- and postharvest disease outbreaks and is based on: (i) the time of the first appearance of air-borne conidia, the first appearance of conidia on the surface of flowers and fruits, and the first quiescent (latent) infections; (ii) the number of conidia on the fruit surface 2 weeks and 1 week before harvest;

(iii) the preharvest incidence of brown rot; and (iv) the mean environmental temperature from pink bud to harvest (Villarino *et al.*, 2012).

Both brown rot and grey mould decays are firm, brownish in colour and develop rapidly (Figs 13.1–13.3). Fruit may be completely rotted after 3–4 days at 20°C. The two pathogens can grow at low temperatures, even at 0°C, although slowly. Conidia of *Monilinia* spp. can germinate at water activity levels as low as 0.95 at 0°C in just 3–4 days (Casals *et al.*, 2009). The two types of decay are not easily differentiated at early stages, but decay caused by *B. cinerea* is of a slightly lighter brown to tan colour than that caused by *Monilinia* spp. At advanced decay stages, the fruit surface is covered with cottony fungal mycelium and conidia. Fungal mycelium and sporulation of *Monilinia* spp. are light brown in colour, whereas those of *B. cinerea* are grey. Brown rot decay of fruit in cold storage often appears black with little or no sporulation. The black-coloured areas (pseudosclerotia) consist of fungal and host tissues (Byrde and Willets, 1977; Batra, 1991). *B. cinerea* produces sclerotia of dense, darkly pigmented mycelium in plant tissue. Pseudosclerotia and sclerotia are types of overwintering or survival stages of these fungi in addition to other mechanisms such as fungal mycelium in cankers or woody tissues.

13.2.2 Rhizopus rot and Gilbertella and Mucor decays

Three fungal species in the Mucoromycota (order Mucorales) that can cause significant postharvest losses on peach and nectarine are *Rhizopus stolonifer* (Ehrenb.:Fr.) Vuill., *Gilbertella persicaria* (E.D. Eddy) Hesselt. and *Mucor piriformis* E. Fisch. (Adaskaveg *et al.*, 2022, 2023). These three fungi are rarely a

Fig. 13.1. Postharvest brown rot of peach fruit caused by *Monilinia fructicola*.

Fig. 13.2. Postharvest grey mould of nectarine fruit caused by *Botrytis cinerea*.

Fig. 13.3. Postharvest brown rot (left) and grey mould (right) on the same peach fruit.

problem preharvest, except in warm, wet climates where infections may occur through injuries or natural cracks on ripening fruit on the tree. The spores of these fungi are ubiquitous in soils and are easily wind disseminated. Infections occur on mature fruit that are injured or bruised, especially during harvest and postharvest handling. *Rhizopus*-infected fruit are first covered with a thick, cottony mycelial layer that rapidly starts sporulating, forming tiny, black, terminal sporangia that contain large numbers of sporangiospores (Fig. 13.4). Infections by *R. stolonifer* of immature fruit develop slowly. Infected mature and ripe fruit decay very rapidly, and an entire fruit may turn into a soft, watery rot within 1–2 days at optimum temperatures (Harvey *et al.*, 1972). Likewise, a cottony mycelial mat, which is thinner than that of *R. stolonifer*, first covers *Gilbertella*-infected fruit. Small, black, glistening sporangia are produced terminally on the mycelium (Fig. 13.5). *Mucor*-infected fruit initially looks similar to Rhizopus rot, but the sporangiophores are often longer and the sporangia are tan to brown in colour. Fruit infections by these pathogens may spread quickly by hyphal contact to healthy fruit (nesting), destroying large numbers of fruit in a basket or box. *R. stolonifer* and *G. persicaria* do not grow at temperatures below 4°C. The optimum temperature range for growth of *R. stolonifer* is 21–27°C with a maximum of 33°C, whereas for *G. persicaria*, the optimum range is 30–33°C with a maximum of 39°C. In contrast, the optimum range for growth of *M. piriformis* is 10–15°C, and the fungus can grow at temperatures near freezing but not at or above 27°C. These temperature requirements indicate that Rhizopus and Gilbertella rot

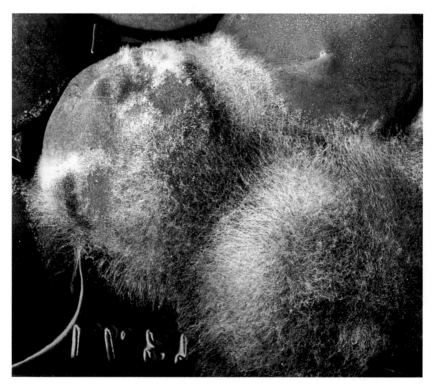

Fig. 13.4. Postharvest Rhizopus rot of peach fruit caused by *Rhizopus stolonifer* with decay spreading by contact to healthy fruit.

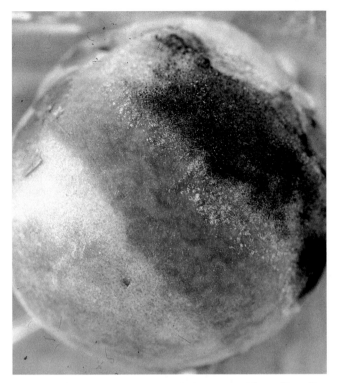

Fig. 13.5. Postharvest Gilbertella rot of peach fruit caused by *Gilbertella persicaria*.

can be managed effectively by fruit storage at appropriate low temperatures (0–2°C), but Mucor decay will still progress under these conditions.

13.2.3   Sour rot

Sour rot of peach, caused by *Geotrichum candidum* Link (teleomorphic synonym *Galactomyces geotrichum* (E.E. Butler and L.J. Petersen) Redhead and Malloch), has only been infrequently reported to cause problems of traditionally handled and marketed fruit (e.g. fruit picked, hydrocooled, transported at low temperatures and displayed at markets) (Adaskaveg *et al.*, 2022). The pathogen can cause decay of many fruits and vegetable crops. Peach fruit that are tree ripened or pre-conditioned (i.e. fruit ripened after harvest for 1–2 days or until a pre-selected firmness is reached and then refrigerated) and ripe fruit are more prone to infection. White-fleshed peaches and nectarines are generally more susceptible to decay. Sour rot may also occur on severely injured, immature fruit, and is prevalent on fruit with split pits. Sour rot-like infections may also be caused by other yeasts and possibly other organisms that are not well characterized.

Symptoms include a brown, watery, soft decay with a thin layer of white mycelial growth on the fruit surface at high relative humidity (Fig. 13.6). The decay may reach the pit and consume the entire fruit. Rotted fruit have a characteristic yeasty to vinegary odour; however, other odours may develop with bacterial contamination, which commonly occurs in the watery decay.

G. candidum is a wound pathogen that decays fruit after spores are deposited into injuries. Punctures, cracks or abrasions may be caused mechanically in the orchard before or during harvest, and insect damage may also function as infection sites. *G. candidum* is widespread and occurs on organic matter in the soil (Harvey *et al.*, 1972) and has been found in peach orchard soils (Yaghmour *et al.*, 2012b) and in dust on fruit surfaces. Inoculum is disseminated by wind-blown dust and can also be transmitted by vinegar flies. The minimum temperature for spore germination, growth and infection of the fungus is about 2°C, with an optimum of 25–27°C and a maximum of 38°C.

13.2.4 Other postharvest decays

Several decays can occur sporadically on senescent peach fruit (Table 13.1) (Adaskaveg *et al.*, 2022, 2023). Alternaria rot caused by *Alternaria alternata* and other closely related *Alternaria* spp. develops on peaches that are stored at low temperatures for long periods. The decay is generally not found on freshly harvested fruit. The lesions are dark green to black, firm, sunken and moist (Fig. 13.7). The decay is shallow initially; however, it may extend into the flesh in a cone shape. At high humidity, older lesions may be covered with olive green, muriform conidia.

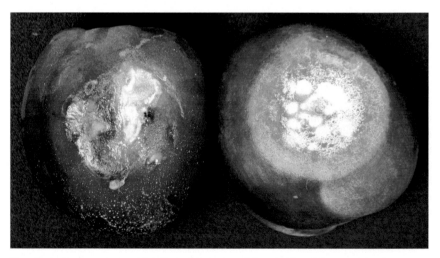

Fig. 13.6. Sour rot caused by *Geotrichum candidum* initiated in wounds of nectarine fruit.

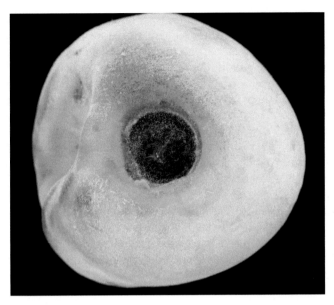

Fig. 13.7. Alternaria decay of peach fruit caused by *Alternaria alternata*.

Anthracnose caused by *Colletotrichum siamense* Prihastuti, L. Cai & K.D. Hyde and *C. fructicola* Prihastuti, L. Cai & K.D. Hyde of the *C. gloeosporioides* species complex, *C. fioriniae* (Marcelino & Gouli) R.G. Shivas & Y.P. Tan and *C. nymphaeae* (Pass.) Aa of the *C. acutatum* species complex, or in rare instances *C. truncatum* (Schwein.) Andrus & W.D. Moore has been reported in areas with high rainfall and warm temperatures during fruit ripening (Grabke *et al.*, 2014; Hu *et al.*, 2015; Lee *et al.*, 2020; Moreira *et al.*, 2020). These species are difficult to identify using morphology alone, and various molecular methods have been developed based on multi-locus sequence analyses (Damm *et al.*, 2009, 2012; Weir *et al.*, 2012). Anthracnose is an important preharvest disease in some areas, but in general, it is a less common postharvest problem. The disease is typically detected at harvest, and infected fruit can be discarded during grading and sorting in the packinghouse. Similar to brown rot, however, quiescent infections can occur on peach and cause losses during marketing. Early symptoms are small, brown spots that become darker and circular, turn orange-brown and are slightly sunken as they age. Lesions initially may be confused with brown rot, but the orange discoloration is distinctive, and lesions expand more slowly and do not extend to more than 5 cm in diameter. The decay is firm to the touch and is often covered with concentric rings of salmon-coloured spore masses. Anthracnose does not develop when storage temperatures are below 4.4°C. The optimum temperature for growth is around 26°C, and no growth occurs above 35°C (Harvey *et al.*, 1972).

Black mould, caused by *Aspergillus niger* and other closely related species (*A. niger* complex), mostly occurs on fruit stored for prolonged periods and develops during ripening at warm temperatures. The optimum temperature for growth is 30–36°C, and the pathogen does not grow below 7°C. The decay generally does not cause serious losses and is rarely found on freshly harvested fruit. Initially, small tan spots enlarge with concentric circles. White mycelium develops in the centre of the lesion and enlarges, followed by production of yellow to then black conidia causing a sooty appearance (Fig. 13.8). Margins of older lesions may have a thin tan area, a thin band of white mycelium and then black sporulation. The decay is localized to a depth of a few millimetres, is brown and can be scooped out leaving a saucer-shaped area of healthy tissue (Harvey *et al.*, 1972).

Blue mould caused by *Penicillium* spp. generally does not occur on freshly harvested and non-injured fruit. *P. expansum* Link. is the main pathogen, but other species have been implicated including *P. crustosum* Thom and *P. chrysogenum* Thom. *Penicillium* spp. are difficult to identify using morphology, and molecular methods have been developed based on β-tubulin, calmodulin, internal transcribed spacer and RNA polymerase II sequences (Visagie *et al.*,

Fig. 13.8. Aspergillus decay or black mould of peach caused by *Aspergillus niger*.

2014). As with decays caused by *Alternaria* and *Cladosporium* spp., blue mould develops on injured fruit during prolonged cold storage. Moist, small, tan spots form on the fruit surface. The epidermis or skin of the fruit can easily slide off if handled. The decay is firm, extends 10–12 mm or more into the fruit, and is easily separated from the healthy tissue. White mycelium develops on the margin, and masses of powdery, blue-green conidia develop at the centre of the lesion (Fig. 13.9). Although blue mould is a minor disease, *P. expansum* can produce the mycotoxin patulin, which has been detected in apricot and peach fruit juices in Italy (Spadaro *et al.*, 2008).

Cladosporium rot caused by *C. herbarum* (Pers.) Link. rarely occurs on peaches but is more common on other stone-fruit crops such as sweet cherry. The decayed area is generally small, with the fungus growing in a visible wound. At first, white mycelium develops on the surface of the injury, but soon the decay turns olive green with production of conidia (Fig. 13.10). Conidia are catenate and micro- or macronematous. Intercalary limoniform macronematous conidia have zero to three distal hila, are ellipsoid to subcylindrical and mostly non-septate, measure 6–16 × 4–6 μm and have roughened walls (micronematous and terminal conidia are smaller, and ramnoconidia are elongated) (Crous *et al.*, 2007). Lesions rarely extend beyond 25 mm in diameter. The distinctive feature of the decay is a grey to black, cone-shaped core that penetrates deep into the flesh of the decay. The margins are brown, moist and easily separated from the healthy tissue (Fig. 13.10).

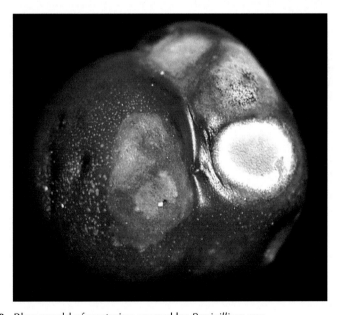

Fig. 13.9. Blue mould of nectarine caused by *Penicillium* spp.

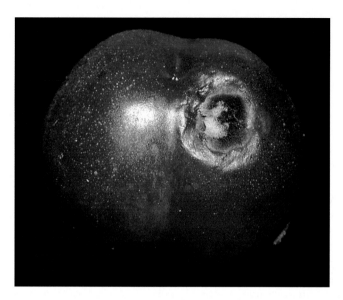

Fig. 13.10. Cladosporium decay of nectarine caused by *Cladosporium herbarum*.

13.3   MANAGEMENT OF POSTHARVEST DECAYS

Management of postharvest decays requires an integrated approach with pre- and postharvest components that focus on maintaining a healthy crop, delaying fruit senescence, avoiding injuries to fruit, sanitation practices and fungicide use (Adaskaveg *et al.*, 2002). Preharvest cultural practices include cultivar selection and proper fertilization, pruning, weed control, irrigation and sanitation (e.g. removal of diseased plant material such as fruit mummies). For brown rot, nectarines are more susceptible than peaches, and fresh-market (i.e. freestone) peaches are typically more susceptible than canning (i.e. cling-stone) peaches. Cultivar differences also exist, although no cultivar is completely resistant. Excess nitrogen fertilization is known to increase susceptibility to brown rot and other diseases. Orchard planting designs, pruning practices and orchard floor (including groundcovers) management can significantly alter the microclimate. Dense tree canopies can restrict air movement, increase relative humidity, prolong leaf wetness duration and decrease pesticide efficacy. Irrigation systems can also affect the incidence of disease. High-angle sprinklers should be avoided to prevent canopies and tree trunks from getting wet. After harvest, it is important to remove all fruit from the trees and from the orchard floor. Additional field sanitation practices include the removal of tree prunings and cleaning of harvest equipment.

Preharvest fungicide treatments are successfully used for the management of brown rot, grey mould and sour rot. Treatments that are applied from 14 to 0 days before harvest help to protect fruit and wounds from infections that occur at harvest time. Fungicides with locally systemic action can also inactivate some of the brown rot and grey mould quiescent infections that only pene-trate the fruit for several cell layers.

Postharvest handling and marketing practices that minimize fruit injuries, utilize sanitation of fruit and equipment (e.g. bins, pick bags and trailers) and involve temperature management of harvested fruit are essential in any post-harvest management programme. For minimizing injuries, shock-absorbing foam pads are used in harvesting, handling and packing equipment (Crisosto and Mitchell, 2002; Thompson *et al.*, 2002a). Pre-washing and hydrocooling with properly adjusted solutions of sodium hypochlorite or other oxidizing materials (e.g. ozone, chlorine dioxide, peroxyacetic acid) are used for fruit sanitation to reduce the inoculum of plant and human pathogens. Harvest bins should be steam-treated or washed with high-pressure water. Other sani-tizing treatments for equipment, but not fruit, include quaternary ammonium compounds.

Temperature management is critical for all postharvest handling sys-tems. Orchard heat is efficiently decreased by hydrocooling (Thompson *et al.*, 2002b). Stone fruits are best kept at 0–2°C, and this temperature should be maintained during handling, storage and transit to the final market. Low tem-peratures maximize the potential life of the fruit and slow or inhibit develop-ment of most fungal pathogens. At 0°C, some decay fungi such as *R. stolonifer* and *G. candidum* will not grow, whereas others such as *Monilinia* spp. and *B. cinerea* are greatly reduced in their growth rate.

Postharvest treatments with antifungal compounds are used to minimize decay from latent and active infections and to prevent the spread of decay between adjacent fruit (i.e. nesting) in packing containers. Currently, post-harvest fungicide treatments are still considered the most effective strategy to maintain postharvest crop health. Without postharvest fungicide use, fruit usually cannot be stored without losses during storage and display-shelf tem-peratures after long-distance transportation and marketing (Adaskaveg *et al.*, 2002, 2022).

The fungicides benomyl, thiophanate-methyl, triforine and iprodione were previously available for use on stone-fruit crops in the USA and were highly effective against brown rot and grey mould (Eckert and Ogawa, 1988). Only iprodione was also effective against Rhizopus rot when mixed with par-affinic oil products as provided by some fruit coatings. Their registrations were withdrawn in the 1980–1990s as a result of the US Environmental Protection Agency's reregistration policy of older pesticides and the Food Quality Protection Act (Adaskaveg *et al.*, 2002). These cancellations were at first considered the end of effective postharvest decay control, but instead they spurred new beginnings in postharvest fungicide registrations. The first 'reduced-risk' fungicide, fludioxonil, was identified and registered in 1997, followed by the 'reduced-risk' fenhexamid and pyrimethanil (Table 13.2). The registration of fludioxonil marked the start of a new era of postharvest safety, and this fungicide ultimately advanced to become the most effective and most widely used broad-spectrum postharvest treatment in the world. Efforts need

Table 13.2. Characteristics of postharvest fungicides registered for peach and nectarine in the USA.

	FRAC code[a]	EPA rating[b]	Efficacy[c]				Compatibility[d]	
			Brown rot	Grey mould	Rhizopus rot	Sour rot	Sodium hypochlorite	Peroxyacetic acid
Fenhexamid[e]	17	RR	+++	+++	–	–	No	Yes
Fludioxonil[f]	12	RR	+++	+++	+++	–	Yes	Yes
Natamycin	48	BP	+++	++	++	–	No	No
Polyoxin D	19	BP	++	+++	–	–	No	Yes
Propiconazole	3	C	+++	++	+++	+++	Yes	Yes
Pyrimethanil[g]	9	RR	+++	+++	–	–	No	Yes
Fludioxonil + propiconazole	12 + 3	RR + C	+++	+++	+++	+++	Yes	Yes

[a]FRAC, Fungicide Resistance Action Committee codes for different modes of action.
[b]US Environmental Protection Agency ratings: RR, reduced risk; BP, biopesticide (biopesticides are exempt from residue tolerance); C, conventional.
[c]Efficacy is based on a rating system: –, not effective; +, moderately effective; ++, very effective; +++, highly effective.
[d]Compatibility with sanitizers when mixed with fungicides in recycling application systems.
[e]Registration of fenhexamid has been suspended due to lack of sales.
[f]Fludioxonil has the broadest spectrum of activity, which also includes many minor decays of peach.
[g]Pending registration.

to be made to educate consumers on the low toxicity of reduced-risk pesticides such as fludioxonil, pyrimethanil and fenhexamid, which are measured in ppm and in some cases fractions of ppm (mg kg^{-1}) for residues on the commodity (Table 13.3). The acute oral toxicity of fludioxonil is nearly 2–30 times lower than that of common food ingredients such as table salt and some preservatives. The advent of marketing pre-ripened peaches and nectarines increased the spectrum of decays that needed to be managed when sour rot became of commercial importance. After testing numerous compounds, the non-reduced-risk fungicide propiconazole was developed in the early 2000s and still is the only highly effective treatment available for sour rot control (McKay *et al.*, 2012; Yaghmour *et al.*, 2012a). The maximum residue limit for propiconazole on fruit is 4 ppm (mg kg^{-1}), but it is effective at a fraction of 1 ppm, and the safety levels are respectable (Table 13.3).

The evolution of postharvest treatments is ongoing. Growing consumer concerns on treating fresh produce with synthetic fungicides led to the identification and purification of natural fermentation compounds with antifungal properties. Examples are natamycin and polyoxin D, which are categorized as 'biopesticides' with exempt-from-residue-tolerance status in the USA. An interesting aspect of natamycin is that resistance has never been selected for in filamentous fungi in over 30 years of use in the food industry. Thus, its use in mixtures with other fungicides (e.g. fludioxonil) can reduce total pathogen population levels to extremely low levels, which makes the probability of selection for resistance to other modes of action in the mixture extremely unlikely.

Currently, five of the six active ingredients of postharvest fungicides available in the USA are in the reduced-risk category that includes biopesticides. The efficacy of these fungicides against the major postharvest decays of peach is presented in Table 13.2. The different modes of action are shown as Fungicide Resistance Action Committee (FRAC) codes. Based on efficacy and FRAC code, mixtures can be prepared in the packing facility to provide broad-spectrum activity, high levels of decay control and resistance management against targeted and non-targeted pathogens. Most fungicides are applied using efficient systems such as controlled-droplet or air-assisted methods that are standard in most packing facilities. These methods do not recycle the fungicide solutions and do not need to be sanitized. However, some postharvest facilities use recirculating fungicide solutions in dip or drench application methods that, based on the US Food Quality Protection Act, need to be sanitized to prevent possible contamination with human bacterial pathogens. Fungicide compatibility with common water sanitizers such as sodium hypochlorite and peroxyacetic acid is provided in Table 13.2.

Increasingly, other 'biological' approaches to manage postharvest diseases of fruit crops are also being explored, and some have been commercialized (Janisiewicz, 2010). These include 'raw' microbial fermentation products, plant extracts including essential oils (Spadaro and Gullino, 2014; Elshafie *et al.*, 2015; Cindi *et al.*, 2016), and biological agents such as various bacteria and yeasts (Tian, 2007). Yeasts and filamentous fungi registered for use on peach are shown in Table 13.4. Mechanisms of action of antagonistic

Table 13.3. Relative toxicity and maximum and targeted residues of selected postharvest fungicides used for managing postharvest decays of peach compared with table salt and sodium nitrite food preservatives.

Toxicity		Fludioxonil	Pyrimethanil	Sodium chloride (table salt)	Natamycin	Fenhexamid	Propiconazole	Sodium nitrite
50% Lethal dose (LD_{50}) or concentration (LC_{50})								
Acute oral	LD_{50} rats	>5,000	4,159–5,971	3,000	2,700	>2,000	2,105	175
	LD_{50} humans	–	–	1,000	–	–	–	–
Acute dermal	LD_{50} rabbits	>2,000	>5,000	10,000	–	>2,000	4,250	Muta-/teratogenic
Inhalation	LC_{50} (g m^{-3} every 4 h)	Non-toxic	1.9	10.2	–	>5	1	5.5
MRL (mg kg^{-1}) of USA/CODEX[b]		5/5	10/4	–	Exempt	10/10	4/4	–
Targeted use residue (mg kg^{-1})[c]		1	2	–	2–3	2	1	–

[a]Toxicity information was obtained from safety data sheets for each chemical. Toxicity is rated from low (fludioxonil) to high (sodium nitrite).
[b]Maximum residue limit (MRL) for the USA and Codex Alimentarius (CODEX) or 'Food Code' for standards, guidelines and codes of practice published by the Food and Agriculture Organization (FAO) and the World Health Organization (WHO).
[c]Targeted use residue is the amount of residual fungicide required to obtain high-level decay control that meets industry standards.

Table 13.4. Biocontrols used to manage major postharvest decays of peach and other stone-fruit crops worldwide[a].

Biocontrol	Strain	FRAC code	Type	Crop	Decay
Aureobasidium pullulans (BioProtect, Botector, BoniProtect; Austria and USA)[b]	DSM 14940 and DSM 14941	BM 02	Yeast	Stone fruits and other crops	Blue mould and grey mould
Candida sake (Candifruit, Spain)	CPA-1	BM 02	Yeast	Stone fruits and other crops	Blue mould and grey mould of temperate fruits and green mould, blue mould, and sour rot of citrus
Metschnikowia fructicola (Shemer, Israel)	NRRL Y-30752	BM 02	Yeast	Stone fruits and other crops	*Penicillium* and Aspergillus decays, grey mould, and Rhizopus rot

[a]The commercial deployment of postharvest biocontrol agents has not had widespread acceptance. This has been attributed to inconsistent performance, the high cost relative to synthetic fungicides, registration hurdles, difficulties in mass production and formulation of the antagonist, and lack of industry adaptations.
[b]BioProtect and Botector are available as preharvest treatments for postharvest decay control, whereas BoniProtect is for postharvest use. Other biocontrol agents that have been discontinued are *Candida oleophila* I-182 (USA) and *Cryptococcus albidus* (South Africa).

microorganisms include: (i) antibiosis; (ii) competition; (iii) parasitism; and (iv) induced host resistance (Droby *et al.*, 2022). Shortcomings of biocontrols have been their inconsistency and lack of high performance, lack of user support, high cost compared with conventional materials, registration hurdles, and formulation and storage issues (Droby *et al.*, 2009). Moreover, the lack of specific information on by-products, secondary metabolites and uncharacterized components that may form during fermentation and that are included in the final biocontrol product is a concern. However, others have argued that biologicals such as *Aureobasidium pullulans*, *Candida sake* and *Metschnikowia fructicola* that are used for decay control of a commodity (Table 13.4) may also provide probiotic activity for the consumer (Karunaratne and Nanayakkara, 2020).

Physical treatments, include heat and ultraviolet irradiation, have been used successfully to stop incipient infections of decay fungi on some crops (Droby *et al.*, 2009; Spadoni *et al.*, 2013). For peaches, changing fruit quality and high energy costs of heating that is followed by cooling of fruit have been limitations. However, treatments with biological agents and plant extracts including essential oils are being evaluated and developed with the goal of broad-spectrum activity and consistent high performance. Combinations of 'biological' treatments are also being tested for a potential additive effect to obtain levels of control acceptable to commercial needs (Jurick and Lichtner, 2022). Another new perspective in the management of postharvest pathogens is the application of antimicrobial peptides and proteins (Marcos *et al.*, 2020). The development of these treatments is still at early experimental stages, and high production costs may limit their agricultural use. An additional long-term goal is breeding for host resistance, and for this, new genetic markers are being utilized (Droby *et al.*, 2022; Jurick and Lichtner, 2022).

REFERENCES

Abate, D., Pastore, C., Gerin, D., de Miccolis Angelini, R.M., Rotolo, C. *et al.* (2018) Characterization of *Monilinia* spp. populations on stone fruit in South Italy. *Plant Disease* 102, 1708–1717.

Adaskaveg, J.E. and Förster, H. (2022) Postharvest decays of stone fruit. In: Adaskaveg, J.E., Förster, H. and Prusky, D. (eds) *Postharvest Pathology of Fruit and Nut Crops.* American Phytopathological Society Press, St. Paul, Minnesota, pp. 193–210.

Adaskaveg, J.E., Förster, H. and Thompson, D.F. (1999) Identification and etiology of visible quiescent infections of *Monilinia fructicola* and *Botrytis cinerea* in sweet cherry fruit. *Plant Disease* 84, 328–333.

Adaskaveg, J.E., Förster, H. and Sommer, N.F. (2002) Principles of postharvest pathology and management of decays of edible horticultural crops. In: Kader, A. (ed.) *Postharvest Technology of Horticultural Crops*, 3rd edn. University of California Agriculture and Natural Resources, Publication No. 3311. University of California, Oakland, California, pp. 163–195.

Adaskaveg, J.E., Förster, H. and Prusky, D.B. (eds) (2022) *Postharvest Pathology of Fruit and Nut Crops*. American Phytopathological Society Press, St. Paul, Minnesota.

Adaskaveg, J.E., Arpaia, M.L., Förster, H. and Mitcham, E.J. (2023) Postharvest disease and insect control. In: Kader, A., Thompson, J.F. and Saltveit, M. (eds) *Postharvest Technology of Horticultural Crops*, 4th edn. Publication No. 21659. University of California, Davis, California (in press).

Bartz, J.A., Fatica, M.K. and Schneider, K.R. (2022) Sanitation practices during post-harvest handling of fresh fruits and vegetables. In: Adaskaveg, J.E., Förster, H. and Prusky, D. (eds) *Postharvest Pathology of Fruit and Nut Crops*. American Phytopathological Society Press, St. Paul, Minnesota, pp. 23–38.

Batra, L.R. (1991) *World Species of Monilinia (Fungi): Their Ecology, Biosystematics and Control*. Mycologia Memoir No. 16. J. Cramer, Berlin.

Byrde, R.J.W. and Willets, H.J. (1977) *The Brown Rot Fungi of Fruit: Their Biology and Control*. Pergamon Press, Oxford, UK.

Casals, C., Viñas, I., Torres, R., Griera, C. and Usall, J. (2009) Effect of temperature and water activity on *in vitro* germination of *Monilinia* spp. *Journal of Applied Microbiology* 108, 47–54.

Cindi, M.D., Soundy, P., Romanazzi, G. and Sivakumar, D. (2016) Different defense responses and brown rot control in two *Prunus persica* cultivars to essential oil vapours after storage. *Postharvest Biology and Technology* 119, 9–17.

Crisosto, C.H. and Mitchell, F.G. (2002) Postharvest handling systems: stone fruits. I. Peach, nectarine, and plum. In: Kader, A. (ed.) *Postharvest Technology of Horticultural Crops*, 3rd edn. University of California Agriculture and Natural Resources, Publication No. 3311. University of California, Oakland, California, pp. 345–350.

Crous, P.W., Braun, U., Schubert, K. and Groenewald, J.Z. (eds) (2007) *The Genus Cladosporium and Similar Dematiaceous Hyphomycetes*. Studies in Mycology, Vol. 58. Centraalbureau voor Schimmelcultures, Utrecht, The Netherlands.

Damm, U., Woudenberg, J.H.C., Cannon, P.F. and Crous, P.W. (2009) *Colletotrichum* species with curved conidia from herbaceous hosts. *Fungal Diversity* 39, 45–87.

Damm, U., Cannon, P.F., Woudenberg, J.H. and Crous, P.W. (2012) The *Colletotrichum acutatum* species complex. *Studies in Mycology* 73, 37–113.

Droby, S., Wisniewski, M., Macarisin, D. and Wilson, C. (2009) Twenty years of post-harvest biocontrol research: is it time for a new paradigm? *Postharvest Biology and Technology* 52, 137–145.

Droby, S., Wisniewski, M. and Norelli, J. (2022) Biological approaches for managing postharvest decay. In: Adaskaveg, J.E., Förster, H. and Prusky, D. (eds) *Postharvest Pathology of Fruit and Nut Crops*. American Phytopathological Society Press, St. Paul, Minnesota, pp. 103–120.

Eckert, J.W. and Ogawa, J.M. (1988) The chemical control of postharvest diseases: deciduous fruits, berries, vegetables and root/tuber crops. *Annual Review of Phytopathology* 26, 433–469.

Elshafie, H.S., Mancini, E., Camele, I., de Martino, L. and de Feo, V. (2015) *In vivo* anti-fungal activity of two essential oils from Mediterranean plants against postharvest brown rot disease of peach fruit. *Industrial Crops and Products* 66, 11–15.

FAOSTAT (2020) Crops and Livestock Products. FAOSTAT Online Database. Food and Agriculture Organization of the United Nations, Rome. Available at: www.fao.org/faostat/en/#data/QC (accessed 8 December 2022).

Fourie, P.H., Holz, G. and Calitz, F.J. (2002) Occurrence of *Botrytis cinerea* and *Monilinia laxa* on nectarine and plum in Western Cape orchards, South Africa. *Australasian Plant Pathology* 31, 197–204.

Gell, I., Cubero, J. and Melgarejo, P. (2007) Two different PCR approaches for universal diagnosis of brown rot and identification of *Monilinia* spp. in stone fruit trees. *Journal of Applied Microbiology* 103, 1364–5072.

Grabke, A., Williamson, M., Henderson, G.W. and Schnabel, G. (2014) First report of anthracnose on peach fruit caused by *Colletotrichum truncatum* in South Carolina. *Plant Disease* 98: 1154.

Harvey, J.M., Smith, W.L. and Kaufman, J. (1972) *Market Diseases of Stone Fruits: Cherries, Peaches, Nectarines, Apricots, and Plums*. Agriculture Handbook 414, Agricultural Research Service, United States Department of Agriculture, Washington, DC.

Hu, M.J., Grabke, A., Dowling, M.E., Holstein, H.J. and Schnabel, G. (2015) Resistance in *Colletotrichum siamense* from peach and blueberry to thiophanate-methyl and azoxystrobin. *Plant Disease* 99, 806–814.

Janisiewicz, W.J. (2010) Quo vadis of biological control of postharvest diseases. In: Prusky, D. and Gullino, M.L. (eds) *Post-Harvest Pathology. Plant Pathology in the 21st Century*. Springer International, Cham, Switzerland, pp. 137–148.

Jurick, W. II and Lichtner, F.J. (2022) Alternative approaches for managing postharvest diseases of fruit crops. In: Adaskaveg, J.E., Förster, H., and Prusky, D. (eds) *Postharvest Pathology of Fruit and Nut Crops*. American Phytopathological Society Press, St. Paul, Minnesota, pp. 121–138.

Karunaratne, A.M. and Nanayakkara, B.S. (2020) Toward probiotic postharvest biocontrol antagonists – appraisal of obstacles. In: Palou, L. and Smilanick, J.L. (eds) *Postharvest Pathology of Fresh Horticultural Produce*. CRC Press, Boca Raton, Florida, pp. 499–519.

Latorre, B.A., Díaz, G.A., Valencia, A.L., Naranjo, P., Ferrada, E.E. *et al.* (2014) First report of *Monilinia fructicola* causing brown rot on stored Japanese plum fruit in Chile. *Plant Disease* 98, 160–160.

Lee, D.M., Hassan, O. and Chang, T. (2020) Identification, characterization, and pathogenicity of *Colletotrichum* species causing anthracnose of peach in Korea. *Mycobiology* 48, 210–218.

Marcos, J.F., Gandía, M., Garrigues, S. and Manzanares, P. (2020) Antifungal peptides and proteins with activity against fungi causing postharvest decay. In: Palou, L. and Smilanick, J.L. (eds) *Postharvest Pathology of Fresh Horticultural Produce*. CRC Press, Boca Raton, Florida, pp. 757–792.

Mari, M., Spadaro, D., Casals, C., Collina, M., de Cal, A. and Usall, J. (2020) Stone fruits. In: Palou, L. and Smilanick, J.L. (eds) *Postharvest Pathology of Fresh Horticultural Produce*. CRC Press, Boca Raton, Florida, pp. 111–140.

McKay, A.H., Förster, H. and Adaskaveg, J.E. (2012) Toxicity and resistance potential of selected fungicides to *Galactomyces* and *Penicillium* spp. causing postharvest fruit decays of citrus and other crops. *Plant Disease* 96, 87–96.

Moreira, R.R., Silva, G.A. and de Mio, L.L.M. (2020) *Colletotrichum acutatum* complex causing anthracnose on peach in Brazil. *Australasian Plant Pathology* 49, 179–189.

Ogawa, J.M. and English, H. (1991) *Diseases of Temperate Tree Fruit and Nut Crops*. Division of Agriculture and Natural Resources, Publication 3345. University of California, Oakland, California.

Ogawa, J.M., Zehr, E.I., Bird, G.W., Ritchie, D.F., Urio, K. and Uyemoto, J.K. (1995) *Compendium of Stone Fruit Diseases.* American Phytopathological Society Press, St. Paul, Minnesota.

Snowdon, A.L. (1990) *A Color Atlas of Post-Harvest Diseases of Fruits and Vegetables, Vol. 1.* CRC Press, Boca Raton, Florida.

Spadaro, D. and Gullino, M.L. (2014) Use of essential oils to control postharvest rots on pome and stone fruit. In: Prusky, D. and Gullino, M.L. (eds) *Post-Harvest Pathology. Plant Pathology in the 21st Century.* Springer International, Cham, Switzerland, pp. 101–110.

Spadaro, D., Garibaldi, A. and Gullino, M.L. (2008) Occurrence of patulin and its dietary intake through pear, peach, and apricot juices in Italy. *Food Additives and Contaminants: Part B* 1, 134–139.

Spadoni, A., Neri, F., Bertolini, P. and Mari, M. (2013) Control of *Monilinia* rots on fruit naturally infected by hot water treatment in commercial trials. *Postharvest Biology and Technology* 86, 280–284.

Thompson, J.F., Mitcham, E.J. and Mitchell, F.G. (2002a) Preparation for fresh market. In: Kader, A. (ed.) *Postharvest Technology of Horticultural Crops*, 3rd edn. University of California Agriculture and Natural Resources, Publication No. 3311. University of California, Oakland, California, pp. 67–79.

Thompson, J.F., Mitchell, F.G., Rumsey, T.R., Kasmire, R.F. and Crisosto, C.H. (2002b) *Commercial Cooling of Fruits, Vegetables, and Flowers.* University of California Agriculture and Natural Resources, Publication No. 21567. University of California, Oakland, California.

Tian, S.P. (2007) Management of postharvest diseases in stone and pome fruit crops. In: Ciancio, A. and Mukerji, K.G. (eds) *General Concepts in Integrated Pest and Disease Management.* Springer, Dordrecht, Netherlands, pp. 131–147.

Villarino, M., Melgarejo, P., Usall, J., Segarra, J., Lamarca, N. and de Cal. A. (2012) Secondary inoculum dynamics of *Monilinia* spp. and relationship to the incidence of postharvest brown rot in peaches and the weather conditions during the growing season. *European Journal of Plant Pathology* 133, 585–598.

Villarino, M., Melgarejo, P. and de Cal, A. (2016) Growth and aggressiveness factors affecting *Monilinia* spp. survival peaches. *International Journal Food Microbiology* 224, 22–27.

Visagie, C.M., Houbraken, J., Frisvad, J.C., Hong, S.-B., Klaassen, C.H.W. *et al.* (2014) Identification and nomenclature of the genus *Penicillium. Studies in Mycology* 78, 343–371.

Wang, J., Guo, L., Xiao, C. and Zhu, X. (2018) Detection and identification of six *Monilinia* spp. causing brown rot using TaqMan real-time PCR from pure cultures and infected apple fruit. *Plant Disease* 102, 1527–1533.

Weir, B.S., Johnston, P.R. and Damm, U. (2012) The *Colletotrichum gloeosporioides* species complex. *Studies in Mycology* 73, 115–180.

Yaghmour, M.A., Bostock, R.M., Adaskaveg, J.E. and Michailides, T.J. (2012a) Propiconazole sensitivity in populations of *Geotrichum candidum*, the cause of sour rot of peach and nectarine, in California. *Plant Disease* 96, 752–758.

Yaghmour, M.A., Bostock, R.M., Morgan, D.P. and Michailides, T.J. (2012b) Biology and sources of inoculum of *Geotrichum candidum* causing sour rot of peach and nectarine fruit in California. *Plant Disease* 96, 204–210.

14

BIOLOGY AND MANAGEMENT OF INSECT PESTS

Nikos T. Papadopoulos[1]*, Brett R. Blaauw[2], Panagiotis Milonas[3] and Anne L. Nielsen[4]

[1]*Department of Agriculture, Crop Production and Rural Environment, University of Thessaly, Volos, Greece; [2]Department of Entomology, University of Georgia, Athens, Georgia, USA; [3]Benaki Phytopathological Institute, Kifissia, Greece; [4]Department of Entomology, Rutgers University, New Brunswick, New Jersey, USA*

14.1 INTRODUCTION

Peach (*Prunus percicae* (L.) Batsch) and nectarine cultivation and production are widely dispersed worldwide. There are two rather broad 'latitudinal zones' that accommodate most peach and nectarine production, one in the northern hemisphere ranging from Egypt (North Africa), north India (Asia) and southern USA (North America) to south Germany (Europe) and north China and Republic of Korea, and the other in the southern hemisphere, including South Africa, Australia, New Zealand, Brazil, Argentina and Chile. Many cultivars have been selected to produce appealing fruit for human consumption, and have been cultivated for different climatic zones, usually receiving significant input to remain productive. Because of the broad distribution of peach orchards among almost all continents, both hemispheres and diverse climatic zones, there is a list of herbivores that have adapted to peach trees and established a close association that often results in severe damage of plants and devastating economic losses for farmers. The list of arthropod pests that feed on peach trees and that can affect the growth and development of the trees as well as fruit production directly is long and includes species in the orders Diptera, Lepidoptera, Thysanoptera, Coleoptera and Hemiptera. A list of the key pests that infest peaches and their relative importance is given in Table 14.1. The wide distribution of peach cultivation, combined with global trade of fresh peach, has contributed to the movement of arthropod pests to different countries and even continents. In addition, climate change has removed climatic

* Email: nikopap@uth.gr

© CAB International 2023. *Peach* (G. Manganaris, G. Costa and C. Crisosto eds)
DOI: 10.1079/9781789248456.0014

Table 14.1. Key peach pests of global and general importance.

Order and family	Species	Damage	Main management approaches
Pests of fruit/reproductive organs			
Coleoptera: Curculionidae	*Conotrachelus nenuphar*	Fruit, stems	Cover sprays targeting adults, biological control targeting soil-dwelling larvae
Diptera: Tephritidae	*Ceratitis capitata* (Wiedemann)	Ripe and ripening fruit	Bait and cover sprays, mass trapping, SIT, inundative release of egg/larvae parasitoids
Diptera: Tephritidae	*Bactrocera zonata* (Saunders)	Ripe and ripening fruit	Bait and cover sprays, male annihilation with methyl eugenol, mass trapping, SIT
Lepidoptera: Torticidae	*Grapholita molesta* (Busck)	Stem borer, fruit surface and flesh	Adult trapping with pheromone traps, insecticide applications, application of CpGV as a biopesticide, mating disruption
	Thaumatotibia (= *Cryptophlebia*) *leucotreta*	Fruit	Adult trapping with pheromone traps, insecticide applications, SIT, mating disruption, biological control (egg parasitoids)
	Platynota idaeusalis (Walker)	Leaves, leaf roller, cosmetic infestation on fruit	Adult trapping with pheromone traps, insecticide applications
Lepidoptera: Gelechidae	*Anarsia lineatella* Zeller	Stem borer, fruit surface and flesh	Adult trapping with pheromone traps, insecticide applications, mating disruption
Hemiptera: Diaspididae	*Pseudaulacaspis pentagona* (Targioni-Tozzetti)	Shoots, branches, fruit	Cover sprays targeting neonate and crawler larvae, application of winter (no longer used in Europe) and summer oils, biological control
	Comstockaspis perniciosus (Comstock)	Shoots, branches, fruit	Cover sprays targeting neonate and crawler larvae, application of winter and summer oils, biological control

Continued

Table 14.1. Continued.

Order and family	Species	Damage	Main management approaches
Hemiptera: Pentatomidae	*Halyomorpha halys* (Stål)	Fruit	Cover sprays or border sprays targeting all life stages, biological control
	Euschistus servus (Say)	Vegetative parts: leaf, buds, fruit	Cover sprays targeting all life stages
Heteroptera: Miridae	*Lygus lineolaris* (Palisot de Beauvois)	Buds, flowers, fruit	Cover sprays targeting all life stages, broadleaf weed management within orchard to minimize alternative hosts
Thysanoptera: Thripidae	*Frankliniella occidentalis* (Pergande)	Mainly fruit, buds, flowers	Weed management within orchard to minimize alternative hosts
Eriophyidae			
Foliage and shoot feeders			
Lepidoptera: Torticidae	*Adoxophyes orana* Fischer von Röeslerstamm	Leaves, shots, fruit (surface scars)	Adult trapping with pheromone traps, insecticide applications
Hemiptera: Diaspididae	*Pseudaulacaspis pentagona* (Targioni-Tozzetti)	Shoots, branches, fruit	Biological control, crawler trapping, insecticide application, summer oils
	Comstockaspis perniciosus (Comstock)	Shoots, branches, fruit	Biological control, crawler trapping, insecticide application, winter and summer oils
Hemiptera: Aphididae	*Myzus persicae* (Sulzer)	Leaf, young shoots	Cover sprays, biological control
Wood borers			
Sesiidae	*Synanthedon exitiosa* (Say)	Wood borer, trunk and roots	Cover sprays, biological control, mating disruption
	Synanthedon pictipes (Grote & Robinson)	Wood borer, scaffolding limbs	Cover sprays, biological control, mating disruption

CpGV, *Cydia pomonella* granulovirus; SIT, sterile insect technique.

barriers for the establishment of these pests in new areas when they arrive. Hence, biological invasions are considered one of the major threats for agricultural production globally, as well as for peach cultivation in a major way. In recent years, there has been an increased emphasis on preventing biological invasions and on developing tools and strategies to alleviate their impact that goes well beyond farmers and involves the trading of fresh fruit commodities and the peach industry in general.

The significant pressure of arthropod pests on peach cultivation is historically connected with intense use of pesticides to control herbivore populations, which has resulted in negative environmental contamination impacts and deterioration of beneficial fauna. There are several initiatives in Europe, North America and elsewhere to reduce the use of pesticides, and there are a number of active ingredients that are currently banned from use. Future pest-management operations in peach and other crops will rely on a much shorter list of pesticides, and therefore more sophisticated approaches and elaborate efforts will be required to keep peach cultivation viable and profitable.

In recent years, there have been many technological advances, such as the global positioning system (GPS), geographical information system (GIS), electronic traps, various sensors to provide real-time data from the field and sophisticated web-based platforms to analyse and share information that have been progressively adopted to manage crops and provide support for pest management. Some of this technology has become affordable for farmers, and a new range of integrated pest management (IPM) tools and algorithms are under development. Management of peach pests may also benefit from these advancements in technology and their adoption.

This chapter elaborates on the key pests for peach and nectarine production, outlining important elements of geographical distribution, pest importance, pest biology and management approaches, and focuses on current and future IPM efforts, with an increased emphasis on modern technologies that can be adopted in the post-pesticides (or reduced use of pesticides) era.

14.2 KEY PEACH PESTS OF GLOBAL AND GENERAL IMPORTANCE

14.2.1 The Mediterranean fruit fly, *Ceratitis capitata* (Wiedemann)

Overview: common names, taxonomy and geographical distribution
The Mediterranean fruit fly, *Ceratitis capitata* (Diptera: Tephritidae), is one of the most important pests for fruit production and trading of horticultural commodities globally (Fig. 14.1). It is an extremely polyphagous species that can infest the fruits of more than 300 plant species (Liquido *et al.*, 1991; White and Elson-Harris 1992; Papadopoulos *et al.*, 2001). Fruits of citrus species and pome and stone fruit, as well as many other tropical and subtropical plant

species, suffer intense infestations of *C. capitata*. Peaches and nectarines are considered among the preferred hosts, and infestations for *C. capitata* can reach high levels of up to 100% of the yield in cases where management is not applied (Fig. 14.2).

Ceratitis capitata originated in the east parts of sub-Saharan Africa and arrived in Mediterranean orchards either through the Nile valley or along the coastal line

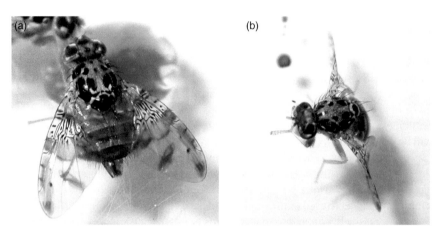

Fig. 14.1. Adult female (a) and male (b) of the Mediterranean fruit fly, *Ceratitis capitata*.

Fig. 14.2. Infestation of peaches by the Mediterranean fruit fly. (a) Oviposition sting on the unripe fruit. (b) Larvae feeding in the fruit flesh.

of western Africa. It was first detected in Spain in 1842, soon after in Portugal and then became dispersed over the Mediterranean countries. In 1905, it was detected in Australia, and by 1906 it was already detected in Brazil and by early 1900 in Hawaii, USA. In the mid-1970s, *C. capitata* was detected in Central America (Carey, 1991; Bonizzoni *et al.*, 2001; Vera *et al.*, 2002; Papadopoulos *et al.*, 2013).

Its current geographical distribution includes almost all of Africa, South and Central America, the Middle East, west Australia and islands of the Pacific, Atlantic and Indian oceans. Frequent detection of *C. capitata* in California, USA, is considered to stem from transient populations (McInnis *et al.*, 2017). However, strong arguments for the existence of established populations have triggered intense debates for decades (Carey, 1991; Papadopoulos *et al.*, 2013; Carey *et al.*, 2017a,b; McInnis *et al.*, 2017; Shelly *et al.*, 2017). Following an intense and costly eradication campaign that has been operating for several decades, *C. capitata* is considered to have been eradicated from almost all parts of Mexico apart from the Guatemalan border. In recent years, *C. capitata* has expanded its geographical distribution to more temperate areas of Europe (Papadopoulos, 2014). *Ceratitis capitata* populations are detected almost every year in countries of central Europe such as Austria, and in the most temperate areas of Mediterranean countries, such as northern Italy and central Macedonia, Greece. Hence, peach, nectarine and other deciduous fruit species (e.g. apple, pear) that grow in these areas are greatly affected. Occurrence of this pest in a specific area may strongly affect trading of fresh fruit commodities, as *C. capitata* is a quarantine pest of major importance for many countries.

Biology: life cycle, longevity and reproduction

Ceratitis capitata is a highly polyphagous, multi-voltine species that can produce many generations every year depending on climatic conditions and the availability of hosts. In tropical and subtropical areas, more than six or seven generations per year can be produced, while in the northern-most limits of its distribution, there may be as few as three (Papadopoulos *et al.*, 2001). The life history traits of *C. capitata* vary depending on temperature conditions, the origin of the population and the host fruit in which the larvae develop. Under constant laboratory conditions (25°C and 60% relative humidity), developmental duration was estimated to be approximately 2, 18 and 11 days for eggs, larvae and pupae, respectively, reared in apples (Papadopoulos *et al.*, 2002). The life expectancy of males and females was estimated at 51 and 60 days, respectively, with fecundity rates of 550 eggs per female. The intrinsic population growth rate (per capita population increase) can range from 0.08 to 1.18 depending on the host fruit that is exploited (Krainacker *et al.*, 1987). Adult longevity, fecundity rates and developmental duration of immature stages may also vary widely depending on the population, as was shown by Diamantidis *et al.* (2011a, b) in a series of elegant studies considering a wide range of global populations.

Damage: symptoms and diagnostics

Females deposit a batch of eggs of variable size in the flesh of ripe or ripening fruit. The number of eggs in each oviposition event may range from one to over nine depending on the host fruit and the physiological state of the female. Oviposition stings may become apparent in some host fruits a few days later, while in other hosts they can remain unnoticed for several weeks. In both peaches and nectarines, it is hard to detect an oviposition sting with the naked eye, and careful examination of the fruit under a stereomicroscope is required. The larvae feed on the flesh of the fruit, which often becomes infected by fungi and bacteria, which intensify the damage and render the fruit unsuitable for consumption and marketing. In several hosts, the oviposition sting alone can cause high rates of fruit dropping.

Infested fruit may escape notice at packinghouses and travel to international markets, thus dispersing propagules over long distances. Interception of infested fruit in cargo shipments is a major element of phytosanitary measures that aim to prohibit invasion of *C. capitata* in novel areas. Identification of an infestation to species level at ports of entry is not an easy task and requires rearing immature stages to adulthood, which may take several days, if not weeks. Reliable morphological keys for fruit fly larvae are not currently available, although there are efforts to develop them. Molecular tools, however, have been developed and can be used to identify immature stages within a few hours (Blacket *et al.*, 2020).

Identification of adults is more straightforward. There are several keys that can be considered and recently the use of mobile apps has also contributed (de Meyer *et al.*, 2022).

Seasonal biology and management

Population growth of the Mediterranean fruit fly depends on climatic conditions, host-fruit availability and the overwintering dynamics, which may define the size and onset of activity of the spring (overwintering) populations in temperate areas. In warmer temperate areas of the northern hemisphere, population densities can be revealed by adult captures in traps, and these peak in summer and autumn. Adult captures are low and sporadic in winter and early spring (Mavrikakis *et al.*, 2000). In cooler, more temperate areas, adult captures begin in summer and peak in autumn. Usually, no adults are detected in such areas during the winter and spring (Papadopoulos *et al.*, 2001). Overwintering is accomplished for all developmental stages that follow a prolonged duration in the warmer southern areas and as larvae within infested fruit in the cooler, more temperate areas. High mortality rates are reported during winter in the most temperate areas where *C. capitata* is found, resulting in extremely low population densities during the spring that cannot be detected and apparently hinder population growth and epidemic population development for months (Papadopoulos *et al.*, 2001).

POPULATION MONITORING. Almost all management approaches against the Mediterranean fruit fly rely on population monitoring with adult trapping. Although several trapping systems exist in local and international markets, the Jackson (delta-type) trap baited with the male-specific attractant trimedlure and several McPhail-type (bucket) traps baited with food attractants such as a combination of ammonium acetate, trimethylamine and putrescine, which attract mainly females but also males, are the most reliable to determine population dynamics (Fig. 14.3) (Epsky *et al.*, 1999; Katsoyannos *et al.*, 1999). The combination of male-targeted and female-targeted trapping systems provides a more reliable information regarding early detection and fluctuations of *C. capitata* populations. *Ceratitis capitata* trapping may serve a wide range of purposes, including pest-management decisions, delimitation of restricted (pest-free and/or low-prevalence) zones, evaluation of the efficacy of implemented control efforts and others. Deployment of trapping devices and their density and spacing should be adjusted depending on the aim of the trap. For making pest-management decisions, it is important to monitor the population in key hosts where population growth and development occur early in the season.

As well as traditional trapping systems, several electronic traps have been proposed to monitor the populations of the Mediterranean fruit fly and that of other fruit flies. Electronic traps are usually based on a 'traditional' type of trap body and lure, and use various approaches to identify and count the attracted and captured insects. Counting sensors include infrared LED emitters and receivers that count flies that 'break' the light beam, micro-cameras that

Fig. 14.3. McPhail-type (a) and Jackson (b) traps baited with the three-component lure (ammonium acetate, putrescine and trimethylamine) and trimedlure, respectively, which are extensively used for population monitoring.

capture frames of adults stuck on a sticky surface and even sensors that can analyse the buzz of attracted adults (Potamitis *et al.*, 2014; Goldshtein *et al.*, 2017; Shaked *et al.*, 2018). Although not yet fully developed, e-traps may provide real-time information regarding the activity of *C. capitata* in the field and may contribute to more economic ways of monitoring and surveillance. Some of the challenges that the technology faces include the development of reliable algorithms to identify captured insects, especially in cases where different sister species are attracted and captured in the same trapping device. Traps using food-based lures that attract both sexes of *C. capitata* and various other non-target species may challenge the identification algorithms.

CULTURAL CONTROL. Infested fruits that drop to the ground, abandoned orchards in a wider area and hosts growing in backyards in suburban areas are considered important for the maintenance and population development of *C. capitata*. Hence, each management programme should include targeted intervention activities to address these areas. Management of populations in abandoned orchards and in urban/suburban areas is more challenging and might require establishment of an area-wide IPM programme. Sanitation of infested fruits that drop to the ground should be always included if it can be economically justified. Collecting and destroying all fruits that remain in the orchard after harvest is important to reduce propagules for future generation of *C. capitata* and should be always considered.

CHEMICAL CONTROL. Bait (e.g. a mixture of attractants with appropriate insecticides) spray applications with appropriate products are often considered as the first approach regarding management of the Mediterranean fruit fly. Bait sprays target adult *C. capitata*, especially females, with the aim of suppressing population densities below the economic injury level. Hence, application of bait sprays should be initiated well before fruit reach the ripening stage. Cover sprays with contact and systemic pesticides could also be considered if bait sprays fail to suppress population densities below the economic injury level. In such cases, special attention should be taken to avoid applications of insecticides close to harvest that may result in unaccepted residues on the fruit. Adult trapping data are a major element in decision making regarding management of *C. capitata* populations. As *C. capitata* is highly polyphagous and multi-voltine, trapping data from a wider area, and alternative and earlier-ripening hosts should be considered to make decisions regarding interventions on a specific crop such as peach and nectarine. Earlier-ripening hosts, such as apricots, figs and pears, should always be considered in a surveillance programme. Establishing economic injury levels for *C. capitata* in peach, nectarine and other hosts is challenging considering the dispersion of the fly from crop to crop and the contribution of alternative hosts to population growth. The susceptibility of different peach cultivars to *C. capitata* infestation may also differ. Cover sprays with appropriate/registered pesticides may be considered, with special attention paid to possible pesticide residues. As stated earlier, *C. capitata* oviposits on ripe or ripening fruit and hence cover insecticide applications should be applied at this phenological stage and very close to harvest. The ban of organophosphates in Europe and elsewhere and the fact that few active ingredients are available to control fruit flies has stimulated research on biopesticides (Di Ilio and Cristofaro, 2021; Benelli *et al.*, 2022), but these are still far from being adopted by growers.

Insecticide resistance has been reported for insecticides used both for bait spray application and for cover sprays. For example, Guillem-Amat *et al.* (2020) reported the detection of spinosad-resistance alleles in field populations of *C. capitata* in Spain. Following the ban on organophosphates, spinosad is the main insecticide used in bait sprays against *C. capitata*. The resistance of Spanish populations of *C. capitata* to pyrethroids has not increased over the last decade (Guillem-Amat *et al.*, 2022), and development of resistance to pyrethroids appears to be rather low in this insect (Voudouris *et al.*, 2018).

NON-CHEMICAL CONTROL. Attract and kill (lure and kill/mass trapping) has long been considered an alternative to the insecticide control approach for *C. capitata* and other fruit flies. Currently, there are many attract-and-kill devices available in local and international markets; however, the efficacy has not been proven using rigorous experimental procedures. All attract-and-kill tactics rely on food attractants and target females before they initiate oviposition. Ammonium acetate, trimethylamine and putrescine are the primary components of the lure for female *C. capitata* (Bali *et al.*, 2021). Additional food baits are also available on the market. A new attractant, Biodelear, has been proposed as a low-cost, non-toxic alternative attractant that can be adopted in mass-trapping efforts against *C. capitata* (Kouloussis *et al.*, 2022). Attract-and-kill and mass-trapping approaches are efficient to suppress low to moderate *C. capitata* populations. Long-term application of attract-and-kill approaches over a wider area may induce a substantial suppression of the *C. capitata* population that can approach the economic injury level (Navarro-Llopis *et al.*, 2013).

BIOLOGICAL CONTROL: FUNCTIONAL BIODIVERSITY, CLASSICAL BIOCONTROL AND ENTOMOPATHOGENS. Although biological control has been considered for control of the Mediterranean fruit fly and other fruit flies, it is not considered a stand-alone method that can suppress the population of *C. capitata* below the economic injury level. There is a list of parasitoids that have been considered for the control of *C. capitata* and mass reared in large facilities, such as *Diachasmimorpha longicaudata* and *Fopius arisanus*. Innundative releases of these parasitoids may be used to suppress populations before the application of a sterile insect release programme (Montoya and Liedo, 2000). Other parasitoids such as *Aganaspis daci* may exist in Mediterranean orchards or can be released innoculatively (Papadopoulos and Katsoyannos, 2003), and can contribute to regulating the populations of *C. capitata* (Fig. 14.4). Besides parasitoids, there has been a recent emphasis on ground-dwelling predators that can reduce the density of larvae that leave fruits to pupate on the ground. The concept

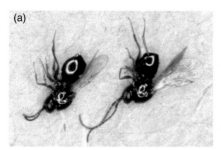

Fig. 14.4. *Aganaspis daci,* an important biological control agent for the Mediterranean fruit fly. (a) Female (left) and male (right; long antennae) adults. (b) A parasitized Mediterranean fruit fly pupa.

of conservation biological control with establishment and management of the ground cover of orchards has been proposed to assist with the control of *C. capitata* (Monzo *et al.*, 2010).

Entomopathogenic nematodes have been also tested experimentally with encouraging results (Kapranas *et al.*, 2021, and references therein). Interestingly, entomopathogenic nematodes have shown significant residual activity that may span up to 4 weeks post-application. In addition, Kapranas *et al.* (2021) showed that entomopathogenic nematodes can attack the larvae of *C. capitata* within infested fruits. Whether a biological control approach using entomopathogenic nematodes is feasible and effective under field conditions still needs to be addressed. Entomopathogenic fungi have been also tested for control of the Mediterranean fruit fly, mainly targeting pupae in the soil (Gazit *et al.*, 2000).

STERILE INSECT RELEASES. The release of sterilized males (known as the sterile insect technique (SIT)) has been applied to contain and eradicate *C. capitata* from North and Central America through costly area-wide campaigns for several decades (Enkerlin *et al.*, 2015). Since 1996, a sterile insect release programme (Medfly PRP) has been established in southern California to address frequent *C. capitata* detection. Indeed, this pre-emptive approach has reduced detection of *C. capitata* (CDFA; https://www.cdfa.ca.gov/plant/PDEP/prpinfo/; accessed 11 January 2023). SIT is also applied to suppress *C. capitata* populations in Valencia, Spain, and in other parts of the world such as Israel. An area-wide approach is essential to assure high efficacy of SIT to suppress *C. capitata* populations.

14.2.2 The peach fruit fly, *Bactrocera zonata* (Saunders)

Overview: common names, taxonomy and geographic distribution
The peach fruit fly, *Bactrocera zonata*, originates from South-East Asia and is considered an important invasive species that threatens peach fruit production

globally (Fig. 14.5). It mainly infests tropical fruits such as mango and guava, but peach is among its preferred hosts. The list of its host fruit exceeds 50 plant species and spans the Cucurbitaceae and Solanaceae. In India, it is considered a highly damaging pest of peaches, with infestation rates reaching high levels up to 100% of fruit (Choudhary *et al.*, 2020). The fly can infest other *Prunus* spp., such as apricot but also citrus fruits, fig and many more. Annual losses caused by the peach fruit fly infestation are estimated to range between US$177 million and US$224 million in Egypt (EPPO, 2010; Mahmoud *et al.*, 2020). *Bactrocera zonata* has dispersed from the ancestral habitats of South-East Asia to Pakistan, Iran, Iraq, the Arab peninsula and North African countries such as Egypt and Libya. It is currently detected and appears to have established in Israel and Lebanon (EPPO, 2010). The peach fruit fly has also invaded La Reunion and Mauritius in the Indian Ocean. Frequent detection of *B. zonata* in California seems to stem from an established population (Papadopoulos *et al.*, 2013), although there is an argument for reinvasion (Carey *et al.*, 2017a,b; Shelly *et al.*, 2017). The fly has also been detected in Florida, but there is no evidence for establishment there. In Europe, *B. zonata* has been detected in Vienna, Austria, almost every year since 2011 (Egartner *et al.*, 2019). Introduction and dispersal in the Middle and Near East and establishment in North Africa pose potential havoc for European countries with a climate friendly for the establishment of *B. zonata*. Most studies of bioclimatic models to predict climatic suitability agree that *B. zonata* has the ability to become established in tropical and subtropical areas (Ni *et al.*, 2012; Zingore *et al.*, 2020). Mediterranean climates are also considered suitable for the persistence and growth of *B. zonata* populations.

Fig. 14.5. Adult female (a) and male (b) of the peach fruit fly, *Bactrocera zonata*.

Biology: life cycle, longevity and reproduction
Similar to *C. capitata*, *B. zonata* is a non-diapausing, multi-voltine species that can produce several generations per year if climatic conditions and host availability are favourable. Embryonic, larval and pupal development range from 1.5 to 10, from 4 to 30 and from 8 to 53 days, respectively, at different constant temperatures ranging from 15 to 35°C (Duyck *et al.*, 2004). However, survival of the immature stages was low, and adult females failed to reach ovarian maturity at the extreme temperatures in this study. *Bactrocera zonata* concludes its development in temperatures ranging from 20 to 35°C, but the optimum range is 25–28°C (Choudhary *et al.*, 2020). The peach fruit fly seems to be able to withstand low humidity better than other tropical fruit flies and hence can persist in relatively dry areas (Zingore *et al.*, 2020; Ben-Yosef *et al.*, 2021). The lifespan of females is approximately 2 months at 25°C and is almost 10 days longer than that of males. Fecundity rates peak at 25°C (440 eggs per female); however, population growth is estimated to maximize at 30°C (intrinsic population growth rate (r_m) = 0.10, doubling time = 7.12 days) (Choudhary *et al.*, 2020).

Damage: symptoms and diagnostics
Like other fruit-feeding tephritids of tropical origin, *B. zonata* females drill a hole and deposit a batch of eggs in the peel or flesh of fruit. Resin exudates may occur in oviposition stings following infestation in peach and other fruits such as mango (Mertens *et al.*, 2021). Usually, a group of larvae feeds in the flesh of the infested fruit, which soon deteriorates and collapses following secondary infestation of microorganisms. Fully grown larvae (L_3) drop out of the fruit and pupate within the surface soil (up to 10 cm depth). Depending on the pervading conditions, adults emerge from pupae one week to several weeks later.

    Identification of adults captured in traps or emerging from pupae is rather straightforward and there are several keys available including electronic ones (White and Elson-Harris, 1992; Drew and Roming, 2016; Plant Health Australia, 2018; Mertens *et al.*, 2021). There are other pests in the genus *Bactrocera* with similar morphological traits that can be confused with adult *B. zonata*, such as *Bactrocera correcta* (Bezzi), although this has a much narrower geographical distribution, which is restricted in South-East Asia. Identification of *B. zonata* at the larval stage is quite challenging. Descriptions of the larval and adult morphological traits as well as detailed adult pictures are provided in EPPO (2013), Plant Health Australia (2018) and Mertens *et al.* (2021). Molecular tools to accurately identify *B. zonata* regardless of the development stage, such as real-time polymerase chain reaction, have been developed (Koohkanzade *et al.*, 2018). Similar to *C. capitata*, infested fruit may be unnoticed and represent a major element of transport and hence invasion of new areas. Thus, molecular identification is important to accurately identify specimens in intercepted fruit at ports of entry to help eliminate the arrival of propagules.

Seasonal biology and management

Year-round population development occurs in the tropics, with host-fruit availability being the main limiting factor. In subtropical areas, capture of adults in traps peaks late in the summer and autumn and follows similar patterns to those of the Mediterranean fruit fly (Bayoumy *et al.*, 2021). Lower temperatures inhibit population growth during the winter and spring months. *Bactrocera zonata* performs better at high temperatures but not well at low ones compared with *C. capitata*. Hence, in temperate areas, winter survival and population growth are expected to be dramatically affected by low winter temperatures. Winter temperature greatly affects the voltinism of the fly. According to FAO/IAEA (2000) the fly can complete seven to nine generations in Egypt. In the drier and hotter areas of east and south Egypt, the number of generations may be reduced to seven per year. The peach fruit fly tends to outcompete and replace *C. capitata* and other fruit flies from warmer coastal areas in La Reunion (Duyck *et al.*, 2006a,b) and Egypt (Elnagar *et al.*, 2010). Both climatic and host availability factors shape the structure of the community of different tephritid species and may explain the interactions among invasive and native species (Charlery de la Masseliere *et al.*, 2017).

POPULATION MONITORING. *Bactrocera zonata* is one of the *Bactrocera* spp. that is strongly attracted to methyl eugenol (4-allyl-1,2-dimethoxybenzene-carboxylate) and hence this is the most commonly used lure for both population monitoring and control approaches using mass trapping, bait stations and male annihilation techniques. Methyl eugenol is sex specific and attracts only males. Sticky delta-type (Jackson) traps can be used to host the dispensers of methyl eugenol. However, bucket (e.g. McPhail-type) traps are also used for population-monitoring purposes. Bucket traps require either an insecticide killing agent or should contain a water solution to kill and retain the attracted insects. Steiner traps can be used for methyl eugenol and are used to trap *B. zonata* males. Dichlorvos (2,2-dichlorovinyl dimethyl phosphate) has been used for decades in McPhail-type traps to kill captured tephritids, but its use is currently restricted or completely banned in most parts of the world. Sticky traps usually have reduced capacity to retain insects, and their performance may be reduced because of dust in dry areas. Inexpensive traps based on transparent plastic bottles have been used by farmers. In a study conducted in Egypt, the International Pheromone McPhail trap followed by the Abdel-Kawi trap outperformed the glass McPhail trap and a farmer-made one (Rizk *et al.*, 2014). Trapping females, which provides important information for control decisions, is more challenging as there is no female-specific attractant available (El-Gendy, 2017). Generic food-based attractants have been considered and are currently used to trap females, such as aqueous solution of torula yeast, and hydrolysed proteins. McPhail traps are used to host the food-based attractants for the peach fruit fly and other tephritids. Selecting the appropriate attractant and trapping strategy is often challenging and should be adapted to local conditions, and should consider the aim of the surveillance. A combination of male- and female-targeted trapping systems should be considered, and female traps should always be included if IPM-related decisions are to be made.

CULTURAL CONTROL.   Similar to other fruit flies, including *C. capitata*, sanitation of infested fruit that drop to the ground of the orchard, and management of unharvested fruit and fruit in abandoned orchards or residential areas are of the utmost importance for reducing the populations of the peach fruit fly. When designing an IPM programme, the high mobility and ability of *B. zonata* to disperse over long distances should be considered. Dispersion of *B. zonata* adults has been estimated to reach 7 km in 1 year based on a recent study by the European Food Safety Authority (Mertens *et al.*, 2021).

CHEMICAL CONTROL.   Several pesticides are registered for the control of *B. zonata* using cover sprays, including organophosphates, pyrethroids and others. Cover sprays with malathion have been used extensively against fruit flies resulting in severe side effects. None the less, bait sprays using malathion as the main pesticide have been used for the control of *B. zonata*, while spinosad-based baits such as GF-120 have also been used extensively (Boulahia-Kheder, 2021). Extensive and continuous use of synthetic insecticides against *B. zonata* should be avoided to reduce the probability of development resistance and possible cross-resistance with other insecticides. For example, use of trichlorfon should be rotated with spinosad or spinetoram to avoid resistance development (Khan *et al.*, 2022).

NON-CHEMICAL CONTROL.   Detection of *B. zonata* in a new area triggers aggressive campaigns to eradicate the pest, including attract-and-kill approaches and, more specifically, the male annihilation technique to eliminate fertile males, followed by foliar bait spray applications using food baits and appropriate insecticides. For established populations, the male annihilation technique is not a common approach to manage *B. zonata*, and in general the fly is not attracted to proteinaceous baits. Recently, Hasnain *et al.* (2022), working towards developing new female-targeted baits, revealed that GF-120 (a standard bait combined with spinosad) was as attractive as a mixture of several compounds such as protein hydrolysate, ammonium acetate, putrescine, trimethylamine and jaggery.

BIOLOGICAL CONTROL: FUNCTIONAL BIODIVERSITY, CLASSICAL BIOCONTROL AND ENTOMOPATHOGENS.   Several biological control agents including fungi, nematodes and parasitoids have been tested and implemented against *B. zonata*. For example, *Metarhizium anisopliae* and *Beauveria bassiana* have shown efficacy against *B. zonata* and other fruit flies (Sookar *et al.*, 2014a; Usman *et al.*, 2021). The application of entomopathogenic nematodes such as *Steinernema feltiae* to fallen fruit infested with *B. zonata* has successfully reduced the adult population in experimental trials (Mahmoud, 2009; Mahmoud *et al.*, 2016).

Likewise, entomopathogenic nematodes of the genus *Heterorhabditis* have shown good efficacy against *B. zonata* (Usman *et al.*, 2021). Fruit fly parasitoids such as *Dirhinus giffardii* (Khan *et al.*, 2020) and *Fopius arisanus* can be also considered in biological control programmes against *B. zonata* (Rousse *et al.*, 2006).

STERILE INSECT RELEASES. SIT technology has recently advanced and is currently implemented to contain and even eradicate *B. zonata* from invaded areas such as the island of Mauritius in the Indian Ocean. However, adoption of this technology at the farmer level for the control of the peach fruit fly is yet to be tested. Combined use of SIT with entomopathogenic fungi has been proposed and tested (Sookar *et al.*, 2014b). This approach considers the use of released sterile males as carriers of entomopathogenic fungi such as *M. anisopliae* and *B. bassiana*, which, through intraspecific interactions, will disperse and infect other individuals.

14.2.3 The Peach Twig Borer, *Anarsia lineatella* (Zeller)

Overview: common names, taxonomy and geographical distribution
The peach twig borer, *Anarsia lineatella* (Lepidoptera: Gelechiidae) is widespread in Central and Southern Europe and North Africa, and eastwards through the Middle East and Turkey to Central Asia and China (Li and Zheng, 1998). *Anarsia lineatella* has been introduced with its host plants to North America where its presence was reported in the middle of the 19th century (Clemens, 1860). It is now widespread in the USA and southern Canada. Records from other areas in North America need to be confirmed because of confusion with morphologically similar species. In Europe, the northern-most occurrence seems to be in northern central Germany. There are two subspecies, *A. lineatella heratella* Amsel, 1967 from Afghanistan and *A. lineatella tauricella* Amsel, 1967 from Turkey (Gregersen and Karsholt, 2017).

The species feeds on a number of Rosaceae, especially *Prunus* spp. *Prunus persica* (L.) Batsch, *P. armeniaca* L., *P. domestica* L. and *P. dulcis* (Mill.) that are considered among its major host plants. In general, the species is oligophagous (Summers *et al.*, 1959; Balachowsky, 1966; Damos and Savopoulou-Soultani, 2008). It is also known to feed on *Malus domestica* Borkh., *Pyrus communis* L., *Prunus cerasus* L. and *Prunus spinosa* L.

The peach twig borer is an important pest of peach, apricot, almond and plum, causing significant damage to young shoots and fruits (Fig. 14.6). Infestation rates in peach and nectarine, which are among the preferred hosts for peach twig borer, can reach a high proportion of the yield in cases where management of the pest is not applied.

Occurrence of *A. lineatella* in a specific area may strongly affect trading of fresh fruit commodities as it is considered a quarantine pest of major importance for many countries.

Fig. 14.6. The peach twig borer, *Anarsia lineatella*. Infestation of a peach fruit.

Biology: life cycle, longevity and reproduction

The peach twig borer is an oligophagous, multi-voltine species that can produce many generations per year depending on climatic conditions and the availability of hosts. In the Mediterranean Basin, it can produce three or four generations in a year, while at the northern-most limits of its distribution, there may be as few as one. The life history traits of *A. lineatella* vary widely depending on temperature conditions. Larval development in constant laboratory conditions on an artificial diet was approximately 25 days at 25°C and only 13 days at 35°C, and total development from egg to adult was 37 days at 25°C and 25 days at 35°C (Damos and Savopoulou-Soultani, 2008). The number of degree-days was 66.6 for egg hatching, 263 for total larval development, 75.7 for pupal development and 400 for egg to adult development. The life expectancy of females was estimated as 10 and 60 days at 25°C and 15°C, respectively, and the lifetime fecundity rate was 127 eggs per female at 25°C. The intrinsic population growth rate (r_m) may range from 0.028 to 0.238, depending on the temperature (Damos, 2013). The optimum temperature for development was estimated to be between 30°C and 35°C (Damos and Savopoulou-Soultani, 2008).

Damage: symptoms and diagnostics

The peach twig borer attacks young twigs in early spring and fruit later in the season. Overwintering larvae exit the hibernacula and bore into a twig from below the pith, hollowing it out and causing exudation of some sap. The twig withers and the larva moves to a new one. Overwintering larvae also attack flowers and buds, feeding on petals and causing significant damage by penetrating ovaries (Mamay *et al.*, 2014). On young trees, repeated death of terminal twigs causes stunted growth and reduced tree vigour. The larvae enter the fruit primarily through the stem end or where the fruit touches another fruit, and feed just under the skin or next to the pit. In mature fruits, the larvae bore inside the fruit and feed on the flesh. Exit holes on fruits from fully grown larvae are visible and can be a source of pathogen infestation. Infested fruits are unsuitable for consumption and marketing.

Identification of *A. lineatella* is based on morphological characteristics of the adults, characterized by a fuscous grey forewing with only a little white and with indistinct black streaks; it appears darker than that of the closely related *Anarsia innoxiella* and has a less fractured pattern on the forewings (Gregersen and Karsholt, 2017). The head capsule of the larva and the prothoracic plate are glistening black, and the body is honey-brown or chestnut brown, with whitish intersegmental divisions. The pinacula are small and black, each with one whitish hair, the anal plate is glistening black, and the prolegs are concolorous with the body (Heckford, 1992).

Seasonal biology and management

Peach twig borer overwinters as first- or second-instar larvae in protected places on the trunk such as in crevices, forming a hibernaculum. In spring, the larvae exit from the hibernacula and feed on buds, flowers and young twigs. Adults of the first generation appear in early to late May in the eastern Mediterranean, Greece and Turkey (Damos and Savopoulou-Soultani, 2008; Mamay *et al.*, 2014), and in April in southern Italy. Depending on the temperature, two or three more generations develop during the summer. In central Greece, the approximate number of degree-days required above a temperature threshold of 10.05°C for the first generation has been estimated at 431, at 661 for the first and second generation respectively, and at 675 for the third. Larvae of the last generation will overwinter in hibernacula.

POPULATION MONITORING. Seasonal abundance and population dynamics are monitored with the use of baited pheromone traps and scouting for twig wilt at the beginning of the season. Shoots damaged by the oriental fruit moth (*Grapholita molesta*) cannot be distinguished from those damaged by *A. lineatella* and need to be opened to determine which pest is present. During bloom, it is important to monitor for peach twig borer larvae and its damage when shoots are emerging, to determine if the pest is active. From bloom onwards, orchards should be monitored for wilt twigs to determine population abundance and time management decisions. Delta traps or wing-style pheromone traps are used for monitoring adult populations. To monitor emerging adults, traps are deployed in peach orchards early in the season. The sex pheromone of *A. lineatella* was isolated and identified in 1975 (Rice and Jones, 1975). The main compounds are (E)-5-decenyl acetate and (E)-5-decen-1-ol at a ratio of 7:1.

CHEMICAL CONTROL. Treatment decisions should be based on careful and appropriate monitoring of the population in the orchard. Degree-day models have been developed that are being used for timing insecticide applications in relation to the targeted developmental stage of *A. lineatella*. Economic injury levels and economic thresholds have been developed and are available for use to decide on insecticide applications against *A. lineatella* (Damos and Savopoulou-Soultani, 2008; Damos and Savopoulou-Soultani, 2010). In IPM programmes, well-timed treatments of environmentally friendly insecticides are preferred. Several insecticides are used for the control of peach twig borer, including spinetoram, spinosad, chlorantraniliprole and deltamethrin.

MATING DISRUPTION. Mating disruption using a dispenser containing the species-specific synthetic pheromone of the peach twig borer has been used since the 1980s. A number of of different pheromone dispensers are available for mating disruption purposes. Dispensers are deployed early in the season before the start of the adult flight to prevent mating and laying of viable eggs. The mating disruption method against the peach twig borer is now widely adopted by peach growers along with simultaneous application for mating disruption of *G. molesta*. In areas where both species are present, the option of using one product for combined mating disruption is available.

BIOLOGICAL CONTROL. There are numerous parasitoids available for *A. lineatella* (Damos and Savopoulou-Soultani, 2008). Parasitization of larvae of the overwintering generation can reach high levels. Conservation practices should be considered to maintain and enhance the activity of the parasitoids in the orchards. For example, application of environmentally friendly insecticides and maintenance of wild vegetation strips are thought to enhance the activity of biological control agents for *A. lineatella* (Perez-Aparicio *et al.*, 2021). Applications of products based on *Bacillus thuringiensis* are available and should target newly hatched larvae.

14.2.4   The oriental fruit moth, *Grapholita molesta* (Busck)

Overview: common names, taxonomy and geographical distribution
The oriental fruit moth, *Grapholita molesta* (Lepidoptera: Tortricidae) is an invasive oligophagous insect species of Asian origin that is currently widely distributed throughout temperate and subtropical regions of the world. Initially, it was spread from north-west China to Japan and Australia, and to Europe, as well as to several states on the east coast of the USA and in California and Washington states in the west. It is also present in Central and South America (Argentina, Brazil and Uruguay) and Africa (Mauritius, Morocco and South Africa) (EPPO, 2022).

The main hosts of the *G. molesta* are *Prunus* spp. (Rosaceace, subfamily Drupaceae), including peach (*Prunus persica*), nectarine, plum (*P. domestica*) and apricot (*P. armeniaca*) (Rothschild and Vickers, 1991). It is considered a pest of major economic importance of peach and nectarine but has also been reported to feed on other plant species. Pome fruits such as apple (*Malus domestica* Borkh) and pear (*Pyrus communis*) are considered secondary hosts (Rice *et al.*, 1972; Rothschild and Vickers, 1991). Switching of oriental fruit moth populations from stone-fruit orchards to pome-fruit orchards during the growing season, particularly after the stone-fruit harvest, has been observed in many parts of its geographical range (Natale *et al.*, 2003; Il'ichev *et al.*, 2007; Myers *et al.*, 2007).

Occurrence of the pest in a specific area may strongly affect trading of fresh fruit commodities as *G. molesta* is considered a quarantine pest of major importance for many countries (e.g. Israel, Mexico).

Biology: life cycle, longevity and reproduction

Oriental fruit moth is an oligophagous, multi-voltine species that can conclude many generations (three to seven) per year depending on climatic conditions and the availability of hosts. In orchards of the Mediterranean coast, four to five generations per year have been estimated. The number of generations completed each year may differ widely, even in orchards located in neighbouring areas. For instance, in Girona, north-east Spain, four generations have been reported, while five were reported in Lleida (Amat *et al.*, 2021).

The life history traits of *G. molesta* varies depending on temperature conditions and the host fruit. At constant temperatures, optimum development is achieved at 25°C (Chen *et al.*, 2019). Larval development lasts for 6–32 days depending on temperature and feeding conditions. The lower developmental threshold for total development is 7.2°C and the upper developmental threshold is 32.2°C. A total of 507.2 degree-days are estimated to be required for development from egg to adult (Croft *et al.*, 1980). However, a higher lower developmental threshold of 9.5°C has been estimated in northern Greece (Damos *et al.*, 2022). The longevity and female fecundity of adults at 25°C in the laboratory has been estimated at 21 days and 160–230 eggs per female, respectively. At 13°C, the oviposition period is extended but fecundity is reduced (approximately 50 eggs per female) compared with that at 25°C (Silva *et al.*, 2011). Adults reared as larvae on peach shoots live longer but lay fewer eggs than those reared on peach fruits. Larval development is faster on peach shoots and immature peach fruit compared with ripe peach fruits. Larval development and adult longevity and fecundity are also affected by host species. On pear fruit, larval development at 25°C lasts 27–31 days and adult longevity lasts 20–25 days, while lifetime fecundity rates range from 128 to 210 eggs per female (Du *et al.*, 2015).

Damage: symptoms and diagnostics

The oriental fruit moth attacks young twigs in early spring and fruits later in the season. Similar to the peach twig borer, larvae of the first generation of *G. molesta* develop on terminal shoots causing wilting. The damage is similar to that caused by the peach twig borer. Larvae enter the young shoots from the top and bore downwards until they find woody tissue. Then they abandon the shoot and infest another one. One larva can destroy up to five shoots to complete its development. The infested shoots wilt and die. In mid-summer, the larvae attack unripe peach fruit, feeding on the surface of the fruit, which may produce gum at the feeding sites. Infested fruit may drop prematurely and are more prone to secondary infection by fungi. On ripe fruits, the spot of larval entry is often unnoticed and hence infested fruits are often detected only after being opened. Infested fruit are unsuitable for consumption and marketing.

Identification of *G. molesta* is based on the morphological characteristics of the adults. Careful inspection of genitalia following laborious dissections is usually required for confirmed identification as the adults are similar to other *Grapholita* spp. The forewing length is 5–6.5 mm with a dull greyish-brown colour and a row of black dots near the apex and termen. The larvae are also similar to other sibling species (Gilligan and Epstein, 2012). Molecular methods are also available to distinguish *G. molesta* larvae from other similar species (Chen and Dorn, 2009).

Seasonal biology and management

Oriental fruit moth overwinters as fully grown larvae in a cocoon in protected places on the trunk, in crevices or in the ground. In spring, the overwintered larvae pupate and adults emerge to lay eggs on leaves or young shoots. In Spain, the adults of the first generation appear in March, and in Italy and Greece in April (Damos *et al.*, 2014; Amat *et al.*, 2021). Three to six generations are completed annually depending on the location. In northern Greece, the approximate number of degree-days required for 50% of the cumulative number of male moths for the first, second and third generations of *G. molesta* were estimated at 654, 785 and 1251, respectively (biofix: 1 March). Later generations overlap, and seem to be characterized by a bimodal seasonal pattern (Damos and Savopoulou-Soultani, 2010). In the USA, peach and *G. molesta* both have the same temperature threshold ($10°C$) so the life cycle is tightly linked and egg laying typically occurs right around petal fall.

POPULATION MONITORING. Seasonal abundance and population dynamics are determined by the use of pheromone-baited traps and scouting for wilted twigs at the start of the season. The sex pheromone for *G. molesta* was identified in 1969 (Roelofs *et al.*, 1969). The main compounds are (Z)-8-dodecenyl acetate, (E)-8-dodecenyl acetate and dodecan-1-ol. Delta traps or wing-style traps are commonly used for adult monitoring of *G. molesta*. Traps should be installed early in the season to monitor emerging adults of the overwintering generation. Besides the original pheromone compounds identified from females that attract males, the addition of terpinyl acetate and acetic acid was found to be critical for efficient monitoring in a mating-disruption-treated orchard (Mujica *et al.*, 2018). In addition to the classical pheromone traps, sophisticated traps based on artificial intelligence and the Internet of Things have been developed for real-time monitoring of male *G. molesta* catches in the orchard (Kim *et al.*, 2011).

CHEMICAL CONTROL. Several insecticides are used for the control of oriental fruit moth (e.g. spinetoram, spinosad, chlorantraniliprole, deltamethrin). Formulations of a granulovirus, which was first isolated from *C. pomonella*, that have isolates selected for *Gydia molesta* are also effective against newly hatched larvae. Treatment decisions should be based on careful and appropriate monitoring of the population in the orchard. Moreover, degree-day models have been developed that are being used for timing insecticide applications in relation to the targeted developmental stage. In California, two degree-day models are used to predict the seasonal phenology of *G. molesta*. In one model, accumulation of degree-days above a threshold of 7.2°C starting from the beginning of the year is used to estimate the appearance of the first adults, whch occurs after 30 degree-days (Croft *et al.*, 1980). Similar degree-days have been estimated for northern Greece (Damos *et al.*, 2022). In other US states, a temperature threshold of 10°C is used for degree-day calculations. In IPM programmes, well-timed treatments of environmentally friendly insecticides should be implemented.

MATING DISRUPTION. Mating disruption using the synthetic pheromone of the oriental fruit moth has been carried out since the 1980s. Different kinds of dispenser are available on the market for use in the field. Dispensers are placed early in the season before the onset of the adult flight to prevent mating and laying of viable eggs. As well as the 'classical' dispensers for mating disruption, microencapsulated formulations have been developed for sprayable application of the pheromone in the field (Il'Ichev *et al.*, 2006). The mating disruption method against *G. molesta* has now been widely adopted by peach growers, along with simultaneous application for the mating disruption of *A. lineatella*.

BIOLOGICAL CONTROL. Numerous natural enemies of *G. molesta* have been recorded (Atanassov *et al.*, 2003). Conservation practices applied in the field maintain and enhance the activity of the parasitoids within orchards. Use of environmentally friendly insecticides and maintenance of wild vegetation strips contribute to the conservation of the natural enemies in the orchard (Aparicio *et al.*, 2021). Lately, the augmentative release of the egg parasitoids *Trichogramma dendrolimi* have shown promising results in decreasing the infestation rates by *G. molesta* in the field (Zhang *et al.*, 2021). Applications of products based on *B. thuringiensis* are available and should target newly hatched larvae. Products based on the *C. pomonella* granulovirus are also available and have been found to be effective against *G. molesta* (Graillot *et al.*, 2017).

14.2.5   The brown marmorated stink bug, *Halyomorpha halys* (Stål)

Overview: common names, taxonomy and geographical distribution
Halyomorpha halys (Hemiptera: Pentatomidae) is a pest of global importance that originates from Asia with populations most prevalent in China, Japan and

the Republic of Korea. In the mid-1990s, it was first classified as an invasive species with aggregating populations detected in eastern Pennsylvania, USA. *Halyomorpha halys* then spread rapidly throughout much of the USA, Canada, Europe and Chile (Leskey and Nielsen, 2018).

The adults of the brown marmorated stink bug are 12–17 mm in length and marbled grey-brown in appearance (Fig. 14.7). Adults can be identified by exposed connexiva, which alternate between black and pale in colour, and the fourth antennal segment, which is white or pale. *Halyomorpha halys* also lacks teeth on the juga and has smooth margins on the antereolateral pronotum, a characteristic that distinguishes it from many Pentatomini (Hoebeke and Carter, 2003). *Halyomorpha halys* eggs are light emerald in colour, and are typically laid in clusters of 28 on the underside of leaves (Nielsen *et al.*, 2008). Each egg is 1.3 mm in diameter and has 30–32 micropylar projections (Hoebeke and Carter, 2003). *Halyomorpha halys* goes through five nymphal instars, all of which are primarily red and black in colour, and the first instars remain aggregated on the egg mass to obtain endosymbionts (Taylor *et al.*, 2014). Instars two to five have the characteristic white fourth antennal segment, as for the adults (Hoebeke and Carter, 2003).

Biology: life cycle, longevity and reproduction

Halyomorpha halys overwinter as non-reproductive diapausing adults in protected areas, primarily human-made shelters, but adults can also be found in cliff outcrops and underneath dead upright trees (Watanabe *et al.*, 1994; Inkley, 2012; Lee *et al.*, 2014). Critical diapause termination occurs at photoperiod of 13.5 h (McDougall *et al.*, 2021). The adults frequently leave the overwintering sites before diapause termination (in early May) and disperse to agricultural areas, with peach orchards among the top early-season crops (Leskey and Nielsen, 2018). Mating and oviposition follow soon after diapause termination, which is estimated at approximately 150 Degree-days (Nielsen *et al.*, 2008). Females lay eggs in clusters of approximately 28 light-green eggs on the underside of leaves. Nymphs hatch and first instars remain aggregated on the egg mass while they obtain endosymbionts on the egg chorion. Development from egg to adult takes 538 Degree-days, and a female can oviposit an average of 244 eggs in her lifetime, an intrinsic

Fig. 14.7. The brown marmorated stink bug, *Halyomorpha halys*. (a) Adult. (b) Eggs. (c) Surface and internal injury on peaches.

rate of increase (r_m) of 0.07 (Nielsen *et al.*, 2008). The five nymphal instars feed on over 170 host plants, primarily in the family Rosaceae ((Nielsen and Hamilton, 2009); www.stopbmsb.org, accessed 12 January 2023). *Halyomorpha halys* feeds primarily on the fruiting structures of plants, with a slight preference for arboreal hosts, and will track plant phenology (Leskey and Nielsen, 2018; Blaauw *et al.*, 2019). Generally, a mixed diet promotes higher fitness, although peach and *Ailanthus altissima* are among the few hosts that the species can complete development on (Acebes-Doria *et al.*, 2016a). Females lay eggs throughout their lifespan, for a period of 4 months post-diapause (Nielsen *et al.*, 2008; Haye *et al.*, 2014), which results in highly overlapping generations. For this reason, it is often difficult to determine the number of generations. Stage-specific modelling has predicted bivoltine populations throughout the USA, with the interaction of photoperiod and climate impacting the population size and phenology of *H. halys* (Nielsen *et al.*, 2017). The largest number of *H. halys* adults typically coincides with peach harvest in mid-July to the end of August.

Damage: symptoms and diagnostics

Halyomorpha halys adults exiting diapause can be found dispersing into peach orchards at approximately $150 \, DD_{14}$ (biofix: 13.5 h photoperiod) in New Jersey, at which time they are beginning to mate and lay eggs (Nielsen *et al.*, 2017). The stage of peach phenology is variable at this time, but feeding at shuck split results in aborted fruit (Nielsen *et al.*, 2011). During feeding, *H. halys* inserts its proboscis into the fruit and injects extra-oral digestive enzymes, killing the plant cells surrounding the wound. This may result in external depressions, sometimes severely deformed, and internal necrosis depending on the plant growth phase (Acebes-Doria *et al.*, 2016a). Early-season feeding by adults in peach can result in gummosis and the highest levels of surface depressions and internal necrosis at harvest (Acebes-Doria *et al.*, 2016a). Feeding when the fruit is taking on water and ripening results in surface depressions that may be corked underneath (Acebes-Doria *et al.*, 2016a). The flesh underneath can even become calcified and sandy in texture (Nielsen *et al.*, 2011). Nymphs can also injure the fruit throughout peach development. The damage varies by insect life stage and peach phenology but results in external depressions and internal necrosis, with a higher number of internal injuries at harvest (Acebes-Doria *et al.*, 2016a). The stylet sheath can frequently be seen during examination and is a helpful way to distinguish injury.

Seasonal biology and management

POPULATION MONITORING. Peach and nectarine are important hosts for all *H. halys* life stages (Acebes-Doria *et al.*, 2016b). Adults and nymphs can be found in peach orchards from approximately shuck split (or mid-May) to harvest. External injury on peach fruit increases throughout the growing season (Blaauw *et al.*, 2015).

Monitoring for *H. halys* has evolved from timed visual samples and limb jarring to pheromone-baited traps. The male-produced aggregation pheromone for *H. halys* is a dual-component lure that is attractive at a 3.5:1 ratio of (3*S*,6*S*,7*R*,10*S*)-10,11-epoxy-1-bisabolen-3-ol to (3*R*,6*S*,7*R*,10*S*)-10,11-epoxy-1-bisabolen-3-ol (Khrimian *et al.*, 2014; Weber *et al.*, 2017). Males and females as well as all mobile immatures are attracted to the pheromone. Two primary baited traps are in use, a black pyramid trap that acts as a trunk mimic and a clear sticky panel trap on a 4-foot wooden stake (Acebes-Doria *et al.*, 2018). The pyramid style trap is more sensitive and may detect adults earlier in the season, but the clear sticky panel trap is inexpensive and allows more traps to be deployed, providing important information on population fluctuations. Both traps have recorded a similar phenology of *H. halys* across geographical regions (Acebes-Doria *et al.*, 2020).

CHEMICAL CONTROL.   Insecticide management of *H. halys* in fruit trees is complex and relies primarily on broad-spectrum pyrethroids and neonicotinoids (Leskey *et al.*, 2012; Leskey and Nielsen 2018). Knockdown and recovery of *H. halys* may occur, especially with pyrethroids (Nielsen *et al.*, 2008; Leskey *et al.*, 2012; Lee *et al.*, 2013). Foliar applications have a short residual activity and must be reapplied frequently to peach trees to combat the continued presence and dispersal by *H. halys*. There is no action or economic threshold for *H. halys* in peach, based on any monitoring method. Frequent application of broad-spectrum insecticides such as pyrethroids has caused secondary pests such as San Jose scale to flare. In peach, insecticides can be applied as full-block, alternate-row and even border-based applications, with the latter two having reduced insecticide application (see Cultural Control, below).

NON-CHEMICAL CONTROL.   There is currently no non-chemical control approach that has been demonstrated to reduce infestation in peach. Hail netting in pear orchards in Italy has significantly reduced injury and may be an acceptable tactic (L. Maistrello, personal communication).

CULTURAL CONTROL.   There are currently no peach varieties that are resistant to stink bug species. Trap crops, specifically a mixture of grain sorghum and sunflower, delay the colonization of *H. halys* on vegetable crops, but this approach has not been evaluated in orchard systems (Mathews *et al.*, 2017). The behavioural response to the aggregation pheromone has been evaluated in two different attract-and-kill approaches. The first uses apple trees baited with high doses of the pheromone, spaced at least 50 m from each other. Attract-and-kill trees are sprayed weekly with insecticide. Injury at harvest is similar to that of whole-block management (Morrison *et al.*, 2019). The second approach uses insecticide-treated netting (Vestergaard-Frandsen, Lausanne, Switzerland) baited with high doses of the aggregation pheromone (Kuhar *et al.*, 2017; Giuseppino *et al.*, 2018), placed on the outside of orchards so that adults attracted to the pheromone will receive a lethal dose of toxicant. These methods are highly effective at killing large numbers of *H. halys* adults but have not been evaluated in peach, where thresholds for injury may be lower than in pome fruit.

Exploiting the perimeter-driven dispersal behaviour of *H. halys* adults has allowed the development of a border spray management tactic in peach (Blaauw *et al.*, 2015, 2016). It was found that 59–80% of adults remained on the border of a peach orchard for 7 days before moving about the orchard. Spatial refinement of insecticide applications to the border plus first full row equates to 25% of a 2.5 ha orchard. Injury at harvest was equal to or less than full-block or alternate-row middle insecticide applications (Blaauw *et al.*, 2015).

BIOLOGICAL CONTROL. Generalist natural enemies have a variable impact on *H. halys* in orchards. Most studies have used sentinel egg masses to monitor biological control services across natural and agricultural ecosystems. Predation on eggs is around 10% (Ogburn *et al.*, 2016), and predators include many generalists, with the top taxa being damsel bugs and katydids. Natural enemy abundance is higher in peach orchards integrating border sprays (A.L. Nielsen, unpublished data). The first discovery of *Trissolcus japonicus* (Hymenoptera: Scelionidae) parasitizing *H. halys* egg masses in North American agriculture was also in peach orchards following border sprays (Kaser *et al.*, 2018). *Trissolcus japonicus* is a co-evolved egg parasitoid of *H. halys* that was detected in the USA in 2014 (Yang *et al.*, 2009; Talamas *et al.*, 2015); however, levels of parasitism in most agricultural habitats remains low.

14.2.6   San Jose scale, *Comstockaspis perniciosus* (Comstock)

Overview: common names, taxonomy and geographical distribution
With nearly worldwide distribution, San Jose scale, *Comstockaspis perniciosus*, is a pest of peach, nectarine, plum and other tree fruits including apple, pear and cherry (Howard and Marlatt, 1896). Adult females are yellow, circular, sac-like insects. They secrete and live beneath a protective covering that is round (1.5 mm), grey-brown and made up of concentric rings surrounding a raised bump near the centre. Adult males are tiny, golden-brown, two-winged insects, about 1 mm long, with a narrow, dark band across the abdomen. They mature under elongated, oval, waxy coverings, about 1 mm long, with a raised, dark bump near one end. The mobile stage of the immatures, known as 'crawlers,' are yellow, somewhat oval, and minute at about 0.25 mm long (Gentile and Summers, 1958; Rice *et al.*, 1982).

Biology: life cycle, longevity and reproduction
Comstockaspis perniciosus overwinters primarily in immature life stages that are nearing adulthood beneath their protective, waxy coverings (Fig. 14.8). During

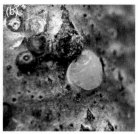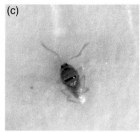

Fig. 14.8. San Jose scale, *Comstockaspis perniciosus*. (a) Under the protective coverings.
(b) An adult female with her protective covering removed. (c) A winged male.

warmer south-east USA winters, mature females also overwinter. The scale
insects remain inactive until sap flow begins in the spring and development
resumes when temperatures reach 10.6°C. The females remain stationary beneath their waxy covering throughout their lives. The males are tiny, winged
insects that seek out female scale insects to mate with (Gentile and Summers,
1958; Rice *et al.*, 1982).

Comstockaspis perniciosus has a very high reproductive potential. Females
produce about ten immatures or 'crawlers' per day for 2–3 weeks, depending
on temperature. Crawlers emerge from under the female scale covering and
move to new sites of infestation on the bark, leaves or fruit. For their size, the
crawlers can walk considerable distances (2–4 m) or be blown by the wind. It is
possible for scale crawlers to be wind blown into orchards from adjacent hedgerows,
wooded borders or orchards. Within 1 day of emerging from under the scale
covering, the crawlers settle, insert their mouth parts into the plant and begin
to feed. Within 2–3 days of settling, the crawlers begin to secrete a white, waxy
material (the 'white cap' stage), which eventually turns black (the 'black cap'
stage). The protected nymphs pass through several moults while maturing before becoming adults (Gentile and Summers, 1958). Serious infestations can
develop between harvest and the onset of winter. This is especially true when
abundant late-summer rainfall promotes succulent growth favourable for scale
development. Management is needed when scale lesions were present on fruit
the previous year or when scales are found on wood during pruning.

Damage: symptoms and diagnostics
Scale insects damage plants directly through feeding injuries by inserting
their piercing/sucking mouth parts and withdrawing nutrients directly from
the plant. In peach, feeding damage can cause leaf chlorosis and twig or limb
dieback, and even the death of trees if scale populations reach high levels. At
high enough populations, *C. perniciosus* covers the branches, creating a bumpy
grey, 'scaly' surface compared with the smooth, brown bark of an uninfested
branch. *Comstockaspis perniciosus* prefers to feed on branches, but at high population levels, finding a good place to feed becomes hard, and the crawlers settle

and feed on the fruit (Rice *et al.*, 1982). This type of feeding injury to the peach fruit produces small, red, measles-like lesions on the skin (Fig. 14.9). Fruit injury from *C. perniciosus* does not damage the flesh of the fruit, but the cosmetic appearance can be rather unappealing.

Seasonal biology and management

POPULATION MONITORING. Adult *C. perniciosus* males can be monitored with pheromone-baited traps (Hoyt *et al.*, 1983). Captures of males can be used to predict the relative densities of the crawlers (Fig. 14.10a) (Badenes-Perez *et al.*, 2002a). The crawler activity timing can be improved by visually monitoring for crawlers on infested trees. Crawler activity can also be assessed by regularly checking new shoot growth on infested trees, as the crawlers will be concentrated on the new growth (Rice *et al.*, 1982). Scale-infested branches should first be wrapped with black electrical tape and then clear double-sided sticky tape (Fig. 14.10b). Using a hand lens or loupe, check the tapes frequently for scale crawlers. Although there is no threshold to initiate spray timing, targeted management is possible according to when the abundance of scale crawlers caught on the tape traps substantially increases.

CHEMICAL CONTROL. Currently, insect growth regulators, such as buprofezin and pyriproxyfen, are effective at managing San Jose scale crawlers, and the systemic nature of spirotetramat can target additional life stages when feeding on plant tissue. In order to preserve the current effectiveness of available insecticides, resistance management programmes should be implemented,

Fig. 14.9. Damage on a peach fruit caused by San Jose scale.

considering the insecticide mode of action classes and rotating when possible (Buzzetti *et al.*, 2015).

NON-CHEMICAL CONTROL.   The most common and effective scale management option is with two horticultural oil applications made during tree dormancy to kill overwintering scales (Badenes-Perez *et al.*, 2002b). Currently, in a commercial orchard, two dormant horticultural oil applications should be applied to every acre, every year. Horticultural oil kills *C. perniciosus* by smothering the scale under their protective coverings, so complete coverage of the tree(s) with dilute sprays is critical in order to reach all the nooks and crannies of the trees to effectively manage them.

CULTURAL CONTROL.   Trees and fruit should be inspected frequently for the presence of scale insects. Scale insects blend in well with peach bark and are generally hard to see, so marking heavily infested trees when scales are detected can help to readily monitor scale development during the season. Additionally, focusing pruning on infested branches and suckers can reduce scale abundance. Good pruning practices and careful tree training can also open up the canopy to facilitate better spray coverage. Burning and proper disposal of removed trees and limbs infested with scale is also important to avoid dispersion of propagules to neighbouring orchards.

BIOLOGICAL CONTROL.   There are several predators and parasitoid wasps that feed on or attack *C. perniciosus* (Nagaraja and Hussainy, 1967; Badenes-Perez *et al.*, 2002b). These natural insect enemies may help reduce scale populations in the south-east USA, but generally they do not provide enough control to prevent

Fig. 14.10. (a) Example of a pheromone-baited sticky card to monitor adult San Jose scale males and (b) a sticky tape 'trap' for sampling San Jose scale crawlers.

damage when scale populations are high. At this time, natural control is only considered a supplement to chemical control. Note that using insecticides, particularly pyrethroids, during the growing season can significantly impact natural enemies, disrupting the natural control of many orchard pests, San Jose scale and mites.

14.2.7 Plum curculio, *Conotrachelus nenuphar* (Herbst)

Overview: common names, taxonomy and geographical distribution
Plum curculio, *Conotrachelus nenuphar* (Coleopter: Curculionidae), is a true weevil and a key insect pest of peach in eastern North America (Fig. 14.11). Adults are approximately 4–6 mm long, brownish-black snout beetles, mottled with lighter grey or brown markings. Their backs are roughened and bear two prominent humps and two smaller humps. The larvae are yellowish-white, legless, brown-headed grubs, approximately 6–9 mm long when fully grown (Quaintance and Jenne, 1912; Johnson *et al.*, 2005c).

Biology: life cycle, longevity and reproduction
Conotrachelus nenuphar is a snout beetle native to North America and is found east of the Rocky Mountains in the USA and Canada. It overwinters as an adult in ground litter or other protected places, both in and around orchards, particularly in nearby woods or fence rows. The overwintered adults become active when mean temperatures reach 10–15.6°C for 3–4 days, and begin moving towards orchards when the maximum temperature reaches 21°C for 2 or more days. This series of temperature events often takes place shortly before or as peaches bloom, especially in the middle and upper south-eastern USA (Quaintance and Jenne, 1912; Johnson *et al.*, 2005c).

Initially, overwintered adults feed on succulent buds, foliage and blooms (Jenkins *et al.*, 2006). The pre-oviposition period following emergence from hibernation may vary from 6 to 17 days, depending on temperature. As the most mature peach fruit reach shuck split, the female begins depositing eggs singly in a hole that she eats in the fruit. The egg hatches on average in about 5 days. The larva feeds on the fruit for 8–22 days. The fully grown larva tunnels out of the peach, enters the soil and constructs a small earthen cell, usually 2.5–8 cm below the surface (Fig. 14.12). After about 2 weeks, the larva transforms into a white pupa and then to an adult. The adults of the first generation usually emerge about 4 weeks after the larvae enter the soil. The complete life cycle, from egg to adult emergence, takes 5–8 weeks, depending on climatic conditions. In the south-eastern USA, *C. nenuphar* usually completes two generations and possibly a partial third generation each year (Johnson *et al.*, 2005c).

Overwintered adults may deposit the first-generation eggs as soon as young peach fruit reach shuck split. Emergence of adults and oviposition may

Fig. 14.11. Adult of the plum curculio, *Conotrachelus nenuphar*.

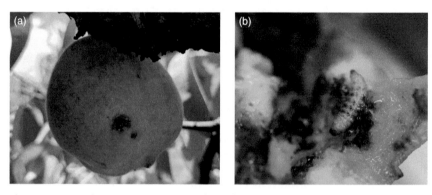

Fig. 14.12. (a) Characteristic crescent-shaped scar on a peach due to the egg-laying behaviour of plum curculio. (b) A plum curculio larva inside a peach.

continue for a period of 6–8 weeks following shuck split. The emergence of the first spring-generation adults usually occurs from late May to July. These adults begin egg laying (second-generation eggs) in early June. Second-generation larvae may be found in peach fruit at harvest time. The adults of the second generation appear in late July or August. Both first- and second-generation adults feed on foliage or fruit until cool weather conditions prevail, when they seek overwintering sites (Johnson *et al.*, 2005c).

Damage: symptoms and diagnostics
Both the adult and larval stages of the plum curculio damage fruit. On nectarines and plums, adult damage consists of tiny circular feeding punctures

or small crescent-shaped oviposition wounds made by females immediately adjacent to egg-laying punctures. On peaches, it is more common to see a 3.2 mm area of shiny fuzz. Teasing away the fuzz will expose a feeding or oviposition scar, possibly an oval, white egg or brown, larval tunnel into the flesh. These feeding and oviposition sites cause conspicuous scarring and malformation as the fruit develop and can provide entry for the brown rot fungus. Feeding damage by adults appears obscure in April but, as the fruit enlarges, plum curculio feeding looks like the injuries caused by cat-facing insects or cold. The larvae tunnel and feed in developing fruit, usually boring to the pit. Most peaches infested by *C. nenuphar* early in the season drop prematurely. Female curculios will deposit eggs whenever fruits are available, but they prefer small, young peaches or peaches within 2 weeks of harvest. Larger peaches, infested after pit hardening begins, generally stay on the tree until ripe, but these infested fruits are of no value due to the flesh damage and/or presence of the grubs (Quaintance and Jenne, 1912; Johnson *et al.*, 2005c).

Seasonal biology and management

POPULATION MONITORING.   Monitoring of *C. nenuphar* adults is commonly done through either circular trunk traps wrapped around the trunk of peach trees to capture adult weevils climbing up the trees or black pyramid traps (Fig. 14.13). Baiting traps with benzaldehyde, plum essence (a mixture of fruit odour extracted from plum) or grandisoic acid (a male-produced aggregation pheromone of plum curculio) can enhance the effectiveness of the pyramid traps as monitoring tools for *C. nenuphar* adult activity (Leskey and Wright 2004, Akotsen-Mensah *et al.*, 2010).

CULTURAL CONTROL.   Keeping the orchard floor closely mowed after harvest affords less protective cover to adults that overwinter in the orchard. Destruction of nearby plum thickets, abandoned peach blocks and other alternative hosts is suggested to reduce migration of *C. nenuphar* into orchards from outside sources (Akotsen-Mensah *et al.*, 2012).

CHEMICAL CONTROL.   *Conotrachelus nenuphar* is the key fruit-infesting insect pest of peaches in the south-eastern USA. Its lengthy emergence and egg-laying periods mandate diligent control. *Conotrachelus nenuphar* is seldom damaging in well-managed orchards. Adult populations are suppressed in the spring by well-timed applications of effective insecticides. Insecticide applications (sprays) provide a protective barrier to prevent overwintering adults from laying the eggs of the first generation. Sprays for *C. nenuphar* control are normally initiated at shuck split. Two or three additional sprays at 10–14-day intervals are needed to ensure control of the overwintered population. Sprays targeting the overwintered generation also provide control of *G. molesta* and suppress stink bugs moving into the orchard. If the egg-laying adults are not controlled effectively, additional applications will be necessary to prevent infested fruit

Fig. 14.13. Examples of a pyramid trap (a) and a circular trunk trap (b) used to monitor plum curculio adult activity.

from second-generation larvae that mature prior to harvest. In infested orchards, special attention should be given to mid- and late-season cultivars by applying insecticide sprays at 6, 4 and 2 weeks before harvest (Lan *et al.*, 2004; Foshee *et al.*, 2008).

Improved knowledge of biology of *C. nenuphar* is allowing greater refinement of control efforts. Once maximum daily temperatures reach 21.1°C for 2 consecutive days from February to early March, the number of accumulating degree-days should start to be monitored. At the pink stage, position two to four pyramid traps per block in the outer row of peach trees adjacent to woodlots or fence rows. At petal fall, begin checking pyramid traps twice weekly for *C. nenuphar* adults. At the same time, inspect 100 fruit along the orchard perimeter for *C. nenuphar* feeding damage. After accumulating 50–100 degree-days (approximately shuck split), growers should expect to start catching adults in pyramid traps or see the first feeding damage on fruit. An insecticide application is recommended if the traps exceed 0.1 adults per trap per day or if damage exceeds 1%. Adult emergence can also be monitored by jarring peach trees along the perimeter over a ground sheet or beating tray. Trees should be jarred in the early morning when the *C. nenuphar* adults are less active and more easily dislodged. Migration of adults into the orchard continues from 50 to 500 degree-days, so this is the period when fruit should be protected by insecticide sprays. Summer adults emerge from the soil after 1000 degree-days (Lan *et al.*, 2004).

BIOLOGICAL CONTROL.   Soil applications of entomopathogenic nematodes, such as *Steinernema riobrave*, in peach orchards targeting the soil-dwelling stage of *C. nenuphar* larvae can result in high levels of control, but the efficacy can depend on soil type and temperature (Shapiro-Ilan *et al.*, 2008; Shapiro-Ilan *et al.*, 2011).

14.2.8 The peachtree borer, *Synanthedon exitiosa* (Say)

Overview: common names, taxonomy and geographical distribution
The peachtree borer, *Synanthedon exitiosa* (Lepidoptera: Sesiidae), sometimes referred to as the 'greater peachtree borer,' is a clearwing moth that is native to much of North America, including the USA and Canada (Armstrong, 1940; Johnson *et al.*, 2005b). The female moth is dark blue, with a broad orange band around the abdomen, with blue, opaque forewings and clear hindwings with opaque blue margins (Fig. 14.14). Males, which are smaller, are dark blue, with several yellow-white stripes around the abdomen and clear wings with dark borders. The larvae are yellowish-white to cream-coloured caterpillars with brown heads and are 2–3 cm long when fully grown (Becker, 1917; Johnson *et al.*, 2005b).

Biology: life cycle, longevity and reproduction
Synantheton exitiosa usually passes the winter as a larva inside its burrow beneath the bark. Some larvae may overwinter in silken coverings (hibernaculae) constructed on the bark outside their burrows. As the larvae mature, they leave their burrows and move to just beneath the soil line within 10 cm of the

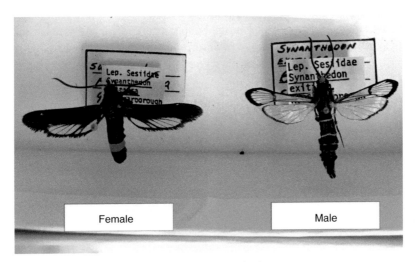

Fig. 14.14. Adult peachtree borer moths, *Synanthedon exitiosa*.

tree trunk and construct silken cocoons in which to pupate. The cocoons are elongated, brownish and about 20 mm long. The pupal stage lasts 3–4 weeks before the adults emerge (Johnson *et al.*, 2005b).

Adult peachtree borers begin to emerge as early as April or May and may be present in orchards up to November. The moths are active during the day, and mating pairs are not an uncommon sight in infested orchards. The females normally mate and begin to lay eggs within a few hours of emerging. Each female lays 200–800 reddish-brown eggs over their lifetime (Johnson *et al.*, 2005b), usually singly in cracks, under loose bark, or near wounds or other rough areas on tree trunks. Eggs hatch in 8–10 days, and the tiny 1 mm long larvae immediately burrow into the bark in the lower part of the tree. Under favourable conditions, the larvae attain a considerable size in a few weeks (Becker, 1917; Johnson *et al.*, 2005b).

Damage: symptoms and diagnostics

Larvae of *S. exitiosa* burrow in and feed on the cambium and inner bark of trees, usually at the base of the trunk from around 8 cm below to 20 cm above the ground line (Becker, 1917; Johnson *et al.*, 2005b). They also feed on large roots that are near the soil surface. Larvae construct and feed in galleries. Masses of accumulating gum, frass and bark chips at the base of the tree are often the first evidence of infestation (Fig. 14.15). Several larvae may develop in one tree. Young trees are particularly susceptible to borers; when infested they are generally weak and grow poorly. Borers easily damage large portions of the vascular tissue in small trees; mortality is common in these instances. Older trees infested by borers may exhibit partial dieback, yellowing of foliage, stunted growth, and loss of vigour and productivity (Becker, 1917).

Seasonal biology and management

POPULATION MONITORING. Pheromone-baited traps deployed in orchards using synthetic sex pheromone to monitor male moths is an effective method

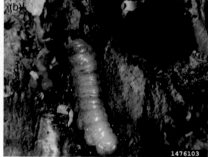

Fig. 14.15. Examples of peachtree borer damage with gumming and frass at the base of a tree (a) and a peachtree borer larva excavated from a peach tree trunk (b).

for detecting infestations and characterizing seasonal flight activity of the peachtree borer (Fig. 14.16) (Yonce and Pate, 1979).

CHEMICAL CONTROL. A drenching trunk spray of long-residual insecticides applied using a handgun is the standard treatment for borer control. This spray establishes a residual insecticide barrier that is lethal to borer larvae for several months. In many south-eastern production areas of the USA, early- and mid-season varieties can also be treated with trunk sprays after harvest, as emergence of this pest continues from June to August.

NON-CHEMICAL CONTROL. Mating disruption for *S. exitiosa* by the deployment of slow-release sex-pheromone dispensers (375 ha⁻¹) placed throughout the orchard canopy, saturating the site's atmosphere with the pheromone, reduces the male moth's ability to find females. This tactic does not kill *S. exitiosa*, but can significantly reduce the number of peachtree borers within a 'disrupted' orchard by minimizing the number of successfully mated females and thus reducing the number of eggs laid within the orchard (Alston *et al.*, 2003).

BIOLOGICAL CONTROL. The application of entomopathogenic nematodes, particularly *Steinernema carpocapsae*, has demonstrated encouraging results for the use of these nematodes as an effective management tool for the peachtree borer. The nematodes are able to actively seek out and attack the borer larvae in the soil or within the tree. When applied in orchards under irrigation or with a sodium polyacrylate polymer gel when irrigation is unavailable, *S. carpocapsae*

Fig. 14.16. A pheromone-baited delta trap with a sticky card liner to monitor peachtree borer adult males.

was able to manage peachtree borer infestations as well as and often better than the standard broad-spectrum insecticide (Shapiro-Ilan *et al.*, 2015).

14.2.9 The lesser peachtree borer, *Synanthedon pictipes* (Grote & Robinson)

Overview: common names, taxonomy and geographical distribution
The lesser peachtree borer, *Synanthedon pictipes* (Lepidoptera: Sesiidae), is native to North America and is an important pest of native and introduced *Prunus* spp., found in most regions east of the Rocky Mountains. *Synanthedon pictipes* adults are metallic, blue-black, clear-winged moths that resemble dark wasps (Fig. 14.17). They have a wingspan of approximately 2–3 cm. Males, except for their pointed abdomen, and females are similar in appearance, with two yellow bands along the abdomen, resembling male peachtree borer moths. The wings are clear except for the dark borders. The larvae are creamy-white caterpillars with dark-brown heads, and are about 2.5 cm long when fully grown (Johnson *et al.*, 2005a).

Biology: life cycle, longevity and reproduction
Synanthedon pictipes overwinters as partially grown larvae beneath the bark, ranging in size from 0.64 to 2.55 cm. They feed periodically during warm spells

Fig. 14.17. Adult lesser peachtree borer moths, *Synanthedon pictipes*. A male is shown on the left and a female on the right.

throughout the winter and complete development in early spring. Prior to pupation, each larva constructs a hibernaculum, a silken, frass-covered protective structure, under the bark near the exit of its gallery. Pupation occurs in the hibernaculum. The mature pupa works its way out of its silken sac and partially through the bark (Johnson *et al.*, 2005a).

Shortly after emerging, the females mate and begin to lay small, reddish-brown eggs along the trunk and limbs. The eggs are usually laid singly in cracks in the bark, frequently in the crotch. Females seem to prefer laying their eggs around wounds or injuries caused by sunscald, winter injury, mechanical injury, broken or cracked limbs, *Cytospora* cankers or existing borer infestations. The eggs hatch in 1–3 weeks, depending on temperature. Upon hatching, the young larvae immediately bore into the bark. The larvae find it more difficult to establish in healthy, undamaged bark. Most larvae feed and develop beneath the bark for about 40–60 days and then pupate to give rise to another brood of moths. Larvae hatching from eggs laid by second-brood moths normally overwinter. A few late first-brood larvae may also overwinter (Johnson *et al.*, 2005a).

Damage: symptoms and diagnostics

Damage to trees is caused by the larval stage. The larval feeding habits and appearance are almost identical to those of the peachtree borer larvae, but lesser borers are usually found higher in the tree (Fig. 14.18). The larvae will attack any above-ground structural wood.

Synanthedon pictipes larvae burrow, feed and develop in the inner bark and cambium, primarily in the upper trunk and large branches. Masses of gum mixed with frass and wood borings normally exude from infested areas. Infestations are most common under loose bark in crotches and around wounds or cankers. 'Bleeding' dark, dead or swollen areas on the trunk or scaffold limbs may indicate infestations. Larval feeding can reduce tree vigour and weaken limbs, and damaged areas may provide entry sites for other pests, such as *Cytospora* canker or shothole borers. In heavily infested trees, large scaffold limbs may be completely girdled by borers and die. *Synanthedon pictipes* infestations frequently worsen as orchards age, because of the wounding inherent in heavy pruning and the overall weakening of trees as they age (Johnson *et al.*, 2005a).

Seasonal biology and management

POPULATION MONITORING. Pheromone-baited traps deployed in the orchards using synthetic sex pheromone to monitor male moths is an effective method for detecting infestations and characterizing seasonal flight activity of *S. pictipes* (Fig. 14.16) (Gentry *et al.*, 1978).

CHEMICAL CONTROL. Broad-spectrum insecticide drenches to scaffold limbs can be applied at bud swell or in the summer. Within-season cover sprays for other insect pests can help suppress *S. pictipes* populations.

Fig. 14.18. (a) Examples of lesser peachtree borer damage with gumming and frass on a scaffolding limb of a tree. (b) A lesser peachtree borer larva excavated from a peach tree limb.

NON-CHEMICAL CONTROL. As for the peachtree borer, mating disruption for the lesser peachtree borer using slow-release sex-pheromone dispensers (375 ha⁻¹) placed throughout the orchard canopy reduces the male moth's ability to find females by saturating the site's atmosphere with the pheromone. This method does not kill the moths but significantly reduces the number of *S. pictipes* by minimizing the number of successfully mated females and thus reducing the number of eggs laid in the orchard (Frank *et al.*, 2020).

CULTURAL CONTROL. *Synanthedon pictipes* larval establishment is closely tied to wounded bark, with infestations more severe where there are injuries from sunburn, cold injury to the undersides of scaffolds, scale, limb breaks, propping or tying wounds, large rough pruning cuts or disease cankers, or where existing *S. pictipes* wounds are already present. Smooth, healthy bark and well-healed pruning cuts at the origin of limbs are less prone to larval infestations. Additionally, *S. pictipes* infestations are more common in poorly managed orchards of low vigour where limb breaks due to overloading with fruit or wind damage are left unattended. Proper canopy management and fruit thinning reduces an orchard's attractiveness to *S. pictipes*. The presence of borers can be determined by inspecting scaffold wounds for the light brown pupal skins (Johnson *et al.*, 2005a).

BIOLOGICAL CONTROL. Curative application of entomopathogenic nematodes has shown promise on established *S. pictipes* infestations when used with an appropriate anti-dessicant, such as a sodium polyacrylate polymer gel, to keep the borer wounds moist for long enough for the nematodes to find the larvae (Shapiro-Ilan *et al.*, 2016).

14.3   IPM OF PEACH PESTS IN THE NEW ERA

The implementation of IPM in crop production has come a long way since its inception (Deping *et al.*, 2019), but there are still ways to improve the approach to ensure the safe, effective and long-lasting control of arthropod pests in peach production. The IPM approach has always relied on novel tactics, and the future of IPM is no different. As we advance innovative tactics and approaches, such as mating disruption, SIT, precision farming and biological control, IPM implementation and adoption will be greatly strengthened while our reliance on chemical pesticides will be substantially reduced.

The concept of mating disruption, which comprises broadcasting synthetic sex pheromones within a crop to interrupt the mate-finding communication and prevent mating of the target pest, is not new to IPM (Miller and Gut, 2015). For decades, mating disruption has been used to successfully manage a variety of insect pests, particularly lepidopteran pests (Evenden, 2016). However, there is an array of management programmes for additional insect pests that may still benefit from this effective technology. For example, sexually reproductive scale insects (Hemiptera: Coccoidea) may be fitting targets for mating disruption, as the females have a limited dispersal ability and adult males have a narrow window of time for mate searching (Franco *et al.*, 2022).

Commercial mating disruption products have been developed for a few hemipteran pests, such as the vine mealybug, *Planococcus ficus* (Signoret) (Hemiptera: Pseudococcidae) (Cocco *et al.*, 2014) and the California red scale, *Aonidiella aurantii* (Maskell) (Hemiptera: Diaspididae) (Vacas *et al.*, 2010). These few successes are encouraging examples that mating disruption may have a high potential for management of additional scale pests, such as the global pest San Jose scale, *C. perniciosus* (Comstock), and the white peach scale, *Pseudaulacaspis pentagona* (Targioni) (Hemiptera: Diaspididae) (Fig. 14.19). Providing that key factors, such as technological advances in pheromone synthesis and formulations, are resolved, mating disruption may play a key role in the future of IPM in peach production.

While mating disruption may be a favourable technique for a variety of key insect pests, it is not a tactic that will be practical for all insect pests of peach. An additional novel technology is SIT, which is a management strategy that uses radiation to generate genetic mutations to produce sterile adult insects in order to prevent successful reproduction with females (Hendrichs and Robinson, 2009). Promising research on SIT has been conducted in peach production. For example, the peach fruit fly, *B. zonata* (Saunders) (Diptera: Tephritidae), is a polyphagous pest of over 50 cultivated and wild plants and is distributed across many regions of the world (Duyck *et al.*, 2004). Sayed *et al.* (2018) demonstrated that irradiation has the potential to effectively sterilize *B. zonata* males, but additional laboratory and field-based research is needed to evaluate SIT as a practical IPM tactic in

Fig. 14.19. Adult female of the white peach scale, *Pseudaulacaspis pentagona*.

peach production. However, SIT is a mature pest-management methodology that is extensively used to suppress the population of the Mediterranean fruit fly, *C. capitata*, in several countries such as South Africa and Spain in an area-wide approach.

Although alternative management practices are ideal for reducing peach pest-management reliance on insecticides, improving the efficacy of chemical pest management is also required. As such, precision agriculture tools in fruit production are likely to have several advantages. For example, laser-guided, variable-rate technology is designed to significantly reduce pesticide use with a positive impact on the environment (Chen *et al.*, 2020). This 'intelligent spray' technology may be more environmentally friendly than a conventional standard sprayer and as effective for management of insect pests, such as oriental fruit moth (*G. molesta*), in peach production. Further advancements in application efficiency may be made by improving the driving paths of sprayer equipment. Gao *et al.* (2020) demonstrated that their algorithm using a colour-depth binocular sensor to acquire images in peach orchards combined with a colour-depth fusion segmentation method based on the leaf wall area of the images could be used to accurately and automatically plan spraying paths in peach orchards. Algorithms can also be used in automated pest surveillance and decision-making techniques. Advancements in real-time data transfer, automated video analysis, sensor technology and data-interpretation software (Cohen *et al.*, 2008) have allowed the development of automated pest surveillance and decision support systems (Potamitis *et al.*, 2017; Ioannou *et al.*, 2019). For example, baited traps modified with optoelectronic sensors can be used to detect, time stamp, GPS tag and identify the collected insect species through the optoacoustic spectrum analysis of their wingbeat, which has been used effectively to automatically monitor tephritid fruit flies (Potamitis *et al.*, 2017). This type of trapping can be used to assist in management decisions, such as the decision support system developed by Ioannou *et al.* (2019), which supports the management of the European cherry fruit fly, *Rhagoletis cerasi* (Diptera: Tephritidae), a major pest of cherry production in Europe, west Asia and North America. Their decision support system implementation reduced the management costs for *R. cerasi* and decreased the overall insecticide exposure to the agroecosystem. With similar technologies also used to monitor lepidopteran pests, such as codling moth, *C. pomonella* (L.) (Lepidoptera: Tortricidae) in pome fruits (Preti *et al.*, 2021), there are opportunities for developing and implementing automatic surveillance systems and decision support tools for key peach pests as well.

Although insecticides will remain the key component of peach pest management, precision application of insecticides through improved timing and efficacy may also help improve natural control of pests within peach orchards. The use of insect natural enemies and entomopathogens for the biological control of insect pests is not a new concept (Simmonds *et al.*, 1976), but it is still a very important and promising tactic for peach production. Applying entomopathogenic fungi, such as *M. anisopliae* and *B. bassiana*, as biopesticides has been demonstrated to be a cost-effective IPM strategy to manage the peach fruit fly, *B. zonata* (Saunders) (Murtaza *et al.*, 2022). Another potential option for pest management is the use of predators of pest species that occur naturally within the agroecosystem. Research has shown that predator functional diversity in polycultures led to greater prey suppression than in monocultures (Greenop *et al.*, 2018). More specifically, Wan *et al.* (2019) determined that natural enemy abundance can increase by 38.1%, with a subsequent 16.9% decrease in insect herbivores in peach orchards with high plant diversification compared with monoculture orchards. As such, there is the potential to enhance orchard plant diversity to further support natural-enemy functional diversity. Denis *et al.* (2021) evaluated the potential of flowering plant species to support key natural enemies in apple and peach orchards. They determined that flowering plants can improve the ecological infrastructure of fruit orchards by supporting the conservation of the biological control agents of key fruit pests. With advancements in pest reproduction interference, precision farming and biological control, and the potential of combining systems approaches and site-focused pest management, the future of IPM in peach production is bright.

14.4 CONCLUSIONS AND FUTURE PERSPECTIVES

Management of insect pests in peach orchards is challenging in many parts of the world. Key pests should be identified in each area and preferably on each farm to start establishing a sound IPM programme. The population dynamics of key pests should be monitored using appropriate methodologies. A few important secondary pests should also be identified and closely monitored. Priority should be given to the collection of reliable data from orchards in terms of seasonality and spatial dispersion of the target pest, infestation rates per cultivar, details of pest control tactics used and economics. In addition, selection of the most reliable system to extract or collect meteorological data should be explored and chosen. Current and historical data from each specific orchard or area should be stored in databases and used to establish longer-term pest-management strategies and to refine current decisions for interventions.

Precision pest-management approaches and algorithms to support interventions should be based on solid data obtained from farms. Development of population growth predictive models for the key pests of peach and nectarine should be tailored to specific areas to support IPM efforts.

ACKNOWLEDGEMENTS

We acknowledge the assistance of George Kyrtisis and Vasileios Rodovitis on pictures regarding adults of *B. zonata* and infestation of *C. capitata* on peaches.

REFERENCES

Acebes-Doria, A.L., Agnello, A.M., Alston, D.G., Andrews, H., Beers, E.H. *et al.* (2020) Season-long monitoring of the brown marmorated stink bug (Hemiptera: Pentatomidae) throughout the United States using commercially available traps and lures. *Journal of Economic Entomology* 113, 159–171.

Acebes-Doria, A.L., Leskey, T.C. and Bergh, J.C. (2016a) Host plant effects on *Halyomorpha halys* (Hemiptera: Pentatomidae) nymphal development and survivorship. *Environmental Entomology* 45, 663–670.

Acebes-Doria, A.L., Leskey, T.C. and Bergh, J.C. (2016b) Injury to apples and peaches at harvest from feeding by *Halyomorpha halys* (Stål) (Hemiptera: Pentatomidae) nymphs early and late in the season. *Crop Protection* 89, 58–65.

Acebes-Doria, A.L., Morrison, W.R., Short, B.D., Rice, K.B., Bush, H.G. *et al.* (2018) Monitoring and biosurveillance tools for the brown marmorated stink bug, *Halyomorpha halys* (Stål) (Hemiptera: Pentatomidae). *Insects* 9: 3.

Akotsen-Mensah, C., Boozer, R. and Fadamiro, H.Y. (2010) Field evaluation of traps and lures for monitoring plum curculio (Coleoptera: Curculionidae) in Alabama peaches. *Journal of Economic Entomology* 103, 744–753.

Akotsen-Mensah, C., Boozer, R.T. and Fadamiro, H.Y. (2012) Influence of orchard weed management practices on soil dwelling stages of plum curculio, *Conotrachelus nenuphar* (Coleoptera: Curculionidae). *Florida Entomologist* 95, 882–889.

Alston, D.G., Reding M.E. and Miller C.A. (2003) Evaluation of three consecutive years of mating disruption for control of greater peachtree borer (*Synanthedon exitiosa*) in peach. In: *Proceedings of the 77th Annual Western Orchard Pest & Disease Management Conference, 5–17 January 2003, Portland, Oregon.*

Amat, C., Bosch-Serra, D., Avilla, J. and Escudero Colomar, L.A. (2021) Different population phenologies of *Grapholita molesta* (Busck) in two hosts and two nearby regions in the NE of Spain. *Insects* 12: 612.

Aparicio, Y., Riudavets, J., Gabarra, R., Agustí, N., Rodríguez-Gasol, N. *et al.* (2021) Can insectary plants enhance the presence of natural enemies of the green peach aphid (Hemiptera: Aphididae) in Mediterranean peach orchards? *Journal of Economic Entomology* 114, 784–793.

Armstrong, T. (1940) The life history of the peach borer, *Synanthedon Exitiosa* Say, in Ontario. *Scientific Agriculture* 20, 557–565.

Atanassov, A., Shearer, P.W. and Hamilton, G.C. (2003) Peach pest management programs impact beneficial fauna abundance and *Grapholita molesta* (lepidoptera: tortricidae) egg parasitism and predation. *Environmental Entomology* 32, 780–788.

Badenes-Perez, F.R., Zalom, F.G. and Bentley, W.J. (2002a) Are San Jose scale (Hom., Diaspididae) pheromone trap captures predictive of crawler densities? *Journal of Applied Entomology* 126, 545–549.

Badenes-Perez, F.R., Zalom, F.G. and Bentley, W.J. (2002b) Effects of dormant insecticide treatments on the San Jose scale (Homoptera: Diaspididae) and its parasitoids *Encarsia perniciosi* and *Aphytis* spp. (Hymenoptera: Aphelinidae). *International Journal of Pest Management* 48, 291–296.

Balachowsky, A.S. (1966) *Entomologie Appliquée à L'Agriculture. Traité. Tome II. Lépidoptères.* Masson et Cie Éditeurs, Paris.

Bali, E.M.D., Moraiti, C.A., Ioannou, C.S., Mavraganis, V. and Papadopoulos, N.T. (2021) Evaluation of mass trapping devices for early seasonal management of *Ceratitis Capita* (Diptera: Tephritidae) populations. *Agronomy* 11: 1101.

Bayoumy, M.H., Michaud, J.P., Badr, F.A.A. and Ghanim, N.M. (2021) Validation of degree-day models for predicting the emergence of two fruit flies (Diptera: Tephritidae) in northeast Egypt. *Insect Science* 28, 153–164.

Becker, G.G. (1917) Notes on the peach-tree borer (*Sanninoidea exitiosa*). *Journal of Economic Entomology* 10, 49–59.

Benelli, G., Ceccarelli, C., Zeni, V., Rizzo, R., Lo Verde, G. *et al.* (2022) Lethal and behavioural effects of a green insecticide against an invasive polyphagous fruit fly pest and its safety to mammals. *Chemosphere* 287: 132089.

Ben-Yosef, M., Verykouki, E., Altman, Y., Nemni-Lavi, E., Papadopoulos, N.T. and Nestel, D. (2021) Effects of thermal acclimation on the tolerance of *Bactrocera zonata* (Diptera: Tephritidae) to hydric stress. *Frontiers in Physiology* 12: 686424.

Blaauw, B.R., Polk, D. and Nielsen, A.L. (2015) IPM-CPR for peaches: incorporating behaviorally-based methods to manage *Halyomorpha halys* and key pests in peach. *Pest Management Science* 71, 1513–1522.

Blaauw, B.R., Jones, V.P. and Nielsen, A.L. (2016) Utilizing immunomarking techniques to track *Halyomorpha halys* (Hemiptera: Pentatomidae) movement and distribution within a peach orchard. *PeerJ* 4, e1997.

Blaauw, B.R., Hamilton, G., Rodriguez-Saona, C. and Nielsen, A.L. (2019) Plant stimuli and their impact on brown marmorated stink bug dispersal and host selection. *Frontiers in Ecology and Evolution* 7: 414.

Blacket, M.J., Agarwal, A., Zheng, L.D., Cunningham, J.P., Britton, D. *et al.* (2020) A LAMP assay for the detection of *Bactrocera tryoni* Queensland fruit fly (Diptera: Tephritidae). *Scientific Reports* 10: 9554.

Bonizzoni, M., Zheng, L., Guglielmino, C.R., Haymer, D.S., Gasperi, G. *et al.* (2001) Microsatellite analysis of medfly bioinfestations in California. *Molecular Ecology* 10, 2515–2524.

Boulahia-Kheder, S. (2021) Advancements in management of major fruit flies (Diptera: Tephritidae) in North Africa and future challenges: a review. *Journal of Applied Entomology* 145, 939–957.

Buzzetti, K., Chorbadjian, R.A. and Nauen, R. (2015) Resistance management for San Jose scale (Hemiptera: Diaspididae). *Journal of Economic Entomology* 108, 2743–2752.

Carey, J.R. (1991) Establishment of the Mediterranean fruit fly in California. *Science* 253, 1369–1373.

Carey, J.R., Papadopoulos, N.T. and Plant, R. (2017a) Tephritid pest populations oriental fruit fly outbreaks in California: 48 consecutive years, 235 cities, 1,500 detections – and counting. *American Entomologist* 63, 232–236.

Carey, J.R., Papadopoulos, N. and Plant, R. (2017b) The 30-year debate on a multi-billion-dollar threat: tephritid fruit fly establishment in California. *American Entomologist* 63, 100–113.

Charlery de la Masseliere, M., Ravigne, V., Facon, B., Lefeuvre, P., Massol, F. *et al.* (2017) Changes in phytophagous insect host ranges following the invasion of their community: long-term data for fruit flies. *Ecology and Evolution* 7, 5181–5190.

Chen, L., Wallhead, M., Reding, M., Horst, L. and Zhu, H. (2020) Control of insect pests and diseases in an Ohio fruit farm with a laser-guided intelligent sprayer. *HortTechnology* 30, 168–175.

Chen, M.H. and Dorn, S. (2009) Reliable and efficient discrimination of four internal fruit-feeding *Cydia* and *Grapholita* species (Lepidoptera: Tortricidae) by polymerase chain reaction-restriction fragment length polymorphism. *Journal of Economic Entomology* 102, 2209–2216.

Chen, Z.-Z., Xu, L.-X., Li, L.-L., Wu, H.-B. and Yu, Y.-Y. (2019) Effects of constant and fluctuating temperature on the development of the oriental fruit moth, *Grapholita molesta* (Lepidoptera: Tortricidae). *Bulletin of Entomological Research* 109, 212–220.

Choudhary, J.S., Mali, S.S., Naaz, N., Mukherjee, D., Moanaro, L. *et al.* (2020) Predicting the population growth potential of *Bactrocera zonata* (Saunders) (Diptera: Tephritidae) using temperature development growth models and their validation in fluctuating temperature condition. *Phytoparasitica* 48, 1–13.

Clemens, B. (1860) Contributions to American lepidopterology. *Proceedings of the Academy of Natural Sciences of Philadelphia* 12, 156–174.

Cocco, A., Lentini, A. and Serra, G. (2014) Mating disruption of *Planococcus ficus* (Hemiptera: Pseudococcidae) in vineyards using reservoir pheromone dispensers. *Journal of Insect Science* 14: 144.

Cohen, Y., Cohen, A., Hetzroni, A., Alchanatis, V., Broday, D. *et al.* (2008) Spatial decision support system for Medfly control in citrus. *Computers and Electronics in Agriculture* 62, 107–117.

Croft, B.A., Michels, M.F. and Rice, R.E. (1980) Validation of a PETE timing model for the oriental fruit moth in Michigan and central California (Lepidoptera: Olethreutidae). *Great Lakes Entomologist* 13, 211–217.

Damos, P. (2013) Demography and randomized life table statistics for peach twig borer *Anarsia lineatella* (Lepidoptera: Gelechiidae). *Journal of Economic Entomology* 106, 675–682.

Damos, P., Bonsignore, C.P., Gardi, F. and Avtzis, D.N. (2014) Phenological responses and a comparative phylogenetic insight of *Anarsia lineatella* and *Grapholita molesta* between distinct geographical regions within the Mediterranean basin. *Journal of Applied Entomology* 138, 528–538.

Damos, P.T. and Savopoulou-Soultani, M. (2008) Temperature-dependent bionomics and modeling of *Anarsia lineatella* (Lepidoptera: Gelechiidae) in the laboratory. *Journal of Economic Entomology*, 1557–1567.

Damos, P.T. and Savopoulou-Soultani, M. (2010) Unique logistic model for simultaneous forecasting of major lepidopterous peach pest complex. In: *IOBC/WPRS Workshop on 'Sustainable protection of fruit crops in the Mediterranean area', Vico del Gargano, Italy*. IOBC/WPRS Bulletin.

Damos, P.T., Soulopoulou, P., Gkouderis, D., Monastiridis, D., Vrettou, M. *et al.* (2022) Degree-day risk thresholds for predicting the occurrence of *Anarsia lineatella*, *Grapholita molesta* and *Adoxophyes orana* in northern Greece peach orchards. *Plant Protection Science* 58, 234–244.

de Meyer, M., Addison, P., Kayenbergh, A. and Virgilio, M. (2022) Electronic multientry identification key for selected fruit flies (Diptera, Tephritidae, Dacinae) of economic importance. *Fruit Fly News* 43, 1–2.

Denis, C., Riudavets, J., Gabarra, R., Molina, P. and Arnó, J. (2021) Selection of insectary plants for the conservation of biological control agents of aphids and thrips in fruit orchards. *Bulletin of Entomological Research* 111, 517–527.

Deping, G., Yongquan, L. and Wenliu, G. (2019) A review of the history and development of integrated pest management (IPM). *Plant Diseases & Pests* 10, 37–40.

Di Ilio, V. and Cristofaro, M. (2021) Polyphenolic extracts from the olive mill wastewater as a source of biopesticides and their effects on the life cycle of the Mediterranean fruit fly *Ceratitis capitata* (Diptera, Tephriditae). *International Journal of Tropical Insect Science* 41, 359–366.

Diamantidis, A.D., Carey, J.R., Nakas, C.T. and Papadopoulos, N.T. (2011a) Ancestral populations perform better in a novel environment: domestication of Mediterranean fruit fly populations from five global regions. *Biological Journal of the Linnean Society* 102, 334–345.

Diamantidis, A.D., Carey, J.R., Nakas, C.T. and Papadopoulos, N.T. (2011b) Populationspecific demography and invasion potential in medfly. *Ecology and Evolution* 1, 479–488.

Drew, R.A.I. and Romig, M.C. (2016) *Keys to the Tropical Fruit Flies (Tephritidae: Dacinae) of South-East Asia*. CAB International, Wallingford, UK.

Du, J., Li, G., Xu, X. and Wu, J. (2015) Development and fecundity performance of oriental fruit moth (Lepidoptera: Tortricidae) reared on shoots and fruits of peach and pear in different seasons. *Environmental Entomology* 44, 1522–1530.

Duyck, P.F., Sterlin, J.F. and Quilici, S. (2004) Survival and development of different life stages of *Bactrocera zonata* (Diptera: Tephritidae) reared at five constant temperatures compared to other fruit fly species. *Bulletin of Entomological Research* 94, 89–93.

Duyck, P.F., David, P. and Quilici, S. (2006a) Climatic niche partitioning following successive invasions by fruit flies in La Reunion. *Journal of Animal Ecology* 75, 518–526.

Duyck, P.F., David, P., Junod, G., Brunel, C., Dupont, R. and Quilici, S. (2006b) Importance of competition mechanisms in successive invasions by polyphagous tephritids in La Reunion. *Ecology* 87, 1770–1780.

EFSA (European Food Safety Authority) (2021) Schenk, M., Mertens, J., Delbianco, A., Graziosi, I. and Vos, S. Pest survey card on *Bactrocera zonata*. *EFSA supporting publication*, EN-1999, 28 pp. doi:10.2903/sp.efsa.2021.EN-1999

Egartner, A., Lethmayer, C., Gottsberger, R.A. and Blümel, S. (2019) Survey on *Bactrocera* spp. (Tephritidae, Diptera) in Austria. *EPPO Bulletin* 49, 578–584.

El-Gendy, I.R. (2017) Host preference of the peach fruit fly, *Bactrocera zonata* (Saunders) (Diptera: Tephritidae), under laboratory conditions. *Journal of Entomology* 14, 160–167.

Elnagar, S., El-Sheikh, M., Hashem, A. and Afia, Y. (2010) Recent invasion by *Bactrocera zonata* (Saunders) as a new pest competing with *Ceratitis capitata* (Wiedemann) in attacking fruits in Egypt. *Aspects of Applied Biology* 104, 97–102.

Enkerlin, W., Gutierrez-Ruelas, J.M., Cortes, A.V., Roldan, E.C., Midgarden, D. *et al.* (2015) Area freedom in Mexico from Mediterranean fruit fly (Diptera: Tephritidae): a review of over 30 years of a successful containment program using an integrated area-wide SIT approach. *Florida Entomologist* 98, 665–681.

EPPO (2010) *Bactrocera zonata*: procedure for official control. *OEPP/EPPO Bulletin* 40, 390–395.

EPPO (2013) PM 7/114 (1) *Bactrocera zonata*. *OEPP/EPPO Bulletin* 43, pp. 412–416.

EPPO (2022) EPPO Global Database, *Grapholita molesta*. Available at: https://gd.eppo.int/taxon/LASPMO/distribution (accessed 27 February 2023).

Epsky, N.D., Hendrichs, J., Katsoyannos, B.I., Vasquez, L.A., Ros, J.P. *et al.* (1999) Field evaluation of female-targeted trapping systems for *Ceratitis capitata* (Diptera: Tephritidae) in seven countries. *Journal of Economic Entomology* 92, 156–164.

Evenden, M. (2016) Mating disruption of moth pests in integrated pest management: a mechanistic approach. In: Jeremy, D.A. and Ring, T.C. (eds) *Pheromone Communication in Moths: Evolution, Behavior, and Application.* University of California Press. Berkeley, California, pp. 365–394.

Foshee, W.G., Boozer, R.T., Blythe, E.K., Horton, D.L. and Burkett, J. (2008) Management of plum curculio and catfacing insects on peaches in Central Alabama: standard crop stage-based vs. integrated pest management-based approaches. *International Journal of Fruit Science* 8, 188–199.

Franco, J.C., Cocco, A., Lucchi, A., Mendel, Z., Suma, P. *et al.* (2022) Scientific and technological developments in mating disruption of scale insects. *Entomologia Generalis* 42, 251–273.

Frank, D.L., Starcher, S. and Chandran, R.S. (2020) Comparison of mating disruption and insecticide application for control of peachtree borer and lesser peachtree borer (Lepidoptera: Sesiidae) in peach. *Insects* 11: 658.

Gao, G., Xiao, K. and Jia, Y. (2020) A spraying path planning algorithm based on colour-depth fusion segmentation in peach orchards. *Computers and Electronics in Agriculture* 173: 105412.

Gazit, Y., Rossler, Y. and Glazer, I. (2000) Evaluation of entomopathogenic nematodes for the control of Mediterranean fruit fly (Diptera : Tephritidae). *Biocontrol Science and Technology* 10, 157–164.

Gentile, A.G. and Summers, F.M. (1958) The biology of San Jose scale on peaches with special reference to the behavior of males and juveniles. *Hilgardia* 27, 269–285.

Gentry, C.C., Holloway, R.L. and Pollet, D.K. (1978) Pheromone monitoring of peachtree borers and lesser peachtree borers in South Carolina. *Journal of Economic Entomology* 71, 247–253.

Gilligan, T.M. and Epstein, M.E. (2012) TortAI: Tortricids of Agricultural Importance. Available at: http://idtools.org/id/leps/tortai/ (accessed 9 December 2022).

Giuseppino, S.P., Bortolotti, P., Roberta, N., Leonardo, M. and Federico, R.P. (2018) Efficacy of long lasting insecticide nets in killing *Halyomorpha halys* in pear orchards. *Outlooks on Pest Management* 29, 70–74.

Goldshtein, E., Cohen, Y., Hetzroni, A., Gazit, Y., Timar, D. *et al.* (2017) Development of an automatic monitoring trap for Mediterranean fruit fly (*Ceratitis capitata*) to optimize control applications frequency. *Computers and Electronics in Agriculture* 139, 115–125.

Graillot, B., Blachere, C., Besse, S., Siegwart, M. and Lopez-Ferber, M. (2017) Host range extension of *Cydia pomonella* granulovirus: adaptation to oriental fruit moth, *Grapholita molesta. Biocontrol* 62, 19–27.

Greenop, A., Woodcock, B.A., Wilby, A., Cook, S.M. and Pywell, R.F. (2018) Functional diversity positively affects prey suppression by invertebrate predators: a meta-analysis. *Ecology* 99, 1771–1782.

Gregersen, K. and Karsholt, O. (2017) Taxonomic confusion around the peach twig borer, *Anarsia lineatella* Zeller, 1839, with description of a new species (Lepidoptera, Gelechiidae). *Nota Lepidopteraologica* 40, 65–85.

Guillem-Amat, A., Sanchez, L., Lopez-Errasquin, E., Urena, E., Hernandez-Crespo, P. and Ortego, F. (2020) Field detection and predicted evolution of spinosad resistance in *Ceratitis capitata. Pest Management Science* 76, 3702–3710.

Guillem-Amat, A., Lopez-Errasquin, E., Castells-Sierra, J., Sanchez, L. and Ortego, F. (2022) Current situation and forecasting of resistance evolution to lambda-cyhalothrin in Spanish medfly populations. *Pest Management Science* 78, 1341–1355.

Hasnain, M., Saeed, S., Naeem-Ullah, U. and Ullah, S. (2022) Development of synthetic food baits for mass trapping of *Bactrocera zonata* S. (Diptera: Tephritidae). Journal of King Saud University 34: 101667.

Haye, T., Abdallah, S., Gariepy, T. and Wyniger, D. (2014) Phenology, life table analysis and temperature requirements of the invasive brown marmorated stink bug, *Halyomorpha halys*, in Europe. *Journal of Pest Science* 87, 407–418.

Heckford, R.J. (1992) *Anarsia lineatella* Zeller (Lepidoptera: Gelechiidae): a larval description. *Entomologist's Gazette* 43: 54.

Hendrichs, J. and Robinson, A. (2009) Sterile insect technique. In: Resh, V.H. and Cardé, R.T. (eds) *Encyclopedia of Insects*, 2nd edn. Academic Press, San Diego, California, pp. 953–957.

Hoebeke, E.R. and Carter, M.E. (2003) *Halyomorpha halys* (Stål) (Heteroptera: Pentatomidae): a polyphagous plant pest from Asia newly detected in North America. *Proceedings of the Entomological Society of Washington* 105, 225–237.

Howard, L.O. and Marlatt, C.L. (1896) The San Jose scale. *US Department of Agriculture Division of Entomology Bulletin* 3: 32.

Hoyt, S.C., Westigard, P.H. and Rice, R.E. (1983) Development of pheromone trapping techniques for male San Jose scale (Homoptera: Diaspididae). *Environmental Entomology* 12, 371–375.

Il'Ichev, A.L., Stelinski, L.L., Williams, D.G. and Gut, L.J. (2006) Sprayable microencapsulated sex pheromone formulation for mating disruption of oriental fruit moth (Lepidoptera: Tortricidae) in Australian peach and pear orchards. *Journal of Economic Entomology* 99, 2048–2054.

Il'ichev, A.L., Williams, D.G. and Gut, L.J. (2007) Dual pheromone dispenser for combined control of codling moth *Cydia pomonella* L. and oriental fruit moth *Grapholita molesta* (Busck) (Lep., Tortricidae) in pears. *Journal of Applied Entomology* 131, 368–376.

Inkley, D.B. (2012) Characteristics of home invasion by the brown marmorated stink bug (Hemiptera: Pentatomidae). *Journal of Entomological Science* 47: 125–130.

Ioannou, C.S., Papanastasiou, S.A., Zarpas, K.D., Miranda, M.A., Sciarretta, A. *et al.* (2019) Development and field testing of a spatial decision support system to control populations of the European cherry fruit fly, *Rhagoletis cerasi*, in commercial orchards. *Agronomy* 9: 568.

Jenkins, D., Cottrell, T., Horton, D.L., Hodges, A. and Hodges, G. (2006) Hosts of plum curculio, *Conotrachelus nenuphar* (Coleoptera: Curculionidae), in central Georgia. *Environmental Entomology* 35, 48–55.

Johnson, D., Cottrell, T. and Horton, D. (2005a) Lesser peachtree borer. In: Horton, D. and Johnson, D. (eds) *Southeastern Peach Growers Handbook, Vol. 1. Georgia Extension Service Handbook*, pp. 270–272.

Johnson, D., Cottrell, T. and Horton, D. (2005b) Peachtree borer. In: Horton, D. and Johnson, D. (eds) *Southeastern Peach Growers Handbook, Vol. 1. Georgia Extension Service Handbook*, pp. 266–269.

Johnson, D., Cottrell, T. and Horton, D. (2005c) Plum curculio. In: Horton, D. and Johnson, D. (eds) *Southeastern Peach Growers Handbook, Vol. 1. Georgia Extension Service Handbook*, pp. 272–274.

Kapranas, A., Chronopoulou, A., Lytra, I.C., Peters, A., Milonas, P.G. and Papachristos, D.P. (2021) Efficacy and residual activity of commercially available entomopathogenic nematode strains for Mediterranean fruit fly control and their ability to infect infested fruits. *Pest Management Science* 77, 3964–3969.

Kaser, J.M., Akotsen-Mensah, C., Talamas, E.J. and Nielsen, A.L. (2018) First Report of *Trissolcus japonicus* parasitizing *Halyomorpha halys* in North American Agriculture. *Florida Entomologist* 101, 680–683.

Katsoyannos, B.I., Heath, R.R., Papadopoulos, N.T., Epsky, N.D. and Hendrichs, J. (1999) Field evaluation of Mediterranean fruit fly (Diptera: Tephritidae) female selective attractants for use in monitoring programs. *Journal of Economic Entomology* 92, 583–589.

Khan, M.A., Kamran, M., Shad, S.A. and Anees, M. (2022) Trichlorfon tolerance, risk assessment, and a cross-resistance trend to four other insecticides in *Bactrocera zonata* (Saunders). *Phytoparasitica* 50, 713–725.

Khan, M.H., Khuhro, N.H., Awais, M., Memon, R.M. and Asif, M.U. (2020) Functional response of the pupal parasitoid, *Dirhinus giffardii* towards two fruit fly species, *Bactrocera zonata* and *B. cucurbitae*. *Entomologia Generalis* 40, 87–95.

Khrimian, A., Zhang, A., Weber, D.C., Ho, H.-Y., Aldrich, J.R. *et al.* (2014) Discovery of the aggregation pheromone of the brown marmorated stink bug (*Halyomorpha halys*) through the creation of stereoisomeric libraries of 1-bisabolen-3-ols. *Journal of Natural Products* 77, 1708–1717.

Kim, Y., Jung, S., Kim, Y. and Lee, Y. (2011) Real-time monitoring of oriental fruit moth, *Grapholita molesta*, populations using a remote sensing pheromone trap in apple orchards. *Journal of Asia-Pacific Entomology* 14, 259–262.

Koohkanzade, M., Zakiaghl, M., Dhami, M.K., Fekrat, L. and Namaghi, H.S. (2018) Rapid identification of *Bactrocera zonata* (Dip.: Tephritidae) using TaqMan real-time PCR assay. *PLoS One* 13: e0205136.

Kouloussis, N.A., Mavraganis, V.G., Damos, P., Ioannou, C.S., Bempelou, E. *et al.* (2022) Trapping of *Ceratitis capitata* using the low-cost and non-toxic attractant Biodelear. *Agronomy* 12: 525.

Krainacker, D.A., Carey, J.R. and Vargas, R.I. (1987) Effect of larval host on life history traits of the Mediterranean fruit fly, *Ceratitis capitata*. *Oecologia* 73, 583–590.

Kuhar, T.P., Short, B.D., Krawczyk, G. and Leskey, T.C. (2017) Deltamethrin-incorporated nets as an integrated pest management tool for the invasive *Halyomorpha halys* (Hemiptera: Pentatomidae). *Journal of Economic Entomology* 110, 543–545.

Lan, Z., Scherm, H. and Horton, D.L. (2004) Temperature-dependent development and prediction of emergence of the summer generation of plum curculio (Coleoptera: Curculionidae) in the southeastern United States. *Environmental Entomology* 33, 174–181.

Lee, D.-H., Wright, S.E. and Leskey, T.C. (2013) Impact of insecticide residue exposure on the invasive pest, *Halyomorpha halys* (Hemiptera: Pentatomidae): analysis of adult mobility. *Journal of Economic Entomology* 106, 150–158.

Lee, D.-H., Cullum, J.P., Anderson, J.L., Daugherty, J.L., Beckett, L.M. and Leskey, T.C. (2014) Characterization of overwintering sites of the invasive brown marmorated stink bug in natural landscapes using human surveyors and detector canines. *PLoS One* 9: e91575.

Leskey, T.C. and Nielsen, A.L. (2018) Impact of the invasive brown marmorated stink bug in North America and Europe: history, biology, ecology, and management. *Annual Review of Entomology* 63, 599–618.

Leskey, T.C. and Wright, S.E. (2004) Monitoring plum curculio, *Conotrachelus nenuphar* (Coleoptera: Curculionidae), populations in apple and peach orchards in the mid-Atlantic. *Journal of Economic Entomology* 97, 79–88.

Leskey, T.C., Lee, D.-H., Short, B.D. and Wright, S.E. (2012) Impact of insecticides on the invasive *Halyomorpha halys* (Hemiptera: Pentatomidae): analysis of insecticide lethality. *Journal of Economic Entomology* 105, 1726–1735.

Li, H. and Zheng, Z. (1998) A taxonomic study on the genus *Anarsia* Zelller from the mainland of China (Lepidoptera: Gelechiidae). *Acta Zoologica Academiae Scientiarum Hungaricae* 43, 121–132.

Liquido, N.J., Shinoba, L.A. and Cunningham, R.T. (1991) *Host Plants of the Mediterranean Fruit Fly (Diptera: Tephritidae): An Annotated World Review*. Entomological Society of America, Lanham, Maryland.

Mahmoud, M.E.E., Mohamed, S.A., Ndlela, S., Abdelmutalab, G.A.A., Khamis, F.M., Bashir, M.A.E. and Ekesi, S. (2020) Distribution, relative abundance, and level of infestation of the invasive peach fruit fly *Bactrocera zonata* (Saunders) (Diptera: Tephritidae) and its associated natural enemies in Sudan. *Phytoparasitica* 48, 589–605.

Mahmoud, M.F. (2009) Susceptibility of the peach fruit fly, *Bactrocera zonata* (Saunders), (Diptera: Tephritidae) to three entomopathogenic fungi. *Egyptian Journal of Biological Pest Control* 19, 169–175.

Mahmoud, Y.A., Ebadah, I.A., Metwally, H. and Saleh, M.E. (2016) Controlling of larvae, pupae and adults of the peach fruit fly, *Bactrocera zonata* (Saund.) (Diptera: Tephritidae) with the entomopathogenic nematode, *Steinernema feltiae*. *Egyptian Journal of Biological Pest Control* 26, 615–617.

Mamay, M., Yanık, E. and Doğramacı, M. (2014) Phenology and damage of *Anarsia lineatella* Zell. (Lepidoptera: Gelechiidae) in peach, apricot and nectarine orchards under semi-arid conditions. *Phytoparasitica* 42, 641–649.

Mathews, C.R., Blaauw, B., Dively, G., Kotcon, J., Moore, J. *et al.* (2017) Evaluating a polyculture trap crop for organic management of *Halyomorpha halys* and native stink bugs in peppers. *Journal of Pest Science* 90, 1245–1255.

Mavrikakis, P.G., Economopoulos, A.P. and Carey, J.R. (2000) Continuous winter reproduction and growth of the Mediterranean fruit fly (Diptera: Tephritidae) in Heraklion, Crete, southern Greece. *Environmental Entomology* 29, 1180–1187.

McDougall, R.N., Ogburn, E.C., Walgenbach, J.F. and Nielsen, A.L. (2021) Diapause termination in invasive populations of the brown marmorated stink bug (Hemiptera: Pentatomidae) in response to photoperiod. *Environmental Entomology* 50, 1400–1406.

McInnis, D.O., Hendrichs, J., Shelly, T., Barr, N., Hoffman, K. *et al.* (2017) Can polyphagous invasive tephritid pest populations escape detection for years under favorable climatic and host conditions? *American Entomologist* 63, 89–99.

Mertens, J., Schenk, M., Delbianco, A., Graziosi, I. and Vos, S. (2021) Pest survey card on *Bactrocera zonata*. EFSA Supporting Publications 18. European Food Safety Authority, Parma, Italy. Available at; https://efsa.onlinelibrary.wiley.com/doi/epdf/10.2903/sp.efsa.2021.EN-1999 (accessed 12 December 2022).

Miller, J.R. and Gut, L.J. (2015) Mating disruption for the 21st century: matching technology with mechanism. *Environmental Entomology* 44, 427–453.

Montoya, P. and Liedo, P. (2000) Biological control of fruit flies (Diptera: Tephritidae) through parasitoid augmentative releases: current status. In: Tan, K.H. (ed.) *Area-Wide Control of Fruit Flies and Other Insect Pests: Joint Proceedings of the International Conference in Area-Wide Control of Insects Pests, 28 May–5 June 1998, and the Fifth International Symposium on Fruit Flies of Economic Importance, 1–5 June 1998*. Penerbit Universiti Sains Malaysia, Penang, Malaysia, pp. 719–723.

Monzo, C., Saabater-Munoz, B., Urbaneja, A. and Castanera, P. (2010) Tracking medfly predation by the wolf spider, *Pardosa cribata* Simon, in citrus orchards using PCR-based gut-content analysis. *Bulletin of Entomological Research* 100, 145–152.

Morrison, W.R. III, Blaauw, B.R., Short, B.D., Nielsen, A.L., Bergh, J.C. *et al.* (2019) Successful management of *Halyomorpha halys* (Hemiptera: Pentatomidae) in commercial apple orchards with an attract-and-kill strategy. *Pest Management Science* 75, 104–114.

Mujica, V., Preti, M., Basoalto, E., Cichon, L., Fuentes-Contreras, E. *et al.* (2018) Improved monitoring of oriental fruit moth (Lepidoptera: Tortricidae) with terpinyl acetate plus acetic acid membrane lures. *Journal of Applied Entomology* 142, 731–744.

Murtaza, G., Naeem, M., Manzoor, S., Khan, H.A., Eed, E.M. *et al.* (2022) Biological control potential of entomopathogenic fungal strains against peach Fruit fly, *Bactrocera zonata* (Saunders) (Diptera: Tephritidae). *PeerJ* 10: e13316.

Myers, C.T., Hull, L.A. and Krawczyk, G. (2007) Effects of orchard host plants (apple and peach) on development of oriental fruit moth (Lepidoptera: Tortricidae). *Journal of Economic Entomology* 100, 421–430.

Nagaraja, H. and Hussainy, S.U. (1967) A study of six species of *Chilocorus* (Coleoptera: Coccinellidae) predaceous on San José and other scale insects. *Oriental Insects* 1, 249–256.

Natale, D., Mattiacci, L., Hern, A., Pasqualini, E. and Dorn, S. (2003) Response of female *Cydia molesta* (Lepidoptera: Tortricidae) to plant derived volatiles. *Bulletin of Entomological Research* 93, 335–342.

Navarro-Llopis, V., Primo, J. and Vacas, S. (2013) Efficacy of attract-and-kill devices for the control of *Ceratitis capitata*. *Pest Management Science* 69, 478–482.

Ni, W.L., Li, Z.H., Chen, H.J., Wan, F.H., Qu, W.W., Zhang, Z. and Kriticos, D.J. (2012) Including climate change in pest risk assessment: the peach fruit fly, *Bactrocera zonata* (Diptera: Tephritidae). *Bulletin of Entomological Research* 102, 173–183.

Nielsen, A.L. and Hamilton, G.C. (2009) Seasonal occurrence and impact of *Halyomorpha halys* (Hemiptera: Pentatomidae) in tree fruit. *Journal of Economic Entomology* 102, 1133–1140.

Nielsen, A.L., Hamilton, G.C. and Matadha, D. (2008) Developmental rate estimation and life table analysis for *Halyomorpha halys* (Hemiptera: Pentatomidae). *Environmental Entomology* 37, 348–355.

Nielsen, A.L., Hamilton, G.C. and Shearer, P.W. (2011) Seasonal phenology and monitoring of the non-native *Halyomorpha halys* (Hemiptera: Pentatomidae) in soybean. *Environmental Entomology* 40, 231–238.

Nielsen, A.L., Fleischer, S., Hamilton, G.C., Hancock, T., Krawczyk, G. *et al.* (2017) Phenology of brown marmorated stink bug described using female reproductive development. *Ecology and Evolution* 7, 6680–6690.

Ogburn, E.C., Bessin, R., Dieckhoff, C., Dobson, R., Grieshop, M. *et al.* (2016) Natural enemy impact on eggs of the invasive brown marmorated stink bug, *Halyomorpha halys* (Stål) (Hemiptera: Pentatomidae), in organic agroecosystems: a regional assessment. *Biological Control* 101, 39–51.

Papadopoulos, N.T. (2014) Fruit fly invasion: historical, biological, economic aspects and management. In: Shelly, T., Epsky, N., Jang, E.B., Reyes-Flores, J. and Vargas, R. (eds) *Trapping And The Detection, Control, and Regulation of Tephritid Fruit Flies*. Springer, Dordrecht, Netherlands, pp. 219–252.

Papadopoulos, N.T. and Katsoyannos, B.I. (2003) Field parasitism of *Ceratitis capitata* larvae by *Aganaspis daci* in Chios, Greece. *Biocontrol* 48, 191–195.

Papadopoulos, N.T., Katsoyannos, B.I., Carey, J.R. and Kouloussis, N.A. (2001) Seasonal and annual occurrence of the Mediterranean fruit fly (Diptera : Tephritidae) in northern Greece. *Annals of the Entomological Society of America* 94, 41–50.

Papadopoulos, N.T., Katsoyannos, B.I. and Carey, J.R. (2002) Demographic parameters of the Mediterranean fruit fly (Diptera: Tephritidae) reared in apples. *Annals of the Entomological Society of America* 95, 564–569.

Papadopoulos, N.T., Plant, R.E. and Carey, J.R. (2013) From trickle to flood: the large-scale, cryptic invasion of California by tropical fruit flies. *Proceedings of the Royal Society B: Biological Sciences* 280, 20131466.

Perez-Aparicio, A., Llorens, J., Rosell-Polo, R.J., Marti, J. and Gemeno, C. (2021) A cheap electronic sensor automated trap for monitoring the flight activity period of moths. *European Journal of Endocrinology* 118, 315–321.

Plant Health Australia (2018) *The Australian Handbook for the Identification of Fruit Flies (version 3.1)*. Plant Biosecurity Cooperative Research Centre (PBCRC) project

Next Generation National Fruit Fly Diagnostics and Handbook PBCRC2147. Plant Health Australia. Canberra.

Potamitis, I., Rigakis, I. and Fysarakis, K. (2014) The electronic McPhail trap. *Sensors* 14, 22285–22299.

Potamitis, I., Rigakis, I. and Tatlas, N.-A. (2017) Automated surveillance of fruit flies. *Sensors* 17: 110.

Preti, M., Favaro, R., Knight, A.L. and Angeli, S. (2021) Remote monitoring of *Cydia pomonella* adults among an assemblage of nontargets in sex pheromone-kairomone-baited smart traps. *Pest Management Science* 77, 4084–4090.

Quaintance, A.L. and Jenne, E.L. (1912) *The Plum Curculio*. Bureau of Entomology, US Department of Agriculture, Washington, DC.

Rice, R.E. and Jones, R.A. (1975) Peach twig borer: field use of a synthetic sex pheromone. *Journal of Economic Entomology* 8, 358–360.

Rice, R.E., Doyle, J. and Jones, R.A. (1972) Pear as a host of the oriental fruit moth in California. *Journal of Economic Entomology* 65, 1212–1213.

Rice, R.E., Flaherty, D.L. and Jones, R.A. (1982) Monitoring and modeling San Jose scale. *California Agriculture* 36, 13–14.

Rizk, M.M.A., Abdel-Galil, F.A., Temerak, S.A.H. and Darwish, D.Y.A. (2014) Factors affecting the efficacy of trapping system to the peach fruit fly (PFF) males, *Bactrocera zonata* (Saunders) (Diptera: Tephritidae). *Archives of Phytopathology and Plant Protection* 47, 490–498.

Roelofs, W.L., Comeau, A. and Selle, R. (1969) Sex pheromone of the oriental fruit moth. *Nature* 224: 723.

Rothschild, G.H.L. and Vickers, R.A. (1991) Biology, ecology and control of the oriental fruit moth. In: van der Geest, L.P.S. and Evenhuis, H.H. (eds). *World Crop Pests: Tortricid Pests*. Elsevier, The Netherlands.

Rousse, P., Gourdon, F. and Quilici, S. (2006) Host specificity of the egg pupal parasitoid *Fopius arisanus* (Hymenoptera: Braconidae) in La Reunion. *Biological Control* 37, 284–290.

Sayed, W.A.A., Farag, S. and Mohamed, S.A. (2018) Evaluation of using silymarin as a radio-protective agent of the peach fruit fly, *Bactrocera zonata* irradiated with gamma radiation. *Egyptian Academic Journal of Biological Sciences* 10, 91–103.

Shaked, B., Amore, A., Ioannou, C., Valdes, F., Alorda, B. *et al.* (2018) Electronic traps for detection and population monitoring of adult fruit flies (Diptera: Tephritidae). *Journal of Applied Entomology* 142, 43–51.

Shapiro-Ilan, D.I., Mizell, R.F., Cottrell, T.E. and Horton, D.L. (2008) Control of plum curculio, *Conotrachelus nenuphar*, with entomopathogenic nematodes: effects of application timing, alternate host plant, and nematode strain. *Biological Control* 44, 207–215.

Shapiro-Ilan, D.I., Leskey, T.C. and Wright, S.E. (2011) Virulence of entomopathogenic nematodes to plum curculio, *Conotrachelus nenuphar*: effects of strain, temperature, and soil type. *Journal of Nematology* 43: 187.

Shapiro-Ilan, D.I., Cottrell, T.E., Mizell, R.F., Horton, D.L. and Zaid, A. (2015) Field suppression of the peachtree borer, *Synanthedon exitiosa*, using *Steinernema carpocapsae*: effects of irrigation, a sprayable gel and application method. *Biological Control* 82, 7–12.

Shapiro-Ilan, D.I., Cottrell, T.E., Mizell, R.F. and Horton, D.L. (2016) Efficacy of *Steinernema carpocapsae* plus fire gel applied as a single spray for control of the lesser peachtree borer, *Synanthedon pictipes*. *Biological Control* 94, 33–36.

Shelly, T.E., Lance, D.R., Tan, K.H., Suckling, D.M., Bloem, K. *et al.* (2017) To repeat: can polyphagous invasive tephritid pest populations remain undetected for years under favorable climatic and host conditions? *American Entomologist* 63, 224–231.

Silva, E.D.B., Kuhn, T.M.A. and Monteiro, L.B. (2011) Oviposition behavior of *Grapholita molesta* Busck (Lepidoptera: Tortricidae) at different temperatures. *Neotropical Entomology* 40, 415–420.

Simmonds, F.J., Franz, J.M. and Sailer, R.I. (1976) History of biological control. In: Huffaker, C.B. and Messenger, P.S. (eds) *Theory and Practice of Biological Control.* Academic Press, London, pp. 17–39.

Sookar, P., Bhagwant, S. and Allymamod, M.N. (2014a) Effect of *Metarhizium anisopliae* on the fertility and fecundity of two species of fruit flies and horizontal transmission of mycotic infection. *Journal of Insect Science* 14: 100.

Sookar, P., Alleck, M., Ahseek, N., Permalloo, S., Bhagwant, S. and Chang, C.L. (2014b) Artificial rearing of the peach fruit fly *Bactrocera zonata* (Diptera: Tephritidae). *International Journal of Tropical Insect Science* 34, S99–S107.

Summers, F.M., Donaldson, D. and Togashi, S. (1959) Control of peach twig borer on almonds and peaches in California. *Journal of Economic Entomology* 52, 637–639.

Talamas, E.J., Herlihy, M.V., Dieckhoff, C., Hoelmer, K.A., Buffington, M. *et al.* (2015) *Trissolcus japonicus* (Ashmead) (Hymenoptera, Scelionidae) emerges in North America. *Journal of Hymenoptera Research* 43, 119–128.

Taylor, C.M., Coffey, P.L., DeLay, B.D. and Dively, G.P. (2014) The importance of gut symbionts in the development of the brown marmorated stink bug, *Halyomorpha halys* (Stål). *PLoS One* 9: e90312.

Usman, M., Wakil, W. and Shapiro-Ilan, D.I. (2021) Entomopathogenic nematodes as biological control agent against *Bactrocera zonata* and *Bactrocera dorsalis* (Diptera: Tephritidae). *Biological Control* 163: 104706.

Vacas, S., Alfaro, C., Navarro-Llopis, V. and Primo, J. (2010) Mating disruption of California red scale, *Aonidiella aurantii* Maskell (Homoptera: Diaspididae), using biodegradable mesoporous pheromone dispensers. *Pest Management Science* 66, 745–751.

Vera, M.T., Rodriguez, R., Segura, D.F., Cladera, J.L. and Sutherst, R.W. (2002) Potential geographical distribution of the Mediterranean fruit fly, *Ceratitis capitata* (Diptera: Tephritidae), with emphasis on Argentina and Australia. *Environmental Entomology* 31, 1009–1022.

Voudouris, C.C., Mavridis, K., Kalaitzaki, A., Skouras, P.J., Kati, A.N. *et al.* (2018) Susceptibility of *Ceratitis capitata* to deltamethrin and spinosad in Greece. *Journal of Pest Science* 91, 861–871.

Wan, N.-F., Ji, X.-Y., Deng, J.-Y., Kiær, L.P., Cai, Y.-M. and Jiang, J.-X. (2019) Plant diversification promotes biocontrol services in peach orchards by shaping the ecological niches of insect herbivores and their natural enemies. *Ecological Indicators* 99, 387–392.

Watanabe, M., Arakawa, R., Shinagawa, Y. and Okazawa, T. (1994) Overwintering flight of brown-marmorated stink bug, *Halyomorpha mista* to the buildings. *Japanese Journal of Sanitary Zoology* 45, 25–31.

Weber, D.C., Morrison, W.R., Khrimian, A., Rice, K.B., Leskey, T.C. *et al.* (2017) Chemical ecology of *Halyomorpha halys*: discoveries and applications. *Journal of Pest Science* 90, 989–1008.

White, I.M. and Elson-Harris, M.M. (1992) *Fruit Flies of Economic Significance Their Identification and Bionomics*. CAB International, Wallingford, UK.

Yang, Z.-Q., Yao, Y.-X., Qiu, L.-F. and Li, Z.-X. (2009) A new species of *Trissolcus* (Hymenoptera: Scelionidae) parasitizing eggs of *Halyomorpha halys* (Heteroptera: Pentatomidae) in China with comments on its biology. *Annals of the Entomological Society of America* 102, 39–47.

Yonce, C.E. and Pate, R.R. (1979) Seasonal distribution of *Synanthedon exitiosa* in the Georgia Peach Belt monitored by pheromone trapping. *Environmental Entomology* 8, 32–33.

Zhang, J., Tang, R., Fang, H., Liu, X., Michaud, J.P. *et al.* (2021) Laboratory and field studies supporting augmentation biological control of oriental fruit moth, *Grapholita molesta* (Lepidoptera: Tortricidae), using *Trichogramma dendrolimi* (Hymenoptera: Trichogrammatidae). *Pest Management Science* 77, 2795–2803.

Zingore, K.M., Sithole, G., Abdel-Rahman, E.M., Mohamed, S.A., Ekesi, S. *et al.* (2020) Global risk of invasion by *Bactrocera zonata*: implications on horticultural crop production under changing climatic conditions. *PLoS One* 15: e0243047.

The Peach Canning Industry

George A. Manganaris[1]*, Thomas M. Gradziel[2], Marina Christofi[1] and Carlos H. Crisosto[2]

[1]*Department of Agricultural Sciences, Biotechnology & Food Science, Cyprus University of Technology, Lemesos, Cyprus; [2]Department of Plant Sciences, University of California, Davis, California, USA*

15.1 INTRODUCTION

Peach (*Prunus persica* (L.) Batsch) is the second most important temperate fruit crop worldwide, being widely consumed as both fresh and after processing, mainly as canned product. According to data from the European Commission, production of peaches and nectarines in 2017 totalled 3.9 million t of which 760,000 t were consumed as processed products. By comparison, 1.0 million t of peaches and nectarines were consumed in the USA in 2016, of which 718,000 t were consumed as fresh and 292,000 t after processing (USDA/NAAS, 2021).

Canning (including the processing for nectarines and juice) accounts for around 80% of all processed peach tonnage in the USA, with other products being frozen sliced and diced products (~15%), dried peaches and baby food purées (~5%). Due to hand labour costs and competition from other countries, canned peaches have decreased in California from 484,820 t in 2007 to approximately 301,460 t in 2017.

Despite representing a significant amount of total peach production, relatively few studies have focused on peaches destined for processing. Canned peaches include canned yellow clingstone peaches, canned spiced yellow clingstone peaches, canned solid-pack yellow clingstone peaches and canned artificially sweetened yellow clingstone peaches (Siddiq *et al.*, 2012). Accordingly, the canned peaches can be sold in many styles, such as halves, quarters, slices, diced or mixed pieces of irregular sizes. In this chapter, the characteristics of processing peach cultivars are presented, along with processing methods and quality determinants.

* Email: george.manganaris@cut.ac.cy

© CAB International 2023. *Peach* (G. Manganaris, G. Costa and C. Crisosto eds) 421
DOI: 10.1079/9781789248456.0015

15.2 HARVEST OPERATION

Most peaches utilized for processing are clingstone types and in almost all cases are characterized by a non-melting flesh. While the clingstone trait represents a hazard for processed fruit because of the greater likelihood of pit fragments being retained in the processed product, it is tightly linked to firm, non-melting flesh that is required for surviving the rigours of bulk transport and processing. The firmer, non-melting flesh also allows cling-stone peaches to be mechanically harvested and canned when they are near their optimal ripeness with full size and optimal colour and flavour. Peaches are mainly hand picked, or they can be mechanically harvested using a shaker and a catching frame system. Following the initial field sorting of green, undersized or damaged fruit, bins are loaded on to trucks and trans-ported to the processor. Prior to processing, the fruit are first washed and sorted by size.

The fruit-pitting operation typically involves cutting the fruit along the suture, followed by application of a torque load to twist the peach halves from the pit, requiring good mesocarp firmness. If the fruit become too soft, they are easily bruised when bulk handled, with additional losses occurring during the pitting and peeling operations. In addition, overly soft flesh may disintegrate during thermal processing. Consequently, high levels of fruit loss can occur during harvesting and processing. While harvesting the fruit at the fully ripe stage increases the damage from soft fruit, it maximizes crop yield as well as fruit colour, sugar content and flavour.

In many cases, because of differences in peach availability in relation to processing-plant capacity and to assure a uniform supply of raw product to the processing plants, peach surplus is stored prior to canning. Storage conditions that maximize postharvest life and protect from fruit flesh browning and decay are like those used for fresh peaches. As a general rule, non-melting flesh geno-types that are destined for canning purposes are characterized by higher tissue retention during their ripening upon removal from cold storage and a higher content of uronic acids, as well as a higher capacity for calcium binding in the water-insoluble pectin fraction compared with melting-flesh cultivars destined for fresh consumption (Manganaris *et al.*, 2006a). In addition, storage tem-perature has a pivotal role and should be near freezing point. Fruit of the refer-ence clingstone peach cultivar 'Andross' is particularly prone to flesh browning symptoms when is cold stored at 5°C for extensive periods (Manganaris *et al.*, 2006b). In certain cases, an even longer pre-canning storage time may be required. Some benefits have been reported on the use of controlled-atmosphere technology (2% O_2 plus 5% CO_2) at 1.1°C. This treatment can maintain higher flesh firmness and retard decay more effectively than air storage. However, controlled-atmosphere storage of fruits picked at the optimum maturity stage produces little benefit over air storage, except under conditions of high disease pressure.

15.3   THE CANNING PROCESS

Common processed forms include peach halves, sliced or diced peach, purée and juice. While the final product and even method of preparation may vary, most processing involves a common sequence of basic events to take place, as described elsewhere (Christofi *et al.*, 2021a). Briefly, fresh peaches are fed into the production line where they are mechanically cut and pitted, followed by inspection for residual pit fragments and sorting of pitted halves through an automatic colour sorting system. Subsequently, peach halves are peeled with 15% caustic soda (sodium hydroxide) and visually inspected/sorted for defects of the peeled halves. Thereafter, the halves are graded based on their size and fed to the appropriate production lines. An additional visual inspection/sorting takes place prior to filling of cans, followed by addition of enough packing medium to fill the interspatial spaces, with the aid of an automatic vacuum filler. The filled cans are then thermally exhausted and introduced into an automatic closing machine for seaming. The hermetically sealed cans are pasteurized in boiling water (97–98°C, 22 min) using a rotary or horizontal line pasteurizer that allows rolling of the can. The can centre is allowed to reach a temperature of around 91–92°C. The canned products are cooled by cold water spraying for 10 min (~40–42°C) to bring the cans to room temperature. A minimum of 6-month storage is typically required to achieve osmotic equilibrium among all components present within the cans.

15.4   FRUIT MATERIAL DESTINED FOR PROCESSING

Non-melting peach cultivars are used for processing as they are more tolerant to pitting and heat treatment during the canning process. In addition, some freestone cultivars (e.g. 'Fairtime', 'Zee Lady' and 'Fay-Elberta') are processed for fruit sections, purée or juice. Many clingstone cultivars developed for canning have been developed by regional breeding programmes. All these cultivars have been bred for good flavour, attractive colour and firmness retention. It is also important that harvested peaches are of uniform size, shape and ripeness. The flesh should be sufficiently firm to survive transport and processing, yet not so firm as to promote pit fragmentation.

A firm flesh is required for successful peeling, pitting and canning (Mallidis and Katsaboxakis, 2002; Crisosto *et al.*, 2007). Later cultivar improvement programmes in the USA, South Africa and Australia led to superior-quality clingstone-type peach cultivars (Byrne, 2002). The dissemination of improved canning peach cultivars, which began in the 1970s, has led to a significant expansion of clingstone peach cultivation (van Nocker and Gardiner, 2014).

New and improved clingstone peach cultivars target improved quality, high yields and resistance to frost, insects and diseases. The University of California at Davis has maintained a processing peach breeding programme since the 1980s with the support of the California peach growers and processors. The current objective of the breeding programme is to develop improved cultivars for the extra-early and early harvest seasons. In Europe, the most important region for processing peaches is Greece, which shares similar cultivars as well as production needs and strategies as California. Below we discuss the main non-melting peach cultivars grown in Greece and the USA.

15.4.1   Greece

The first cultivars used for canning in Greece were 'Elberta' and later 'Red Haven', but the final product was of inferior quality due to low flesh firmness. During the period 1980–2000, the market was dominated by three cultivars ('Andross', 'Catherina' and 'Everts') that had overlapping ripening periods. In particular, 'Andross' is the predominant peach cultivar grown in Greece, and is a key worldwide non-melting peach for canning (Christofi *et al.*, 2022). Consequently, the canning industry is facing an oversupply of peach fruits for canning when these predominant cultivars are ripening. Therefore, there is an urgent request by the industry to expand the harvest season with cultivars with appreciable qualitative attributes for canning. Towards the expansion of the processing peach harvest season, both early- and late-ripening cultivars were introduced. The main non-melting peach cultivars currently grown in Greece are 'Romea', 'Catherina', 'Fortuna', 'Loadel', 'PI-A37', 'Andross', 'PI-IB42', 'Everts' and 'PI-E45' (Drogoudi and Tsipouridis, 2007), while some new releases are being cultivated with promising results (Fig. 15.1). Among them, 'Romea' is the earliest-producing processing peach cultivar grown in Greece although its yield is relatively low compared with the late-ripening cultivars. 'Catherina' is a high-yield cultivar with good fruit quality and is the most favoured cultivar by growers due to its stable production under unfavourable climatic conditions (i.e. spring frost or rain during flowering). 'Mirel®' is a patented mid-ripening cultivar, recently released from France which is characterized by high production volumes, good fruit quality and good tree-holding ability. 'Fercluse®' is a mid-season patented cultivar with a more orange-coloured fruit flesh. 'Everts' is a late-ripening cultivar with yellow-coloured fruit flesh. 'Ferlate®' is a late-ripening patented French cultivar also characterized by orange-coloured flesh. 'VLG' is late-ripening (mid-September) cultivar with unknown origin, characterized by satisfactory production that can potentially offer the option to extend the canning season. Figure 15.1 shows the harvesting period for each cultivar and the qualitative properties of the harvested fruit.

15.4.2   USA

Most of the processing peach industry is concentrated in the Central Valley of California where the relatively long and largely rain-free growing season

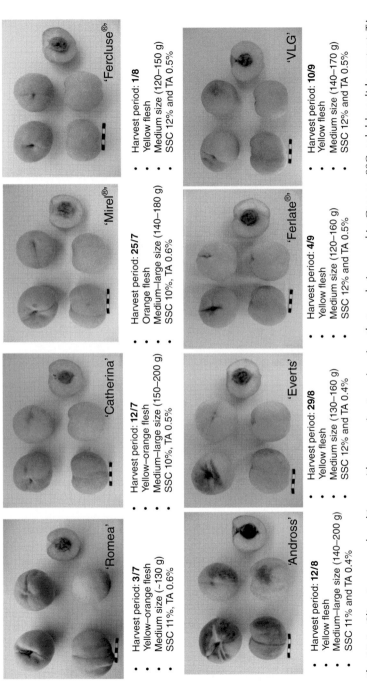

Fig. 15.1. Clingstone peach cultivars with successive tree ripening that are being used in Greece. SSC, soluble solids content; TA, titratable acidity. Photographs courtesy of M. Christofi.

'Fercluse®'
- Harvest period: **1/8**
- Yellow flesh
- Medium size (120–150 g)
- SSC 12% and TA 0.5%

'VLG'
- Harvest period: **10/9**
- Yellow flesh
- Medium size (140–170 g)
- SSC 12% and TA 0.5%

'Mirel®'
- Harvest period: **25/7**
- Orange flesh
- Medium–large size (140–180 g)
- SSC 10%, TA 0.6%

'Ferlate®'
- Harvest period: **4/9**
- Yellow flesh
- Medium size (120–160 g)
- SSC 12% and TA 0.5%

'Catherina'
- Harvest period: **12/7**
- Yellow–orange flesh
- Medium–large size (150–200 g)
- SSC 10%, TA 0.5%

'Everts'
- Harvest period: **29/8**
- Yellow flesh
- Medium size (130–160 g)
- SSC 12% and TA 0.4%

'Romea'
- Harvest period: **3/7**
- Yellow–orange flesh
- Medium size (~130 g)
- SSC 11%, TA 0.6%

'Andross'
- Harvest period: **12/8**
- Yellow flesh
- Medium–large size (140–200 g)
- SSC 11% and TA 0.4%

Fig. 15.2. Peaches of the commercial processing cultivar 'Kader' demonstrating the required traits of uniform size, shape and flesh colour free from red anthocyanin staining and pit fragments. Reproduced from Gradziel and Marchand (2019b).

favours high quality and productivity (Gradziel and McCaa, 2008). Many cultivars have been developed to provide the processing plants with a continuous supply of raw fruit possessing a uniform golden-yellow flesh colour free from red anthocyanin staining and pit fragments having fruit with 10% or greater soluble solids content (SSC) and a diameter of 60 mm or more (Gradziel and Marchand, 2019a,b). In the major peach-processing production regions, cultivars are developed within or alongside breeding programmes for fresh-market fruit as the genetics and breeding methodologies are identical (Zhang *et al.*, 1996; Layne, 1997).

The cultivar improvement programmes differ primarily in the breeding objectives and consequently the selection criteria, as fresh-market and processing cultivar ideotypes differ in important and sometimes dramatic ways (Fresnedo-Ramirez *et al.*, 2016). A major difference between processing and fresh-market cultivars is that processing peaches require greater fruit retention and uniformity for bulk handling and processing, and greater yields to compensate for the generally lower price for the raw product (Minas *et al.*, 2018).

In addition, as most flavour volatiles are lost during processing, processing peach breeding programmes must emphasize other components of peach flavour. While some fresh-market freestone cultivars such as 'O'Henry' are also processed in California, their soft melting flesh and typical high levels of water-soluble anthocyanin pigments result in poor processed quality unless processing is applied by rapid freezing or drying. As in Greece, most US processed peach cultivars utilize the clingstone non-melting fruit flesh trait as the firm flesh is more resistant to physical damage during bulk fruit harvest, transport, processing and fruit slicing (Carles, 1984).

To be successful, a cultivar must be compatible with the nuances of every component of the processing pathway. An example is seen in the cultivar 'Kader' (Fig. 15.2), which was recently released to fill the production gap between 'Carson' and 'Bowen' resulting from the rejection by processors of the previous cultivar 'Dixon' due to its high incidence of pit fragments and red-staining of flesh (Gradziel and Marchand, 2019a,b). As discussed previously, a critical requirement for processing peaches is the ability to maintain a tree-ripe flesh firmness of 30 N or greater throughout harvest, transport and processing. With the ongoing requirements for greater labour and harvest efficiency, an emerging need is the ability of future varieties to maintain tree-ripe fruit firmness levels for a week or more after the initial tree-ripe stage to allow more efficient once-over harvesting of the entire orchard, either by hand or by mechanized harvesting (Gradziel and Marchand, 2019a).

15.5 ATTRIBUTES OF CANNING PRODUCTS

The flavour, texture and colour of fresh and processed products are parameters critical to consumers' acceptance and thus the product's market success. These parameters are most often characterized by humans (sensory science) and machines (instrumental analysis) (Ross, 2009). Qualitative and sensorial, including textural, attributes, as well as phytochemical properties of canned peach products are described in the following sections.

15.5.1 Qualitative and sensorial attributes

A bright yellow colour with no red coloration in the pit cavity, a firm texture and a good flavour are desired. Peaches with fruit sizes larger than 60 mm and yellow-gold fruit flesh are the most desirable, but the fruit pit cavity should not develop a pink to red colour from the formation of red anthocyanins. Anthocyanins are very heat labile and subject to browning in canning operations; this has led to the selection for flesh of canning peaches that is anthocyanin free (Bassi and Monet, 2008). In many peach cultivars, there are cells near the stone that contain anthocyanin (red pigment) and when the fruit is halved, the appearance is often of red rays extending out from the centre, which contains the pit (flesh bleeding). Leaking of red pigment into the juice that turns brown when cooked creates consumer dissatisfaction. These bleeding symptoms are intensified during cold storage. In a few cultivars, red-brown bleeding into the juices also originates from the skin, which can also be rich in anthocyanins that leak into the juice. During the breeding process, most of these fruit types are not selected for commercial use.

Fruit should be almost free from defects or blemishes such as decay, worms, worm holes and split pits, and free from damage caused by scab, bacterial spot, other diseases, insects, bruises or other damage. Fruit should be uniform in size and symmetry for the intended use.

The canning industry is dominated by yellow-fleshed peach cultivars. However, currently, peach cultivars that lead to a canned product with an orange colour are being promoted into the market as an alternative option.

A colour intensity index for assessment of canned peaches has been developed (Delwiche, 1989). The evaluation scale was structured with the following scores: 1–2, light green; 3–4, slightly yellow; 5–6, yellow; 7–8, light orange; and 9–10, dark orange. Figure 15.3 presents categorization of seven peach cultivars in three distinct colour groups according to the mean scores

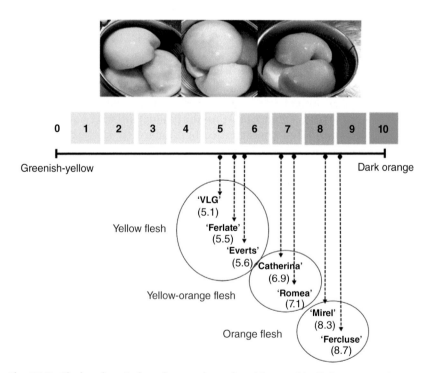

Fig. 15.3. Flesh colour index of canned peach cultivars with distinct properties. Adapted from Christofi *et al.* (2021b).

of cultivars: (i) yellow for 'VLG', 'Ferlate®' and 'Everts'; (ii) yellow–orange for 'Catherina' and 'Romea'; and (iii) orange for 'Mirel®' and 'Fercluse®' (Christofi *et al.*, 2021b).

The SSC of fruit is not a strict requirement because the syrup can be adjusted at the end of the canning process resulting in processed products with an SSC ranging from 10% (light) to 22% (extra-heavy). The packing medium may differ, ranging from light syrup (15.4°Brix, pH 3.56) to pear or grape juice. An alternative filling medium targeting less calorific content is diluted clarified concentrated grape juice (12.3°Brix, pH 3.66) (Christofi *et al.*, 2021a).

Endocarp or pit splitting (at the carpel suture) or shattering (radial fractures) may affect both early- and late-ripening cultivars, depending on orchard conditions. It is reported that cultural practices to improve fruit size (e.g. supplemental irrigation, thinning, girdling) may also increase the incidence of endocarp splitting or shattering when carried out improperly (Crisosto and Costa, 2008). These two undesired endocarp modifications are a commercial problem because of the potential risk to consumers, especially children, in eating the stone fragments. In the canning industry, these undesirable fruits are not processed because the fragments are difficult to be eliminated from processed fruits.

Throughout the canning process, large amounts of fruit are damaged during mechanical pitting operations. Fruit disintegration is thus referred to as pitting damage because it is produced during the pitting operation in the canning process. A strong correlation between fruit firmness measured in the weakest position on the fruit and pitting damage has been established (Crisosto *et al.*, 2001; Valero *et al.*, 2003, 2006). The percentage of 'Andross', 'Carson' and 'Ross' fruit with pitting damage increased as fruit firmness fell below 18 N; this value is called the critical pitter threshold. Thus, peach canning quality is horticulturally determined by reaching a minimum flesh colour, an adequate fruit size and firmness greater than 18 N to maximize consumer quality and avoid losses (Metheney *et al.*, 2002).

Regarding sensorial attributes, only recently, a standardized vocabulary ('consensus language') was developed for the determination and quantification of 15 parameters for odour, appearance, texture, taste and flavour through quantitative descriptive analysis. A complete list of all sensorial attributes, definitions and reference standards used in the training sessions is available in a recent work by Christofi *et al.* (2021a) and is provided in Table 15.1.

Fruit packed in diluted clarified grape juice concentrate, aiming towards a less calorific content product, demonstrated an inferior consumer perception regarding bitterness, astringency and 'off-flavour' and thus is not recommended for commercial usage (Christofi *et al.*, 2021b).

15.5.2 Textural properties

A key quality attribute of peach, either as fresh fruit or processed product, is texture because this sensory property is directly linked to the perception

Table 15.1. List of sensory attributes, definitions and reference standards used in the training sessions of the sensory panel.

Attributes		Definitions	Reference standards (intensity)
Odour	Peach aroma	Intensity of characteristic aroma of commercial canned peach (none to very)	None: mineral water Very: freshly prepared commercial canned peach was blended to purée
Appearance	Peach colour	Based on colour scale given, define the colour of peach half (greenish yellow to dark orange)	A colour scale was constructed and given
	Colour uniformity	Based on scale given, define the degree of colour uniformity (non-uniform to uniform)	Images were taken and given as an example of the two ends
	Brightness	Based on scale given, define the glossy surface showing bright reflection (less shiny to very shiny)	Images were taken and given as an example of the two ends
	Residual peel	The skin remaining after peeling with caustic soda (none to very)	–
	Blemished fruit	Something spoils the appearance of peach half, which is otherwise aesthetically perfect (none to very)	Images were taken and given as an example of the two ends
Texture	Hardness	Force required to compress and deform 75% of the half centre, using your thumb (low to very hard)	Low: cream cheese Moderate: soft pitted canned olive Very hard: raw groundnut
	Difficulty in chewiness	The degree of difficulty observed during chewing of peach half (low to very)	Try the appropriate piece of area (as shown in additional part) and evaluate the samples based on the chewing time (<3 s: low and >12 s: very)
Taste	Sweetness	Taste characteristic of sucrose (low to very)	Low: freshly prepared canned peach purée with sucrose (4.5 g l^{-1}) Very: freshly prepared canned peach purée with sucrose (30 g l^{-1})

Attribute	Definition	Reference standards
Acidity	Taste characteristic of citric fruits (low to very)	Low: freshly prepared canned peach purée with 50% more water Very: freshly prepared canned peach purée with citric acid (0.5 g l^{-1})
Bitterness	Taste characteristic of caffeic acid (none to very)	None: mineral water Very: freshly prepared aqueous solution with caffeine (0.5 g l^{-1})
Astringency	Sensation of dryness on the palate or muscle contraction (squeeze lips) caused by some substances like tannins (none to very)	None: mineral water Very: freshly prepared canned peach purée with tannic acid (0.5 g l^{-1})
Peach flavour	Typical flavour of commercial canned peach (low to very)	Low: freshly prepared canned peach purée with 50% more water Very: freshly prepared canned peach was blended with peach nectar
Fruitiness	The characteristic fruity note associated with fruits besides peach (low to very)	Low: freshly prepared aqueous solution with isoamyl acetate (5.0 mg l^{-1}) Very: freshly prepared aqueous solution with isoamyl acetate (40 mg l^{-1})
Off-flavour	Intensity of atypical flavour perceived in your mouth after chewing and is often associated with deterioration of the product such as overcooked (none to very)	None: commercial canned peach Very: commercial canned peach half was baked in oven for 20 min (180°C)

in the mouth during chewing (Harker *et al.*, 2010) and represents the most strongly correlated attribute with descriptive sensory attributes (Contador *et al.*, 2016). Texture sensory attributes of peach fruit is an important quality parameter that is substantially affected during canning. As indicated previously, the optimum maturity of peaches for canning purposes is when the fruit is yellow coloured and still firm (Siddiq *et al.*, 2012).

A toolkit to evaluate the textural properties of canned peaches using large deformation mechanical testing has recently been developed (Christofi *et al.*, 2021a). Textural properties can effectively be evaluated with a TA-XT Plus texture analyser by applying three discrete large deformation tests: (i) a puncture test with a flat cylindrical probe; (ii) a texture profile analysis (TPA) with a flat compression plunger; and (iii) a Kramer shear cell test with a bladed fixture. This provides a total of nine textural properties, namely, puncture firmness (individual halves), Kramer hardness (applied in a complex mixture of peach slices), TPA hardness (central section of halves), fragmentation, consistency, cohesiveness, springiness, chewiness and total hardness (Fig. 15.4).

15.5.3 Content of bioactive compounds

Fresh clingstone-type peaches undergo thermal processing as a preservation method. Despite its economic importance, little information exists regarding the fate of phytochemicals after the canning process and contradictory results have been reported. The concentration of total carotenoids was found to be significantly reduced except for zeaxanthin (which increased during storage), while phenolic compounds were less affected by pasteurization of fresh peaches (Oliveira *et al.*, 2012). Other studies showed that thermal processing

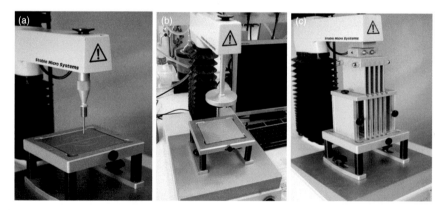

Fig. 15.4 TA-XT Plus texture analyser for application of three discrete large deformation tests: (a) a puncture test with a flat cylindrical probe; (b) a texture profile analysis with a flat compression plunger; and (c) a Kramer shear cell test with a bladed fixture. Photographs courtesy of M. Christofi.

and storage have negative impacts on peach procyanidins (Hong *et al.*, 2004; Techakanon *et al.*, 2016).

Although thermal processing of fruits and vegetables is thought to reduce the content of bioactive components, one study showed that clingstone processed peaches contained higher quantities of bioactive compounds than fresh clingstone peaches (Campbell and Padilla-Zakour, 2013). Moreover, Durst and Weaver (2013) reported that canned peaches are nutritionally equivalent to fresh, and a fourfold increase in vitamin C in canned peaches was recorded. Further studies are needed to clarify the nutritional value of canned peaches and how this is affected by the processing protocol and the initial raw material.

Recently, Christofi *et al.* (2022) employed an array of comprehensive and highly complementary protocols to determine the phytochemical properties of an array of non-melting peach cultivars. The results demonstrated that the phytochemical profile is primarily genotype dependent, as is their ability to retain bioactive compounds upon canning and storage. 'Andross' presented the highest levels of zeaxanthin and lutein, which were relatively unaffected by the canning storage steps compared with β-carotene. Therefore, as well as its proven exceptional agronomical properties and yield performance, 'Andross' appears to possess appreciably high contents of bioactive compounds (Christofi *et al.*, 2022). However, soluble phenolic compounds, such as neochlorogenic acid, chlorogenic acid, procyanidin B1 and catechin registered a dramatic decrease following canning. Remarkably, the use of grape juice as a filling medium, an additional source of polyphenols, resulted in a reduction in losses of bioactive compounds, possibly by balancing out diffusion processes between the fruit tissue and packing medium (Oliveira *et al.*, 2012).

15.6   CONCLUSIONS

The world canning industry is facing an oversupply of peach fruits for canning; thus, in the long term, canned peach consumption should be increased to benefit producers and consumers. In the short term, a large crop needs to be rapidly processed. However, many producers have facilities with limited capacities in relation to their seasonal crop, causing processing delays that affect fruit flavour. These processing delays reduce consumer quality, affecting canned peach marketing and consumption. Therefore, special attention should be focused on breeding programmes to release high-quality cultivars. These cultivars should ripen uniformly within the canopy or be able to remain on the tree while maintaining fruit firmness and high flavour sensory characteristics (see Chapter 4, this volume). Selection of freestone and non-melting cultivars may reduce the potential for undesirable pit fragments in canned fruit. Advancement of storage technologies to protect and maintain peach quality prior to processing should be pursued, combined with the development of processing equipment to ensure safe pit removal and eating quality.

REFERENCES

Bassi, D. and Monet, R. (2008) Botany and taxonomy. In: Layne, D.R. and Bassi, D. (eds) *The Peach: Botany, Production and Uses*. CAB International, Wallingford, UK, pp. 1–37.

Byrne, D.H. (2002) Peach breeding trends: a world wide perspective. *Acta Horticulturae* 592, 49–59.

Campbell, O.E. and Padilla-Zakour, O.I. (2013) Phenolic and carotenoid composition of canned peaches (*Prunus persica*) and apricots (*Prunus armeniaca*) as affected by variety and peeling. *Food Research International* 54, 448–455.

Carles, L. (1984) Clingstone peaches, fruits specially adapted for the manufacture of fruits in syrup. *Arboriculture Fruitière* 31, 47–48.

Christofi, M., Mourtzinos, I., Lazaridou, A., Drogoudi, P., Tsitlakidou, P. *et al.* (2021a) Elaboration of novel and comprehensive protocols towards determination of textural properties and other sensorial attributes of canning peach fruit. *Journal of Texture Studies* 52, 228–239.

Christofi, M., Mauromoustakos, A., Mourtzinos, I., Lazaridou, A., Drogoudi, P. *et al.* (2021b) The effect of genotype and storage on compositional, sensorial, and textural attributes of canned fruit from commercially important non-melting peach cultivars. *Journal of Food Composition & Analysis* 103: 104080.

Christofi, M., Pavlou, A., Lantzouraki, D.Z., Tsiaka, T., Myrtsi, E. *et al.* (2022) Profiling phytochemicals in fresh and canned fruit of non-melting peach cultivars: impact of genotype and canning process on their content. *Journal of Food Composition & Analysis* 114: 104734.

Contador, L., Díaz, M., Hernández, E., Shinya, P. and Infante, R. (2016) The relationship between instrumental tests and sensory determinations of peach and nectarine texture. *European Journal of Horticultural Science* 81, 189–196.

Crisosto, C.H. and Costa, G. (2008) Preharvest factors affecting peach quality. In: Layne, D.R. and Bassi, D. (eds) *The Peach: Botany, Production and Uses*. CAB International, Wallingford, UK, pp. 536–549.

Crisosto, C.H., Slaughter, D., Garner, D. and Boyd, J. (2001) Stone fruit critical bruising thresholds. *Journal of the American Pomological Society* 55, 76–81.

Crisosto, C.H., Valero, C. and Slaughter, D.C. (2007) Predicting pitting damage during processing in Californian clingstone peaches using color and firmness measurements. *American Society of Agricultural and Biological Engineers* 23, 189–194.

Delwiche, M.J. (1989) Maturity standards for processing clingstone peaches. *Journal of Food Engineering* 10, 269–284.

Drogoudi, P.D. and Tsipouridis, C.G. (2007) Effects of cultivar and rootstock on the antioxidant content and physical characters of clingstone peaches. *Scientia Horticulturae* 115, 34–39.

Durst, R.W. and Weaver, G.W. (2013) Nutritional content of fresh and canned peaches. *Journal of the Science of Food and Agriculture* 93, 593–603.

Fresnedo-Ramirez, J., Frett, T.J., Sandefur, P.J., Salgado-Rojas, A., Clark, J.R. *et al.* (2016) QTL mapping and breeding value estimation through pedigree-based analysis of fruit size and weight in four diverse peach breeding programs. *Tree Genetics & Genomes* 12: 25.

Gradziel, T. and Marchand, S. (2019a) 'Vilmos' peach: a processing clingstone peach expressing a novel 'stay-ripe' trait with improved harvest quality, ripening in the 'Andross' maturity season. *HortScience* 54, 2078–2080.

Gradziel, T. and Marchand, S. (2019b) 'Kader' peach: a processing clingstone peach with improved harvest quality and disease resistance, ripening in the 'Dixon' maturity season. *HortScience* 54, 754–757.

Gradziel, T.M. and McCaa, J.P. (2008) Processing peach cultivar development. In: Layne, D.R. and Bassi, D. (eds) *The Peach: Botany, Production and Uses.* CAB International, Wallingford, UK, pp. 175–192.

Harker, F.R., Redgwell, R.J., Hallett, I.C., Murray, S.H. and Carter, G. (2010) Texture of fresh fruit. *Horticultural Reviews* 20, 121–224.

Hong, Y.J., Barrett, D.M. and Mitchell, A.E. (2004) Liquid chromatography/mass spectrometry investigation of the impact of thermal processing and storage on peach procyanidins. *Journal of Agricultural and Food Chemistry* 52, 2366–2371.

Layne, R.E.C. (1997) Peach and nectarine breeding in Canada: 1911 to 1995. *Fruit Varieties Journal* 51, 218–228.

Mallidis, C.G. and Katsaboxakis, C. (2002) Effect of thermal processing on the texture of canned apricots. *International Journal of Food Science and Technology* 37, 569–572.

Manganaris, G.A., Vasilakakis, M., Diamantidis, G. and Mignani, A. (2006a) Diverse metabolism of cell wall components of melting and non-melting peach genotypes during ripening after harvest or cold storage. *Journal of the Science of Food and Agriculture* 86, 243–250.

Manganaris, G.A., Vasilakakis, M., Diamantidis, G. and Mignani, I. (2006b) Cell wall physicochemical aspects of peach fruit related to internal breakdown symptoms. *Postharvest Biology and Technology* 39, 69–74.

Metheney, P.D., Crisosto, C.H. and Garner, D. (2002) Developing canning peach critical bruising thresholds. *Journal of the American Pomological Society* 56, 75–78.

Minas, I.S., Tanou, G. and Molassiotis, A. (2018) Environmental and orchard bases of peach fruit quality. *Scientia Horticulturae* 235, 307–322.

Oliveira, A., Pintado, M. and Almeida, D.P.F. (2012) Phytochemical composition and antioxidant activity of peach as affected by pasteurization and storage duration. *LWT – Food Science and Technology* 49, 202–207.

Ross, C.F. (2009) Sensory science at the human–machine interface. *Trends in Food Science and Technology* 20, 63–72.

Siddiq, M., Liavoga, A. and Greiby, I. (2012) Peaches and nectarines. In: Sinha, N.K., Sidhu, J.S., Barta, J., Wu, J.S.B. and Cano, M.P. (eds) *Handbook of Fruits and Fruit Processing*, 2nd edn. Wiley, Ames, Iowa, pp. 535–549.

Techakanon, C., Gradziel, T., Zhang, L. and Barrett, D. (2016) Effects of peach cultivar on enzymatic browning following cell damage from high pressure processing. *Journal of Agricultural and Food Chemistry* 64, 7606–7614.

USDA/NAAS (2021) Peaches. United States Department of Agriculture National Agricultural Statistics Service, Washington, DC. Available at: https://quick-stats.nass.usda.gov/results/46E21842-F8F0-33BF-A287-8FDC2DBFEE16?pivot=short_desc (accessed 12 December 2022).

Valero, C., Crisosto, C.H., Metheney, D. and Bowerman, E. (2003) Developing critical pitter thresholds for canning peaches using a nondestructive firmness sensor. *Acta Horticulturae* 604, 811–815.

Valero, C., Crisosto, C.H. and Slaughter, D. (2006) Relationship between nondestructive firmness measurements and commercially important ripening fruit stages for peaches, nectarines, and plums. *Postharvest Biology and Technology* 44, 248–253.

van Nocker, S. and Gardiner, S.E. (2014) Breeding better cultivars, faster: applications of new technologies for the rapid deployment of superior horticultural tree crops. *Horticulture Research* 1: 14022.

Zhang, G.R., Zong, X.P., Shen, Y.S., Wang, Z.Q., Zuo, Q.Y. and Zhu, G.R. (1996) Zhenghuang 5, a new late canning peach cultivar. *Journal of Fruit Science* 13, 130–131.

Index

Note: Page numbers in **bold** type refer to **figures** and page numbers in *italic* type refer to *tables*.